VOLUME FOUR HUNDRED AND THIRTY-EIGHT

METHODS IN ENZYMOLOGY

Small GTPases in Disease, Part A

METHODS IN ENZYMOLOGY

Editors-in-Chief

JOHN N. ABELSON AND MELVIN I. SIMON

Division of Biology
California Institute of Technology
Pasadena, California

Founding Editors

SIDNEY P. COLOWICK AND NATHAN O. KAPLAN

VOLUME FOUR HUNDRED AND THIRTY-EIGHT

METHODS IN ENZYMOLOGY

Small GTPases in Disease, Part A

EDITED BY

WILLIAM E. BALCH
The Scripps Research Institute
Department of Cell Biology
La Jolla, CA, USA

CHANNING J. DER
Lineberger Comprehensive Cancer Center
University of North Carolina
Chapel Hill, NC, USA

ALAN HALL
Chair Cell Biology Program
Memorial Sloan-Kettering Cancer Center
New York, NY, USA

AMSTERDAM • BOSTON • HEIDELBERG • LONDON
NEW YORK • OXFORD • PARIS • SAN DIEGO
SAN FRANCISCO • SINGAPORE • SYDNEY • TOKYO
Academic Press is an imprint of Elsevier

ELSEVIER

Academic Press is an imprint of Elsevier
525 B Street, Suite 1900, San Diego, California 92101-4495, USA
84 Theobald's Road, London WC1X 8RR, UK

For information on all Elsevier Academic Press publications
visit our Web site at www.books.elsevier.com

ISBN: 978-0-12-373968-1

PRINTED IN THE UNITED STATES OF AMERICA
08 09 10 11 9 8 7 6 5 4 3 2 1

CONTENTS

CONTRIBUTORS

Bruno Antonny
Institut de Pharmacologie Moléculaire et Cellulaire, Centre National de la Recherche Scientifique, Valbonne, France

Susan D. Arden
Cambridge Institute for Medical Research, University of Cambridge, Cambridge, United Kingdom

William E. Balch
The Institute for Childhood and Neglected Diseases and Department of Cell Biology, The Scripps Research Institute, La Jolla, California

Dafna Bar-Sagi
Department of Biochemistry, New York University School of Medicine, New York, New York

Martin O. Bergo
Wallenberg Laboratory, Department of Medicine, Sahlgrenska University Hospital, Gothenburg, Sweden

Jennifer L. Bromberg-White
Laboratory of Cancer and Developmental Cell Biology, Van Andel Research Institute, Grand Rapids, Michigan

Folma Buss
Cambridge Institute for Medical Research, University of Cambridge, Cambridge, United Kingdom

Paola Caprari
Neural Development Group, Mouse Cancer Genetics Program, National Cancer Institute, Frederick, Maryland

Mark Carrington
Department of Biochemistry, University of Cambridge, Cambridge, United Kingdom

Sergio D. Catz
Division of Biochemistry, Department of Molecular and Experimental Medicine, The Scripps Research Institute, La Jolla, California

Jacqueline Cherfils
Laboratoire d'Enzymologie et Biochimie Structurales, Centre National de la
Recherche Scientifique, Gif-sur-Yvette, France

Margarita V. Chibalina
Cambridge Institute for Medical Research, University of Cambridge, Cambridge,
United Kingdom

Yuchen Chien
Department of Cell Biology, University of Texas Southwestern Medical Center,
Dallas, Texas

Christopher M. Counter
Department of Radiation Oncology and Department of Pharmacology and Cancer
Biology, Durham, North Carolina

Jon M. Davison
Department of Pathology, Johns Hopkins University School of Medicine,
Baltimore, Maryland

Adam Denley
Department of Molecular and Experimental Medicine, The Scripps Research
Institute, La Jolla, California

Nicholas S. Duesbery
Laboratory of Cancer and Developmental Cell Biology, Van Andel Research
Institute, Grand Rapids, Michigan

Edo Elstak
Department of Cell Biology, University Medical Center Utrecht, Utrecht, The
Netherlands

Pedro F. Esteban
Neural Development Group, Mouse Cancer Genetics Program, National Cancer
Institute, Frederick, Maryland

Mark C. Field
Department of Pathology, University of Cambridge, Cambridge, United Kingdom

Catalin M. Filipeanu
Department of Pharmacology and Experimental Therapeutics, Louisiana State
University Health Sciences Center, New Orleans, Louisiana

Loren G. Fong
Division of Cardiology, Department of Internal Medicine, University of California,
Los Angeles, California

Tomo Funaki
Department of Molecular Immunology, Institute of Development, Aging and
Cancer, Tohoku University, Sendai, Japan

Bianka L. Grosshans
Novartis Institutes for Biomedical Research, Basel, Switzerland

Marco Gymnopoulos
Department of Molecular and Experimental Medicine, The Scripps Research Institute, La Jolla, California

Jonathan R. Hart
Department of Molecular and Experimental Medicine, The Scripps Research Institute, La Jolla, California

David Horn
London School of Hygiene and Tropical Medicine, London, United Kingdom

Sebastian Höpfner
Kanzlei Dr. Volker Vossius, Munich, Germany

Darren M. Hutt
Department of Cell Biology, The Scripps Research Institute, La Jolla, California

Nobuyuki Ichijo
Department of Molecular Immunology, Institute of Development, Aging and Cancer, Tohoku University, Sendai, Japan

Hao Jiang
Department of Molecular and Experimental Medicine, The Scripps Research Institute, La Jolla, California

John Kendrick-Jones
MRC Laboratory of Molecular Biology, Cambridge, United Kingdom

Tom Kirchhausen
Department of Cell Biology, Harvard Medical School, and IDI Immune Research Institute, Boston, Massachusetts

Shunsuke Kon
Department of Molecular Immunology, Institute of Development, Aging and Cancer, Tohoku University, Sendai, Japan

Steven D. Leach
Department of Surgery, Oncology and Cell Biology, Johns Hopkins University School of Medicine, Baltimore, Maryland

Corinne M. Linardic
Department of Pediatrics, Duke University, Durham, North Carolina, and Department of Pharmacology and Cancer Biology, Durham, North Carolina

Eric Macia
L'Institut de Pharmacologie Moléculaire et Cellulaire, CNRS, Valbonne, France, and Department of Cell Biology, Harvard Medical School, and IDI Immune Research Institute, Boston, Massachusetts

Sally Martin
Institute for Molecular Bioscience and Centre for Microscopy and Microanalysis, University of Queensland, Brisbane, Queensland, Australia

Ultan McDermott
Center for Molecular Therapeutics, Massachusetts General Hospital Cancer Center and Harvard Medical School, Charlestown, Massachusetts

Jun Miyoshi
Department of Molecular Biology, Osaka Medical Center for Cancer and Cardiovascular Diseases, Osaka, Japan

Sohini Mukherjee
University of Texas Southwestern Medical Center, Dallas, Texas

Waka Natsume
Department of Molecular Immunology, Institute of Development, Aging and Cancer, Tohoku University, Sendai, Japan

Maaike Neeft
Department of Cell Biology, University Medical Center Utrecht, Utrecht, The Netherlands

Noriyuki Nishimura
Department of Biochemistry, Institute of Health Biosciences, The University of Tokushima Graduate School, Tokushima, Japan

Peter Novick
Department of Cell Biology, Yale University School of Medicine, New Haven, Connecticut

Kim Orth
University of Texas Southwestern Medical Center, Dallas, Texas

Kazuhiro Osanai
Department of Respiratory Medicine, Kanazawa Medical University, Kahokugun, Ishikawa, Japan

Arun Pal
Max Planck Institute of Molecular Cell Biology and Genetics, Dresden, Germany

Robert G. Parton
Institute for Molecular Bioscience and Centre for Microscopy and Microanalysis, University of Queensland, Brisbane, Queensland, Australia

Seung Woo Park
Department of Internal Medicine, Yonsei University College of Medicine, Seoul, South Korea

Henry E. Pelish
Makato Life Sciences, Inc., Boston, Massachusetts, and Department of Cell Biology, Harvard Medical School, and IDI Immune Research Institute, Boston, Massachusetts

Paul A. Randazzo
Laboratory of Cellular and Molecular Biology, National Cancer Institute, Bethesda, Maryland

Katherine A. Rauen
Department of Pediatrics, Division of Medical Genetics and UCSF Helen Diller Family, Comprehensive Cancer Center and Cancer Research Institute, University of California, San Francisco, California

Jerry M. Rhee
Department of Surgery, Oncology and Cell Biology, Johns Hopkins University School of Medicine, Baltimore, Maryland

Rhys C. Roberts
MRC Laboratory of Molecular Biology, Cambridge, United Kingdom

Pablo Rodriguez-Viciana
UCSF Helen Diller Family, Comprehensive Cancer Center and Cancer Research Institute, University of California, San Francisco, California

Ayuko Sakane
Department of Biochemistry, Institute of Health Biosciences, The University of Tokushima Graduate School, Tokushima, Japan

Takuya Sasaki
Department of Biochemistry, Institute of Health Biosciences, The University of Tokushima Graduate School, Tokushima, Japan

Masanobu Satake
Department of Molecular Immunology, Institute of Development, Aging and Cancer, Tohoku University, Sendai, Japan

Tatsuhiro Sato
Department of Microbiology, Immunology and Molecular Genetics, Jonsson Comprehensive Cancer Center, Molecular Biology Institute, University of California, Los Angeles, California

Jeffrey Settleman
Center for Molecular Therapeutics, Massachusetts General Hospital Cancer Center and Harvard Medical School, Charlestown, Massachusetts

Fedor Severin
A. N. Belozersky Institute of Physico-Chemical Biology, Moscow State University, Russia

Sreenath V. Sharma
Center for Molecular Therapeutics, Massachusetts General Hospital Cancer Center and Harvard Medical School, Charlestown, Massachusetts

Peter van der Sluijs
Department of Cell Biology, University Medical Center Utrecht, Utrecht, The Netherlands

Yoshimi Takai
Division of Molecular and Cellular Biology, Department of Biochemistry and Molecular Biology, Kobe University Graduate School of Medicine/Faculty of Medicine, Kobe, Japan

Fuyuhiko Tamanoi
Department of Microbiology, Immunology and Molecular Genetics, Jonsson Comprehensive Cancer Center, Molecular Biology Institute, University of California, Los Angeles, California

Kenji Tanabe
Department of Molecular Immunology, Institute of Development, Aging and Cancer, Tohoku University, Sendai, Japan

Sonia Terrillon
Department of Biochemistry, New York University School of Medicine, New York, New York

Lino Tessarollo
Neural Development Group, Mouse Cancer Genetics Program, National Cancer Institute, Frederick, Maryland

Akiko Umetsu
Department of Microbiology, Immunology and Molecular Genetics, Jonsson Comprehensive Cancer Center, Molecular Biology Institute, University of California, Los Angeles, California

Thijs van Vlijmen
Department of Cell Biology, University Medical Center Utrecht, Utrecht, The Netherlands

Dennis R. Voelker
Program in Cell Biology, Department of Medicine, National Jewish Medical and Research Center, Denver, Colorado

Peter K. Vogt
Department of Molecular and Experimental Medicine, The Scripps Research Institute, La Jolla, California

Annika M. Wahlstrom
Wallenberg Laboratory, Department of Medicine, Sahlgrenska University Hospital, Gothenburg, Sweden

Toshio Watanabe
Department of Molecular Immunology, Institute of Development, Aging and Cancer, Tohoku University, Sendai, Japan

Michael A. White
Department of Cell Biology, University of Texas Southwestern Medical Center, Dallas, Texas

Marnix Wieffer
Department of Cell Biology, University Medical Center Utrecht, Utrecht, The Netherlands

Guangyu Wu
Department of Pharmacology and Experimental Therapeutics, Louisiana State University Health Sciences Center, New Orleans, Louisiana

Hye-Young Yoon
Laboratory of Cellular and Molecular Biology, National Cancer Institute, Bethesda, Maryland

Stephen G. Young
Division of Cardiology, Department of Internal Medicine, University of California, Los Angeles, California

Jean-Christophe Zeeh
Laboratoire d'Enzymologie et Biochimie Structurales, Centre National de la Recherche Scientifique, Gif-sur-Yvette, France

Mahel Zeghouf
Laboratoire d'Enzymologie et Biochimie Structurales, Centre National de la Recherche Scientifique, Gif-sur-Yvette, France

Marino Zerial
Max Planck Institute of Molecular Cell Biology and Genetics, Dresden, Germany

Li Zhao
Department of Molecular and Experimental Medicine, The Scripps Research Institute, La Jolla, California

Fuguo Zhou
Department of Pharmacology and Experimental Therapeutics, Louisiana State University Health Sciences Center, New Orleans, Louisiana

Methods in Enzymology

VOLUME 275. Viral Polymerases and Related Proteins
Edited by LAWRENCE C. KUO, DAVID B. OLSEN, AND STEVEN S. CARROLL

VOLUME 276. Macromolecular Crystallography (Part A)
Edited by CHARLES W. CARTER, JR., AND ROBERT M. SWEET

VOLUME 277. Macromolecular Crystallography (Part B)
Edited by CHARLES W. CARTER, JR., AND ROBERT M. SWEET

VOLUME 278. Fluorescence Spectroscopy
Edited by LUDWIG BRAND AND MICHAEL L. JOHNSON

VOLUME 279. Vitamins and Coenzymes (Part I)
Edited by DONALD B. MCCORMICK, JOHN W. SUTTIE, AND CONRAD WAGNER

VOLUME 280. Vitamins and Coenzymes (Part J)
Edited by DONALD B. MCCORMICK, JOHN W. SUTTIE, AND CONRAD WAGNER

VOLUME 281. Vitamins and Coenzymes (Part K)
Edited by DONALD B. MCCORMICK, JOHN W. SUTTIE, AND CONRAD WAGNER

VOLUME 282. Vitamins and Coenzymes (Part L)
Edited by DONALD B. MCCORMICK, JOHN W. SUTTIE, AND CONRAD WAGNER

VOLUME 283. Cell Cycle Control
Edited by WILLIAM G. DUNPHY

VOLUME 284. Lipases (Part A: Biotechnology)
Edited by BYRON RUBIN AND EDWARD A. DENNIS

VOLUME 285. Cumulative Subject Index Volumes 263, 264, 266–284, 286–289

VOLUME 286. Lipases (Part B: Enzyme Characterization and Utilization)
Edited by BYRON RUBIN AND EDWARD A. DENNIS

VOLUME 287. Chemokines
Edited by RICHARD HORUK

VOLUME 288. Chemokine Receptors
Edited by RICHARD HORUK

VOLUME 289. Solid Phase Peptide Synthesis
Edited by GREGG B. FIELDS

VOLUME 290. Molecular Chaperones
Edited by GEORGE H. LORIMER AND THOMAS BALDWIN

VOLUME 291. Caged Compounds
Edited by GERARD MARRIOTT

VOLUME 292. ABC Transporters: Biochemical, Cellular, and Molecular Aspects
Edited by SURESH V. AMBUDKAR AND MICHAEL M. GOTTESMAN

VOLUME 293. Ion Channels (Part B)
Edited by P. MICHAEL CONN

Rab1b Silencing Using Small Interfering RNA for Analysis of Disease-Specific Function

Darren M. Hutt* *and* William E. Balch*,†

Contents

Abstract

Rab1 GTPase is a critical component required for endoplasmic reticulum (ER)-to-Golgi as well as intra-Golgi trafficking. It is required for the proper recruitment of tethering factors to mediate vesicle docking and subsequent fusion to the target membrane compartment. Much is known about the role of Rab1 in ER-to-Golgi trafficking through overexpression of dominant negative mutation that inhibit GTP binding or GTPase activity of the protein, as well as through the use of antibodies to inhibit endogenous protein activity. These techniques have allowed for the establishment of a central role for Rab1 in trafficking in the early secretory pathway; however, the use of these techniques is limited. The introduction of antibodies relies on permeabilization of the cell membrane for their introduction. The use of dominant negative mutations relies on the mutation overwhelming the endogenous protein activity without removing it from the cell. The advent of siRNA to silence genes of interest provides a means to overcome this limitation. Here we describe optimal conditions for the efficient silencing of Rab1b using siRNA to analyze its role in disease.

* Department of Cell Biology, The Scripps Research Institute, La Jolla, California
† The Institute for Childhood and Neglected Diseases, The Scripps Research Institute, La Jolla, California

Methods in Enzymology, Volume 438
ISSN 0076-6879, DOI: 10.1016/S0076-6879(07)38001-4

1. INTRODUCTION

With over 70 mammalian isoforms, Rab GTPases represent the largest of the five subfamilies of the Ras superfamily of proteins (Pereira-Leal and Seabra, 2001) with each isoform thought to mediate a specific trafficking step. Rab GTPases, as their Ras counterparts, cycle between an active membrane–associated, GTP-bound, and inactive cytosolic, GDP-bound states, and are therefore ideal molecular switches for regulating the orderly progression of vesicles through membranous compartments of the eukaryotic exocytic and endocytic pathways. Following activation by the guanine nucleotide exchange factor (GEF), Rab GTPases mediate the recruitment of effector proteins, which include tethering and fusion factors, to mediate the docking and fusion of trafficking vesicles with its targeted acceptor membrane compartment (Gurkan *et al.*, 2005; Pfeffer and Aivazian, 2004). Following membrane docking and fusion, a GTPase-activating protein (GAP) initiates the GTP hydrolysis of Rab proteins, resulting in inactivation of the protein as well as the generation of a binding target for the GDP dissociation inhibitor (GDI). The latter mediates the extraction of Rab GTPases from the acceptor membrane compartment and directs it's recycling to the donor compartment.

Rab1 had been reported to localize to both the endoplasmic reticulum (ER) and the Golgi compartments (Plutner *et al.*, 1991), suggesting that it might mediate the trafficking between these compartments. *In vitro* reconstitution experiments in yeast indicate that the docking of ER-derived vesicles is sensitive to the GTPase, Ypt1p, which in turn regulates Uso1p-dependent tethering of donor vesicles to a target membrane (Cao *et al.*, 1998). This suggests that Ypt1p mediates tethering of ER-derived vesicles prior to membrane fusion. This observation is consistent with the proposed role of Rab1, the mammalian ortholog of Ypt1p, in mediating the docking of ER-derived vesicle with the cis-Golgi compartment. Rab1 has been shown to recruit the cytosolic tethering factors p115 to the vesicle (Allan *et al.*, 2000) and to interact with the Golgi membrane proteins GM130 and GRASP (Moyer *et al.*, 2001) to mediate the docking process. In addition, antibodies specific for Rab1 were able to block ER to Golgi trafficking as well as intra-Golgi trafficking in semipermeable cells (Plutner *et al.*, 1991). These data support a critical role for Rab1 in ER to Golgi trafficking.

In order to characterize the role of Rab1 in trafficking various cargo proteins through the endomembrane compartments of the cell, we have relied on the availability of dominant negative mutations and antibodies to inhibit the function of this protein. The use of these tools is somewhat limited, namely, that the overexpression of mutant Rab1 is dependent on efficiency of transfection in the cell line of interest as well as the ability of the given mutant to overcome the endogenous Rab1 and act as a dominant negative. In addition, the introduction of antibodies into the cytosol of cells has proven to be a useful tool to block endogenous protein function. However, this invasive technique requires microinjection or the permeabilization of the plasma membrane, which renders the given experiment into an artificial system since the normal homeostasis of the cell is compromised.

In recent years, the use of small inhibitory RNA (siRNA) molecules as a silencing tool has grown exponentially. This technique calls for the introduction of a short RNA sequence, typically 19 nucleotides in length, which is directed to a sequence-specific region in a target gene without off-target silencing of other potentially critical genes. The siRNA is introduced using various techniques, including cationic lipids, which is reminiscent of standard transfection techniques used for the introduction of plasmid DNA. There is now an extensive literature and a large number of companies that design siRNA probes that have minimal "off-target" (e.g., nonspecific) effects that can be used for silencing studies.

In order to ascertain that Rab1b knockdown resulted not only in decreased levels of endogenous Rab1, but also in impaired ER to Golgi trafficking, we monitored the maturation of the cystic fibrosis transmembrane conductance regulator (CFTR), the gene mutated in cystic fibrosis (Kerem et al., 1989; Riordan et al., 1989; Rommens et al., 1989). The gene codes for a 1480–amino-acid chloride channel that localizes to the apical plasma membrane of epithelial cells (Anderson et al., 1991a,b). As CFTR is processed in the ER, it acquires its N-linked core glycosylation. This glycosylation is further processed to an endoglycosidase H–resistant form as it matures in the Golgi compartment. These two glycoforms of the protein migrate at 135 and 160 kDa, respectively, on SDS-PAGE, and are referred to as band B and C, respectively. This differential migration of the ER and Golgi forms of the protein allows for a direct measure of the trafficking and maturation of CFTR by Western blotting (Cheng et al., 1990). Although there are currently more than 1500 mutations linked to cystic fibrosis, greater than 90% of patients have at least one copy of the CFTR gene containing a three–base-pair deletion resulting in the loss of phenylalanine at position 508 (ΔF508). This mutation results in a trafficking-deficient form of CFTR that is restricted to the ER, and therefore is only detectable on SDS-PAGE as the band B glycoform (Cheng et al., 1990).

Herein, we describe the methodologies required for the efficient silencing of Rab1b in a lung cell line and an analysis of its effect on the trafficking of both the wildtype CFTR and ΔF508 CFTR.

2. MATERIALS FOR siRNA SILENCING OF Rab1B

Purified siRNA duplex–targeting human Rab1b was purchased from Ambion (cat. no. AM16104), and the sense strand sequence was GAUCC-GAACCAUCGAGCUGtt. The nontargeting siRNA sense strand sequence used for these experiments was GCGCGCUUUGUAGGAUUCtt and was purchased from Dharmacon. HiPerFect transfection reagent was purchased from Qiagen. Rab1 was detected using a previously characterized polyclonal antibody (p68) (Plutner *et al.*, 1991), and CFTR was detected using the M3A7 ascites raised against the nucleotide-binding domain 2 (NBD2) of human CFTR, which was obtained from J. R. Riordan, PhD (University of North Carolina).

Human bronchial epithelial cells (HBEs) and cystic fibrosis bronchial epithelial cells (CFBE41o-) were cultured in growth media: α-MEM culture media (Gibco cat. no. 12000–063) supplemented with 100 U/ml penicillin, 100 μg/ml streptomycin, 10% FBS, 2 mM L-glutamine, and 1 μg/ml blasticidin or 2 μg/ml puromycin, respectively.

3. PROCEDURE FOR TRANSFECTION OF siRNA

1. One day prior to transfection HBE or CFBE, cells are seeded at a density of 1.0×10^5 cells per well of a 12-well dish, and cultured overnight in growth media. This cell density will allow cells to be confluent 3 days post-transfection.
2. On the day of transfection, remove the growth media and replace with 1.1 ml of fresh growth media.
3. Dilute the siRNA to a 12× working concentration in 100 μl of α-MEM without serum and add 6 μl of HiPerFect reagent. Vortex the mixture at 75% maximum for 5 s. This mixture is for each well of a 12-well culture dish that is to be transfected.
4. Incubate the mixture for 10 min at room temperature to generate transfection complexes.
5. Add the mixture drop-wise to the well containing 1.1 ml of growth media, and incubate for 48 h at 37°/5% CO_2.
6. Following the initial 48-h incubation, replace the growth media containing transfection complexes with 1.1 ml of fresh growth media.
7. Prepare transfection complexes as in steps 3 to 5.
8. Culture cells for an additional 48 h.
9. Cells are harvested following 96 h for analysis.

4. TITRATION OF OPTIMAL siRNA CONCENTRATION

In order to avoid transfection-related toxicity, it is necessary to optimize the siRNA concentration for efficient silencing of Rab1b. HBE cells were transfected with a complex composed of 6 μl of HiPerFect transfection reagent and 5, 15, or 50 nM final concentration of Rab1b siRNA or 50 nM final concentration of the scrambled control siRNA as described above. One set of cells was transfected once with the indicated siRNA concentration and cultured for 2 days post-transfection, while the second set of cells was transfected twice with the indicated concentration of siRNA at 2-day intervals and cultured for 2 days after the final transfection as described above (Fig. 1.1). These results show that cells which were only transfected with a single hit of siRNA only had a maximal knockdown of Rab1 of 20.8 $\pm$ 0.8% at the highest concentration of siRNA (50 nM)

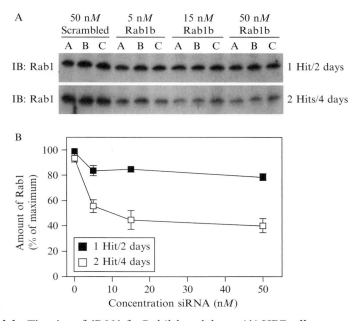

Figure 1.1   Titration of siRNA for Rab1b knockdown. (A) HBE cells were transfected with the indicated siRNA concentration for one hit (upper panel) or two hits (lower panel) and cultured for 2 days post-transfection. Rab1 was detected on Western blot using the polyclonal antibody, p68. (B) Quantitative analysis of Rab1 knockdown in HBE cells. The amount of Rab1 in HBE cells following transfection with 5, 15, or 50nM siRNA for one hit (solid squares) or two hits (open squares) is quantified as a percent of Rab1 in HBE cells transfected with 50nM scrambled siRNA. Error bars represent the standard error of the mean ($n = 3$).

relative to scrambled control. When cells were transfected with two hits of Rab1b siRNA at this same concentration, the knockdown of Rab1 was 59.6 ± 8.5. These data reveal that the optimal conditions for efficient silencing of Rab1b in HBE cells include two sequential transfections with complexes composed of 50 nM final concentration of siRNA mixed with 6 μl of HiPerFect. The optimal volume of HiPerFect to be used for each well was previously determined for other siRNA targets (data not shown).

5. SDS-PAGE ANALYSIS OF RAB1 KNOCKDOWN

Lysis Buffer: 50 mM Tris-HCl, pH 7.4, 150 mM NaCl, 1% Triton X-100, and 2 mg/ml of EDTA-free protease-inhibitor cocktail (Roche cat. no. 13129100)
SDS Sample Buffer: 120 mM, Tris-HCl, pH 7.0, 30% glycerol, 6% SDS, 0.6% bromophenol blue, and 100 mM DTT

Following the knockdown procedure described above, cells were transferred to ice and incubated in lysis buffer on ice for 30 min with occasional rocking. The lysates were collected and centrifuged at 14,000 $\times$ g for 20 min at 4° and the supernatants were collected. The protein concentration in each sample was determined by Bradford assay (Pierce product no. 1856209) using BSA (Pierce product no. 23209) as a standard. Equal amounts of protein (30 μg) for each sample were supplemented with 6$\times$ SDS Sample Buffer and boiled for 15 min prior to loading. The proteins are separated on a 12% SDS-PAGE run and subjected to Western blotting with polyclonal anti-Rab1 antibody p68 (Plutner *et al.*, 1991). The detection was done by ECL and the band intensities compared by quantitative analysis using an Alpha Innotech Fluor SP apparatus and accompanying software (Figs. 1.2 and 1.3; see also Fig. 1.1).

6. SDS-PAGE ANALYSIS OF CFTR

Analysis of CFTR was performed by SDS-PAGE as described for Rab1; however, samples were incubated at 37° for 15 min in SDS sample buffer instead of boiling since CFTR irreversibly aggregates at high temperatures and will not be detected by SDS-PAGE. These samples were separated on 8% SDS-PAGE in order to resolve the ER and Golgi glycoforms from one another. CFTR was detected using the M3A7 ascites raised against the NBD2 domain of CFTR, and band intensities were analyzed as for Rab1. Relative intensities were normalized to percent of maximal intensity and replicates were averaged and compared using a two-tailed t-test (see Figs. 1.2 and 1.3).

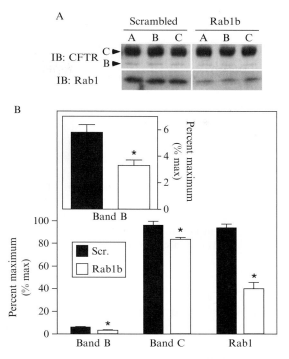

Figure 1.2 Knockdown of Rab1b affects stability and trafficking of wildtype CFTR. (A) Western blot analysis of CFTR in cells transfected with 50 nM Rab1b or scrambled siRNA. (B) Quantitative analysis of CFTR stability and trafficking in cells transfected with 50nM Rab1b (open bar) or scrambled (solid bar) siRNA. The levels of bands B and C are expressed as a percent of the most intense signal. ($n = 3$, $\star p = 0.05$) *Inset:* Magnification of the band B signal to illustrate the effect of Rab1b siRNA.

In order to ensure that the knockdown of Rab1 resulted in the predicted ER–to–Golgi trafficking defect, we assessed the trafficking and stability of both wildtype and ΔF508 CFTR. A 60% knockdown of Rab1b in cells expressing wildtype CFTR caused a 43.1 ± 6.9% decrease in the levels of band B (ER) and a concomitant 19.3 ± 3.1% decrease in band C (post-Golgi) (see Fig. 1.2A and B). Upon ER export of CFTR, there is no accumulation within the Golgi compartment nor in endocytic vesicular storage pools; hence the band C glycoform represents surface CFTR, which has been previously shown to exhibit a half life on the order of 24 h or more (Lukacs *et al.*, 1993). This suggests that any effects on ER–to–Golgi trafficking resulting from Rab1 knockdown would have minimal effect on the surface pool of wildtype CFTR unless these conditions persisted for prolonged periods of time. This observation is confirmed by the fact that we observed a ∼20% reduction in the band C glycoform while a 50% reduction

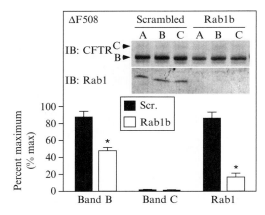

Figure 1.3 Knockdown of Rab1b affects stability of ΔF508 CFTR. Quantitative analysis of ΔF508 CFTR in cells transfected with Rab1b (open bar) or scrambled (closed bar) siRNA. The levels of bands B and C are expressed as a percent of the most intense signal ($n = 3$, $\star p = 0.05$).

in the band B glycoform is seen. The decreased levels of the ER glycoform of wildtype CFTR is associated with both a decrease in Rab1b synthesis reflecting its loss of message and rapid proteosomal degradation of CFTR, a pathway that prevents accumulation of wildtype CFTR that fails to exit the ER. In order to more completely ascertain the effect of Rab1b depletion on cell surface stability, a pulse chase experiment would be required to follow both trafficking kinetics (conversion from B to C) and the kinetics of degradation of the band C glycoform.

The data above confirm that the knockdown of Rab1b using siRNA is able to result in a defect in ER-to-Golgi trafficking of wildtype CFTR in human cell lines. This is supported by the effect of Rab1b knockdown on the stability of ΔF508 CFTR, which is unable to traffic to the Golgi compartment and mature to the band C glycoform. Under conditions of maximal Rab1b knockdown, there is a $45.2 \pm 5.0\%$ decrease in the band B form of ΔF508 CFTR (see Fig. 1.3).

7. DISCUSSION

We have shown that transfection complexes composed of 50 nM siRNA and 6 μl of HiPerFect transfection reagent presented to cells in two separate hits over a 4-day culture period provide the optimal silencing of Rab1 in human bronchial epithelial cells. Although the concentration of siRNA and the amount of transfection reagent, as well as the length of the post-transfection incubation should be optimized for a given siRNA and cell line as per manufacturer suggestions, similar protocols can be used to successfully silence a number of other proteins in these same cell lines as well as in human embryonic kidney cells (HEK293) (data not shown). Although some variation in efficiency of silencing is observed for knockdown of different targets, with the correct siRNA probes, it is not unusual to obtain more than 50% of control, often a value sufficient to elicit a meaningful biological effect. More efficient knockdowns (>80%) are likely to yield more pronounced biological effects, and can be observed pending design and choice of siRNA probe. Although monitoring steady-state levels of a protein is sufficient to indicate impact on cell physiology, a protein with a long half-life could have altered functionality in response to multiple epigenetic modifications (phosphorylation, acetylation, etc.), or in the case of membrane-bound proteins, redistribution to different compartments yielding a different functionality(s). Thus, a second measure of efficacy of knockdown is to follow the protein of interest using a pulse-chase protocol. For example, monitoring the kinetic of synthesis of Rab1b would yield a direct measure of translating message. Moreover, in addition to monitoring the steady state levels of both the B and C glycoforms of CFTR, it would have also been useful to follow maturation of *de novo* synthesized protein via a pulse chase experiment to obtain a more direct read-out of the ER, Golgi, and cell surface trafficking events. The least reliable method in terms of defining knockdown of function is to follow the message level by quantitative RT-PCR given the unknown half-life of a particular target. However, it has the advantage of being rapid, and does not require additional target-specific reagents. Quantitative RT-PCR can, at minimum, indicate efficacy of the siRNA probe being used, and is useful to monitor the presence of various closely related isoforms (e.g., Rab1a) that could confound interpretation of results if they display overlap in function.

In general, it should be kept in mind that alteration of a particular protein activity such as Rab1b by siRNA silencing may have multiple indirect effects on linked pathways over the typical time course of an siRNA experiment. Consequently, results should be interpreted with caution and validated by alternative approaches, and through rescue of silencing by supplementation with a silencing-resistant cDNA construct with altered code encoding the target protein where possible.

REFERENCES

Allan, B. B., Moyer, B. D., and Balch, W. E. (2000). Rab1 recruitment of p115 into a cis-SNARE complex: Programming budding COPII vesicles for fusion. *Science* **289,** 444–448.

Anderson, M. P., Gregory, R. J., Thompson, S., Souza, D. W., Paul, S., Mulligan, R. C., Smith, A. E., and Welsh, M. J. (1991a). Demonstration that CFTR is a chloride channel by alteration of its anion selectivity. *Science* **253,** 202–205.

Anderson, M. P., Rich, D. P., Gregory, R. J., Smith, A. E., and Welsh, M. J. (1991b). Generation of cAMP-activated chloride currents by expression of CFTR. *Science* **251,** 679–682.

Cao, X., Ballew, N., and Barlowe, C. (1998). Initial docking of ER-derived vesicles requires Uso1p and Ypt1p but is independent of SNARE proteins. *EMBO J.* **17,** 2156–2165.

Cheng, S. H., Gregory, R. J., Marshall, J., Paul, S., Souza, D. W., White, G. A., O'Riordan, C. R., and Smith, A. E. (1990). Defective intracellular transport and processing of CFTR is the molecular basis of most cystic fibrosis. *Cell* **63,** 827–834.

Gurkan, C., Lapp, H., Alory, C., Su, A. I., Hogenesch, J. B., and Balch, W. E. (2005). Large-scale profiling of Rab GTPase trafficking networks: The membrome. *Mol. Biol. Cell* **16,** 3847–3864.

Kerem, B., Rommens, J. M., Buchanan, J. A., Markiewicz, D., Cox, T. K., Chakravarti, A., Buchwald, M., and Tsui, L. C. (1989). Identification of the cystic fibrosis gene: Genetic analysis. *Science* **245,** 1073–1080.

Lukacs, G. L., Chang, X. B., Bear, C., Kartner, N., Mohamed, A., Riordan, J. R., and Grinstein, S. (1993). The delta F508 mutation decreases the stability of cystic fibrosis transmembrane conductance regulator in the plasma membrane. Determination of functional half-lives on transfected cells. *J. Biol. Chem.* **268,** 21592–21598.

Moyer, B. D., Allan, B. B., and Balch, W. E. (2001). Rab1 interaction with a GM130 effector complex regulates COPII vesicle cis—Golgi tethering. *Traffic* **2,** 268–276.

Pereira-Leal, J. B., and Seabra, M. C. (2001). Evolution of the Rab family of small GTP-binding proteins. *J. Mol. Biol.* **313,** 889–901.

Pfeffer, S., and Aivazian, D. (2004). Targeting Rab GTPases to distinct membrane compartments. *Mol. Cell Biol.Nat. Rev. Mol. Cell. Biol.* **5,** 886–896.

Plutner, H., Cox, A. D., Pind, S., Khosravi-Far, R., Bourne, J. R., Schwaninger, R., Der, C. J., and Balch, W. E. (1991). Rab1b regulates vesicular transport between the endoplasmic reticulum and successive Golgi compartments. *J. Cell Biol.* **115,** 31–43.

Riordan, J. R., Rommens, J. M., Kerem, B., Alon, N., Rozmahel, R., Grzelczak, Z., Zielenski, J., Lok, S., Plavsic, N., and Chou, J. L. (1989). Identification of the cystic fibrosis gene: Cloning and characterization of complementary DNA. *Science* **245,** 1066–1073.

Rommens, J. M., Iannuzzi, M. C., Kerem, B., Drumm, M. L., Melmer, G., Dean, M., Rozmahel, R., Cole, J. L., Kennedy, D., and Hidaka, N. (1989). Identification of the cystic fibrosis gene: Chromosome walking and jumping. *Science* **245,** 1059–1065.

CHAPTER TWO

Rab8-Optineurin-Myosin VI: Analysis of Interactions and Functions in the Secretory Pathway

Margarita V. Chibalina,* Rhys C. Roberts,† Susan D. Arden,* John Kendrick-Jones,† and Folma Buss*

Contents

Abstract

The small GTPase Rab8 has been shown to regulate polarized membrane trafficking pathways from the TGN to the cell surface. Optineurin is an effector protein of Rab8 and a binding partner of the actin-based motor protein myosin VI. We used various approaches to study the interactions between myosin VI and its binding partners and to analyze their role(s) in intracellular membrane trafficking pathways. In this chapter, we describe the use of the mammalian two-hybrid assay to demonstrate protein–protein interactions and to identify

* Cambridge Institute for Medical Research, University of Cambridge, Cambridge, United Kingdom
† MRC Laboratory of Molecular Biology, Cambridge, United Kingdom

Methods in Enzymology, Volume 438
ISSN 0076-6879, DOI: 10.1016/S0076-6879(07)38002-6

binding sites. We describe a secretion assay that was used in combination with RNA interference technology to analyze the function of myosin VI, optineurin, and Rab8 in exocytic membrane trafficking pathways.

1. INTRODUCTION

The Rab proteins are a large family of small GTPases that participate in and regulate intracellular membrane trafficking pathways (Zerial and McBride, 2001). For example, Rab8 plays an important role in exocytic membrane traffic from the Golgi complex to the plasma membrane in polarized epithelial cells (Huber *et al.*, 1993b), in photoreceptor cells (Moritz *et al.*, 2001), and in polarized neurons (Huber *et al.*, 1993a). It is localized in vesicles at/around the trans-Golgi network, in recycling endosomes, and at the plasma membrane in membrane ruffles (Ang *et al.*, 2003; Peranen *et al.*, 1996). Rab8 binds to optineurin (Hattula and Peranen, 2000), an adapter protein associated with the Golgi complex, which is known to bind to the actin-based motor protein myosin VI (Sahlender *et al.*, 2005), and to huntingtin, the protein mutated in Huntington's disease (Faber *et al.*, 1998). The characterization of Rab8 effector proteins and complexes is required to establish the precise role of Rab8 in exocytic membrane trafficking pathways. Furthermore, this approach will help us to understand the role of Rab8 and its effector proteins in a number of diseases.

Rab8 is linked by optineurin to myosin VI, which has been shown to be involved in endocytic and exocytic membrane trafficking pathways. Mutations in the myosin VI gene in mouse and humans are linked to deafness (Ahmed *et al.*, 2003; Avraham *et al.*, 1995) and to hypertrophic cardiomyopathy (Mohiddin *et al.*, 2004), and the myosin VI knockout mouse also has neurological abnormalities (Osterweil *et al.*, 2005). The specific intracellular functions of myosin VI are mediated by a range of interacting proteins (Buss *et al.*, 2004). Optineurin was identified as a myosin VI–binding partner in a yeast two-hybrid screen, and was shown to link myosin VI to the Golgi complex and to mediate its function in the secretory pathway (Au *et al.*, 2007; Sahlender *et al.*, 2005). This interaction between myosin VI and optineurin may be of medical interest, since mutations in the human optineurin gene cause primary open–angle glaucoma (Rezaie *et al.*, 2002). Optineurin is a 67-kDa protein that contains two leucine zippers and one zinc finger separated by large stretches of coiled coil. This myosin VI binding partner is the cellular target of the adenoviral protein Ad E3–14.7 K, which inhibits tumor necrosis factor alpha (TNF-A)–induced cytolysis, an important defense mechanism to protect cells against viral infection (Li *et al.*, 1998). Optineurin also links Rab8 to huntingtin, the protein mutated in the progressive neurodegenerative disorder, Huntington's disease (Hattula and Peranen, 2000). Although a wealth of data suggests that polyglutamine-mediated aggregation of huntingtin is a major cause of the disease, new evidence suggests that loss of normal huntingtin function may also play a role in Huntington's disease pathogenesis. Wildtype huntingtin is present at/around the TGN and on vesicles, and is suggested to play a role in endocytosis and exocytosis (Velier *et al.*, 1998) (DiFiglia *et al.*, 1995).

Therefore, to understand the function of Rab8, optineurin, myosin VI, and huntingtin in exocytic membrane-trafficking pathways, it is of vital importance to establish how they interact in defined protein complexes.

This chapter describes the experimental approaches we have used to identify and verify myosin VI–binding partners and methods to test their function in the secretory pathway.

2. IDENTIFICATION OF MYOSIN VI–BINDING PARTNERS

The two-hybrid approach in yeast has been used extensively during the past decade as a tool to identify and characterize macromolecular interactions. The yeast two-hybrid system is an *in vivo* system that identifies interacting proteins by the reconstitution of active transcription factor complexes (Chien *et al.*, 1991; Fields and Song, 1989). A protein of interest is fused to a DNA-binding domain of a transcription factor, while a second protein is fused to an activating domain. If the proteins of interest interact, an active transcription factor is formed and expression of specific reporter gene occurs. The yeast two-hybrid system can be used in two main ways: to screen a library for novel interacting proteins (Stephens and Banting, 2000) or to further characterize binding of known interactors, such as by mapping interacting domains (Eugster *et al.*, 2000). As a first step to identify myosin VI–binding partners, we performed a yeast two-hybrid screen. The myosin VI tail was used as a bait to screen a human umbilical vein epithelial cells cDNA library. To minimize self-activation of reporter gene expression by the myosin VI tail, we used the GAL4-based ProQuest Two-Hybrid System (Invitrogen), in which HIS3 reporter sensitivity was titrated with increasing concentrations of 3-amino-1,2,4-triazole (3AT), as described in the ProQuest Instruction Manual (Invitrogen). Using this method, 1 million yeast transformants were screened, yielding 14 different positive interactors.

3. VERIFICATION OF YEAST TWO-HYBRID SCREEN RESULTS

Since one of the major problems with the yeast two-hybrid system is the isolation of false positive interactions, we used a range of different *in vitro* and *in vivo* assays to verify "true" myosin VI interacting proteins. First, we used a mammalian two-hybrid system to demonstrate that the interactions occur in mammalian cells, where the expressed proteins are more likely to be in their native conformation. Furthermore, since myosin VI is not expressed in yeast, the mammalian cell is more likely to express, for example, regulatory kinases or phosphatases that might be involved in controlling the interaction between myosin VI and its binding partner. Next, we identified and mutated the exact binding site and showed for several binding partners, such as optineurin, Dab2, and GIPC, that highly specific binding to the C-terminal tail of myosin VI depends on a short amino acid sequence (Sahlender *et al.*, 2005; Spudich *et al.*, 2007). In the case of Dab2, a single point mutation can abolish binding. To further test whether the interactions between the proteins were direct or part of a larger protein complex, we performed GST pull-down assays of bacterial expressed and purified myosin VI tail and *in vitro* translated optineurin or *in vitro* translated Rab8. In addition, we immunoprecipitated myosin VI/optineurin–containing protein complexes under native conditions from a variety of tissue culture cells to verify the interaction *in vivo*. We also performed localization studies of myosin VI and its binding partners by indirect immunofluorescence to establish whether both proteins are present on the same intracellular compartment, such as, for example, the Golgi complex or intracellular vesicles. Finally, we performed functional studies on siRNA knockdown cells or on cells overexpressing dominant negative mutants, to see whether myosin VI and its proposed binding partner were functioning in the same intracellular pathway and whether the ablation leads to similar phenotypes.

Since GST pull-down assays and coimmunoprecipitation are standard procedures described elsewhere in great detail (see, for example, Johansson and Olkkonen, 2005; Monier and Goud, 2005), we will focus first of all on the mammalian two-hybrid assay and its use to verify and further dissect protein–protein interaction. To identify and characterize proteins involved in constitutive secretion, we have developed a SEAP assay, which is described later.

4. MAMMALIAN TWO-HYBRID ASSAY

The principles of the mammalian two-hybrid system are the same as for the yeast two-hybrid system, namely that the protein of interest is fused to a DNA-binding domain and the potential interacting protein is fused to an activating domain of a transcription factor. If the proteins of interest interact, an active transcription factor is formed and expression of the luciferase reporter occurs. Figure 2.1 illustrates the components of the mammalian two-hybrid assay. Four plasmids carrying two proteins of interest fused to

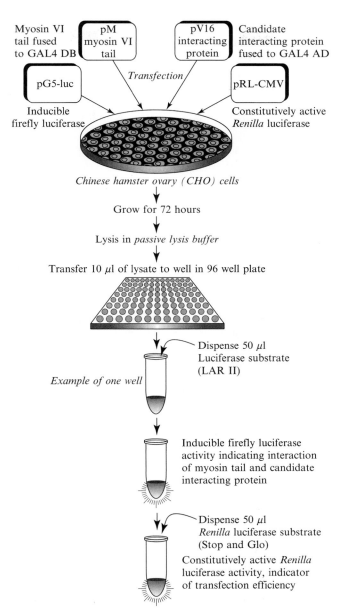

Figure 2.1 Schematic flow diagram of the mammalian two-hybrid assay. Both plasmids coding for the two proteins of interest fused to transcription factor domains (pM and pV16) are cotransfected with an inducible reporter gene (pG5-luc) and a constitutively expressed transfection control (pRL-CMV) into CHO cells. After 72 h, the cells

transcription factor domains, an inducible reporter gene and a constitutively expressed transfection control, are cotransfected into Chinese hamster ovary (CHO) cells. pG5luc (Promega) codes for the inducible reporter—firefly luciferase under control of the GAL4 regulatory element. *Renilla* luciferase is coded by pRL-CMV (Promega); it is constitutively expressed, and thus provides measurement of transfection efficiency. This allows variations in vector uptake to be taken into account when assaying induced firefly luciferase activity. The bait vector, pM (Clontech), codes for the GAL4 DNA-binding domain fused to N-terminus of the myosin VI tail, and the prey vector, pV16 (Clontech), expresses the myosin VI–binding partner optineurin fused to the activating domain derived from the VP16 protein of herpes simplex virus. When two proteins interact, the reconstituted active transcription factor binds to the GAL4 regulatory DNA element in the pG5luc plasmid and the firefly luciferase is expressed. A *Not*I site was introduced into the vectors pM and pV16 downstream from the *Sal*I site in the multiple cloning sites (KJ Patel, MRC-LMB, Cambridge, UK). The modified vectors were referred to as pM★ and pV16★. In this way, cDNA from the ProQuest yeast two-hybrid system was directly subcloned into the mammalian two-hybrid vectors.

The relative luciferase activity was determined by calculating the ratio of inducible (firefly) to constitutively active (*Renilla*) luciferase activity for cells transfected with MyoVI-tail/pM★ and optineurin/pV16★, divided by the ratio of inducible to constitutively active luciferase from cells transfected with MyoVI-tail/pM★ and empty pV16★. A significant interaction was deemed to occur when the relative luciferase activity was greater than 25.

$$\text{Activity} = \frac{\textit{Renilla}/\text{firefly from MyoVI-tail}/\text{pM}^*\text{and optineurin}/\text{pV16}}{\textit{Renilla}/\text{firefly from MyoVI-tail}/\text{pM}^*\text{and pV16 empty}}$$

Initially, we used the mammalian two-hybrid assay to confirm the interaction between myosin VI and optineurin in mammalian cells *in vivo*. As a next step, we used deletion mutants and site-directed mutagenesis to map the exact binding site for optineurin on myosin VI. Optineurin binds to the RRL motif (aa 1107–1109) in the globular tail domain of myosin VI. Mutation in the WWY, the Dab2 binding motif in the myosin VI tail, did not affect optineurin binding (Figure 2.2).

---

are lysed and the lysates assayed in 96-well plates using the Dual-Luciferase reporter Assay System, which supplies the LARII and Stop&Glo reagents as substrates to assay the firefly and *Renilla* luciferase, respectively.

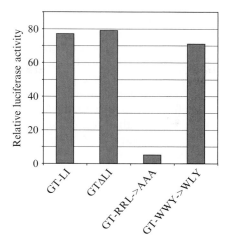

Figure 2.2 Mammalian two-hybrid assay to characterize interaction of optineurin and myosin VI. Optineurin in pVP16 vector was transfected into CHO cells together with pM vector encoding globular tail of myosin VI (aa 1034–1276) (GT-LI), the deletion mutant lacking the large insert (GTΔLI), or the tails carrying point mutations RRL→AAA (aa 1107–1109) or WWY→WLY (aa 1183–1185). The luciferase activity was assayed as described in the text. The graph shows the result from a single representative experiment.

5. PROTOCOL FOR MAMMALIAN TWO-HYBRID ASSAY

The mammalian two-hybrid assay was performed in CHO (Chinese hamster ovary) cells obtained from the European Collection of Animal Cell Cultures. CHO cells were cultured in F-12 HAM medium supplemented with 10% fetal calf serum, 2 mM L-glutamine, 100 U/ml penicillin, and 100 μg/ml streptomycin at 37° and 5% CO_2. For transfection, the cells were seeded out and grown to about 50 to 60% confluency in 35-mm tissue culture dishes. The cells were transfected with highly purified plasmid DNA using FuGENE™ transfection reagent (Roche Diagnostics, UK), according to the manufacturer's instructions. Each dish was cotransfected with 0.8 μg pG5luc, 0.8 μg pRL-CMV, 1 μg MyoVI-tail/pM★, and 1 μg optineurin/pV16★ using 12 μl of FuGENE™. As a negative control background luciferase activity was measured in cells transfected with MyoVI-tail/pM★ and empty pV16★ together with both luciferase plasmids. As a positive control, MyoVI-tail/pM★ and Dab2/pV16★ were used. Cells were grown for 48 to 72 h with a daily change of medium. To measure the luciferase activity, the Dual-Luciferase Reporter Assay System (Promega) was used. After a single wash in PBS, the cells were lysed in 500 μl of Passive Lysis Buffer (Promega) for 30 min at room temperature on a shaker. The lysate including the cell debris was collected, and used immediately for quantitation of the luminescent signal or frozen at −70° for later use. The Dual-Luciferase Reporter Assay was performed according to the instructions in the manual (Promega). The measurements were performed on an Orion Microplate Luminometer (Berthold Detection Systems). For the assay 10 μl samples of the cell lysates were pipetted onto a white 96-well plate. Fifty microliters of the luciferase assay reagent (LARII) was mixed with the samples, and the firefly luciferase luminescence measured immediately. Into the same well, 50 μl of the Stop & Glo Reagent was dispensed. This reagent quenches the firefly reaction and activates the *Renilla* luciferase, which gives the second measurement.

5.1. Notes

1. A negative control must be done in each experiment as the transfection efficiency and luciferase activity may vary significantly from experiment to experiment.
2. As a standard control, an insert swap between pM and pVP16 is recommended.
3. We have used a luminometer that allows automatic substrate dispensing and processes the samples successively, one at a time. If only manual substrate addition and measurements are possible, measuring more than few samples at a time is not recommended, because the luminescence fades very fast.
4. Although the Promega manual recommends the use of relatively small quantities of a control reporter vector—that is, ratios of up to 1:50 for coreporter vector:experimental vector—we found that reproducible measurements can only be obtained when this ratio is at least 1:5.
5. In our experience, CHO cells gave higher luciferase activity than HeLa cells, possibly because of a combination of higher transfection rates and higher expression levels.
6. When luciferase activity is low, more concentrated lysate can be prepared by scraping cells into a smaller amount of lysis buffer followed by vigorous pipetting. In this way, the volume of lysis buffer used for a 35-mm dish can be reduced to 200 μl.

6. SEAP Secretion Assay

Myosin VI and optineurin are present at the Golgi complex, and both proteins have been shown to play a role in maintenance of Golgi morphology and constitutive secretion (Sahlender *et al.*, 2005; Warner *et al.*, 2003) (Figure 2.3). To measure constitutive secretion, we expressed a *s*ecreted form of *a*lkaline *p*hosphatase (SEAP) in tissue culture cells, and quantified

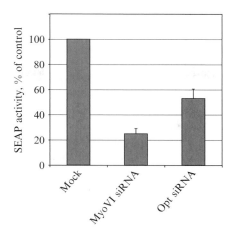

Figure 2.3   SEAP assay to measure constitutive secretion in myosin VI and optineurin knockdown cells. A stable HeLa cell line expressing SEAP was transfected twice with siRNA specific for myosin VI (MyoVI) or optineurin (Opt). The activity of alkaline phosphatase secreted into the culture medium was measured 48 h after the last transfection. The results shown are mean (plus/minus standard deviation) from three independent experiments.

the release of the enzyme into the tissue culture medium. The SEAP used is an engineered variant of the placental alkaline phosphatase (PLAP), which is normally anchored to the plasma membrane via a phosphatidylinositol-glycan (Berger *et al.*, 1988b). The engineered version has a C-terminal truncation, and is therefore efficiently secreted and no longer linked to the plasma membrane (Berger *et al.*, 1988a). In contrast to other forms of alkaline phosphatases that are found in bones and intestine, SEAP is heat resistant and survives heat inactivation up to 65°; thus, its activity can be measured in the presence of endogenous alkaline phosphatases. SEAP was originally described as a reporter enzyme to measure gene expression in eukaryotic cells (Berger *et al.*, 1988a). The pSEAP2-control vector expresses SEAP under control of the SV40 early promoter and can be purchased from Clontech.

7. Measurement of SEAP Secretion in Fibroblasts from Myosin VI Knockout Mouse

To measure secretion in cells lacking myosin VI, fibroblasts were isolated from the myosin VI knockout mouse (Snell's waltzer mouse). A litter of newborn mice was genotyped using a PCR-based method, as described in Self *et al.* (1999). Skin and muscle tissue was cut into small pieces and incubated in 0.25% trypsin for 15 to 30 min at 37°. After gentle pipetting, large tissue lumps were allowed to fall to the bottom of the tube, and cell suspension was collected and the cells plated in growth medium (DMEM supplemented with 10% fetal calf serum, 2 mM of glutamine, and 60 μM of 2-mercaptoethanol). The cells were split every 2 to 3 days, and after several months a number of immortal cell lines from wildtype and Snell's waltzer mice were generated. For transient transfections with the pSEAP2-control plasmid, the cells were grown on 35-mm dishes to 50 to 60% confluency and transfected with 2 μg of plasmid using FuGENE^{TM} according to the manufacturer's instructions. The transfection efficiency of the pSEAP2-control plasmid was normalized by cotransfection of a second plasmid expressing GFP. Since the transfection efficiency of the mouse fibroblasts was generally very low, usually around 5%, for detection we used a chemiluminescent assay based on the substrate CSPD (disodium 3-(4-methoxyspiro(1,2-diosethane-3,2''-(5'-chloro)tricyclo decan-4-yl)phenyl phosphate) (Clontech). Samples of the culture medium were taken every 24 h for a total of 4 days and assayed in the chemiluminescent SEAP assay according to the manufacturer's handbook (Great EscAPe SEAP User Manual, Clontech). The assay was performed in 96-well, flat-bottom microtiter plates suitable for plate luminometers. In brief, 15-μl aliquots of the culture medium were diluted with 45 μl of dilution buffer and heated up to 65° for 30 min. After cooling the samples to room temperature, 60 μl of assay buffer and 60 μl of diluted CSPD substrate were added, and 15 to 60 min later measurements were performed with a plate luminometer (Berthold Detection Systems, Orion Luminometer). Growth medium taken from untransfected cells was used for baseline measurement.

8. MEASUREMENT OF SEAP SECRETION IN siRNA KNOCKDOWN CELLS

To study protein secretion in siRNA knockdown cells, stable cell lines expressing SEAP were generated. The SEAP cDNA was subcloned from pSEAP2-control plasmid into the mammalian expression vector pIRES-neo2 using *Nhe*I and *Hpa*I restriction enzymes. The resulting construct was transfected into two 35-mm culture dishes of HeLa cells using FuGENE. Two days after transfection, the cells were trypsinized and transferred to a 10-cm culture dish. Selection commenced the next day with 500 μg/ml of the antibiotic G418. After 1 to 2 weeks, single surviving clones were picked and transferred to a 24-well plate. Twenty to 30 clones were tested for SEAP secretion, and about 10 clones were selected with a range of levels of SEAP secretion. Selected clones were grown on and stocks were frozen in 10% DMSO, and 25% fetal calf serum in RPMI, and kept in liquid nitrogen.

HeLa cells stably expressing SEAP were routinely cultured in the presence of 500 μg/ml of G418, but medium with no antibiotic was used during knockdown experiments. For siRNA knockdown, SEAP-expressing cells were plated on 6- or 24-well plates and transfected with myosin VI or optineurin siRNA duplexes twice (on days 1 and 3), and medium was changed on day 4 and SEAP was let to accumulate in the medium for a further 1 to 2 days. On the day of the assay, small aliquots of culture medium were collected and used for assay immediately or were frozen at −70°. The basic protocol for the SEAP activity assay was taken from www. genetherapysystems.com and slightly modified as described. Growth medium, which did not have contact with cells, was used for baseline measurement.

9. PROTOCOL FOR SEAP ASSAY

Growth medium from control or siRNA-treated SEAP expressing cells was collected and heated to 65° for 30 min. In a 96-well plate, 10 μl of 0.05% Zwittergent (Calbiochem) (0.5% stock in water, dilute 1:10 before the assay) was mixed with 20 μl of heated culture medium (or 20 μl of heated growth medium for baseline measurements) followed by 200 μl of PNPP substrate. To prepare the substrate, PNPP (para-nitrophenyl phosphate disodium salt, Sigma 104 phosphatase substrate) was dissolved in 1 M diethanolamine, pH 9.8, and 1 mM MgCl$_2$ to 1 mg/ml immediately before the assay. The reaction was incubated for 20 to 30 min at room temperature in the dark and the absorption measured at 405 nm in a spectrometer such as the Anthos HTII plate reader or similar.

9.1. Notes

1. SEAP secretion data have to be normalized to the cell number, as cell density may vary dramatically among cells treated with different siRNA oligos.
2. We have extensively tested the HeLa cell lines stably expressing SEAP and found that they behave as normal HeLa cells in all standard applications such as transfection, siRNA knockdowns or immunofluorescence. They can be successfully used for assaying secretion after transfection of different constructs (e.g., dominant negative tail of myosin VI or Rab8Q67L mutant), given that the transfection efficiency is high enough so that the decrease/increase in secretion in transfected cells is not

masked by bulk SEAP secretion by cells, which do not express the construct of interest.

3. Since only small amounts of culture medium are needed for the assay and the cells themselves are left intact, the cells from the secretion assay can also be used, for example, for Western blotting to monitor levels of protein depletion/expression or for immunofluorescence if plated on coverslips.

4. The SEAP is constitutively secreted and thus always present in the culture medium. In order to see small changes in secretion after manipulating the cells, some extra work may be required to find suitable experimental conditions for each protein/construct of interest. For example, HeLa clones with different levels of SEAP secretion can be tested and SEAP can be allowed to accumulate in freshly changed medium for different times—from 6 h to 3 days.

ACKNOWLEDGMENTS

This work was funded by a Wellcome Trust Senior Fellowship (F.B.) and supported by the Medical Research Council. The Cambridge Institute for Medical Research is the recipient of a strategic award from the Wellcome Trust.

REFERENCES

Ahmed, Z. M., Morell, R. J., Riazuddin, S., Gropman, A., Shaukat, S., Ahmad, M. M., Mohiddin, S. A., Fananapazir, L., Caruso, R. C., Husnain, T., Khan, S. N., Griffith, A. J., et al. (2003).Mutations of MYO6 are associated with recessive deafness, DFNB37. *Am. J. Hum. Genet.* **72,** 1315–1322.

Ang, A. L., Folsch, H., Koivisto, U. M., Pypaert, M., and Mellman, I. (2003). The Rab8 GTPase selectively regulates AP-1B-dependent basolateral transport in polarized Madin-Darby canine kidney cells. *J. Cell Biol.* **163,** 339–350.

Au, J., Puri, C., Ihrke, G., Kendrick-Jones, J., and Buss, F. (2007). Myosin VI is required for sorting of AP-1B dependent cargo to the basolateral domain in polarised MDCK cells. *J. Cell Biol.* **177,** 103–114.

Avraham, K. B., Hasson, T., Steel, K. P., Kingsley, D. M., Russell, L. B., Mooseker, M. S., Copeland, N. G., and Jenkins, N. A. (1995). The mouse Snell's waltzer deafness gene encodes an unconventional myosin required for structural integrity of inner ear hair cells. *Nat. Genet.* **11,** 369–375.

Berger, J., Hauber, J., Hauber, R., Geiger, R., and Cullen, B. R. (1988a). Secreted placental alkaline phosphatase: A powerful new quantitative indicator of gene expression in eukaryotic cells. *Gene* **66,** 1–10.

Berger, J., Howard, A. D., Brink, L., Gerber, L., Hauber, J., Cullen, B. R., and Udenfriend, S. (1988b). COOH-terminal requirements for the correct processing of a phosphatidylinositol-glycan anchored membrane protein. *J. Biol. Chem.* **263,** 10016–10021.

Buss, F., Spudich, G., and Kendrick-Jones, J. (2004). Myosin VI: Cellular fuctions and motor properties. *Annu. Rev. Cell Dev. Biol.* **20,** 649–676.

Chien, C. T., Bartel, P. L., Sternglanz, R., and Fields, S. (1991). The two-hybrid system: A method to identify and clone genes for proteins that interact with a protein of interest. *Proc. Natl. Acad. Sci. USA* **88,** 9578–9582.

DiFiglia, M., Sapp, E., Chase, K., Schwarz, C., Meloni, A., Young, C., Martin, E., Vonsattel, J. P., Carraway, R., and Reeves, S. A. (1995). Huntingtin is a cytoplasmic protein associated with vesicles in human and rat brain neurons. *Neuron* **14,** 1075–1081.

Eugster, A., Frigerio, G., Dale, M., and Duden, R. (2000). COP I domains required for coatomer integrity, and novel interactions with ARF and ARF-GAP. *EMBO J.* **19,** 3905–3917.

Faber, P. W., Barnes, G. T., Srinidhi, J., Chen, J., Gusella, J. F., and MacDonald, M. E. (1998). Huntingtin interacts with a family of WW domain proteins. *Hum. Mol. Genet.* **7,** 1463–1474.

Fields, S., and Song, O. (1989). A novel genetic system to detect protein-protein interactions. *Nature* **340,** 245–246.

Hattula, K., and Peranen, J. (2000). FIP-2, a coiled-coil protein, links Huntingtin to Rab8 and modulates cellular morphogenesis. *Curr. Biol.* **10,** 1603–1606.

Huber, L. A., de Hoop, M. J., Dupree, P., Zerial, M., Simons, K., and Dotti, C. (1993a). Protein transport to the dendritic plasma membrane of cultured neurons is regulated by rab8p. *J. Cell Biol.* **123,** 47–55.

Huber, L. A., Pimplikar, S., Parton, R. G., Virta, H., Zerial, M., and Simons, K. (1993b). Rab8, a small GTPase involved in vesicular traffic between the TGN and the basolateral plasma membrane. *J. Cell Biol.* **123,** 35–45.

Johansson, M., and Olkkonen, V. M. (2005). Assays for interaction between Rab7 and oxysterol binding protein related protein 1L (ORP1L). *Methods Enzymol.* **403,** 743–758.

Li, Y., Kang, J., and Horwitz, M. S. (1998). Interaction of an adenovirus E3 14.7-kilodalton protein with a novel tumor necrosis factor alpha-inducible cellular protein containing leucine zipper domains. *Mol. Cell Biol.* **18,** 1601–1610.

Mohiddin, S. A., Ahmed, Z. M., Griffith, A. J., Tripodi, D., Friedman, T. B., Fananapazir, L., and Morell, R. J. (2004). Novel association of hypertrophic cardiomyopathy, sensorineural deafness, and a mutation in unconventional myosin VI (MYO6). *J. Med. Genet.* **41,** 309–314.

Monier, S., and Goud, B. (2005). Purification and properties of rab6 interacting proteins. *Methods Enzymol.* **403,** 593–599.

Moritz, O. L., Tam, B. M., Hurd, L. L., Peranen, J., Deretic, D., and Papermaster, D. S. (2001). Mutant rab8 impairs docking and fusion of rhodopsin-bearing post-Golgi membranes and causes cell death of transgenic Xenopus rods. *Mol. Biol. Cell* **12,** 2341–2351.

Osterweil, E., Wells, D. G., and Mooseker, M. S. (2005). A role for myosin VI in postsynaptic structure and glutamate receptor endocytosis. *J. Cell Biol.* **168,** 329–338.

Peranen, J., Auvinen, P., Virta, H., Wepf, R., and Simons, K. (1996). Rab8 promotes polarized membrane transport through reorganization of actin and microtubules in fibroblasts. *J. Cell Biol.* **135,** 153–167.

Rezaie, T., Child, A., Hitchings, R., Brice, G., Miller, L., Coca-Prados, M., Heon, E., Krupin, T., Ritch, R., Kreutzer, D., Crick, R. P., and Sarfarazi, M. (2002). Adult-onset primary open-angle glaucoma caused by mutations in optineurin. *Science* **295,** 1077–1079.

Sahlender, D. A., Roberts, R. C., Arden, S. D., Spudich, G., Taylor, M. J., Luzio, J. P., Kendrick-Jones, J., and Buss, F. (2005). Optineurin links myosin VI to the Golgi complex and is involved in Golgi organization and exocytosis. *J. Cell Biol.* **169,** 285–295.

Self, T., Sobe, T., Copeland, N. G., Jenkins, N. A., Avraham, K. B., and Steel, K. P. (1999). Role of myosin VI in the differentiation of cochlear hair cells. *Dev. Biol.* **214,** 331–341.

Spudich, G., Chibalina, M. V., Au, J. S., Arden, S. D., Buss, F., and Kendrick-Jones, J. (2007). Myosin VI targeting to clathrin-coated structures and dimerization is mediated by binding to Disabled-2 and PtdIns(4,5)P(2). *Nat. Cell. Biol.* **9,** 176–183.

Stephens, D. J., and Banting, G. (2000). The use of yeast two-hybrid screens in studies of protein: Protein interactions involved in trafficking. *Traffic* **1,** 763–768.

Velier, J., Kim, M., Schwarz, C., Kim, T. W., Sapp, E., Chase, K., Aronin, N., and DiFiglia, M. (1998). Wild-type and mutant huntingtins function in vesicle trafficking in the secretory and endocytic pathways. *Exp. Neurol.* **152,** 34–40.

Warner, C. L., Stewart, A., Luzio, J. P., Steel, K. P., Libby, R. T., Kendrick-Jones, J., and Buss, F. (2003). Loss of myosin VI reduces secretion and the size of the Golgi in fibroblasts from Snell's waltzer mice. *EMBO J.* **22,** 569–579.

Zerial, M., and McBride, H. (2001). Rab proteins as membrane organizers. *Nat. Rev. Mol. Cell. Biol.* **2,** 107–117.

CHAPTER THREE

Characterization of Rab27a and JFC1 as Constituents of the Secretory Machinery of Prostate-Specific Antigen in Prostate Carcinoma Cells

Sergio D. Catz

Contents

Abstract

Prostate-specific antigen (PSA) and prostate-specific acid phosphatase (PSAP) are produced by prostate carcinoma cells. Their secretion has implications in both prostate cancer diagnosis and progression. The mechanisms involved in PSA and PSAP secretion in response to androgens have remained relatively unknown. The small GTPase Rab27a regulates exocytosis in several tissues. Here, we present methods for the characterization of Rab27a and its effector JFC1/Slp1 as key components of the secretory machinery that regulates exocytosis in prostate carcinoma cells.

Division of Biochemistry, Department of Molecular and Experimental Medicine, The Scripps Research Institute, La Jolla, California

Methods in Enzymology, Volume 438
ISSN 0076-6879, DOI: 10.1016/S0076-6879(07)38003-8

25

1. INTRODUCTION

Rab27a is a small GTPase involved in exocytosis in several tissues (Tolmachova *et al.*, 2004). Rab27a, as well as other Rab proteins, operates through specific Rab effectors that regulate multiple steps of vesicle transport including exocytosis. These effectors can regulate vesicular trafficking in many different ways. For example, they can facilitate vesicle transport from the site of vesicle formation to the acceptor membrane by interacting with the cytoskeleton and/or with motor components (Strom *et al.*, 2002; Wu *et al.*, 2002). Other effectors assist with the docking of the Rab-containing vesicle to the plasma membrane during exocytosis. Generally, these effectors contain a Rab-binding domain (RBD) and an adaptor domain that recognizes membrane lipids or proteins at the docking site. JFC1/Slp1 (synaptotagmin-like protein 1), a Rab27a-binding protein identified in our laboratory (McAdara-Berkowitz *et al.*, 2001), associates with target membranes through the C2 domains distributed in its C-terminus (Catz *et al.*, 2002) and with Rab27a through its RBD in the N-terminus (Strom *et al.*, 2002). We found significant expression of JFC1 in tissues that have a secretory function, with the highest level of expression observed in the prostate (McAdara-Berkowitz *et al.*, 2001). Cell fractionation analysis showed that JFC1 is partially and peripherally associated with cellular membranes (Catz *et al.*, 2002; Munafo *et al.*, 2007). The C2A domain of JFC1 localizes to the plasma membrane of various cell types when expressed as an EGFP chimera (Catz *et al.*, 2002; Johnson *et al.*, 2005a; Munafo *et al.*, 2007). It has been suggested that the binding of this domain to the plasma membrane is based, at least in part, through binding to 3′-phosphoinositides (Catz *et al.*, 2002). The RBD of JFC1-lacking C2 domains localized to punctate structures that resemble vesicles when overexpressed in prostate carcinoma cells (Johnson *et al.*, 2005a). However, those structures were not distributed in the proximity of the plasma membrane, supporting the idea that if JFC1 is involved in docking, its C2 domains will be important for vesicle tethering or docking to the plasma membrane (Johnson *et al.*, 2005a).

The major cell type within the normal prostatic epithelium is the secretory luminal cell (Catz and Johnson, 2003). These cells produce prostate secretory proteins and express the androgen receptor that renders the cells androgen-dependent. Secretory prostate cells are the cell of origin of most human prostate adenocarcinomas (Denmeade *et al.*, 1996). They produce and secrete prostate-specific antigen (PSA) and prostatic acid phosphatase (PSAP). PSA is a kallikrein (Yousef and Diamandis, 2002) with serine protease activity. The expression and secretion of PSA are regulated by androgens in normal prostate secretory epithelial cells and by diverse signaling pathways in cells that became androgen-independent

(Lee *et al.*, 2003; Wen *et al.*, 2000). PSA plasma level is often elevated in prostate cancer and is broadly used as a blood-borne diagnostic marker of the disease. Furthermore, PSA secretion may have pathological connotations because it degrades extracellular matrix proteins (Webber *et al.*, 1995), and this property has been associated with invasiveness by prostate carcinoma cells. PSAP, which is involved in the regulation of phospho-tyrosine proteins (Lee *et al.*, 2003) and is produced and secreted exclusively by the prostate epithelium, has been used as an indicator of hormonal therapy effectiveness (Huggins and Hodges, 1972; van Steenbrugge *et al.*, 1983).

Pharmacological intervention during prostate cancer may differentially and independently regulate prostate marker secretion and cell proliferation (Dixon *et al.*, 2001). Therefore, the identification of the molecular machinery involved in the secretion of PSA and PSAP and the characterization of its regulation are essential steps for the understanding of the pathophysiology of prostate cancer and for the analysis of its progression based on the detection of secreted prostatic markers. Here, we describe the cellular biology approaches used to study and characterize the role of Rab27a and JFC1 in the secretory mechanism of prostate carcinoma cells.

2. METHODS

2.1. Materials

2.1.1. Cell culture

The human prostate carcinoma cell line LNCaP-FGC, which is referred to in this report as LNCaP, was obtained from American Type Culture Collection (Rockville, MD). RPMI 1640 medium was obtained from Invitrogen (Carlsbad, CA). Charcoal-stripped fetal bovine serum was from Gemini Bio-products (Sacramento, CA). The nonaromatizable androgen 6α-fluorotestosterone was obtained from BIOMOL International (Plymouth Meeting, PA).

2.1.2. Antibodies

The anti-JFC1 antibody was raised by inoculating rabbits with the N-terminal peptide MAHGPKPETEGLLDLS conjugated to keyhole limpet hemocyanin (Chiron Mimotopes, San Diego, CA) and was purified using a Montage antibody purification kit with PROSEP-A medium (Millipore, MA). The polyclonal antibody against Rab27a was raised using recombinant GST (glutathione S-transferase)-Rab27a, the RiBi Adjuvant System (RiBi Immunochem Research, Hamilton, MT) and Imject® alum (Pierce, Rockford, IL) to immunize New Zealand White rabbits. The antibodies raised against PSAP and PSA of mouse and rabbit origin were from NeoMarkers (Union City, CA) and the goat anti-PSA antibody was from R & D Systems (Minneapolis, MN). The anti–VAMP-2 (vesicle-associated membrane protein-2) and anti-LAMP–2 (lysosome-associated membrane protein-2) were from Santa Cruz Biotechnology (Santa Cruz, CA); the anti-EEA1 (early endosome antigen 1) was from Transduction Laboratories (San Jose, CA).

2.2. Immunofluorescence analysis of endogenous proteins in prostate carcinoma cells

In the analysis of molecules with potential secretory functions, it is important to determine the subcellular localization of the endogenous factors of interest in relationship to the cargo proteins involved in the exocytic process. To help elucidate the function of these molecules, their subcellular localization should be studied in association with known markers of cellular organelles. In our studies, we utilized confocal microscopy to analyze the distribution of the cargo proteins PSA and PSAP in relationship to the secretory proteins Rab27a and JFC1, using VAMP-2, LAMP-2, and EEA1 as markers for secretory vesicles, lysosomes, and early endosomes, respectively.

2.2.1. Solutions

Washing solution: Phosphate-buffered saline (PBS, Invitrogen, CA)
Fixing solution: 3.7% (w/v) paraformaldehyde in de-ionized water
Permeabilization solution: 0.01% saponin in PBS
Blocking solution: 1.5% bovine serum albumin in PBS

2.2.2. Method

For confocal microscopy analysis, cells are grown on coverslips. In our experiments, we use Lab-Tek eight-well chambered 1.0 Borosilicate cover-glass (Nalge Nunc International, IL; cat. no. 155411). Coverglass is pre-treated with poly-L-lysine (Sigma, MO) at 0.01% in deionized, sterile water for 10 min at room temperature (21°). Poly-L-lysine solution is removed and coverslips are dried for 30 min. LNCaP cells are seeded at 70% confluence (5×10^4 cells per well in a total volume of 400 μl). The whole process should be performed in a laminar flow hood. For immunofluorescence analysis, the medium is carefully removed and cells are washed once with PBS for 30 sec at RT. PBS is replaced with 200 μl of fixing solution. To minimize modifications in the epitopes, the incubation period in the presence of fixative must be limited to 10 min at room temperature. Samples are washed twice with 400 μl of PBS for 5 min. Next, the cells are permeabilized with 0.01% saponin for 10 min and blocked with 1.5% BSA in PBS. These two steps may be combined. Dilutions of the appropriate primary antibodies and pre-bleed sera or nonspecific IgG from the same species as negative controls are prepared in blocking solution containing 0.01% saponin at a dilution of 1:200. Antibodies (200 μl solution) are added overnight at 4°. After incubation with the primary antibodies, the samples are washed three times with PBS for 15 min, and combinations of the secondary antibodies (488) and/or (594) Alexa-Fluor–conjugated, donkey anti-rabbit, anti-mouse, and/or anti-goat (Molecular Probes, CA) are added at a dilution of 1:300 in blocking solution plus saponin. Samples are incubated further for 2 h at room temperature (21°). Cells are washed twice with PBS, and washing solution is replaced with mounting solution Fluoromount-G (Southern Biotechnology, Birmingham, AL). Samples are stored in the dark at 4° until analyzed by laser-scanning confocal microscopy. If nuclear visualization is desired, samples are incubated with 100 ng/ml DAPI (4′,6-diamidino-2-phenylindole) for 5 min at 21° before the final wash. For visualization, fluorescence associated with Alexa-Fluor-594-labeled secondary antibodies is excited using the 568-nm laser line and collected using a standard Texas Red filter. Fluorescence associated with Alexa-Fluor-488-labeled secondary antibodies is visualized using the 488-nm laser line and collected using a standard FITC filter set.

2.3. Functional analysis of exocytosis in LNCaP cells

2.3.1. Cell culture and transfection of LNCaP cells

LNCaPs are prostate secretory carcinoma cells that express the androgen receptor and grow slowly in the absence of androgens. Because of the nature of our experiments, we cultured the cells in the absence of hormone. The cells were cultured in T-75 culture flasks using RPMI containing 10% fetal bovine serum, 50 units/ml of penicillin, 50 μg/ml streptomycin, at 37° in 5% CO_2. To further reduce the possible presence of androgens in the serum, cells were transferred to charcoal-stripped fetal bovine serum 24 h before experiments in which androgen stimulation was required. We used LNCaP cells with no more than 40 passages. Long-passage LNCaP cells become androgen-independent and lose PSAP expression (Lee *et al.*, 2003). For secretion studies, LNCaP cells are transfected by nucleofection (Amaxa, Germany). Cells are trypsinized for 5 min, washed in culture media, counted and resuspended in Amaxa transfection solution V at 2 × 10^6 cells per 90 μl. At this point, all elements necessary for transfection and post-transfection recovery must be readily available so that cells do not remain in transfection solution for more than 15 min. Using a gel-loading tip, 90 μl of solution V containing LNCaP cells is transferred to a transfection cuvette. Next, 10 μl of a solution containing 2–10 μg of the expression vector of interest is added to the cells and mixed by pipetting. Cells are transfected using an Amaxa nucleofector I apparatus and pulsed using the electrical setting T01. Immediately after pulsation, 200 μl of RPMI medium are added to the cells, and 145 μl of the new cell suspension are then transferred to each of both experimental and control wells in six-well plates containing 1.5 ml of RPMI with 10% charcoal-stripped fetal bovine serum. Cells are recovered for 24 h.

2.3.2. Study of subcellular distribution of JFC1 domains in prostate carcinoma cells

Rab effectors are heterogeneous molecules that are considered to be specific for particular transport systems (Segev, 2001). These effectors can regulate vesicular trafficking in different ways. For instance, Rab effectors may assist with the docking of the Rab-containing vesicle to the acceptor membrane. These effectors contain a Rab-binding domain (RBD) and an adaptor domain that recognizes structures at the docking site. An example of this kind of operational molecule is Rabphilin-3a, a Rab3a effector that associates with membrane phospholipids through its C2 domains (Yamaguchi et al., 1993). Similarly, the Rab27a-binding protein JFC1 contains an RBD in its amino terminal domain and tandem C2 domains in its carboxy-terminus. The observation that the C2A domain, but not the C2B domain, of JFC1 localizes exclusively at the plasma membrane (Catz et al., 2002) suggests that if JFC1 is responsible for the docking of Rab27a vesicles, its C2A domain would play a central role in this mechanism. To shed light on the putative role of the Rab27a effector JFC1 in the secretory pathway of prostate carcinoma cells, we expressed various JFC1 domains as EGFP (enhanced-green fluorescent protein) chimeras in LNCaP cells and analyzed their distribution by confocal microscopy. The various steps in the cloning of the constructs were performed by standard techniques and all constructs were verified by sequencing using an automated fluorescent dye–terminator sequencer. The *jfc1* DNA fragments cloned into the *Eco*RI/*Sal*I sites of the pEGFP-C2 polylinker were generated by PCR using *jfc1* cDNA as template and *pfu* polymerase (Stratagene, La Jolla, CA). To generate the C2A domain insert, we used the following oligonucleotides containing *Eco*RI or *Sal*I sites (underlined): GGG$\underline{GAATTC}$AGCCTGTCAGGCGACGCGGAGGCG-GTG (coordinates $+766$ to $+792$) and GGG$\underline{GTCGAC}$CCGCCCCAGT-CCCACGTGTCCAGGGG (coordinates $+1131$ to $+1108$). To generate the triple mutant EGFP-C2A ($K^{318,322,323}E$), we used the QuikChange® Multi Site-Directed Mutagenesis Kit (Stratagene), the primer GCAAGC-GCGAGACGGCGGTGGAGGAACGGAATCTG and EGFP-C2A as template. The chimera of EGFP and the RBD domain of JFC1 encoding amino acids 1 to 110 was generated by PCR using the QuikChange Multi Site-Directed Mutagenesis Kit to generate a stop codon in EGFP-full length JFC1.

For direct fluorescence analysis of the distribution of the various JFC1 truncation and mutant chimeras, LNCaP cells are transfected by nucleofection as described above using 5 μg of the expression vector of interest. Before transfections, sterile, Poly-L-lysine-treated Lab-Tek eight-well, chambered coverslips containing 300 μl of RPMI media per well should be readily available. The whole process should be performed in a laminar flow hood. Transfected cells (5×10^4) are seeded in the wells and recovered for 24 to 48 h. Cells are carefully washed with PBS to avoid cellular resuspension, fixed with 3.7% paraformaldehyde for 10 min at RT and washed again twice with PBS for 15 min. PBS is then removed and replaced with 200 μl of mounting medium fluoromount-G (Southern Biotechnology, Birmingham, AL). Samples are ready for analysis and can be stored in the dark at 4° for 10 days. The expression of different JFC1 domains as EGFP chimeras helped us to determine that the C2A domain of JFC1, a phosphoinositide-binding domain (Catz *et al.*, 2002), localizes exclusively at the plasma membrane in LNCaP prostate carcinoma cells (Fig. 3.1). We also showed by mutagenesis analysis that the lysine stretch present in the C2A domain of JFC1 is essential to maintain plasma localization. In similar experiments, we demonstrated that the RBD domain of JFC1 which lacks C2 domains localizes at punctate structures that resemble vesicles, which are not in close proximity to the plasma membrane. These results established the basis for the functional analysis of the secretory role of JFC1 in prostate carcinoma cells, described in the following.

2.3.3. Analysis of PSA and PSAP exocytosis in LNCaP cells

C2 domains frequently direct plasma membrane localization of secretory proteins in many cell types (Catz *et al.*, 2002; Chapman and Davis, 1998). We hypothesized that if JFC1 plays a role in the docking of Rab27a-containing vesicles at the plasma membrane through its C2A domain,

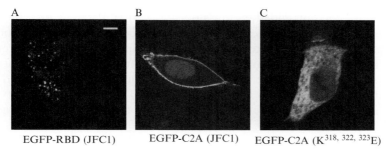

EGFP-RBD (JFC1)   EGFP-C2A (JFC1)   EGFP-C2A ($K^{318, 322, 323}E$)

Figure 3.1   Subcellular localization of JFC1 domains in LNCaP prostate carcinoma cells. The chimaeras consisting of the EGFP-Rab-binding domain (RBD) of JFC1 (A), the EGFP-C2A domain of JFC1 (B), or the C2A domain with mutations in the lysine stretch (EGFP-C2A-$K^{318,322,323}E$) (C), were expressed in LNCaP cells. The cells were fixed 48 h after transfection and were analyzed by confocal microscopy. Scale bars, 5 μm. (Reproduced in part from Johnson, J. L., Ellis, B. A., Noack, D., Seabra, M. C., and Catz, S. D. (2005). The Rab27a binding protein JFC1 regulates androgen-dependent secretion of prostate-specific antigen and prostate-specific acid phosphatase. *Biochem. J.* 391, 699–710, with permission from the publisher Portland Press Ltd.)

the presence of this domain would compete with endogenous JFC1 for a docking point at the plasma membrane and should impair secretion. To test our hypothesis, we evaluated the secretory function of LNCaP cells exogenously expressing the C2A domain of JFC1. For these experiments, LNCaP cells were harvested from T-75 (80% confluent) flasks 2 days after passage and transfected with 10 μg of various expression vectors using the Amaxa nucleofector exactly as described above. For the analysis of the participation of JFC1 in the secretion of PSA and PSAP, we used the expression vectors for myc-tagged full-length JFC1, myc-tagged C2A domain of JFC1 or the pCMV-tag 2 empty vector (Stratagene, La Jolla, CA). For the analysis of the participation of Rab27a in the secretion of PSA, we utilized the expression vector pcDNA3.1-Rab27a (obtained from the UMR cDNA resource center, www.cDNA.org) or the constitutively active mutant Rab27aQ^{78}L, which was generated using the primer 5′-TGGGACACAGCAGGGCTG-GAGAGGTTTCGTAGCT-3′ and the QuikChange® Multi Site-Directed Mutagenesis kit (Stratagene).

Transfected cells are transferred to six-well plates containing 1.5 ml of RPMI with 10% charcoal-stripped fetal bovine serum. After 12 h, the medium is replaced with fresh RPMI 1640 medium containing 100 nM of 6α-fluorotestosterone (Biomol International, Philadelphia, PA) or carrier (DMSO). The final concentration of DMSO should be 0.5% v/v or lower. Testosterone-containing medium can be prepared before hand and stored at 4° in the dark until use. Cells are incubated in the presence of androgens for 24 or 48 h. The supernatants are collected and kept at −20° until analyzed for the presence of PSAP and PSA. Appropriate sample dilution should be empirically calculated. For the conditions described here and using ELISA kits from Alpha Diagnostics International (San Antonio, TX) a 1:100 dilution is a recommended starting point.

To further analyze protein expression and intracellular concentration of PSA and PSAP, the cells are trypsinized for 5 min, harvested, washed once with RPMI medium, counted, and lysed by incubating in 200 μl of 2% Nonidet P40 in PBS for 30 min on ice. Cell debris is eliminated by centrifugation at 16,000 × for 10 min at 4°, and the clear supernatants are kept at −20° until use. For the analysis of protein expression, 10 μg of protein lysate is resolved by electrophoresis using 10% NuPAGE gels and Mops SDS running buffer (Invitrogen, CA). Proteins are transferred to 0.2-μm pore-size nitrocellulose membranes (Invitrogen, CA) by standard procedures. The antibodies used to determine the level of expression of endogenous or overexpressed JFC1 and Rab27a were described above. For the detection of myc-tagged proteins, the monoclonal antibody clone 9B11 raised against the peptide EQKLISEEDL from Cell Signaling Technology (Boston, MA) works well. For detection, we used the secondary antibodies anti-rabbit IgG or anti-mouse IgG conjugated to horseradish peroxidase at 1:5000 (Caltag Laboratories, Carlsbad, CA). For visualization, we used the chemiluminescence substrate system LumiGlo (Millipore/Upstate Biotechnology, Charlottesville, VA) and Kodak BioMax Light Film (Kodak, New Haven, CT).

2.3.4. Use of siRNA to downregulate Rab27a

Specific downregulation of a gene product by RNA interference is frequently used to reveal the particular function of the protein of interest. To elucidate whether Rab27a plays a central role in exocytosis, we utilized an siRNA gene-targeting approach to downregulate this small GTPase. We used this approach to demonstrate that Rab27a is a key component of the secretory machinery of granulocytes (Munafo et al., 2007); a similar approach was applied to LNCaP prostate carcinoma cells. The sequences of the oligonucleotides used to efficiently and specifically down-regulate Rab27a are: 5′-ggagagguuucguagcuua and 5′-ccagugua-cuuuaccaaua (Dharmacon, CO). For transfections, 2×10^6 cells are resuspended in 90 μl of solution V (Amaxa, Germany). Either specific siRNA oligonucleotides or control siRNAs are added to reach a final oligonucleotide concentration of 20 μM in a final volume of 100 μl. Due to possible nonspecific interference, it is recommended that at least two specific siRNA oligonucleotides and two nonspecific controls are used in the assays. The cells are transfected using the electrical setting of T01, transferred to 1.5 ml RPMI, and recovered for 48 h to 72 h. At this point, the expression of Rab27a is analyzed by Western blot following standard procedures. Functional analysis of the secretion of PSA and PSAP by LNCaP prostate carcinoma cells is then performed exactly as described above.

2.4. Functional analysis of the Rab27a-JFC1 secretory mechanism in difficult-to-transfect cells

To overcome the difficulties presented by hard to transfect cells, permeabilization can be used to allow the access of modulatory factors to target intracellular molecules. Several permeabilization methods have been employed on a variety of cell types to elucidate diverse cellular mechanisms (Brown et al., 2003; Chen et al., 2005; Munafo et al., 2007; Plutner et al., 1992). In our experiments, we use SLO (streptolysin-O), a bacterial pore-forming toxin, to establish a semi-intact assay system for the analysis of granule secretion. SLO is a thiol-activated toxin produced by Gram-positive bacteria. It binds to cholesterol-containing membranes, oligomerizes into ring-shaped structures and forms pores of around 30 nm (Palmer et al., 1998). Brown et al. (2003) demonstrated that permeabilization of human cells with SLO results in excellent maintenance of intracellular structure when evaluated by electron microscopy. To demonstrate that the Rab27a-JFC1 machinery is involved in the exocytosis of secretory organelles, we used a cellular approach based on the principle that the intracellular presence of the C2A domain of JFC1 would compete with endogenous JFC1 for a docking site on the plasma membrane and thereby impair secretion (Munafo *et al.*, 2007).

2.4.1. Purification of recombinant proteins

2.4.1.1. Cloning JFC1 or JFC1-truncation constructs were engineered by standard techniques (Sambrook *et al.*, 1989). JFC1 full-length, truncation 256–377 containing the C2A domain (C2A) and 256–562 containing both C2 domains (C2A-C2B) were amplified from wildtype JFC1 cDNA with *pfu* polymerase (Stratagene, CA) using 5′ primers that contained a *Sal*I site (underlined) and 3′-antisense primers that contained a *Not*I site (underlined). The 5′ primers used for amplification were GGG<u>GTCGAC</u>C CATGCCCCAGAGGGGCCACCCATCGCAA for wildtype JFC1 and GGG<u>GTCGAC</u>CCAGCCTGTCAGGCGACGCGGAGGCGGTG for truncations 256–377 and 256–562, whereas the 3′ primers were (GGG <u>GCGGCCGC</u>GCGCCCCAGTCCCACGTGTCCAGGGG) for the truncation 256–377 and GGG<u>GCGGCCGC</u>CGTCCTGGGGGCCAGGTT GGTTCT for wildtype and truncation 256–562. PCR products were cloned into pGEX-6P-1 (Amersham Biosciences, NJ). Constructs were sequenced using an automated fluorescent dye–terminator sequencer.

2.4.1.2. Preparation of recombinant fusion proteins

Buffers:

Lysis buffer: 0.2% Igepal, 0.2 m*M* PMSF (phenylmethylsulphonyl fluoride) in PBS containing complete protease inhibitor cocktail (Roche, Germany)

Elution buffer: 100 m*M* Tris-HCl pH 8.0 containing 20 m*M* glutathione and 300 m*M* NaCl

Dialysis buffer: 50 m*M* Tris/HCl pH 7.4 containing 100 m*M* NaCl, 5 m*M* KCl, 1 m*M* EDTA and 1 m*M* PMSF

Recombinant fusion proteins composed of an upstream glutathione *S*-transferase (GST) linked to a downstream JFC1 (GST-JFC1), and truncations C2A (GST-C2A) or C2A-C2B (GST-C2AB) are purified by affinity chromatography on glutathione-agarose beads by standard methods as described by the manufacturer (Amersham Biosciences, NJ). For the production of the recombinant proteins, the final concentration of IPTG (Isopropyl *β*-D-1-thiogalactopyranoside) used to induce protein expression is 0.01 m*M*. Induction is carried out at RT (21°) for 4 h. This condition helps minimize the presence of these recombinant proteins in inclusion bodies. Thereafter, all procedures are performed at 4° and bacteria are kept in iced at all times. After induction, bacteria is spun down and resuspended in ice-cold lysis buffer. Aliquots (30 ml) are sonicated three times for 1 min with a 30-s rest interval using a Misonix 3000 Ultrasonicator. Lysates are spun down at 16,000 × g for 10 min and supernatants are incubated with 1 ml of prewashed glutathione-agarose beads for 1 h. Beads are then washed twice with 30 ml lysis buffer and once with PBS for 15 min. Elution, using 5 ml of elution buffer per each milliliter of glutathione-agarose beads, is performed with rotation for 10 min at room temperature (21°). Relatively high NaCl concentration in the elution buffer is essential to achieve efficient elution of these recombinant proteins. Proteins are dialyzed overnight against 4 liters of dialysis buffer using Slide-A-Lyzer Dialysis Cassettes 10K MWCO (Pierce, IL). Buffer is changed, and proteins are dialyzed against 4 liters of buffer for 2 more h. Small aliquots of the recombinant proteins are flash frozen in liquid nitrogen and stored at −80° until used. Protein concentration is determined by the Bradford method (Bio-Rad, Hercules, CA) using BSA as standard.

2.4.2. Functional assays in permeabilized cells

Permeabilization buffer: 20 mM Hepes/Na$^+$ pH 7.5, 140 mM KCl, 1 mM MgSO$_4$, 1 mM ATP, 1 mM GTP, 0.1% BSA and Complete protease inhibitor cocktail (Roche, Germany).

SLO solution should be freshly prepared by dissolving the lyophilized powder (Sigma, MO) in PBS, and then activated with 10 mM dithiothreitol for 1 h at 37° before use (Imai *et al.*, 2004). Cells (2.5 × 10^6) are washed twice with PBS, resuspended in 100 μl of permeabilization buffer, transferred to a 1.5-ml eppendorf tube containing 5 units of SLO in the presence of 2 μM GST, GST-C2A(JFC1), or GST-JFC1, and incubated for 5 min at 37°. The cells are stimulated with phorbol ester (PMA; 0.1 μg/ml) for 20 min. The reactions are stopped by transferring the samples to ice and immediately centrifuging at 16,000 × g for 5 min at 4°. Supernatants and cellular pellets are stored at −20° until analyzed by ELISA.

2.5. Protein–protein interaction analysis

In studies involving Rab proteins and their effectors, characterization of the interactions between endogenous proteins by immunoprecipitation and confocal microscopy analysis is essential and should be performed provided that specific antibodies are available. We have used these approaches to show that endogenous JFC1 interacts and colocalizes with Rab27a (Johnson *et al.*, 2005a; Munafo *et al.*, 2007). However, other approaches including co-immunoprecipitation of tagged proteins and pull-down assays have been widely utilized to characterize the molecular details of the Rab–Rab effector interaction (Fukuda and Kanno, 2005; Strom *et al.*, 2002). The technical particulars of these experiments applied to the interaction of Rab27a with JFC1 are discussed below.

2.5.1. Analysis of interaction between JFC1 and Rab27a by coimmunoprecipitation assays

The utilization of mutant proteins in coimmunoprecipitation assays is useful to recognize molecular mechanisms of protein–protein interaction as well as identify dominant negative molecules with potential applications in functional assays. In our analysis of the binding between JFC1 and Rab27a, we utilized the mutant JFC1-W^{83}S. It contains a serine substitution for the tryptophan located in the sequence T^{80}GDWFQ^{85} of JFC1, which is partially homologous to the conserved SGAWFF motif in Rabphilin3a that is part of the contact interface of this molecule with Rab3a (Ostermeier and Brunger, 1999). This mutant, as we show in Fig. 3.2 and in reference (Johnson et al., 2005b), loses ability to bind to Rab27a.

Radio immunoprecipitation assay buffer (RIPA) is made of 50 mM Tris–HCl, 1% Nonidet-P40, 0.1% Na-deoxycholate, 150 mM NaCl, 1 mM ethylene-diaminetetraacetic acid (EDTA), 1 mM PMSF, 1 mM Na$_3$VO$_4$ and 1 mM NAF, pH 7.4)

The various steps in the cloning of the constructs were performed by standard techniques. The myc-tag vectors for the expression of the JFC1 mutant W83 to serine was generated using the QuickChange$^{®}$ Multi Site–Directed Mutagenesis Kit (Stratagene, CA) and the primer CCTGA

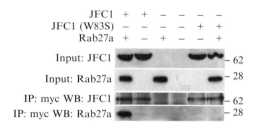

Figure 3.2   Rab27a is coimmunoprecipitated with JFC1 but not with the mutant JFC1-W^{83}S. Cells were transiently transfected with 5 μg of the expression vectors pmyc-JFC1, pmyc-JFC1-(W^{83}S) or empty vector (pCMV-Tag 3B), and with the expression vector pcDNA3.1-Rab27a or with the corresponding empty vector, by nucleofection. The cells were lysed in radio–immunoprecipitation (IP) assay buffer 48 h after transfection, and 10% of the sample was saved to establish the level of transfection (Input). For this purpose, JFC1 and Rab27a were detected using an anti-myc and a polyclonal anti-Rab27a antibody, respectively. The IP reactions were carried out by the addition of anti-myc agarose to the cleared lysate supernatant with overnight incubation at 4°. Immunopellets were washed, and proteins were resolved by gel electrophoresis and transferred to nitrocellulose membranes. Immunoprecipitated Rab27a and JFC1 were determined by Western blot (WB) analysis using polyclonal antibodies raised against Rab27a and JFC1, respectively. (From Johnson, J. L., Pacquelet, S., Lane, W. S., Eam, B., and Catz, S. D. (2005). Akt regulates the subcellular localization of the Rab27a-binding protein JFC1 by phosphorylation. *Traffic* **6**, 667–681, with permission from the publisher Blackwell Synergy.)

CAGGGGACTCGTTCCAGGAAGCAC. For Rab27a and JFC1 coimmunoprecipitation experiments, we used the human cervical carcinoma cell line Hela S3 (American Type Culture Collection, Rockville, MD), which was maintained in DMEM supplemented with 10% fetal calf serum and 0.45% glucose (Invitrogen, Carlsbad, CA) at 37° in 5% CO_2. Cells (3×10^5) are plated in 35-mm culture dishes in 2 ml of DMEM medium overnight to achieve a density of 70% confluence. The cells are transiently transfected with 5 μg of the expression vectors pmyc-JFC1, pmyc-JFC1 ($W^{83}S$), or empty vector (pCMV-Tag 3B), and with the expression vector pcDNA3.1-Rab27a or with the corresponding empty vector, using FuGene6 (Roche Diagnostics Corp., Indianapolis, IN) and following the manufacturer's recommendations. After 48 h, cells are washed with PBS, 1 ml of RIPA buffer is added, and cells are scraped off and kept on ice for 30 min. Immunoprecipitation is carried out at 4°. Lysates are spun down at $16,000 \times g$ for 10 min. An aliquot of each sample ($\sim$10%) is saved to establish the level of expression of exogenous proteins in the lysates by Western blot analysis. Lysates are precleared by incubation with 30 μl of protein A-agarose for 30 min. Immunoprecipitation is carried out by the addition of 40 μl of anti-myc agarose (Sigma-Aldrich, St. Louis, MO) to the cleared lysate supernatant and rotated overnight. Immunopellets are washed twice with RIPA buffer and once with PBS for 15 min. Proteins were resolved by gel electrophoresis and transferred to nitrocellulose membranes. The presence of Rab27a and JFC1 was determined by Western blot using specific polyclonal antibodies.

2.5.2. Pull-down assays using *in vitro*–translated proteins

2.5.2.1. Preparation of in vitro–translated [^{35}S]-Rab27a
Another useful way to analyze the characteristics of the interactions between Rab proteins and effectors is through *in vitro* pull-down assays. In our experiments, we took this approach to analyze the Rab27a-binding properties of phosphorylated JFC1. In this experiment, we used recombinant GST-JFC1 to pull down *in vitro*–translated [^{35}S]-Rab27a. To generate radioactive Rab27a, we used the vector pBK-CMV expressing wildtype Rab27a under the control of the T3 promoter to produce *in vitro*–translated protein labeled with translational-grade [^{35}S]-methionine (Amersham Biosciences, NJ). To this end, the TNT-coupled transcription and translation system (Promega, Madison, WI) is used following the manufacturer's recommendations. *In vitro* translated [^{35}S]-Rab27a was loaded with GTPγS (Calbiochem, San Diego, CA) or GDP (Sigma, St. Louis, MO) by dialyzing the radiolabeled protein against high EDTA-containing buffer and subsequently incubating the recombinant protein with either 10 mM GTPγS or 10 mM GDP in $MgCl_2$-containing buffer. For this purpose, the whole *in vitro*–translated reaction (100 μl) containing [^{35}S]-Rab27a is dialyzed against nucleotide exchange buffer (Christoforidis and Zerial, 2000) containing

20 mM Hepes, 100 mM NaCl, 10 mM EDTA, 5 mM MgCl$_2$, 1 mM dithiothroitol (DTT) and 1 mM GTPγS or 1 mM GDP, pH 7.4 for 1 h at room temperature using a Microdialyzer™ System 100 (Pierce, IL). Next, the buffer is exchanged for nucleotide stabilization buffer lacking EDTA, but containing 10 mM MgCl$_2$ and 10 mM GTPγS or GDP (Christoforidis and Zerial, 2000). Samples are dialyzed for another hour.

2.5.2.2. Phosphorylation of GST-JFC1 The kinase buffer is comprised of 50 mM Hepes, pH 7.4, 10 mM magnesium chloride, 1 mM DTT, and 1 mM ATP. Phosphorylation of recombinant JFC1 by recombinant active Akt (Upstate Biotechnology, NY) is performed by incubating a reaction mixture containing 3 μg of recombinant GST-JFC1 or GST control (purified as described above) and 0.4 pmol of active Akt in kinase buffer in a total volume of 50 μl for 30 min at 37°. In parallel reactions, 10 μCi of [γ-^{33}P]-ATP is included and labeled proteins are visualized by autoradiography as control of the phosphorylation reaction.

2.5.2.3. Binding reaction The binding buffer is comprised of 20 mM Tris-HCl pH 7.4, 80 mM NaCl, 4 mM MgCl$_2$, 1 mM DTT, 0.25% Nonidet-P40, and 1 mM GTP or 1 mM GDP. For binding reactions, 20 μl of the *in vitro*–translation reaction containing GDP- or GTPγS-loaded [^{35}S]-Rab27a are incubated in the presence of 1.5 μg of phosphorylated or unphosphorylated GST-JFC1 or equimolar amount of GST in binding buffer rotating overnight at 4° in a final volume of 800 μl. The protein complexes are pulled down by the addition of 50 μl pre-washed GSH-Agarose (Amersham Biosciences, NJ). The samples are rotated for 1 h at 4°, washed four times for 15 min with binding buffer containing 1 mM GDP or GTPγS, spun down at 4000 × g for 10 min and boiled in 1× sample buffer. Samples are resolved by gel electrophoresis, and presence of radiolabeled-Rab27a in the pellets is detected by autoradiography using Kodak Biomax MR film (Eastman Kodak Company).

3. Concluding Remarks

In recent years, there has been increasing evidence of the central role played by small GTPases in vesicular transport systems and, in particular, on the role of Rab27a and effectors in the process of exocytosis. At least 11 Rab27a-interacting proteins have been identified. How these molecules coordinate various steps in the exocytic process including docking, priming and fusion remains relatively unknown. Here, we presented a range of methodologies that helped elucidate the role of JFC1 and Rab27a in the secretion of prostate-specific markers. They are potentially useful for analyzing other Rab effector functions in similar cellular systems.

ACKNOWLEDGMENTS

These studies are supported by US Public Health Service grant AI-024227, and by the Sam and Rose Stein Endowment Fund.

REFERENCES

Brown, G. E., Stewart, M. Q., Liu, H., Ha, V. L., and Yaffe, M. B. (2003). A novel assay system implicates PtdIns(3,4)P(2), PtdIns(3)P, and PKC delta in intracellular production of reactive oxygen species by the NADPH oxidase. *Mol. Cell* **11**, 35–47.

Catz, S. D., and Johnson, J. L. (2003). BCL-2 in prostate cancer: A minireview. *Apoptosis* **8**, 29–37.

Catz, S. D., Johnson, J. L., and Babior, B. M. (2002). The C2A domain of JFC1 binds to 3′-phosphorylated phosphoinositides and directs plasma membrane association in living cells. *Proc. Natl. Acad. Sci. USA* **99**, 11652–11657.

Chapman, E. R., and Davis, A. F. (1998). Direct interaction of a Ca^{2+}-binding loop of synaptotagmin with lipid bilayers. *J. Biol. Chem.* **273**, 13995–14001.

Chen, C. Y., Sakisaka, T., and Balch, W. E. (2005). Use of Hsp90 inhibitors to disrupt GDI-dependent Rab recycling. *Methods Enzymol.* **403**, 339–347.

Christoforidis, S., and Zerial, M. (2000). Purification and identification of novel Rab effectors using affinity chromatography. *Methods* **20**, 403–410.

Denmeade, S. R., Lin, X. S., and Isaacs, J. T. (1996). Role of programmed (apoptotic) cell death during the progression and therapy for prostate cancer. *Prostate* **28**, 251–265.

Dixon, S. C., Knopf, K. B., and Figg, W. D. (2001). The control of prostate-specific antigen expression and gene regulation by pharmacological agents. *Pharmacol. Rev.* **53**, 73–91.

Fukuda, M., and Kanno, E. (2005). Analysis of the role of Rab27 effector Slp4-a/Granuphilin-a in dense-core vesicle exocytosis. *Methods Enzymol.* **403**, 445–457.

Huggins, C., and Hodges, C. V. (1972). Studies on prostatic cancer. The effect of castration, of estrogen and androgen injection on serum phosphatases in metastatic carcinoma of the prostate. *CA Cancer J. Clin.* **22**, 232–240.

Imai, A., Yoshie, S., Nashida, T., Shimomura, H., and Fukuda, M. (2004). The small GTPase Rab27B regulates amylase release from rat parotid acinar cells. *J. Cell Sci.* **117**, 1945–1953.

Johnson, J. L., Ellis, B. A., Noack, D., Seabra, M. C., and Catz, S. D. (2005a). The Rab27a binding protein JFC1 regulates androgen-dependent secretion of prostate-specific antigen and prostate-specific acid phosphatase. *Biochem. J.* **391**, 699–710.

Johnson, J. L., Pacquelet, S., Lane, W. S., Eam, B., and Catz, S. D. (2005b). Akt regulates the subcellular localization of the Rab27a-binding protein JFC1 by phosphorylation. *Traffic* **6**, 667–681.

Lee, M. S., Igawa, T., Yuan, T. C., Zhang, X. Q., Lin, F. F., and Lin, M. F. (2003). ErbB–2 signaling is involved in regulating PSA secretion in androgen-independent human prostate cancer LNCaP C-81 cells. *Oncogene* **22**, 781–796.

McAdara-Berkowitz, J. K., Catz, S. D., Johnson, J. L., Ruedi, J. M., Thon, V., and Babior, B. M. (2001). JFC1, a novel tandem C2 domain-containing protein associated with the leukocyte NADPH oxidase. *J. Biol. Chem.* **276**, 18855–18862.

Munafo, D. B., Johnson, J. L., Ellis, B. A., Rutschmann, S., Beutler, B., and Catz, S. D. (2007). Rab27a is a key component of the secretory machinery of azurophilic granules in granulocytes. *Biochem. J.* **402,** 229–239.

Ostermeier, C., and Brunger, A. T. (1999). Structural basis of Rab effector specificity: Crystal structure of the small G protein Rab3A complexed with the effector domain of rabphilin-3A. *Cell* **96,** 363–374.

Palmer, M., Harris, R., Freytag, C., Kehoe, M., Tranum-Jensen, J., and Bhakdi, S. (1998). Assembly mechanism of the oligomeric streptolysin O pore: The early membrane lesion is lined by a free edge of the lipid membrane and is extended gradually during oligomerization. *EMBO J.* **17,** 1598–1605.

Plutner, H., Davidson, H. W., Saraste, J., and Balch, W. E. (1992). Morphological analysis of protein transport from the ER to Golgi membranes in digitonin-permeabilized cells: Role of the P58 containing compartment. *J. Cell Biol.* **119,** 1097–1116.

Sambrook, J., Maniatis, T., and Fritsch, E. F. (1989). "Molecular cloning: A laboratory manual." Cold Spring Harbor Laboratory Press, Cold Spring Harbor, NY.

Segev, N. (2001). Ypt/rab GTPases: Regulators of protein trafficking. *Sci. STKE* **2001,** RE11.

Strom, M., Hume, A. N., Tarafder, A. K., Barkagianni, E., and Seabra, M. C. (2002). A family of Rab27-binding proteins. Melanophilin links Rab27a and myosin Va function in melanosome transport. *J. Biol. Chem.* **277,** 25423–25430.

Tolmachova, T., Anders, R., Stinchcombe, J., Bossi, G., Griffiths, G. M., Huxley, C., and Seabra, M. C. (2004). A general role for Rab27a in secretory cells. *Mol. Biol. Cell* **15,** 332–344.

van Steenbrugge, G. J., Blankenstein, M. A., Bolt-de, V. J., Romijn, J. C., Schroder, F. H., and Vihko, P. (1983). Effect of hormone treatment on prostatic acid phosphatase in a serially transplantable human prostatic adenocarcinoma (PC-82). *J. Urol.* **129,** 630–633.

Webber, M. M., Waghray, A., and Bello, D. (1995). Prostate-specific antigen, a serine protease, facilitates human prostate cancer cell invasion. *Clin. Cancer Res.* **1,** 1089–1094.

Wen, Y., Hu, M. C., Makino, K., Spohn, B., Bartholomeusz, G., Yan, D. H., and Hung, M. C. (2000). HER-2/neu promotes androgen-independent survival and growth of prostate cancer cells through the Akt pathway. *Cancer Res.* **60,** 6841–6845.

Wu, X. S., Rao, K., Zhang, H., Wang, F., Sellers, J. R., Matesic, L. E., Copeland, N. G., Jenkins, N. A., and Hammer, J. A., III (2002). Identification of an organelle receptor for myosin-Va. *Nat. Cell Biol.* **4,** 271–278.

Yamaguchi, T., Shirataki, H., Kishida, S., Miyazaki, M., Nishikawa, J., Wada, K., Numata, S., Kaibuchi, K., and Takai, Y. (1993). Two functionally different domains of rabphilin-3A, Rab3A p25/smg p25A-binding and phospholipid- and Ca^{2+}-binding domains. *J. Biol. Chem.* **268,** 27164–27170.

Yousef, G. M., and Diamandis, E. P. (2002). Human tissue kallikreins: A new enzymatic cascade pathway? *Biol. Chem.* **383,** 1045–1057.

CHAPTER FOUR

In Vitro Assays to Characterize Inhibitors of the Activation of Small G Proteins by Their Guanine Nucleotide Exchange Factors

Jean-Christophe Zeeh,* Bruno Antonny,[†] Jacqueline Cherfils,* and Mahel Zeghouf*

Contents

Abstract

Guanine nucleotide exchange factors (GEFs) are essential regulators of the spatiotemporal conditions of small GTP-binding protein (SMG) activation. Their cellular activities combine the biochemical stimulation of GDP/GTP exchange, which leads to the active conformation of the SMG, to the detection of upstream signals and, in some cases, interaction with downstream effectors. Inhibition of GEF activities by small molecules has become recently a very active field, both for understanding biology with the tools of chemistry and because GEFs are emerging as therapeutic targets. The natural compound brefeldin A (BFA) was the first inhibitor of a GEF to be characterized, and several inhibitors of SMG activation have since been discovered using a variety of screening

* Laboratoire d'Enzymologie et Biochimie Structurales, Centre National de la Recherche Scientifique, Gif-sur-Yvette, France
† Institut de Pharmacologie Moléculaire et Cellulaire, Centre National de la Recherche Scientifique, Valbonne, France

Methods in Enzymology, Volume 438
ISSN 0076-6879, DOI: 10.1016/S0076-6879(07)38004-X

methods. An essential step toward their use in basic research or as leads in therapeutics is the characterization of their mechanism of inhibition. GEFs function according to a multistep mechanism, involving transient ternary (nucleotide-bound) and binary (nucleotide-free) intermediates. This mechanism thereby offers many opportunities for blockage, but a thorough analysis is necessary to define the inhibition mechanism and the steps of the reaction that are affected by the inhibitor. Here, based on the case study of how BFA inhibits the activation of Arf activation by Sec7 domains, we describe a flow-chart of assays to decipher the mechanism of inhibitors of the activation of SMGs by their GEFs.

1. GEFs and Diseases: Inhibiting the Exchange Reaction by Small Molecules

In many human diseases, one or several signaling pathways controlled by small GTP-binding proteins (SMGs) are either upregulated or acquire a critical role in the context of the pathology. This was first recognized for activating mutations of the p21Ras oncogene in many cancers (reviewed in Wittinghofer and Waldmann, 2000), but many more examples have since been documented such as in cardiovascular diseases (reviewed in Laufs and Liao, 2000), mental retardation (reviewed in Ramakers, 2002), and infections (reviewed in Aktories and Barbieri, 2005) and possibly insulin resistance (Hafner *et al.*, 2006). For instance, members of the Rho family have been shown to play critical roles in the morphology and motility changes of cancer cells (reviewed in Sahai and Marshall, 2002), and some of their guanine nucleotide exchange factors (GEFs) are oncogenes (reviewed in Rossman *et al.*, 2005). Recently, loss-of-function mutations in BIG2, a GEF for the Arf family (Sheen *et al.*, 2004), and gain-of-function mutations in SOS1, a GEF for Ras (Roberts *et al.*, 2007; Tartaglia *et al.*, 2007), have been shown to cause congenital disorders. Viruses and bacteria that take command of cellular pathways by injecting their own GEFs into host cells, such as the *Legionella* ArfGEF RalF (Nagai *et al.*, 2002) and the *Salmonella* RacGEF SopE (Hardt *et al.*, 1998), or by blocking a cellular ArfGEF, such as the viral peptide 3A (Wessels *et al.*, 2006), illustrate the pivotal role of these regulators in diseases. Because GEFs are essential for both the activation and signaling specificity of SMGs, their inhibition by small molecules could in principle block the activation of promiscuous SMGs, specifically in the conditions associated with disease, while leaving normal functions unaffected (reviewed in Zeghouf *et al.*, 2005).

Activation of SMGs by GEF-stimulated GDP/GTP exchange is a multistep reaction (Fig. 4.1), thus offering more than one target conformation for the action of drugs. The kinetics of the exchange reaction have been

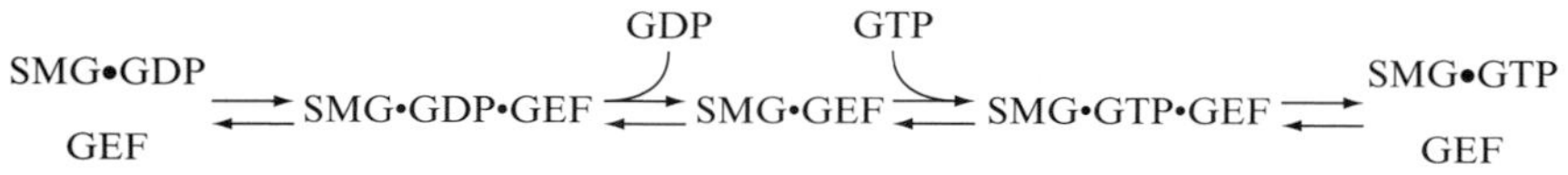

$$\text{SMG•GDP} \underset{\text{GEF}}{\overset{}{\rightleftarrows}} \text{SMG•GDP•GEF} \overset{\text{GDP}}{\rightleftarrows} \text{SMG•GEF} \overset{\text{GTP}}{\rightleftarrows} \text{SMG•GTP•GEF} \underset{\text{GEF}}{\overset{}{\rightleftarrows}} \text{SMG•GTP}$$

Figure 4.1 Schematic representation of the GEF-stimulated nucleotide exchange reaction.

described in detail for only a few systems, including the activation of Ran by RCC1 (Klebe *et al.*, 1995), Ras by cdc25 (Lenzen *et al.*, 1998), and representative Rab/GEF pairs (Esters *et al.*, 2001), yet the same principles apply to all SMG/GEF pairs (reviewed in Cherfils and Chardin, 1999). The exchange reaction initiates by the formation of a low-affinity complex between the GEF and the SMG to which GDP is still tightly bound. This intermediate decreases the affinity for GDP and converts into a nucleotide-free SMG/GEF intermediate of high affinity. Entry of GTP and dissociation of the active, GTP-bound SMG involve the reverse sequence of events.

The first known inhibitor of a GEF is a fungal macrocyclic lactone, brefeldin A (earlier called decumbin), which inhibits the activation of Arf proteins by their GEFs carrying a Sec7 catalytic domain. BFA was discovered in the late 1950s as the product of a *Penicillum* strain isolated from spoiled corn and first described for its toxicity in animals and plants associated with a lack of antibiotic activity (Singleton *et al.*, 1958). Its antitumoral and antiviral properties were subsequently documented (Tamura *et al.*, 1968), long before its target and mechanism were known. BFA was shown in the early 1990s to inhibit the activation of Arf1 in cells (Donaldson *et al.*, 1992; Helms and Rothman, 1992) and has since been instrumental in understanding the molecular basis of traffic at the Golgi (reviewed in Jackson, 2000). Elucidation of its target and mechanism awaited the discovery of the GEFs for Arfs (Chardin *et al.*, 1996; Peyroche *et al.*, 1996), which rapidly revealed an unusual uncompetitive mechanism in which BFA targets the Arf-GDP/GEF complex, and stabilizes it in an abortive conformation unable to release GDP (Peyroche *et al.*, 1999). Subsequent crystallographic analysis of BFA-trapped Arf-GDP/Sec7 intermediates (Mossessova *et al.*, 2003; Renault *et al.*, 2003), together with structures of unbound Arf (Amor *et al.*, 1994; Goldberg, 1998) and of a nucleotide-free Arf/GEF complex (Goldberg, 1998), revealed how BFA hijacks conformational rearrangements of Arf-GDP relative to its GEF. New inhibitors of GEF activity have recently been discovered using different screens. New ArfGEF inhibitors have been screened using RNA aptamers (Mayer *et al.*, 2001), aptamer displacement screens (Hafner *et al.*, 2006) and structure-based *in silico* screening (Viaud *et al.*, 2007). The importance of RhoGEFs in cancer and other human diseases also fostered the exploration of novel screening methods, and several inhibitors have recently been discovered (Blangy *et al.*, 2006; Gao *et al.*, 2004; Nassar *et al.*, 2006; Schmidt *et al.*, 2002). Taking BFA and its inhibition of Arf activation by its GEFs as a case study because of its exhaustive biochemical and structural coverage, we define here a flowchart of assays aimed at characterizing the mechanism of action of candidate GEF inhibitors *in vitro*. The first part summarizes the initial inhibition assays for the identification of GEF inhibitors. We then discuss the use of Michaelis-Menten formalism for the analysis of the GDP/GTP exchange mechanism and its inhibition, and how this can be used to analyze the inhibition profiles of candidate inhibitors. In cases where the mechanism is not of competitive nature, assays to refine the mechanism will then be described.

2. INITIAL SCREEN OF INHIBITORS OF SMG ACTIVATION

Numerous techniques are available to study the activation of SMGs. Among them, fluorescence spectroscopy is a rapid and sensitive approach, which allows for real-time monitoring and determination of the kinetic parameters of the nucleotide exchange reaction. After excitation at a specific wavelength (λ_{ex}), a fluorescent group emits light at a longer wavelength (λ_{em}), and the resulting fluorescent signal is dependent of the local environment of the molecule. In the case of Arf proteins, the conformational change of two tryptophans that occurs during the nucleotide exchange reaction results in a large increase of tryptophan fluorescence.

A typical nucleotide exchange assay is illustrated in Fig. 4.2, where [Δ17] Arf1, a soluble N-terminally truncated form of Arf1, was activated by the catalytic Sec7 domain of its GEF ARNO. Tryptophan fluorescence measurements were performed mainly as described in Zeeh *et al.* (2006) using $\lambda_{ex} = 292$ nm, $\lambda_{em} = 340$ nm, and slits of 5 nm with a Varian Cary Eclipse spectrofluorometer. Typical reaction conditions were 1 μM of [Δ17]Arf1-GDP and 50 nM of Sec7 domain in 700 μl of reaction buffer (50 mM Tris-HCl, pH 8, 50 mM NaCl, 2 mM MgCl$_2$, and 2 mM β-mercaptoethanol). The protein solution was first equilibrated for 5 min in a thermostated cuvette at 37 °C under constant stirring, until a constant fluorescent signal was reached. The exchange reaction was then initiated by addition of 10 μM of GTP and an apparent activation rate constant k$_{obs}$ of 0.039 s^{-1} was determined by fitting the fluorescence change to a single exponential. One can subsequently examine the GTP-to-GDP GEF-stimulated exchange by injecting a large excess of GDP (200 μM) when a plateau has been reached. Under these conditions, a k$_{obs'}$ of 0.0077 s^{-1} for the GTP-to-GDP exchange was determined by fitting the fluorescence decay to a single exponential. Spontaneous exchange activity was similarly measured except that the Sec7 domain was omitted and a large excess of GTP (100 μM) was used.

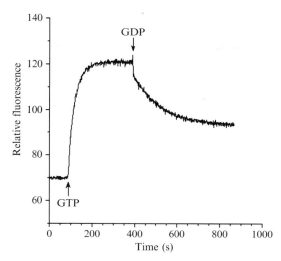

Figure 4.2 Tryptophan fluorescence–based nucleotide exchange assay. 1 μM of [Δ17] Arf1-GDP was incubated with 50 nM of ARNO Sec7 domain at 37 °C. Exchange reaction was initiated by successive injections of 10 μM of GTP and 200 μM of GDP. k_{obs} values were determined from the exponential fit of the change in fluorescence.

Intrinsic tryptophan fluorescence is not always suitable to monitor the exchange reaction, as for Ras, which lacks tryptophan residue. An alternative approach is to use GDP or GTP analogs labeled with a fluorescent group sensitive to its environment. Among them, guanine nucleotides carrying a N-methylanthranyloyl (or mant) group covalently attached to their ribose hydroxyls have been widely used, as delineated by Wittinghofer and colleagues (John *et al.*, 1990, reviewed in Lenzen *et al.* 1995). The mant fluorescent group is small and, in general, disturbs neither the binding of nucleotides nor the exchange reaction. Nevertheless, this point has to be assessed for each SMG/GEF pair studied, since crystal structures show that the ribose hydroxyls of GTP or GTP analogs often interact with the SMG. Mant guanine nucleotides (mGDP or mGTP) can be excited at a wavelength remote from that where proteins and nucleotides absorb, and they display a large increase in fluorescence intensity upon binding to SMG. Moreover, they can be either synthesized in-house as previously described (John *et al.*, 1990) or purchased. In some cases, it might be interesting to use fluorescence resonance energy transfer (FRET) between tryptophan and mant nucleotide to follow the exchange reaction. FRET is a nonradiative energy transfer between two fluorescent groups having overlapping energy spectra. Here, the tryptophan residue is excited and acts as a donor to induce the fluorescence of the acceptor mant group.

FRET is extremely sensitive to the distance between the two fluorophores, and occurs only when the mant nucleotide is bound to the SMG, thus providing another way to follow and quantify SMGs activation.

The use of mGTP is exemplified in Fig. 4.3A. The exchange reaction assay was performed as described above with 2 μM of [Δ17]Arf1-GDP and 200 nM of Sec7 domain in 800 μl of reaction buffer. The reaction was initiated by addition of 10 μM of mGTP (Jena Bioscience) and simultaneously monitored with mant ($\lambda_{ex} = 360$ nm, $\lambda_{em} = 440$ nm) and FRET ($\lambda_{ex} = 292$ nm, $\lambda_{em} = 440$ nm) fluorescence. In both cases, a large increase of fluorescence is observed and a k_{obs} of 0.012 and 0.011 s^{-1} were respectively determined from the mant and the FRET recordings. Those values are very similar to the k_{obs} of 0.013 s^{-1} calculated from tryptophan fluorescence signal obtained under the same conditions (not shown). Note that the protein concentrations in Fig. 4.3A are different from those in Figs. 4.2 and 4.3B, thus precluding a direct comparison of the calculated k_{obs}. It should be noted that the FRET-based approach provides a better signal-to-noise ratio (about a twofold increase) than the direct excitation of the mant nucleotide. Alternatively, the exchange reaction can be carried out using SMG preloaded with mGDP, thus allowing for use of large excess of nonfluorescent nucleotide in the assay. In the experiment presented in Fig. 4.3B, [Δ17] Arf1-GDP was first preloaded with mGDP by incubating the protein with a 25-fold molar excess of mGDP in 50 mM Tris-HCl, pH 8, 50 mM NaCl, 2 mM β-mercaptoethanol and 5 mM EDTA for 40 min in the dark and at room temperature. The reaction was stopped by desalting on a prepacked NAP-5 column previously equilibrated with reaction buffer which contains 2mM Mgcl$_2$. The exchange reaction was carried out using 1 μM [Δ17] Arf1-mGDP, 50 nM of Sec7 domain, and 200 μM of GTP in 700 μl of reaction buffer. The observed fluorescence changes led to the determination of k_{obs} values of 0.035, 0.038, and 0.034 s^{-1} from the tryptophan, mant, and FRET signal, respectively.

The above activation assays provide a straightforward *in vitro* screen to probe small molecules effects on SMGs activation. Initially, one can simply compare the k_{obs} and/or $k_{obs'}$ obtained in the absence or in the presence of 100 μM of the compound of interest after 2 to 5 min of incubation with the protein solution. As a control, it is important to test the effect of the solvent in which the compound is prepared. For example, we found that ethanol, which is commonly used as a solvent for small molecules, inhibits Arf1 activation by the Sec7 domain of ARNO when used at above about 1% concentration. Before starting a detailed analysis of an apparently "active" molecule, it is also necessary to check its absorption, excitation and emission spectra, as well as its solubility in the reaction buffer, since the physicochemical properties of the molecule itself might interfere with the fluorescence assay (Feng and Shoichet, 2006). At this point, it is also important to conduct some pilot experiments that could give insights into

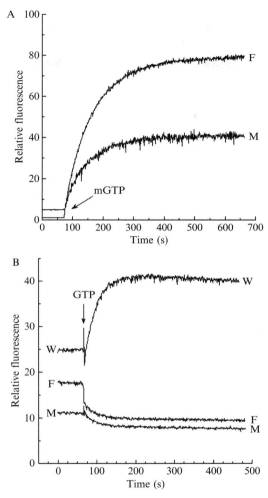

Figure 4.3 Nucleotide exchange assay using mant guanine nucleotides. (A) 2 μM of [Δ17]Arf1–GDP and 200 nM of ARNO Sec7 domain were incubated at 37 °C before addition of 10 μM of mGTP to initiate the exchange reaction. (B) Nucleotide exchange assay using a preloaded SMG: 1 μM of [Δ17]Arf1-mGDP was activated by 50 nM of ARNO Sec7 domain in the presence of 200 μM of GTP at 37 °C. Note that we did not observe significant differences within 10-to-200-μM nucleotide concentration range for identical protein conditions. F, FRET fluorescence; M, mant; W, tryptophan.

the mechanism of inhibition. Thus, instead of simply carrying out kinetics experiments at increasing concentration of the drug, one should test whether the inhibition is weakened or strengthened when a higher concentration of SMG is used. The first case is a good indication of a competitive mechanism. The second case suggests a more complex mechanism. Note also that the order of additions is important. Thus, if Arf1–GDP is first incubated with BFA and GTP before the addition of Sec7 domain to trigger activation, one observes complex activation kinetics. The reaction starts with the same rate as that observed in the absence of BFA, but then slows down within a few seconds. This delay is due to the fact that BFA does not bind to a single protein species but rather to a transient complex between Arf1–GDP and Sec7 domain. Simple (monoexponential) kinetics requires Arf1–GDP and Sec7 domain to be incubated with BFA for a few minutes before the addition of GTP. After active compounds have been identified, the comparison of their apparent inhibition constant KI_{app} can be useful to estimate their relative efficiencies, as exemplified with the study of BFA and its analog BFC in (Zeeh *et al.*, 2006). Sec7-stimulated exchange assays were performed as described above, in the presence of increasing concentrations of the inhibitor, except that the reaction is initiated by addition of 100 μM of GTP. KI_{app} was determined from the hyperbolic fit of k_{obs} values plotted as a function of the inhibitor concentration. An illustration is presented in Fig. 4.4, where we used intrinsic tryptophan fluorescence to analyze BFA inhibition of [Δ17]Arf1 activation by a BFA–sensitive ARNO mutant, called ARNO⁴ᴹ. From these data, we calculated a KI_{app} of 12 μM for BFA whereas BFC had a KI_{app} of 200 μM. This result highlighted the importance of the hydroxyl at the C7 position, which is absent in BFA,

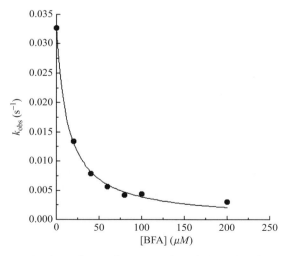

Figure 4.4 Determination of KI_{app} by tryptophan fluorescence kinetics. k_{obs} values were measured as described in Figure 4.3 in the presence of various concentration of BFA. KI_{app} was determined by hyperbolic fit of k_{obs} values plotted as a function of inhibitor concentration.

and is consistent with the presence of a hydrogen bond between this hydroxyl and Tyr190 of the Sec 7 domain in the Arf-GDP/ARNO^{4M}/BFA complex (Renault *et al.*, 2003) and previous mutagenesis studies of ARNO (Peyroche *et al.*, 1999; Robineau *et al.*, 2000). It should be noted that, for BFA, we obtained identical KI_{app} values when monitoring the reaction with tryptophan, mant, or FRET fluorescence (unpublished data).

3. DETERMINATION OF INHIBITION PROFILE BY USING MICHAELIS-MENTEN FORMALISM

Despite the absence of a chemical reaction, Arf1 activation by ARNO can be analyzed as an enzymatic reaction in the presence of a high excess of GTP to prevent the reverse exchange (Fig. 4.5). Considering ARNO as the enzyme (E) and Arf1-GDP as the substrate (S), we can apply the Michaelis-Menten formalism to the overall GEF-stimulated exchange reaction since the formation of the ternary complex ARNO-Arf1-GDP (ES) is an obligatory intermediate.

This has been used to investigate in details the GEF-stimulated guanine nucleotide exchange for Ran/RCC1 (Klebe *et al.*, 1995), Ras/Cdc25 (Lenzen *et al.*, 1995), Arf1/ARNO (Béraud-Dufour *et al.*, 1998), RhoA/GDS (Hutchinson and Eccleston, 2000), or Ypt52/Vps9 (Esters *et al.*, 2001).

This approach was used to establish the BFA inhibition profile of [Δ17] Arf1 activation by the Sec7 domain of ARNO (Peyroche *et al.*, 1999). Initial rates of the ARNO-stimulated exchange reaction were determined graphically from the slopes of fluorescence recordings considering the signal as proportional to the [Δ17]Arf1-GTP concentration. Initial rates V were plotted as a function of the substrate, i.e. [Δ17]Arf1-GDP, at several fixed concentrations of BFA and fitted nonlinearly to Michaelis-Menten Eq. (4.1):

$$V = \frac{k_{cat}[\text{Sec7}][\text{Arf1-GDP}]}{(K_m + [\text{Arf1-GDP}])} \quad (4.1)$$

It should be noted that in the absence of BFA, the Sec7 domain of ARNO could not be saturated by [Δ17]Arf1-GDP even using a 100:1 ratio

$$\text{ARNO} + \text{Arf1-GDP} \underset{k_{-1}}{\overset{k_1}{\rightleftharpoons}} \text{Arf1-GDP-ARNO} \overset{\text{GTP}}{\underset{k_{cat}}{\longrightarrow}} \text{ARNO} + \text{Arf1-GTP}$$

E   S k_{-1} ES k_{cat} E   P

Figure 4.5 Schematic representation of the Michaelis–Menten formalism applied to the activation of a SMG by its GEF.

between the two proteins, which excludes the determination of absolute k_{cat} and K_m values for the Arf1 activation by ARNO (Béraud-Dufour *et al.*, 1998). Nevertheless, the authors found that BFA affects both k_{cat} and K_m by the same factor of $1+[I]/K_I$ (Fig. 4.6A). Accordingly, the corresponding double-reciprocal plot ($1/V$ as a function of $1/[\Delta 17]$Arf1-GDP), also known as Lineweaver–Burk linearization, led to a typical pattern of parallel lines (Fig. 4.6B). This result was consistent with the uncompetitive mechanism depicted in Fig. 4.6C, where BFA targets the Arf1-GDP/Sec7 intermediate. This mechanism has been deciphered in great detail, and clearly demonstrated by additional biochemical and structural data (Peyroche *et al.*, 1999; Renault *et al.*, 2003; Robineau *et al.*, 2000).

Although double-reciprocal plots are an easy means to obtain an estimate of the inhibition profile of a compound, systematic weighting errors are introduced during the required data manipulations. Curve-fitting programs, such as Grafit (Erithacus software), EZ-Fit (Perrella Scientific, Inc.), Graph-Pad (GraphPad Software, Inc.), or Sigmaplot (Systat Software, Inc.), should thus be preferred for an accurate determination of inhibitor modality and kinetic parameters from raw data. First, experimental data are fitted according to equations that correspond to several inhibition types (the competitive, noncompetitive, and uncompetitive inhibition to begin with). The goodness of fit is then statistically tested to determine which model best describes the experimental data.

4. COMPLEMENTARY ASSAYS TO REFINE NONCOMPETITIVE AND UNCOMPETITIVE MECHANISMS

In order to refine the mode of action of noncompetitive and uncompetitive inhibitors, it is necessary to assay the effect of the compound on the discrete exchange steps delineated in Fig. 4.1, as well as possible effects on spontaneous nucleotide dissociation. Fluorescent spectroscopy assays described above can readily map which discrete step of the GEF-stimulated exchange reaction is blocked by the inhibitor, by taking advantage of the fact that in most cases, GEFs stimulate the exchange reaction regardless of the nature of the leaving and incoming nucleotide (see Fig. 4.2). For instance, GEF-stimulated exchange of GDP for mGDP (or mGDP for GDP) involves only the SMG-GDP/GEF and SMG/GEF intermediates of the exchange scheme in Fig. 4.1. An effect of the compound on this reaction indicates that it acts on the exchange reaction prior to the binding of GTP. Conversely, GEF-stimulated exchange of GTP for mGTP (or mGTP for GTP) can be used to probe whether an inhibitor acts on the second part of the exchange reaction. It should be noted, however, that in some cases the GEF might not stimulate the dissociation of GTP. This was

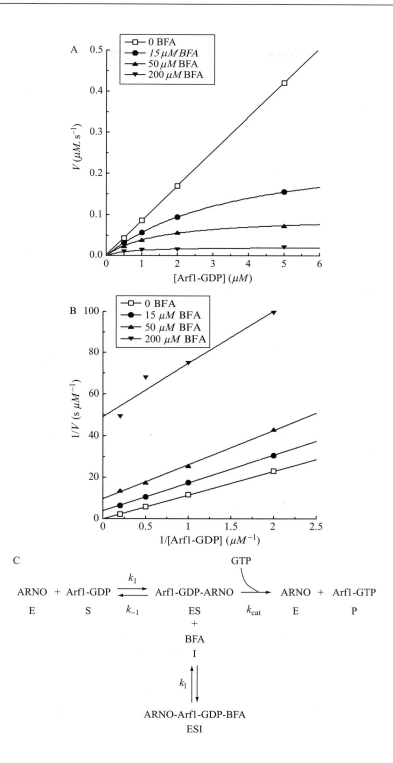

the case that we observed with the Sec7 domain of ARNO, which stimulates GTP/GDP exchange from Arf1, but not from the closely related Arf6 (unpublished data). Fluorescence kinetics were also used to analyze the effect of BFA on the entry of GTP in the nucleotide-free Arf/Sec7 complex. Such an effect could be ruled out by comparing the k_{obs} obtained with or without inhibitor for the binding of GTP to a preformed nucleotide-free complex (Robineau *et al.*, 2000).

Alternatively, noncompetitive and uncompetitive inhibitors can act by stabilizing SMG–GEF interactions, rather than preventing them, as exemplified by BFA. Fluorescence anisotropy provides a direct means to analyze the effect of inhibitors on the dissociation constant of protein complexes (reviewed in Hill and Royer, 1997), and it is also well suited to the expected micromolar affinity range for the SMG-GDP/GEF exchange reaction intermediate. This assay is based on the fact that the polarization of the light emitted by fluorescent probe when excited with a plane-polarized light increases as the size of the complex to which it is attached increases. Accordingly, formation of a complex between a fluorescently labeled SMG and its GEF increases the anisotropy signal, from which a dissociation constant can be derived by correlation between concentration of the GEF and anisotropy. For this type of experiment, Arf1-GDP is labeled with Alexa 488 succinimidyl ester (Molecular Probes, excitation and emission wavelengths of $\lambda_{ex} = 494$ nm and $\lambda_{em} = 520$ nm), which reacts with protein amino groups. Labeling is performed essentially as recommended by the manufacturer (50 mM bicarbonate buffer at pH 8.3, 50 to 75 μM Arf1-GDP, 1 mg/ml Alexa 488, room temperature), except that the incubation is reduced to 1 h to favor unique labeling of the amino-terminus. Excess dye is removed on a desalting PD 10 column (GE-Healthcare). The final Arf1-GDP concentration and the labeling ratio are determined according to manufacturer instructions.

A typical fluorescence anisotropy experiment is shown in Fig. 4.7. Assays were performed in 50 mM Tris-HCl, pH 8, 200 mM NaCl, 2 mM MgCl$_2$, and 2 mM β-mercaptoethanol. A large excess of GDP (10 mM) was added to favor the ternary nucleotide-bound Arf-GDP/GEF complex, together

---

Figure 4.6 Determination of the uncompetitive inhibition profile of BFA. (A) Effect of BFA on GEF-stimulated Arf1 activation. Nucleotide exchange assays were performed with 200 nM of a BFA-sensitive Sec7 domain mutant (ARNO $^{F190Y\ A191S}$) and various concentrations of [Δ17]Arf1-GDP in the presence of the indicated concentrations of BFA. Initial velocity was determined graphically from tryptophan fluorescence recordings and plotted as a function of [Δ17]Arf1-GDP concentration. (From Peyroche, A., Antonny, B., Robineau, S., Acker, J., Cherfils, J., and Jackson, C. L. (1999). Brefeldin A acts to stabilize an abortive ARF-GDP-Sec7 domain protein complex: Involvement of specific residues of the Sec7 domain. *Mol. Cell* **3**, 275–285.) (B) The parallel lines pattern obtained after Lineweaver-Burk linearization of the saturation curves shown in A, is typical of uncompetitive inhibition. (C) Model of the uncompetitive inhibitor BFA mechanism on Arf1 activation by ARNO.

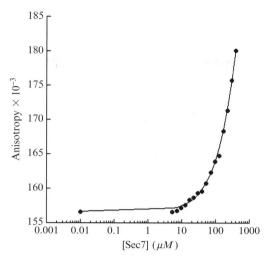

Figure 4.7 Determination of the affinity of the Arf1-GDP/Sec7 complex. The anisotropy-binding profile was obtained from the titration of 10 nM of Alexa 488-labeled [Δ17] Arf1-GDP with the ARNO Sec7 domain using a serial dilution format (see text for details).

with BSA (0.1 mg/ml) and glycerol (10%) to avoid nonspecific interactions. The anisotropy blank was measured under these conditions using unlabeled Arf1-GDP and the maximal GEF concentration. It should be noted that an increase in anisotropy of at least 10 milliA (10^{-3} anisotropy units) between the extreme points (no GEF and maximal GEF concentration) is recommended, and that anisotropy values above 320 milliA, which is the maximal theoretical value for the Alexa 488 dye, indicate protein aggregation. Anisotropy profiles were measured in the serial dilution format as described in Grillo *et al.* (1999), maintaining Alexa 488-labeled Arf1-GDP at a concentration of 10 nM in the cuvette and decreasing the concentration of the GEF by taking out aliquots (one-third of the total volume) from the Arf1-GDP/GEF solution and replacing them with a solution containing labeled Arf-GDP alone. The dissociation constant is determined using a biological equation solver, such as the BioEqs software (Royer and Beechem, 1992, http://abcis.cbs.cnrs.fr/BIOEQS/). It should be noted that, due to solubility limits of the unlabeled GEF, a high concentration plateau might not be reached, thus yielding a less well-defined value for the dissociation constant.

5. CONCLUSION

This flowchart of assays was recently applied to the characterization of a new inhibitor of the activation of Arf by BFA-insensitive GEFs that was discovered using structure-based screening (Viaud *et al.*, 2007).

It allowed the deciphering of a noncompetitive mechanism of inhibition, which was supported by additional biochemical and NMR spectroscopy assays. In the case of fast reactions, the same flowchart can be used in a stopped-flow apparatus to study pre–steady-state kinetics (Hemsath and Ahmadian, 2005). These fluorescence assays are also amenable to automation in 96-well format for high-throughput screening because of their high sensitivity and simplicity (Rojas *et al.*, 2003), and might provide important tools for target-based drug discovery.

ACKNOWLEDGMENTS

We are grateful to Alain Chavanieu and Julien Viaud (Centre de Biochimie Structurale [CBS], Centre National de la Recherche Scientifique [CNRS], Montpellier, France), whose collaboration has fueled the establishment of the flowchart of GEF inhibition assays presented in this review, and to Cathy Royer (CBS, CNRS, Montpellier, France) for teaching us the fundamentals of fluorescence anisotropy. This work was supported by grants from the CNRS (J.C.), the Agence National de la Recherche ANR-PCV (Physique et Chimie du Vivant) (J.C.), and the Association pour la Recherche contre le Cancer (J.-C.Z.). All authors are members of the CNRS research consortium GDR2823.

REFERENCES

Aktories, K., and Barbieri, J. T. (2005). Bacterial cytotoxins: Targeting eukaryotic switches. *Nat. Rev. Microbiol.* **3,** 397–410.

Amor, J. C., Harrison, D. H., Kahn, R. A., and Ringe, D. (1994). Structure of the human ADP-ribosylation factor 1 complexed with GDP. *Nature* **372,** 704–870.

Béraud-Dufour, S., Robineau, S., Chardin, P., Paris, S., Chabre, M., Cherfils, J., and Antonny, B. (1998). A glutamic finger in the guanine nucleotide exchange factor ARNO displaces Mg2+ and the beta-phosphate to destabilize GDP on ARF1. *EMBO J.* **17,** 3651–3659.

Blangy, A., Bouquier, N., Gauthier-Rouviere, C., Schmidt, S., Debant, A., Leonetti, J. P., and Fort, P. (2006). Identification of TRIO-GEF1 chemical inhibitors using the Yeast Exchange Assay. *Biol. Cell.* **98,** 511–522.

Chardin, P., Paris, S., Antonny, B., Robineau, S., Beraud-Dufour, S., Jackson, C. L., and Chabre, M. (1996). A human exchange factor for ARF contains Sec7- and pleckstrin-homology domains. *Nature* **384,** 481–484.

Cherfils, J., and Chardin, P. (1999). GEFs: Structural basis for their activation of small GTP-binding proteins. *Trends Biochem. Sci.* **24,** 306–311.

Donaldson, J. G., Finazzi, D., and Klausner, R. D. (1992). Brefeldin A inhibits Golgi membrane-catalysed exchange of guanine nucleotide onto ARF protein. *Nature* **360,** 350–352.

Esters, H., Alexandrov, K., Iakovenko, A., Ivanova, T., Thoma, N., Rybin, V., Zerial, M., Scheidig, A. J., and Goody, R. S. (2001). Vps9, Rabex-5 and DSS4: Proteins with weak but distinct nucleotide-exchange activities for Rab proteins. *J. Mol. Biol.* **310,** 141–156.

Feng, B. Y., and Shoichet, B. K. (2006). A detergent-based assay for the detection of promiscuous inhibitors. *Nat. Protoc.* **1,** 550–553.

Gao, Y., Dickerson, J. B., Guo, F., Zheng, J., and Zheng, Y. (2004). Rational design and characterization of a Rac GTPase–specific small molecule inhibitor. *Proc. Natl. Acad. Sci. USA* **101,** 7618–7623.

Goldberg, J. (1998). Structural basis for activation of ARF GTPase: Mechanisms of guanine nucleotide exchange and GTP-myristoyl switching. *Cell* **95,** 237–248.

Grillo, A. O., Brown, M. P., and Royer, C. A. (1999). Probing the physical basis for trp repressor–operator recognition. *J. Mol. Biol.* **287,** 539–554.

Hafner, M., Schmitz, A., Grune, I., Srivatsan, S. G., Paul, B., Kolanus, W., Quast, T., Kremmer, E., Bauer, I., and Famulok, M. (2006). Inhibition of cytohesins by SecinH3 leads to hepatic insulin resistance. *Nature* **444,** 941–944.

Hardt, W. D., Chen, L. M., Schuebel, K. E., Bustelo, X. R., and Galan, J. E. (1998). *S. typhimurium* encodes an activator of Rho GTPases that induces membrane ruffling and nuclear responses in host cells. *Cell* **93,** 815–826.

Helms, J. B., and Rothman, J. E. (1992). Inhibition by brefeldin A of a Golgi membrane enzyme that catalyses exchange of guanine nucleotide bound to ARF. *Nature* **360,** 352–354.

Hemsath, L., and Ahmadian, M. R. (2005). Fluorescence approaches for monitoring inter-actions of Rho GTPases with nucleotides, regulators, and effectors. *Methods* **37,** 173–182.

Hill, J. J., and Royer, C. A. (1997). Fluorescence approaches to study of protein-nucleic acid complexation. *Methods Enzymol.* **278,** 390–416.

Hutchinson, J. P., and Eccleston, J. F. (2000). Mechanism of nucleotide release from Rho by the GDP dissociation stimulator protein. *Biochemistry* **39,** 11348–11359.

Jackson, C. L. (2000). Brefeldin A revealing the fundamental principles governing mem-brane dynamics and protein transport. *Subcell. Biochem.* **34,** 233–272.

John, J., Sohmen, R., Feuerstein, J., Linke, R., Wittinghofer, A., and Goody, R. S. (1990). Kinetics of interaction of nucleotides with nucleotide-free H-ras p21. *Biochemistry* **29,** 6058–6065.

Klebe, C., Prinz, H., Wittinghofer, A., and Goody, R. S. (1995). The kinetic mechanism of Ran—Nucleotide exchange catalyzed by RCC1. *Biochemistry* **34,** 12543–12552.

Laufs, U., and Liao, J. K. (2000). Targeting Rho in cardiovascular disease. *Circ. Res.* **87,** 526–528.

Lenzen, C., Cool, R. H., Prinz, H., Kuhlmann, J., and Wittinghofer, A. (1998). Kinetic analysis by fluorescence of the interaction between Ras and the catalytic domain of the guanine nucleotide exchange factor Cdc25Mm. *Biochemistry* **37,** 7420–7430.

Lenzen, C., Cool, R. H., and Wittinghofer, A. (1995). Analysis of intrinsic and CDC25-stimulated guanine nucleotide exchange of p21ras-nucleotide complexes by fluorescence measurements. *Methods Enzymol.* **255,** 95–109.

Mayer, G., Blind, M., Nagel, W., Bohm, T., Knorr, T., Jackson, C. L., Kolanus, W., and Famulok, M. (2001). Controlling small guanine–nucleotide-exchange factor function through cytoplasmic RNA intramers. *Proc. Natl. Acad. Sci. USA* **98,** 4961–4965.

Mossessova, E., Corpina, R. A., and Goldberg, J. (2003). Crystal structure of ARF1★Sec7 complexed with Brefeldin A and its implications for the guanine nucleotide exchange mechanism. *Mol. Cell* **12,** 1403–1411.

Nagai, H., Kagan, J. C., Zhu, X., Kahn, R. A., and Roy, C. R. (2002). A bacterial guanine nucleotide exchange factor activates ARF on *Legionella* phagosomes. *Science* **295,** 679–682.

Nassar, N., Cancelas, J., Zheng, J., Williams, D. A., and Zheng, Y. (2006). Structure-function based design of small molecule inhibitors targeting Rho family GTPases. *Curr. Top. Med. Chem.* **6,** 1109–1116.

Peyroche, A., Antonny, B., Robineau, S., Acker, J., Cherfils, J., and Jackson, C. L. (1999). Brefeldin A acts to stabilize an abortive ARF-GDP-Sec7 domain protein complex: Involvement of specific residues of the Sec7 domain. *Mol. Cell* **3,** 275–285.

Peyroche, A., Paris, S., and Jackson, C. L. (1996). Nucleotide exchange on ARF mediated by yeast Gea1 protein. *Nature* **384,** 479–481.

Ramakers, G. J. (2002). Rho proteins, mental retardation and the cellular basis of cognition. *Trends Neurosci.* **25**, 191–199.

Renault, L., Guibert, B., and Cherfils, J. (2003). Structural snapshots of the mechanism and inhibition of a guanine nucleotide exchange factor. *Nature* **426**, 525–530.

Roberts, A. E., Araki, T., Swanson, K. D., Montgomery, K. T., Schiripo, T. A., Joshi, V. A., Li, L., Yassin, Y., Tamburino, A. M., Neel, B. G., and Kucherlapati, R. S. (2007). Germline gain-of-function mutations in SOS1 cause Noonan syndrome. *Nat. Genet.* **39**, 70–74.

Robineau, S., Chabre, M., and Antonny, B. (2000). Binding site of brefeldin A at the interface between the small G protein ADP-ribosylation factor 1 (ARF1) and the nucleotide-exchange factor Sec7 domain. *Proc. Natl. Acad. Sci. USA* **97**, 9913–9918.

Rojas, R. J., Kimple, R. J., Rossman, K. L., Siderovski, D. P., and Sondek, J. (2003). Established and emerging fluorescence-based assays for G-protein function: Ras-superfamily GTPases. *Comb. Chem. High Throughput Screen* **6**, 409–418.

Rossman, K. L., Der, C. J., and Sondek, J. (2005). GEF means go: Turning on RHO GTPases with guanine nucleotide-exchange factors. *Nat. Rev. Mol. Cell Biol.* **6**, 167–180.

Royer, C. A., and Beechem, J. M. (1992). Numerical analysis of binding data: Advantages, practical aspects, and implications. *Methods Enzymol.* **210**, 481–505.

Sahai, E., and Marshall, C. J. (2002). RHO-GTPases and cancer. *Nat. Rev. Cancer* **2**, 133–142.

Schmidt, S., Diriong, S., Mery, J., Fabbrizio, E., and Debant, A. (2002). Identification of the first Rho-GEF inhibitor, TRIPalpha, which targets the RhoA-specific GEF domain of Trio. *FEBS Lett.* **523**, 35–42.

Sheen, V. L., Ganesh, V. S., Topcu, M., Sebire, G., Bodell, A., Hill, R. S., Grant, P. E., Shugart, Y. Y., Imitola, J., Khoury, S. J., Guerrini, R., and Walsh, C. A. (2004). Mutations in ARFGEF2 implicate vesicle trafficking in neural progenitor proliferation and migration in the human cerebral cortex. *Nat. Genet.* **36**, 69–76.

Singleton, V. L., Bohonos, N., and Ullstrup, A. J. (1958). Decumbin, a new compound from a species of *Penicillium*. *Nature* **181**, 1072–1073.

Tamura, G., Ando, K., Suzuki, S., Takatsuki, A., and Arima, K. (1968). Antiviral activity of brefeldin A and verrucarin A. *J. Antibiot. (Tokyo)* **21**, 160–161.

Tartaglia, M., Pennacchio, L. A., Zhao, C., Yadav, K. K., Fodale, V., Sarkozy, A., Pandit, B., Oishi, K., Martinelli, S., Schackwitz, W., Ustaszewska, A., Martin, J., *et al.* (2007). Gain-of-function SOS1 mutations cause a distinctive form of Noonan syndrome. *Nat. Genet.* **39**, 75–79.

Viaud, J., Zeghouf, M., Barelli, H., Zeeh, J.-C., Padilla, A., Guibert, B., Chardin, P., Royer, C., Cherfils, J., and Chavanieu, A. (2007). Structure-based discovery of an inhibitior of Arf activation by Sec7 domains through targeting of protein-protein complexes. *Proc. Natl. Acad. Sci. USA* **104**, 10370–10375.

Wessels, E., Duijsings, D., Niu, T. K., Neumann, S., Oorschot, V. M., de Lange, F., Lanke, K. H., Klumperman, J., Henke, A., Jackson, C. L., Melchers, W. J., and van Kuppeveld, F. J. (2006). A viral protein that blocks Arf1-mediated COP-I assembly by inhibiting the guanine nucleotide exchange factor GBF1. *Dev. Cell* **11**, 191–201.

Wittinghofer, A., and Waldmann, H. (2000). Ras—A molecular switch involved in tumor formation. *Angew. Chem. Int. Ed.* **39**, 4192–4214.

Zeeh, J. C., Zeghouf, M., Grauffel, C., Guibert, B., Martin, E., Dejaegere, A., and Cherfils, J. (2006). Dual specificity of the interfacial inhibitor brefeldin a for arf proteins and sec0007 domains. *J. Biol. Chem.* **281**, 11805–11814.

Zeghouf, M., Guibert, B., Zeeh, J. C., and Cherfils, J. (2005). Arf, Sec7 and Brefeldin A: A model towards the therapeutic inhibition of guanine nucleotide-exchange factors. *Biochem. Soc. Trans.* **33**, 1265–1268.

CHAPTER FIVE

ANALYSIS OF SMALL GTPASE FUNCTION IN TRYPANOSOMES

Mark C. Field,* David Horn,‡ *and* Mark Carrington†

Contents

Abstract

Trypanosomatids are protozoan parasites, of interest due to both their disease burden and deeply divergent position within the eukaryotic lineage. The African trypanosome, *Trypanosoma brucei*, has emerged as a very amenable model system, with a considerable toolbox of methods available, including inducible overexpression, RNA interference, and a completed genome. Here we describe some of the special considerations that need to be addressed when studying trypanosome gene function, and in particular small GTPases; we provide protocols for transfection, RNA interference, overexpression and basic transport assays, in addition to an overview of available vectors, cell lines, and strategies.

1. OVERVIEW

Trypanosomatids comprise a protist lineage that includes many infectious species alongside numerous free-living taxa. Trypanosomes infect plants, arthropods, fish, and mammals; the latter category is of the most

* Department of Pathology, University of Cambridge, Cambridge, United Kingdom
† Department of Biochemistry, University of Cambridge, Cambridge, United Kingdom
‡ London School of Hygiene and Tropical Medicine, London, United Kingdom

Methods in Enzymology, Volume 438
ISSN 0076-6879, DOI: 10.1016/S0076-6879(07)38005-1

interest as these species are responsible for significant morbidity and mortality in humans and domestic animals (Barrett *et al.*, 2003). There are three major trypanosome lineages involved, the African trypanosome, *Trypanosoma brucei* sp., the South American trypanosome, *T. cruzi*, and the Leishmanias, both Old and New World. These organisms are responsible for a spectrum of disease affecting the majority of the human population across Africa, South America, Europe, and Asia. Genome sequencing indicates a remarkable degree of relatedness between these three lineages (Berriman *et al.*, 2005).

Small GTPases are highly important regulators of function in trypanosomes, and because of the absence of heterotrimeric GTPases from trypanosomatid genomes, Ras-like GTPases likely shoulder a larger proportion of the signal transduction burden than in higher eukaryotes. Over 40 small GTPases are encoded by the *T. brucei* genome (Field, 2005); considerable work has been done on the Rab subfamily (Field and Carrington, 2004), and a lesser amount on the Arf proteins (Price *et al.*, 2005), but essentially nothing is known of the functions of the GTPases more closely related to the Ras and Rho families (Field, 2005). Interest in intracellular trafficking is mainly prompted by the fact that the surfaces of trypanosomes are dominated by GPI-anchored proteins and glycoconjugates, and that in *T. brucei* endocytosis is likely a component of the immune evasion machinery (Field *et al.*, 2007). We remain almost completely ignorant of the downstream molecules that interact with trypanosome small GTPases; informatics is able to identify candidate molecules—for example, the nine TBC domain–containing proteins encoded in the trypanosome genome are likely Rab GTPase activators—but no information is available on function (A. J. O'Reilly, C. Gabernet-Castello, and M. C. Field, unpublished).

Much of the analytical tool kit exploited by the wider community is applicable to trypanosome GTPases, such as overexpression of dominantly active forms of the proteins, use of two hybrid libraries for screening for interacting partners, and epitope tagging, and these topics and strategies have been well discussed elsewhere. However, several aspects are unique; in particular, the trypanosome transcriptional machinery is predominantly polycistronic, and the basal transcriptional apparatus appears rather divergent (Ivens *et al.*, 2005). Further, a dismally low transfection efficiency, at less than 10^{-6} transformants/μg DNA for bloodstream forms, which has so far resisted all attempts at significant improvement, means that creation of transgenic trypanosomes must be done in a carefully considered manner, and that forward genetic screening methods are currently very cumbersome; a recent advance in technology bodes well for increased efficiency (Burkard *et al.*, 2007). The consequence is that specific vector systems have been developed, and reliable constitutive and inducible systems are now available primarily for reverse genetics approaches. The latter require use of engineered cell lines harboring the tetracycline repressor. Due to the diploid genome and the comparative cumbersome nature of inducible expression systems in trypanosomes, initial analysis of small GTPase function by gene deletion is not recommended; at least three rounds of transfection are required—to remove both endogenous alleles and to provide a conditional copy in the case of essentiality—and under some circumstances the ability to control expression of inducible ectopic genes can prove challenging. Further, such approaches result in overexpression, as induction requires T7 or RNA pol I–driven transcription. As an overall strategy, we favor a combination of knockdown by RNAi, coupled with constitutive overexpression of wildtype or mutant forms of the relevant GTPase; such approaches have been highly successful (Hall *et al.*, 2004, 2005a,b,c).

This article focuses on analysis of small GTPases in *T. brucei* on account of this system being more technically advanced than other trypanosomatids; we will consider relevant vectors and their use and describe the various reporter lines that are required for specific approaches. We also describe general assays for analysis of endocytosis and exocytosis.

2. CULTURING

Trypanosomes require specialized media for culturing. These can be made *de novo*, and are straightforward to make, but media can also be purchased direct as custom orders from several companies (e.g., Invitrogen, Sigma, JRH Biosciences). In addition, workers may wish to contact the authors as the community places bulk orders with companies on an occasional basis. The most commonly used laboratory strains of *T. brucei* are derived from the Lister 427 isolate. The procedures outlined below have been optimized using Lister 427, and it may be necessary to modify some of these conditions for the growth of other isolates in culture.

For the bloodstream form, *T. brucei* HMI-9 is used (Hirumi and Hirumi, 1989). To the basic medium, add the following per liter: 3 g sodium bicarbonate (Sigma), 10 ml penicillin/streptomycin (100 × 5000 U to 5000 μg/ml, Life Tech, #15070-063), and 14 μl β-mercaptoethanol. Filter, sterilize, and store at 4° for up to 6 months. Add 10% heat inactivated (56°, 30 min) fetal calf serum before use. Grow cells at 37° with 5% CO_2, in nonadherent culture flasks, with vented caps. Do not allow the cells to exceed 2 × 10^6/ml.

For procyclic forms, the routine culture medium is SDM-79 and can be made up from its components (Brun and Schonenberger, 1979). When making SDM-79, add the following fresh materials: 2 g/l of sodium hydrogen carbonate. Once made, the basal medium is stable for several months at 4°. To complete the medium, 3 ml of 2.5 mg/ml hemin dissolved in 50 mM sodium hydroxide and 100 ml fetal bovine serum are added per liter. Once complete, the medium is stable for 1 month at $4°$. Plastic tissue culture flasks are used: 10 ml culture in a 25-cm^2 flask, 30 ml in a 75-cm^2 flask, and 70 ml in a 175-cm^2 flask. Lids are adjusted to allow free gaseous exchange. Cultures are maintained at $27°$; it is not necessary to provide a CO_2 source. Procyclic trypanosomes do not grow well if diluted to less than $1 \times 10^5/\text{ml}$ and reach a stationary phase at around $4 \times 10^7/\text{ml}$. Routine cultures are maintained between $5 \times 10^5/\text{ml}$ and $1 \times 10^7/\text{ml}$, a 20-fold range that can be conveniently maintained by making a 1-in-20 dilution every 2 or 3 days, depending on the strain. Subculturing using a dilution of more than 1 in 20 is not routinely used, as there can be a long lag before growth resumes.

The procyclic cell line Lister 427 29-13 (containing pLEW29 and pLEW13) or the *Single Marker Bloodstream* (SMB, *T7RNAP::TETR:: NEO*) line are routinely used for the expression of tetracycline-inducible transgenes and tetracycline-inducible RNAi. Both contain integrated transgenes that constitutively express the tet repressor and T7 RNA polymerase. The 29-13 cell line does not grow as well as the wildtype; the doubling time is increased and cultures contain a higher percentage of cells with aberrant morphologies. An alternative with superior morphology and similar inducibility is the PTT line, also derived from 427; again the antibiotic selection must be maintained at all times. The morphology of procyclics can often be improved by rapid passage, and if this is unsuccessful, by cloning.

If performing RNAi knockdown experiments is intended, then additional consideration is advisable in testing fetal bovine serum batches. Normally, batches require testing for procyclic forms as occasional batches do not support growth. However, in addition, some batches of serum may be contaminated with oxytetracycline at sufficient concentration to induce low levels of RNAi. If the RNAi is deleterious to growth, and then there is rapid selection for cells that cannot perform the desired ablation for one of several reasons. This may be avoided by testing serum batches for the ability to support the growth of a Lister 427 29-13 or SMB cell line containing a rapidly lethal RNAi construct such as eIF4A1 or clathin heavy chain, respectively.

3. TRANSFECTION AND CLONING OF T. BRUCEI

The production of recombinant trypanosomes is dependent on antibiotic selection. The antibiotics routinely used at present are geneticin (G418), hygromycin, phleomycin (zeocin), blasticidin, and puromycin (Tables 5.1 and 5.3). Antibiotics can be stored as $100\times$ or $1000\times$ stocks in sterile water. If you suspect an issue with antibiotic selection, such as antibiotic resistance in the absence of expression of the expected transgene

Table 5.1 Plasmid-Based Vectors Commonly Used for Manipulation of *Trypanosoma brucei* Gene Expression

Name	Features	Integration site	Applications	Selection[a]	Reference
p2T7[b]	T7 promoter/TET operator	rRNA spacer	RNA interference	H	La Count et al., 2000
pXS5[c]	rRNA promoter	rRNA spacer	Constitutive expression in BSF	GPH	Alexander et al., 2002
pXS2	Procyclin promoter	Tubulin intergenic	Constitutive expression in PCF	G	Bangs et al., 1996
pLEW100	Procyclin promoter	rRNA spacer	Inducible expression in BSF or PCF	P	Wirtz et al., 1999
pRPa^GFP	rRNA promoter	rRNA spacer	Inducible expression in 2T1 cells	ZH	Alsford et al., 2005
pRPa^Con	rRNA promoter	rRNA spacer	Constitutive expression in 2T1 cells	ZH	Alsford and Horn, unpublished
pRPa^SL	rRNA promoter	rRNA spacer	RNAi in 2T1 cells	ZH	Alsford and Horn, unpublished

[a] G, G418/neomycin; H, hygromycin; P, phleomycin; Z, zeocin.
[b] A new version of this construct, p2T7^TABlue, with increased efficiency for PCR cloning and selection is described in Subramanium et al. (2006).
[c] A version of both pXS2 and pXS5, designated pXS219 and pXS519, with a few additional cloning sites, is also available from the Field lab.

or unexpectedly high transfection efficiencies, it may be advisable to titer the concentration of the antibiotic, as the effective concentration may be influenced by the batch of fetal calf serum. To do this, it is suggested that a concentration range of 0.25 to 4.0× of the concentration given in Table 5.1 is assayed for efficient killing of cultures. For procyclics select the concentration that restricts growth to less than 2×10^6/ml, and causes cell death (loss of motility) within 8 days. Phleomycin sometimes takes up to 10 days to kill the cells. For bloodstream forms, efficient killing should be obtained after 2 days. Antibiotic-mediated killing does not occur with stationary phase cells.

BSF trypanosomes are comparatively challenging to transfect as the efficiency is extremely low. Therefore adherence to the protocol is essential, as even a small decrease in efficiency can compromise the experiment. It is recommended that all transfection experiments include a positive control to ensure that the protocol has been correctly carried out, either a strong RNAi (e.g., clathrin heavy chain/eIF4A) or a cytoplasmic eGFP expression vector. The protocol described in the following is for the p2T7 series of RNAi vectors, but can be adapted for other constructs by altering the restriction enzyme used for linearization, and the selecting antibiotic, as required.

For transfection, a Bio-Rad Gene pulser II is used; other models can work but the pulse conditions will need to be parameterized for the specific instrument. For inducible expression, the SMB cell line described by Wirtz *et al.* (1999) is typically used, while a wildtype 427 line is preferable for a simple overexpression experiment, as the genetic background is unmanipulated. Use 10 μg of digested DNA for each electroporation, and usually perform two independent procedures per construct. The DNA must be completely linearized to prevent contamination with circular plasmid, which can both mis-integrate and provide a transient transfection. For p2T7, digest the DNA overnight with twofold excess of *Not*I, and check the plasmid digestion on an agarose gel with undigested p2T7^{TAblue} as a control. It is most convenient if the electroporation is performed in the morning. At least 2.5×10^7 cells are required for each electroporation. Harvest the cells by centrifugation, at 1000×g for 10 min at ambient temperature and resuspend them in ~25 ml of cytomix[1] at 37° in a falcon tube. Pellet again and resuspend at ~6 × 10^7/ml in cytomix. While the cells are pelleting, aliquot sterilized linear DNA (10 μg in 10 μl) into microfuge tubes; the DNA is best sterilized by ethanol precipitation prior to use. Mix 0.45 ml of the cells with DNA, and transfer to a 2-mm gap electrocuvette and administer one pulse (1.4 kV, 25 μF). Five to 15% of the cells

---

[1] Cytomix: 2 mM EGTA, pH 7.6, 5 mM MgCl$_2$, 120 mM KCl, 120 mM, 0.5% glucose (dextrose), 0.15 mM CaCl$_2$, 0.1 mg/ml BSA, 10 mM K$_2$HPO$_4$/KH$_2$PO$_4$, pH 7.6, 1 mM hypoxanthine, 25 mM HEPES, pH 7.6. Adjust to pH 7.6 with KOH, filter sterilize, and store at 4° (adapted from Hoff, *et al.*, 1992).

should survive the pulse. After electroporation, place all contents of the cuvette into 36 ml of HMI-9 in a 25-cm flask, and 6 hr later add 25 μl of a mixture of HygromycinB/G418[2] to each flask and transfer to 24-well plates. For transfection of wildtype 427 cells, the addition of G418 is omitted. On day 6, expand the positive wells and freeze samples[3] as required. It is usually unnecessary to consider cloning of bloodstream trypanosomes, as the frequency of transfection will normally ensure that each well that grows up is a clone. Cloning by limiting dilution, that is, plating a theoretical 0.5 cells per well, is also possible should the need to clone arise. No adaptations to media are required for this procedure.

Procyclic cells are substantially more efficient to transfect, and hence for some purposes may be the stage of choice. Preparation of DNA is essentially the same as for bloodstream cells. For transfection into the PCF 29-13 cell line (*T7RNAP::TETR::NEO::HYG*), 20 μg of linearized plasmid is added to 2×10^7–log phase cells that have been resuspended in a total volume of 0.45 ml of Opti-MEM (Invitrogen). Thoroughly mix the sample with a P200 Gilson pipette, and then transfer to a precooled 2-mm gap electro-cuvette (Bio-Rad), and chill on ice for 5 min prior to electroporation. Deliver a single pulse from a Bio-Rad Gene Pulser II electroporator (1.5 kV, 25 μF). Immediately following electroporation, transfer the cells into 10 ml of SDM79/10%FCS containing 25 μg/ml G418 and 25 μg/ml hygromycin, and leave the culture to recover at 27° overnight. Approximately 16 h following electroporation, add phleomycin at 2.5 μg/ml to select for transformants. Drug-resistant cell lines typically grow within 2 to 3 weeks. For transfection of wildtype 427 cells, the addition of G418 and hygromycin is omitted.

Procyclic trypanosomes can also be cloned by limiting dilution. Prepare 25 ml each of cells at one cell per milliliter and 0.2 cell per milliliter. Plate out as 24 1-ml cultures in a 24-well tissue culture plate, and inspect from 8 days onward up to 14 days. If eight wells or less contain trypanosomes on one of the plates, the well can be considered clonal. Cloning presents particular problems as procyclic trypanosomes do not tolerate dilution well, and either the use of conditioned medium or increased fetal bovine serum concentration and growth in CO_2 is required. Conditioned medium is prepared from a mid–log phase culture by centrifugation to remove most of the cells and filtering through a 0.22-μm filter to remove residual cells and to resterilize. For cloning, use a mix of 60% fresh medium/40%

---

[2] Drug selection mix: hygromycin B stock; 5 mg/ml, G418 stock; 10 mg/ml. Store at −20°. For 10 flasks, make a working stock of 200 μl hygromycin B and 80 μl G418; 25 μl in each flask gives a final concentration of 2.5 μg/ml and 2 μg/ml, respectively.

[3] Cryopreservation mix: Mix cells in HMI9 plus 10% glycerol (sterile). Place in a suitable cryo-box or polystyrene box sealed with tape and place at −70° overnight. Move vials to a liquid nitrogen vessel for long-term storage.

conditioned medium. The alternative is to increase the fetal bovine serum to 15% and to grow in 5% CO_2. If this latter approach is used, then the selective concentration of any antibiotic may require retitering.

4. VECTOR SYSTEMS FOR OVEREXPRESSION AND RNA INTERFERENCE

In common with most biological systems, trypanosomes have specific elements that are required for gene transcription, and these elements must be included in ectopic expression systems. In addition, for inducible expression the T7 polymerase/tet-on system that has been frequently exploited has been successfully imported into trypanosomes (Wirtz and Clayton, 1995); this feature is particularly valuable for RNA interference, where tight control of potentially lethal dsRNAs is essential.

Unique features of trypanosome transcription include the complete lack of characterized RNA polymerase II promoters for protein–coding genes, the requirement for *trans*-splicing, and the importance of the 3′ end in mRNA stability. For ectopic expression, these considerations have been incorporated into many vector designs (Fig. 5.1A). The pXS2 and pXS5 vectors (Alexander *et al.*, 2002; Bangs *et al.*, 1996) exploit RNA polymerase I promoters, together with intergenic regions from high–abundance transcripts/protein products, which provide the essential signals for *trans*-splicing and polyadenylation. pXS2 and pXS5 are also targeted to repetitive arrays within the genome, the tubulin locus on chromosome I, and the rRNA repeats. The expression cassettes are built into pBluescript backbones for easy propagation in *Escherichia coli*. In our lab, these vectors are highly reliable, producing between 3- and 10-fold overexpression of small GTPases compared to endogenous levels (Pal *et al.*, 2003), and such expression appears to be highly stable over a great number of generations. Overexpression of other constructs is also possible (Alexander *et al.*, 2002; Bangs *et al.*, 1996; Chung *et al.*, 2004). One possible complication with pXS2 is that variable transcriptional read-through occurs into the body of chromosome I, and in some cell lines increased levels of both tubulin mRNA and protein are observed, indicating that read-through transcripts can be processed and translated, although no obvious phenotype results (Dhir and Field, 2004; V. Koumandou and M. C. Field, unpublished data). It is not clear if this is a general phenomenon or specific to pXS2, and certainly care should be taken when selecting a vector to avoid potential artifacts. It should also be borne in mind that targeting to these arrays has the potential to result in differential expression depending on the precise context of the integration event. For this reason, we recommend that at least three clones or lines should be analyzed to ensure that the desired overexpression has been achieved.

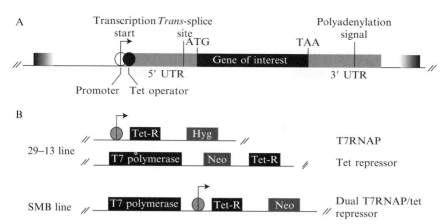

Figure 5.1 Features of expression systems in trypanosomes. (A) General features of trypanosome ectopic expression cassettes. Due to the polycistronic expression of most trypanosome protein–coding genes, a specialized set of vectors has been developed. These include a promoter (open oval), typically the Pol I promoter for VSG, procyclin, or rRNA, 5′ sequence derived from a well-characterized gene (light gray), such as tubulin, which provides a 5′ *trans*-splicing site for addition of the 39 nucleotide miniexon, and a 3′ sequence from a stable transcript (light gray), which also donates the polyadenylation site. In some instances, these promoters have been modified to include the Tet operator to facilitate conditional expression. The 3′ end sequence is chosen carefully as this region also contains elements responsible for control of mRNA levels, and while these elements are poorly characterized, the behavior of such elements for the most part is predictable in terms of mRNA levels and developmental regulation. In addition, the promoter may be the T7 polymerase promoter, with the tet repressor binding site (operator, black oval) to facilitate inducible expression. Regions of more than 200 base pairs are normally used to facilitate faithful integration into the genome (gradient gray) and the gene of interest is shown as a black box. The double slash (//) indicates region of the construct that is not shown. The elements are not shown to scale. (B) Features of the tet-on inducible system for 90-13, SMB, and 29-13 lines. The 90-13 and 29-13 lines incorporate two elements, one carrying the tet repressor and the other the T7 RNA polymerase. Production of T7 pol and the downstream tet repressor ORF relies on read-through transcription. Production of tet repressor is also facilitated through a second element, which uses a crippled T7 promoter, with about 10% of normal transcriptional activity (gray oval). The SMB line was created by a two-step procedure to remove one of the selectable markers. The resulting cassette contains both the T7 polymerase and the tet repressor, and is maintained by neomycin/G418 selection. Again, transcription of the T7 polymerase relies on endogenous transcription, while the tet repressor is transcribed primarily from a 10% T7 promoter (gray oval). All of these elements are integrated into the tubulin repeat array on chromosome I, except the tet repressor construct in 29-13, which is in the RNP1 locus. The elements are not shown to scale; selectable markers are shown in a light gray box, and ORFs required for functioning of the transcription system in black boxes. The double slash (//) indicates sequence features not shown. (Based on data in Wirtz, E., (1999).)

The expression of transgenes is occasionally deleterious to the cell, a problem that can be overcome with a tightly regulated inducible promoter. This technology requires the use of a cell line engineered to produce T7 RNA polymerase and the tet repressor; several of these are now available and selections are described in Tables 5.1 and 5.2. It should be noted that the inducible cell lines are suitable for both overexpression and RNAi experiments, as the basic expression cassettes are common (see Fig. 5.1). In all instances the various combinations of antibiotic must be continually maintained as expression of the T7 polymerase is not well tolerated and tends to be lost rapidly in the absence of selection.

A transgene cloned into the pLEW100 vector is expressed from an EP procyclin promoter that itself is regulated by two tet repressor binding sites (Wirtz *et al.*, 1999). The promoter is activated by the addition of tetracycline or doxycycline to the culture medium, the concentration can be used to titer expression, although this is technically challenging to achieve in a reproducible manner (Wirtz *et al.*, 1999). Routinely, an open reading frame is cloned into the HindIII and BamHI sites of pLEW100; the resultant chimeric mRNA has the EP procyclin 5'UTR, the inserted open reading frame, and a modified aldolase 3'UTR. Prior to electroporation, pLEW100-derived plasmids are linearized with the restriction enzyme NotI, and then targeted to the nontranscribed spacer in one of the rRNA gene loci. At maximum expression levels using 1 μg/ml tetracycline, the transgene is usually expressed at about the level of an abundant cytoplasmic protein, such as the translation initiation factor eIF4A1 (Dhalia *et al.*, 2006). The promoter is silent in the absence of tetracycline and pLEW100, and its derivatives have been used for the regulated expression of several lethal dominant mutants (Dhalia *et al.*, 2006). Several labs have developed derivatives of pLEW100 to express transgenes with N- or C-terminal tags; epitope tags for detection and fluorescent proteins tag for subcellular localization. One such set of vectors is described at http://web.mac.com/mc115/iWeb/mclab/resources.html. The ability to use doxycycline also provides the possibility of using such systems in animal models, specifically mice and rats, as induction can be achieved by adding doxycycline to drinking water (Lecordier *et al.*, 2005).

A number of second-generation vectors and cell lines are available or are in development. These provide a range of options, a variety of epitope tags for N- or C-terminal tagging and improvements in ease of plasmid construction and transfection efficiency. A few examples are outlined below.

The p2T7 RNAi vector was modified for direct cloning of PCR products (Alibu *et al.*, 2005). This p2T7[TAblue] vector has been used to screen more than 200 genes on *T. brucei* chromosome I (Subramanium *et al.*, 2006), and is now available with a variety of selectable marker genes. The cloning site comprises a *LacZ* stuffer flanked by *Eam*1105I sites engineered for T-A cloning. The vector is digested with *Eam*1105I at 37° for

Table 5.2 Some Common *Trypanosoma brucei* Cell Lines Used in Expression and/or RNA Interference Studies

Name	Features	Applications	Used markers	Reference
927	Genome strain	Not commonly used	None	van Deursen et al., 2001
Lister 427	Common lab strain	Overexpression, knockout	None	Melville et al., 2000
SMB (328.114)	Bloodstream line	Inducible expression/RNAi	G418	Wirtz et al., 1999
29-13[a]	Procyclic line	Inducible expression/RNAi	G418 and Hygro	Wirtz et al., 1999
90-13	Bloodstream line	Inducible expression/RNAi	G418 and Hygro	Wirtz et al., 1999
PTT	Procyclic line	Inducible expression/RNAi	G418 and Hygro	Bastin, P, unpublished
PTH	Procyclic line	Inducible expression (not T7)	Hygro	Bastin et al., 1999
PTP	Procyclic line	Inducible expression (not T7)	Phleo	Wickstead et al., 2003
2T1	Bloodstream line with tagged *rRNA* spacer	Inducible/constitutive expression/RNAi	Phleo and Puro[b]	Alsford et al., 2005

[a] If this strain is not cultured carefully, it can have poor morphology that may be problematic for morphometric and cell cycle analysis. The PTT cell line is superior for this type of analysis, and is essentially identical in all other regards.
[b] The *PAC* gene is lost from the genome upon correct integration of the expression/RNAi vector.

Table 5.3   Selectable Markers in Use for Transfection of Trypanosomes

	Concentration (μg/ml)	
Antibiotic	Procyclic form	Bloodstream form
Geneticin (neomycin/G418)	15	1
Hygromycin	25	5
Phleomycin (zeocin/bleomycin)	2.5	1
Nourseothricin	100	25
Puromycin	1.0	0.1
Blasticidin	10	10

Note: These concentrations are based on experiments with *Trypanosoma brucei brucei* Lister 427 lines cultured in HMI-9 (bloodstream form) or SDM79 (procyclic form), and hence may require adapting for other strains, subspecies or media.

1 to 2 h in preparation for ligation with RNAi target fragments with 3′ A-overhanging ends. We typically use *Taq* DNA polymerase for target fragment amplification. In *E. coli*, on plates containing X-gal and IPTG, uncut or re-ligated vector-containing cells form blue colonies, while the desired recombinants generate white colonies.

Targeted integration at a tagged *rRNA* spacer locus in the 2T1 cell line has three significant advantages (Alsford *et al.*, 2005). Transformation efficiency is increased approximately four-fold, such that multiple clones are derived using ∼5 μg of linearized DNA. Faithful double-crossover integration is rapidly verified by screening for puromycin-sensitivity in multiwell plates, and a *PAC* gene is removed from the genome upon correct integration. And finally, all clones analyzed to date exhibit equivalent expression; consequently, screening multiple clones for differential expression is not required (D. Horn and S. Alsford, unpublished data). Recovery of cells with expression cassettes integrated at the correct locus is ensured since complementing portions of the selectable marker are located on the vector and within the 2T1 genome. We typically use a 2T1 strain with a tagged *rRNA* spacer to support the highest expression level because, in most cases, it is desirable to express the maximum quantity of recombinant mRNA or dsRNA in the case of RNAi. There are now a number of 2T1-compatible expression constructs available. These are mostly driven by the *rRNA* promoter, and therefore do not require parallel expression of T7 RNA polymerase; the vector series includes constructs for constitutive or inducible expression and for RNAi. The cloning sites are versatile, and they are linearized for integration using *Asc*I.

For Tet-inducible expression of a tagged protein, we typically allow 24 h induction prior to analysis at which point expression is usually maximal with rare cases of toxicity resulting from overexpression. The situation is different for RNAi experiments, where it is important to carefully consider the timeframe between Tet-addition (RNAi-induction) and specific phenotype assay(s). "Growth curve" analysis can act as a guide if a growth defect is seen. In these circumstances, it is desirable to carry out specific phenotype assays prior to, or immediately after, the defect is seen to reveal primary rather than secondary phenotypes. Monitoring mRNA and/or protein knockdown is also important and can act as a guide where no growth defect is seen. If more rapid and efficient knockdown is required, this can be achieved with "stem-loop" dsRNA vectors, probably due to the more stable intramolecular interaction, but assembly of these vectors is more complex. It is also important to note that the RNAi effect may be lost over time using any of these systems. If RNAi is deleterious to growth, cells that lose a component of the RNAi machinery will come to predominate in the population following an extended period of induction (Chen *et al.*, 2003).

5. CLONING AND MANIPULATION OF TRYPANOSOME SMALL GTPASE SEQUENCES

Trypanosome small GTPases fully conserve the overall primary structure of the Ras superfamily, retaining G boxes one to five, the hypervariable region, and, in most cases, the C-terminal prenylation motif. Some deviation from this structure is apparent for several GTPases (see Field, 2005 and Fig. 5.2), but no functional analysis is available on these factors at present.

Construction and overexpression of either wildtype or mutant forms of trypanosome small GTPases is best achieved by constitutive ectopic expression, and provides a convenient method for both localization and analysis of function. The pXS2 or pXS5 vector series have proven to be highly reliable for this purpose, and in the vast majority of cases transformation and overexpression are readily achieved; no specific selection of host strain is required for these vectors. The constitutive approach is preferred over inducible systems here as precise control of the levels of induction, together with epigenetic effects, as the parasite adjusts to an increase or decrease in activity of the transgene make reproducible control of the system difficult using induction; a serious issue if detailed analysis of function is intended, for example, by trafficking assays. Recent work indicates that manipulation of Rab5 expression levels also influences the expression of clathrin heavy chain (CLH), without altering the levels of CLH mRNA, indicating a post-translational effect that would be challenging to control in an inducible format (Hall *et al.*, 2004; S. Natesan, V. Koumandou, and M. C. Field, unpublished data). Monitoring of overexpression is important, as occasional mistargeting or other factors can compromise function of the introduced transgene—this is

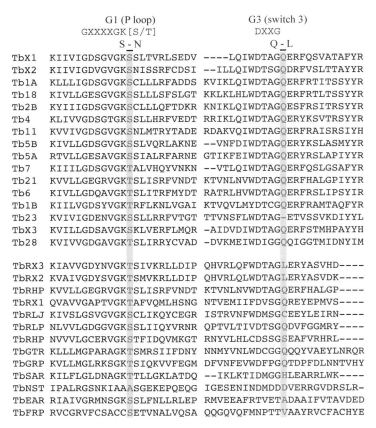

<pre>
                G1 (P loop)   G3 (switch 3)
             GXXXXGK[S/T]   DXXG
                 S - N   Q - L
   TbX1   KIIVIGDSGVGKSSLTVRLSEDV ----LQIWDTAGQERFQSVATAFYR
   TbX2   KIIVIGDVGVGKSNISSRFCDSI --ILLQIWDTSGQDRFVSLTTAYYR
   Tb1A   KLLLIGDSGVGKSCLLLRFADDS KVIKLQIWDTAGQERFRTITSSYYR
   Tb18   KIVLLGESGVGKSSLLLSFSLGT KKLKLHLWDTAGQERFRTLTSSYYR
   Tb2B   KYIIIGDSGVGKSCLLLQFTDKR KNIKLQIWDTAGQESFRSITRSYYR
   Tb4   KLIVVGDSGTGKSSLLHRFVEDT RRIKLQIWDTAGQERYKSVTRSYYR
   Tb11   KVVIVGDSGVGKSNLMTRYTADE RDAKVQIWDTAGQERFRAISRSIYH
   Tb5B   KIVLLGDSGVGKSSLVQRLAKNE --VNFDIWDTAGQERYKSLASMYYR
   Tb5A   RTVLLGESAVGKSSIALRFARNE GTIKFEIWDTAGQERYRSLAPIYYR
   Tb7   KIIILGDSGVGKTALVHQYVNKN --VTLQIWDTAGQERFQSLGSAFYR
   Tb21   KVVLLGEGRVGKTSLISRFVNDT KTVNLNVWDTAGQERFHALGPIYYR
   Tb6   KIVLLGDQAVGKTSLITRFMYDT RATRLHVWDTAGQERFRSLIPSYIR
   Tb1B   KIILVGDSYVGKTRFLKNLVGAI KTVQVLMYDTCGQERFRAMTAQFYR
   Tb23   KVIVIGDENVGKSSLLRRFVTGT TTVNSFLWDTAG-ETVSSVKDIYYL
   TbX3   KVILLGDSAVGKSKLVERFLMQR -AIDVDIWDTAGQERFSTMHPAYYH
   Tb28   KVIVVGDGAVGKTSLIRRYCVAD -DVKMEIWDIGGQQIGGTMIDNYIM

   TbRX3   KIAVVGDYNVGKTSIVKRLLDIP QHVRLQFWDTAGLERYASVHD----
   TbRX2   KVAIVGDYSVGKTSMVKRLLDIP QHVRLQLWDTAGLERYASVDK----
   TbRHP   KVVLLGEGRVGKTSLISRFVNDT KTVNLNVWDTAGQERFHALGP----
   TbRX1   QVAVVGAPTVGKTAFVQMLHSNG NTVEMIIFDVSGQREYEPMVS----
   TbRLJ   KIVSLGSVGVGKSCLIKQYCEGR ISTRVNFWDMSGCEEYLEIRN----
   TbRLP   NLVVLGDGGVGKSSLIIQYVRNR QPTVLTIVDTSGQDVFGGMRY----
   TbRHP   NVVVLGCERVGKSTFIDQVMKGT RNYVLHLCDSSGSEAFVRHRL----
   TbGTR   KLLLMGPARAGKTSMRSIIFDNY NNMYVNLWDCGGQQQYVAEYLNRQR
   TbGRP   KVLLMGLRKSGKTSIQKVVFEGM DFVNFEVWDFPGQTDPFDLNNTVHY
   TbSAR   KILFLGLDNAGKTTLLGKLATDQ --IKLKTIDMGGHLEARRLWK----
   TbNST   IPALRGSNKIAAASGEKEPQEQG IGESENINDMDDDVERRGVDRSLR-
   TbEAR   RIAIVGRMNSGKSSLFNLLRLEP RMVEEAFRTVETADAAIFVTAVDED
   TbFRP   RVCGRVFCSACCSETVNALVQSA QQGQVQFMNPTTVAAYRVCFACHYE
</pre>

Figure 5.2   Key residues in trypanosome Rab and signaling class GTPases. Portions of a ClustalX alignment are shown highlighting the GVGKS (G1) and WDTAGQ (G3) boxes that are highly conserved between small GTPases—the critical residues are shaded. The complete Rab family together with the majority of the signaling GTPases are included (see Ackers *et al.*, 2005; and Field, 2005). These regions may be exploited for the construction of S- to N-dominant negative "GDP-locked," and Q to L constitutively active "GTP-locked" forms of the proteins for overexpression studies (e.g., Pal *et al.*, 2003). A minority of the GTPases are divergent within the WDTAGQ region, but most Ras-, Rho-, Rac-, and Rab-related GTPases retain the consensus sequence.

most conveniently done either by introducing an N-terminal tag, using an antibody raised against the recombinant protein, or by qRT-PCR.

By contrast, RNAi experiments are always conducted using inducible systems, and hence must be performed in the SMB, 29-13, or equivalent cell background. It is critical that the expression level of the targeted gene of interest is monitored. This is required both for ensuring specificity, that is, that the intended target has been suppressed and to eliminate off-target effects, and also for monitoring the progress of the emergence of a phenotype. In bloodstream stage cells, effects from RNAi normally manifest within 1 to 2 days, but longer times of up to 5 days are not uncommon (see Subramaniam *et al.*, 2006, and trypanofan.path.cam.ac.uk for examples), and in procyclics knockdown can be even more protracted. Monitoring knockdown by Western blot is preferable, as it is the level of protein that is important for phenotype, but in the absence of such a reagent, qRT-PCR is rapid, convenient, and accurate; however, such data need to be considered carefully as the half-life of small GTPases in trypanosomes can vary considerably. For the majority of trypanosome small GTPases, RNAi-mediated knockdown results in severe growth inhibition (Dhir and Field, 2004; Hall *et al.*, 2004, 2005a, b; K Abbassi, and M. C. Field, unpublished data).

Construction of expression constructs can be achieved by PCR very readily. As trypanosome open reading frames lack introns, recovery of sequences is most easily achieved by amplification direct from genomic DNA, and subcloning into a convenient vector, followed by the required mutagenesis. It is recommended that a prolonged hot start for up to 30 min is employed when using trypanosome genomic DNA, but because of the small size of the GTPase open reading frame, a proofreading polymerase is not necessary and Taq can be used, providing that the resulting construct is fully sequenced. Sequences can be retrieved from GeneDB at www.genedb.org/genedb/tryp/. As the genome strain (TREU 927) is not the same as the common laboratory strain, it is to be expected that a small number of single nucleotide polymorphisms will be detected between the database sequence and that obtained by PCR from strain 427 DNA. The vast majority are noncoding, and any concerns may be allayed by also sequencing the PCR product directly. There is no significant base bias or abnormal codon usage in the African trypanosome, and hence design of mutagenic oligonucleotides can follow standard protocols. In addition, expression of trypanosome small GTPases as GST–fusion proteins is also standard, and there is no need for specialized host strains for production of recombinant material, either for immunization or for biochemistry on purified protein; however, we do recommend the use of BL21 (DE5) or Rosetta strains of *E. coli* for expression to improve both yield and solubility. Yields of 1 mg/liter of the GTPase can be easily attained.

6. Trafficking Assays to Monitor Exocytosis and Endocytosis

The analysis of bulk exocytosis in trypanosomes is comparatively straightforward as the GPI-anchored surface antigen, variant surface glycoprotein (VSG), is very highly expressed, and represents ~80% of exported protein. For endocytosis only, the transferrin receptor represents a well-characterized system for receptor-mediated endocytosis, while a second assay, uptake of the mannose-binding lectin concanavalin A is also useful. As VSG represents 90% of surface molecules and is bound by conA, this assay mainly reports on VSG uptake. Using the lectin has the advantage that VSG-specific antibodies are not required, which for the majority of VSG isoforms are unavailable. Protocols for analysis by fluorescence microscopy can be found in Field *et al.* (2004).

Export of VSG to the cell surface uses accessibility of surface VSG to GPI-phospholipase C (GPI-PLC) (Allen *et al.*, 2003; Bangs *et al.*, 1986); the supernatant following GPI-PLC autodigestion contains soluble VSG (sVSG) but internal (mf) VSG remains in the pellet fraction. Take 5×10^7 mid–log phase BSF cells, wash in labeling medium (DMEM without methionine or cysteine) prewarmed to $37°$. After centrifugation ($800 \times g$, 10 min), resuspend in 1 ml labeling medium, and incubate for 15 min to starve the cells. Pulse label at $37°$ for 5 to 10 min with [^{35}S] methionine/cysteine ProMix (Amersham Biosciences, >1000 Ci/mm) to a final concentration of 200 μCi/ml, and chase by adding prewarmed complete HMI9. Withdraw 1-ml aliquots and place on ice. Pellet cells in a microfuge ($20,000 \times g$, 20 s, $4°$), wash once in ice-cold PBS/1 mg/ml BSA, and resuspend in 920 μl of hypotonic lysis buffer (10 mM Tris-HCl, pH 7.5). Incubate for 5 min on ice followed by 10 min at $37°$ to enable GPI-PLC autodigestion. Separate mfVSG from sVSG by centrifuging for 10 min in a microfuge ($20,000 \times g$, $4°$), and remove 900 μl of supernatant (sVSG) to a new eppendorf. Wash the pellet (mfVSG) with 1 ml ice-cold hypotonic buffer, and resuspend in 1 ml ice-cold solubilization buffer (50 mM Tris-HCl, pH 7.5, 150 mM NaCl, 1% NP-40), and incubate on ice for 25 min. Add 90 μl of 10$\times$ the solubilization buffer to the supernatant to bring samples to equivalence. Clear the lysates by centrifugation for 15 min ($20,000 \times g$, $4°$) and transfer to new microfuge tubes. Include 5 mM iodoacetamide, 0.1 mM N a-p-tosyllysine chloromethyl ketone (TLCK), and 1 μg/ml leupeptin as peptidase inhibitors. Add 10 μl of a 50% slurry of ConA-sepharose 4B in Con A wash buffer (10 mM Tris-HCl (pH 7.5), 150 mM NaCl, 1 mM $CaCl_2$ and 1 mM $MnCl_2$) to each sample, and rotate for 1 h at $4°$. Pellet the beads and wash three times in ice-cold Con A wash buffer, and then once in ice-cold PBS. Following the final wash, resuspend the beads in SDS-PAGE, reducing sample buffer at 1×10^5 cell equivalents per microliter. Analyze by electrophoresis and autoradiography on 12% polyacrylamide gels.

Quantitation of FITC-transferrin (Molecular Probes) accumulation, that is, receptor-mediated endocytosis, may be conveniently performed using smaller numbers of cells. Serum (including transferrin) must be removed prior to the assay by washing the cells at least once in serum-free HMI9 medium containing 1% BSA. The cells can be resuspended in HMI9/BSA at 1×10^7 ml^{-1} and preincubated at 37° for 20 min. FITC-transferrin (100 μg/ml) is added, and cells are incubated at 37°. Uptake is stopped by either the addition of 1 ml ice-cold HMI9 medium or by withdrawing an aliquot from the labeling culture and adding this to 1 ml ice-cold HMI9 medium. Cells are then washed once in PBS at 4° before fixing for 1 h at 4° in 4% paraformaldehyde, and processing for immunofluorescence (Field et al., 2004). The procedure for conA accumulation is essentially the same. Following washing, biotin or FITC-conjugated conA (100 μg/ml) (Vectalabs) is added and the cells incubated at 37° for up to 1 h. Uptake is stopped by placing cells on ice at relevant time points. Labeled cells are washed in HMI9/BSA at 4°, then fixed with 4% paraformaldehyde as described above. For biotinylated probes, the lectin is visualized using FITC-streptavidin (Molecular Probes). Quantitation for both of these assays is done by defining regions of interest in captured digital images and determining the fluorescence intensity. Typically, at least 20 individual cells need to be examined for each data point. The assay has the distinct advantage that analysis is at the single-cell level; as RNAi knockdown is not synchronous, and cell populations are therefore heterogeneous, this flexibility is critical and allows interrogation of cells specifically where a morphological defect or loss of a marker protein (probed by co-staining with an antibody, for example) can be used to verify the degree of knockdown.

ACKNOWLEDGMENTS

Work in our laboratories is supported by program and project grants from the Wellcome Trust. We are grateful to the many laboratories whose efforts have contributed to the emergence of a sophisticated toolbox for genetic manipulation of trypanosomes and to their generosity in making such tools and protocols available to the community. Development of next-generation p2T7 vectors was accomplished as part of trypanoFAN, a Wellcome Trust–funded functional genomics initiative.

REFERENCES

Ackers, J. P., Dhir, V., and Field, M. C. (2005). A bioinformatic analysis of the RAB genes of Trypanosoma brucei. Mol. Biochem. Parasitol. 141(1), 89–97.

Alexander, D. L., Schwartz, K. J., Balber, A. E., and Bangs, J. D. (2002). Developmentally regulated trafficking of the lysosomal membrane protein p67 in Trypanosoma brucei. J. Cell Sci. 115, 3253–3263.

Allen, C. L., Goulding, D., and Field, M. C. (2003). Clathrin-mediated endocytosis is essential in Trypanosoma brucei. Embo J. 22(19), 4991–5002.

Alibu, V. P., Storm, L., Haile, S., Clayton, C., and Horn, D. (2005). A doubly inducible system for RNA interference and rapid RNAi plasmid construction in Trypanosoma brucei. Mol. Biochem. Parasitol. 139, 75–82.

Alsford, S., Kawahara, T., Glover, L., and Horn, D. (2005). Tagging a *T. brucei* RRNA locus improves stable transfection efficiency and circumvents inducible expression position effects. *Mol. Biochem. Parasitol.* **144**(2), 142–148.

Bangs, J. D., Andrews, N. W., Hart, G. W., and Englund, P. T. (1986). Post-translational modification and intracellular transport of a trypanosome variant surface glycoprotein. *J. Cell Biol.* **103**, 255–263.

Bangs, J. D., Brouch, E. M., Ransom, D. M., and Roggy, J. L. (1996). A soluble secretory reporter system in *Trypanosoma brucei*. Studies on endoplasmic reticulum targeting. *J. Biol. Chem.* **271**, 18387–18393.

Barrett, M. P., Burchmore, R. J., Stich, A., Lazzari, J. O., Frasch, A. C., Cazzulo, J. J., and Krishna, S. (2003). The trypanosomiases. *Lancet* **362**(9394), 1469–1480.

Bastin, P., MacRae, T. H., Francis, S. B., Matthews, K. R., and Gull, K. (1999). Flagellar morphogenesis: Protein targeting and assembly in the paraflagellar rod of trypanosomes. *Mol. Cell Biol.* **19**(12), 8191–8200.

Berriman, M., Ghedin, E., Hertz-Fowler, C., Blandin, G., Renauld, H., Bartholomeu, D. C., Lennard, N. J., Caler, E., Hamlin, N. E., Haas, B., Bohme, U., Hannick, L., *et al.* (2005). The genome of the African trypanosome *Trypanosoma brucei*. *Science* **309**, 416–422.

Brun, R., and Schonenberger, M. (1979). Cultivation and *in vitro* cloning or procyclic culture forms of *Trypanosoma brucei* in a semi-defined medium. *Acta Trop.* **36**, 289–292.

Burkard, G., Fragoso, C. M., and Roditi, I. (2007 Jun). Highly efficient stable transformation of bloodstream forms of Trypanosoma brucei. *Mol. Biochem. Parasitol.* **153**(2), 220–223.

Chen, Y., Hung, C. H., Burderer, T., and Lee, G. S. (2003). Development of RNA interference revertants in *Trypanosoma brucei* cell lines generated with a double stranded RNA expression construct driven by two opposing promoters. *Mol. Biochem. Parasitol.* **126**, 275–279.

Chung, W. L., Carrington, M., and Field, M. C. (2004). Cytoplasmic targeting signals in *trans*-membrane invariant surface glycoproteins of trypanosomes. *J. Biol. Chem.* **279**, 54887–54895.

Dhalia, R., Marinsek, N., Reis, C. R. S., Katz, R., Muniz, J. R. C., Standart, N., Carrington, M., and de Melo Neto, O. P. (2006). The two eIF4A helicases in *Trypanosoma brucei* are functionally distinct. *Nucl. Acids Res.* **34**, 2495–2507.

Dhir, V., and Field, M. C. (2004). TbRAB23; a nuclear–associated Rab protein from *Trypanosoma brucei*. *Mol. Biochem. Parasitol.* **136**, 297–301.

Dhir, V., Goulding, D., and Field, M. C. (2004). TbRAB1 and TbRAB2 mediate trafficking through the early secretory pathway of *Trypanosoma brucei*. *Mol. Biochem. Parasitol.* **137**, 253–265.

Field, M. C., and Carrington, M. (2004). Intracellular membrane transport systems in *Trypanosoma brucei*. *Traffic* **5**, 905–913.

Field, M. C., Natesan, S. K., Gabernet-Castello, C., and Koumandou, V. L. (2007 Jun). Intracellular trafficking in trypanosomatids. *Traffic* **8**(6), 629–639.

Field, M. C. (2005). Signalling the genome: The Ras-like small GTPase family of trypanosomatids. *Trends Parasitol.* **21**, 447–450.

Field, M. C., Allen, C. L., Dhir, V., Goulding, D., Hall, B. S., Morgan, G. W., Veazey, P., and Engstler, M. (2004). New approaches to the microscopic imaging of *Trypanosoma brucei*. *Microsc. Microanal.* **10**, 621–636.

Hall, B. S., Pal, A., Goulding, D., and Field, M. C. (2004). Rab4 is an essential regulator of lysosomal trafficking in trypanosomes. *J. Biol. Chem.* **279**, 45047–45056.

Hall, B., Allen, C. L., Goulding, D., and Field, M. C. (2005a). Both of the Rab5 subfamily small GTPases of *Trypanosoma brucei* are essential and required for endocytosis. *Mol. Biochem. Parasitol.* **138**, 67–77.

Hall, B. S., Pal, A., Goulding, D., Acosta-Serrano, A., and Field, M. C. (2005b). *Trypanosoma brucei*: TbRAB4 regulates membrane recycling and expression of surface proteins in procyclic forms. *Exp. Parasitol.* **111**, 160–171.

Hall, B. S., Smith, E., Langer, W., Jacobs, L. A., Goulding, D., and Field, M. C. (2005c). Developmental variation in Rab11-dependent trafficking in *Trypanosoma brucei*. *Eukaryot. Cell* **4,** 971–980.

Hirumi, H., and Hirumi, K. (1989). Continuous cultivation of *Trypanosoma brucei* blood stream forms in a medium containing a low concentration of serum protein without feeder cell layers. *J. Parasitol.* **75,** 985–989.

Ivens, A. C., Peacock, C. C., Worthey, E. A., Murphy, L., Aggarwal, G., Berriman, M., Sisk, E., Rajandream, M. A., Adlem, E., Aert, R., Anupama, A., Apostolou, Z., *et al.* (2005). The genome of the kinetoplastid parasite, *Leishmania major*. *Science* **309,** 436–442.

La Count, D. J., Bruse, S., Hill, K. L., and Donelson, J. E. (2000). Double-stranded RNA interference in *Trypanosoma brucei* using head-to-head promoters. *Mol. Biochem. Parasitol.* **111,** 67–76.

Lecordier, L., Walgraffe, D., Devaux, S., Poelvoorde, P., Pays, E., and Vanhamme, L. (2005). *Trypanosoma brucei* RNA interference in the mammalian host. *Mol. Biochem. Parasitol.* **140,** 127–131.

Melville, SE, Leech, V., Navarro, M., and Cross, G. A. (2000). The molecular karyotype of the megabase chromosomes of *Trypanosoma brucei* stock 427. *Mol. Biochem. Parasitol.* **111,** 261–273.

Pal, A., Hall, B. S., Jeffries, T. R., and Field, M. C. (2003). Rab5 and Rab11 mediate transferrin and anti-variant surface glycoprotein antibody recycling in *Trypanosoma brucei*. *Biochem. J.* **374,** 443–451.

Price, H. P., Panethymitaki, C., Goulding, D., and Smith, D. F. (2005). Functional analysis of TbARL1, an N-myristoylated Golgi protein essential for viability in bloodstream trypanosomes. *J. Cell Sci.* **118,** 831–841.

Subramaniam, C., Veazey, P., Redmond, S., Hayes-Sinclair, J., Chambers, E., Carrington, M., Gull, K., Matthews, K., Horn, D., and Field, M. C. (2006). Chromosome-wide analysis of gene function by RNA interference in the african trypanosome. *Eukaryot. Cell* **5,** 1539–1549.

van der Hoff, M. J., Moorman, A. F., and Lamers, W. H. (1992). Electroporation in 'intracellular' buffer increases cell survival. *Nucl. Acids Res.* **20,** 2902.

van Deursen, F. J., Shahi, S. K., Turner, C. M., Hartmann, C., Guerra-Giraldez, C., Matthews, K. R., and Clayton, C. E. (2001). Characterisation of the growth and differentiation *in vivo* and *in vitro*—of bloodstream—form *Trypanosoma brucei* strain TREU 927. *Mol. Biochem. Parasitol.* **112,** 163–171.

Wickstead, B., Ersfeld, K., and Gull, K. (2003). The frequency of gene targeting in *Trypanosoma brucei* is independent of target site copy number. *Nucl. Acids Res.* **31,** 3993–4000.

Wirtz, E., and Clayton, C. (1995). Inducible gene expression in trypanosomes mediated by a prokaryotic repressor. *Science* **268,** 1179–1183.

Wirtz, E., Leal, S., Ochatt, C., and Cross, G. A. (1999 Mar 15). A tightly regulated inducible expression system for conditional gene knock-outs and dominant-negative genetics in *Trypanosoma brucei*. *Mol. Biochem. Parasitol.* **99**(1), 89–101.

WEB RESOURCES

Useful sites for details on vectors, protocols and genome data:

http://trypanofan.path.cam.ac.uk. Maintained by Mark Field. A functional genomics website describing the data obtained from RNAi analysis of T. brucei chromosome I, together with several protocols.

http://tryps.rockefeller.edu/. Maintained by George A. M. Cross. This laboratory developed a number of important aspects of transfection and inducible expression, and the website offers extensive details on a number of available vectors and protocols.

http://homepages.lshtm.ac.uk/~ipmbdhor/dhhome.htm. Maintained by David Horn. Contains details of a number of second-generation vectors.

http://web.mac.com/mc115/iweb/mclab/home.html. Maintained by Mark Carrington. Contains details of a number of second-generation vectors designed using the pLEW100 inducible system.

http://www.genedb.org/genedb/tryp/. Maintained by the Sanger Institute. An excellent and well-curated resource for trypanosome genome annotation and sequence data.

http://www.izb.unibe.ch/res/seebeck/taggingvectors.php. Maintained by Thomas Seebeck. Provides details of vectors for in situ tagging. Only suitable for procyclic (insect) form due to transfection efficiency.

CHAPTER SIX

Use of Dynasore, the Small Molecule Inhibitor of Dynamin, in the Regulation of Endocytosis

Tom Kirchhausen,*,1 Eric Macia,*,‡ *and* Henry E. Pelish*,†

Contents

Abstract

The large GTPase dynamin is essential for clathrin-dependent coated-vesicle formation. Dynasore is a cell-permeable small molecule that inhibits the GTPase activity of dynamin1, dynamin2 and Drp1, the mitochondrial dynamin. Dynasore was discovered in a screen of ~16,000 compounds for inhibitors of

* Department of Cell Biology, Harvard Medical School, and IDI Immune Research Institute, Boston, Massachusetts
† Makoto Life Sciences, Inc., Boston, Massachusetts
‡ L'Institut de Pharmacologie Moléculaire et Cellulaire, CNRS, Valbonne, France
1 Corresponding author

Methods in Enzymology, Volume 438
ISSN 0076-6879, DOI: 10.1016/S0076-6879(07)38006-3

the dynamin2 GTPase. Dynasore is a noncompetitive inhibitor of dynamin GTPase activity and blocks dynamin-dependent endocytosis in cells, including neurons. It is fast acting (seconds) and its inhibitory effect in cells can be reversed by washout. Here we present a detailed synthesis protocol for dynasore, and describe a series of experiments used to analyze the inhibitory effects of dynasore on dynamin *in vitro* and to study the effects of dynasore on endocytosis in cells.

1. INTRODUCTION

Dynamin functions in membrane tubulation and fission of budding vesiculo-tubular structures. It is essential for clathrin-dependent endocytosis from the plasma membrane, for the fission of plasma membrane caveolae to form free transport vesicles, and for vesicle formation at the trans-Golgi network (Cao *et al.*, 2000; Corda *et al.*, 2002; Nichols, 2003; Takei *et al.*, 2005). It also appears to participate in actin comet formation and transport of macropinosomes and in the function of podosomes, probably by interaction with actin-binding proteins. A related role in membrane fission has also been assigned to homolog proteins of dynamin (Dnm1 in mammalian cells and Drp1 in yeast) in the biogenesis of mitochondria and peroxisomes (Koch *et al.*, 2005; Schrader, 2006).

2. DYNAMIN

Dynamin (for recent reviews, see Kirchhausen, 1999; Praefcke and McMahon, 2004; Thompson and McNiven, 2001; Wiejak and Wyroba, 2002; Yang and Cerione, 1999) is a multidomain protein of $\sim$100 kDa containing a GTPase module, a lipid-binding pleckstrin homology (PH) domain, a GTPase effector domain (GED), and a proline/arginine-rich C-terminal segment (PRD) containing amino-acid sequences that bind to the SH3 domains of other proteins. Dynamin is unusual among GTPases because its affinity for GDP and GTP is rather low (10 to 25 μM) when compared to small and heterotrimeric GTPases, and because its intrinsic GTP hydrolysis rate is high (1 to 2 min^{-1}) and dramatically increases by polymerization ($>$100 min^{-1}) (referred to here as the intrinsic GAP activity). Because of its kinetic parameters and the high levels of intracellular GTP (1 mM), dynamin is thought to be only transiently in the GDP bound state (1 to 10 ms) (Sever *et al.*, 2000a). Pure dynamin spontaneously forms rings and spirals in conditions of low ionic strength, and decorates tubulin microtubules and lipid vesicles with helices of similar dimensions. Any condition leading to self-assembly or ring formation also leads to stimulation of the dynamin GTPase activity. A useful and simple trick to stimulate the

GTPase activity of dynamin is to mix it with GST–Grb2 (containing its two SH3 domains) (Barylko *et al.*, 1998). Using this approach, we screened for interfering small molecules and identified one compound that we named dynasore (Macia *et al.*, 2006). In cells, dynasore inhibits clathrin-mediated endocytosis at two distinct steps, the transition from a half-formed ("U" pit) to fully formed pit and from a fully formed pit ("O" pit) to an endocytic vesicle (Fig. 6.1).

There are several models to explain the role of dynamin in membrane tubulation and fission of budding vesiculotubular structures (reviewed in

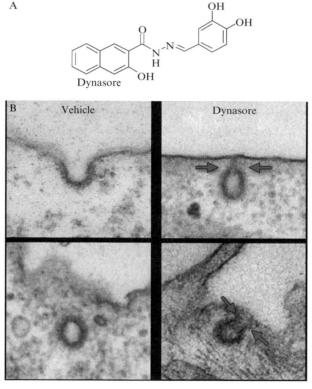

Figure 6.1   (A) Chemical structure of dynasore. (B) Effect of dynasore on clathrin-coated structures. The figure shows representative images of clathrin–coated structures of cells treated with DMSO (vehicle) or with 80 μM dynasore. The upper and lower left panels illustrate the appearance of endocytic coated pits and coated vesicles; the upper and lower right panels show the appearance of "U" and "O" shape-coated pits associated with the plasma membrane in cells treated with dynasore. The gray arrows highlight the extent of the constriction states observed upon treatment with dynasore. (From Macia, E., Ehrlich, M., Massol, R., Boucrot, E., Brunner, C., and Kirchhausen, T. (2006). Dynasore, a cell-permeable inhibitor of dynamin. *Dev. Cell* **10**, 839–850.) (See color insert.)

Kelly, 1999; Kirchhausen, 1999; McNiven, 1998; Sever *et al.*, 2000b; Yang and Cerione, 1999). These models range from viewing dynamin strictly as a mechanochemical enzyme to considering it as a regulatory protein for the recruitment of the downstream enzymatic partner(s) responsible for fission. Viewed as a mechanochemical enzyme, dynamin self-assembles around the neck of the budding pit, and then undergoes a conformational change in response to GTP binding and/or GTP hydrolysis. It is assumed that the coordinated change in conformation of ring elements leads to neck constriction and scission, and many variants to this model have been proposed. These models are strongly influenced by the results from *in vitro* self-assembly studies, either alone or in the presence of flexible and inflexible lipid scaffolds (Stowell *et al.*, 1999; Sweitzer and Hinshaw, 1998; Zhang and Hinshaw, 2001). Dynamin viewed as a regulatory GTPase stems from studying the effects by overexpression of dynamin mutants defective in self-assembly and/or intrinsic GAP activity (Sever *et al.*, 1999). Based on the observation that dynR725A and dynK694A maintain or even stimulate the endocytic rate of receptor-mediated uptake of transferrin, it has been proposed that dynamin-GTP, rather than GTP hydrolysis, facilitates vesicle budding. The opposite view is held by McMahon and coworkers who analyzed the effect of overexpression of several point mutants of dynamin's GTPase effector (GED) and GTPase domains and found that dynamin oligomerization and GTP binding alone are not sufficient for endocytosis *in vivo*. They concluded that efficient GTP hydrolysis and an associated conformational change are also required (Marks *et al.*, 2001). When tested using microtubules or lipid tubes as assembly scaffolds, the intrinsic GAP activity of these mutants is about the same as with wildtype dynamin (Marks *et al.*, 2001; Sever *et al.*, 1999). Thus, it is possible that *in vivo*, the dynamin mutants assembled around membrane necks and displayed relatively "normal" GTPase activity.

3. DYNAMIN AND THE ACTIN CYTOSKELETON

Dynamin, alone or in combination with amphiphysin, can form membrane tubes of dimensions similar to those on collars of deeply invaginated clathrin coated pits (Takei *et al.*, 1999). This was the first indication that a coated pit might not be a required template for dynamin function. Dynamin colocalizes with actin in growth cones (Torre *et al.*, 1994), and binds to a number of proteins involved in the regulation of actin cytoskeleton. They include profilin, cortactin, syndapin (a partner of N-WASP), and SH3-domain containing proteins like Abp1 linking cortical actin with endocytosis (Kessels *et al.*, 2001; McNiven *et al.*, 2000; Qualmann *et al.*, 1999; Witke *et al.*, 1998). Presently, it is not clear how these interactions are in any way related to a possible link between actin and clathrin-based endocytosis, as suggested by the partial inhibition of receptor-mediated endocytosis induced on the depolymerization of actin with latrunculin or cytocholasin (Boucrot *et al.*, 2006). Dynamin is found in actin comets involved in intracellular movement of macropinosomes and of *Listeria monocytogenes* in infected cells (Lee and De Camilli, 2002; Orth *et al.*, 2002). It is also found in podosomes (Ochoa *et al.*, 2000), narrow membrane invaginations similar in diameter to the elongated necks of coated pits emanating from the plasma membrane; these membranes are surrounded by actin and are positioned perpendicular to the substratum. Overexpression of dynamin mutants defective in GTP binding and hydrolysis (dynK44A) or lacking the C-terminal PRD segment decreased the intracellular motility of macropinosomes and *Listeria* linked to actin comets (Lee and De Camilli, 2002; Orth *et al.*, 2002). It is not known whether the efficient linkage of dynamin and actin, or of its function in this context, requires dynamin assembly and intrinsic GAP activity.

4. THE "CHEMICAL GENETICS" DISCOVERY APPROACH

In the last decade, a number of laboratories have engaged in medium- and high-throughput phenotype-based screens of libraries of chemical compounds in an approach dubbed "chemical genetics." The stated goal is to identify small molecules that disrupt the function of proteins or protein complexes (Gura, 2000). The Institute of Chemistry and Cell Biology (ICCB) at Harvard Medical School, now the ICCB-Longwood, is a major screening center for this approach. Using its facilities, we have identified a compound, now called secramine, that blocks membrane traffic from the trans–Golgi network to the plasma membrane (Pelish *et al.*, 2001, 2006). A number of other compounds, such as Exo1, Exo2, BLT1, and vacuolin (Cerny *et al.*, 2004; Feng *et al.*, 2003, 2004; Nieland *et al.*, 2002), inhibit unique traffic steps along the secretory pathway or within endosomes. These small molecules were found by "forward chemical genetics," that is, use of a cell-based screen for inhibition or activation of a particular cellular process, followed by molecular target identification (Fig. 6.2). In the "reverse chemical genetics" approach, one first searches for compounds that inhibit or activate *in vitro* proteins known to be involved in a defined process, followed by studies to determine the phenotypic effects in cells and/or organisms. In comparison to "forward chemical genetics," this approach has the significant advantage of bypassing the target identification step, and was used for the discovery of dynasore (see Fig. 6.1A).

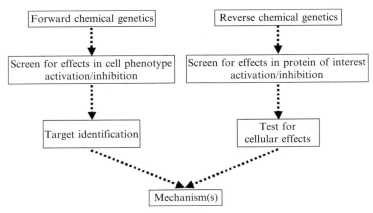

Figure 6.2   Strategy for the chemical genetics discovery approach. The reverse chemical genetics approach was used to discover dynasore. Dynasore is cell permeable and interferes with all functions known to be associated with dynamin. (From Macia, E., Ehrlich, M., Massol, R., Boucrot, E., Brunner, C., and Kirchhausen, T. (2006). Dynasore, a cell-permeable inhibitor of dynamin. *Dev. Cell* **10**, 839–850; and Newton, A. J., Kirchhausen, T., Murthy, V. N. (2006). Inhibition of dynamin completely blocks compensatory synaptic vesicle endocytosis. *Proc. Natl. Acad. Sci. USA* **103**, 17955–17960.)

 ## 5. WHY DO WE NEED INTERFERING SMALL MOLECULES?

Interfering small molecules allow researchers to freeze biological processes at interesting points. This is particularly useful in the investigation of transient phenomena, such as membrane traffic. Much of the recent progress in understanding protein trafficking pathways has been achieved using approaches based on genetic dissection and morphological and biochemical analysis. However, the dynamic nature of these events (Cole *et al.*, 1996) makes it particularly difficult to use slow techniques such as genetic deletion and immunological depletion to study them. Temperature-sensitive (ts) mutants have in some cases proved helpful, but the number of proteins for which ts mutants exist is not large, and the effect can take several hours to be observable. Fast-acting chemical agents would be an ideal way to probe the dynamics of these complex systems.

5.1. Acute interference with membrane traffic

Only two ts mutants, one for dynamin and another for the ϵ-subunit of COPI, are available for studies in mammalian cells (Damke *et al.*, 1995; Guo *et al.*, 1994). The dynamin mutant displays an interfering phenotype within minutes of transfer to the nonpermissive temperature, while the ϵ-mutant requires several hours before having a strong effect. Use of the dynamin ts mutant was instrumental in unraveling the enormous capacity of the endocytic pathway to accommodate perturbation. In less than 1 h after temperature shift, the rate of fluid phase uptake returns to normal levels (Damke et al., 1995). This example illustrates the value of studying the effect of rapid perturbations in complex systems; the same can be said of studies involving the dramatic and acute effects of brefeldin A on the integrity of the Golgi complex, which led to our current views concerning the regulated traffic between the endoplasmic reticulum (ER) and the Golgi and the biogenesis of the Golgi (Pelletier et al., 2000; Ward et al., 2001). There are a few other chemicals that act in the endocytic and the secretory pathway, such as wortmanin (Kundra and Kornfeld, 1998; Spiro et al., 1996), ilimaquinone (Takizawa et al., 1993), and monensin (Pless and Wellner, 1996). Nevertheless, the chemical tools available are both few in number and unsatisfactory in terms of specificity for a particular protein target or cellular pathway. This deficiency exists because most of the available chemical tools were discovered serendipitously and not through directed screens. In addition to our directed screening approach, Robinson, McCluskey and coworkers screened for and identified a class of small molecules that inhibit the GTPase activity of dynamin1 in vitro (Hill et al., 2005). The inhibitors they identified, dimeric tyrphostins, bear some resemblance to dynasore, as both display at least one benzenediol. Unlike dynasore, however, no in vivo data has been reported for these compounds.

6. SYNTHESIS OF DYNASORE

We identified dynasore in a screen of $\sim$16,000 compounds (part of the Diverset E, Chembridge Library) for inhibition of the GST-Grb2-stimulated GTPase activity of dynamin2 (Macia et al., 2006) (assay described below). Here we describe our synthesis of dynasore (Fig. 6.3). Our approach is based on the strategy of Ling et al. (2001) for the synthesis of benzoic acid arylidenehydrazides. Dynasore ($C_{18}H_{14}N_2O_4$, molecular

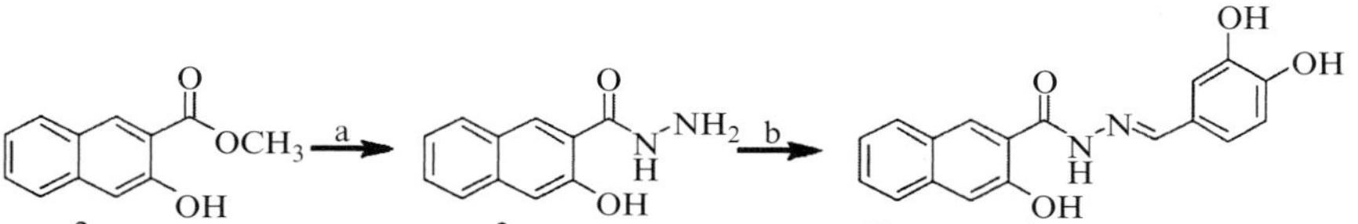

Figure 6.3 Scheme for the synthesis of dynasore. (a) H_4N_2, CH_3OH, $65°$, 43% yield. (b) 3,4-dihydroxy-benzaldehyde, CH_3CO_2H, CH_3CH_2OH, $78°$, 85% yield.

weight 322.31 g/mol) (1) is easily synthesized on gram scale in two steps from commercially available methyl 3-hydroxy-2-naphthoate (2) without the need for column chromatography.

Both reactions were performed in oven-dried glassware under a positive pressure of argon. Starting materials and reagents were purchased from commercial suppliers and used without further purification. ^{1}H and ^{13}C NMR spectra were recorded on a Varian INOVA500 or Mercury400 spectrometer. Chemical shifts for proton and carbon resonance are reported in parts per million (δ) relative to DMSO (δ 2.49 and 39.5, respectively). Tandem high-pressure liquid chromatography/mass spectral (LCMS) analyses were performed on a Waters Platform LCZ mass spectrometer in electrospray ionization (ES) mode. Samples were passed through a Symmetry C18 column using a gradient of 85% water/0.1% formic acid and 15% acetonitrile/0.1% formic acid to 100% acetonitrile/0.1% formic acid in 5 min.

The dynasore synthesis begins with the conversion of methyl 3-hydroxy-2-naphthoate (2) to 3-hydroxyl-2-naphtoylhydrazine (3). Hydrazine (2.3 ml, 5.0 equivalent) was added to a solution of methyl 3-hydroxy-2-naphthoate (2) (3 g, 14.8 mmol, 1.0 equivalent) in methanol (50 ml) at room temperature. The mixture was refluxed overnight at 65°. Upon cooling, brown needles formed. The solid was collected on a filter, washed with cold methanol, and dried to yield 3-hydroxyl-2-naphtoylhydrazine (3) (1.28 g, yield of 43%). The ^{1}H NMR (400 MHz, (CD3)$_2$SO) analysis follows: δ 8.44 (s, 1H), 7.81 (d, J = 8.0 Hz, ^{1}H) 7.71 (d, J = 8.1 Hz, ^{1}H), 7.45 to 7.49 (m, ^{1}H), 7.30 to 7.34 (m, ^{1}H), 7.26 (s, ^{1}H); 13C NMR (100 MHz, (CD3)$_2$SO): δ 167.0, 155.0, 135.8, 129.0, 128.6, 128.0, 126.6, 125.8, 123.6, 118.1, 110.6; and LCMS (ES+) calculated for $C_{11}H_{10}N_2O_2$ (M-H$^+$) was 203.07 (found 203.23).

We subsequently converted 3-hydroxyl-2-naphtoylhydrazine (3) into dynasore (3-hydroxy-naphthalene-2-carboxylic acid (3,4-dihydroxy-benzylidene)-hydrazide). Ethanol (50 ml) and acetic acid (0.4 ml) were added to 3 (1.28 g, 6.33 mmol, 1 equivalent) and 3,4-dihydroxy-benzaldehyde (0.87 g, 6.33 mmol, 1 equivalent). Upon heating to 78°, 3 and 3,4-dihydroxy-benzaldehyde dissolved. A new precipitate subsequently formed. The solution was refluxed overnight at 78°. Upon cooling, the precipitate was collected on a filter, washed with cold ethanol, and dried to yield pure dynasore (1.74 g, yield of 85%). The ^{1}H NMR (500 MHz, (CD3)$_2$SO) analysis follows: δ 11.80 (s, ^{1}H), 11.41 (s, ^{1}H), 9.44 (s, ^{1}H), 9.31 (s, ^{1}H), 9.29 (s, ^{1}H) 8.45 (s, ^{1}H), 8.24 (s, ^{1}H), 7.89 (d, J = 8.3 Hz, ^{1}H) 7.75 (d, J = 8.3 Hz, ^{1}H), 7.50 (dd, J = 7.6, 7.6 Hz, ^{1}H), 7.35 (dd, J = 7.6, 7.6, ^{1}H), 7.31 (s, ^{1}H), 7.28 (s, ^{1}H), 6.97 (d, J = 6.3 Hz); ^{13}C NMR (100 MHz, (CD3)$_2$SO): δ 163.6, 154.4, 149.2, 148.2, 145.8, 135.8, 130.0, 128.7, 126.8, 125.8, 125.5, 123.8, 120.9, 120.0, 115.6, 112.8, 110.6; and LCMS (ES+) calculated for $C_{18}H_{14}N_2O_4$ (M-H$^+$) was 323.10 (found 323.02).

7. STORAGE CONDITIONS FOR DYNASORE

Dynasore is stored as a dry solid under argon in the dark at $-20°$. Dynasore can also be stored at $-20°$ or $-80°$ in the dark (no need to flash freeze) as a 200-mM solution in DMSO under argon. Aliquots of 10 to 20 μl are stored in 0.5-ml microcentrifuge tubes. After adding argon and closing the cap, the microcentrifuge tubes are sealed with parafilm. To avoid the capture of moisture, the DMSO aliquots of dynasore are warmed up to room temperature before opening. The aqueous solution of dynasore will appear light yellow and the working final concentration for *in vivo* experiments is $\sim$80 μM (0.2% DMSO final), which typically results in a greater than 90% block in endocytosis. An appropriate volume of the stock solution of dynasore is added to the working solution, and mixed by gentle tumbling or by up-and-down pipetting. Dynasore binds to serum proteins and loses activity. We thus use dynasore dissolved in media lacking albumin or serum. Typically we use DMEM $\pm$10% Nuserum, PBS with glucose, or PBS with glucose and 10% Nuserum. Nuserum is a synthetic, low-protein alternative to traditional serum. For *in vivo* experiments, we first rinse the cells four to five times with serum-free medium (e.g., 3-ml sequential washes for cells seeded in a 12-well plate).

8. EXPRESSION, PURIFICATION, AND STORAGE OF DYNAMIN

8.1. Protein expression

We express human dynamin in SF9 insect cells (*Spodoptera frugiperda*, GIBCO-BRL, Gaithersburg, MD) grown in SF-900 II SFM (GIBCO-BRL) essentially as described (Damke *et al.*, 2001). Using the Bac-to-Bac baculovirus expression system (GIBCO-BRL), a full-length, cDNA encoding human dynamin1 containing a 6-His-tag at the N-terminus is subcloned into the baculovirus vector pFastBac. A bacmid is generated after transposition in *Escherichia coli*, and several independent clones are selected to transfect Sf9 cells using the transfection reagent CellFECTIN (Invitrogen, Carlsbad, CA).

For transfection, 10^6 SF9 cells are seeded in each well of a six-well plate containing 2 ml of Sf-900 II SFM medium supplemented with 0.5X penicillin/streptomycin. Prior to transfection, the cells are permitted to attach to the bottom of the plates for at least 1 h at $27°$. Transfection proceeds as follows: (1) 5 μl of mini-prep bacmid (0.1 to 0.4 mg/ml) is added to $100\mu l$ of Sf 900 II SFM without antibiotics (solution A). (2) Separately, 6 μl

CellfFECTIN is added to 100 μl Sf-900 II SFM without antibiotics (solution B). (3) Solutions A and B are mixed and incubated at room temperature for 15 to 45 min. (4) The cells are washed once with 2 ml of Sf-900 II SFM without antibiotics. (5) Next, 0.8 ml of Sf-900 II SFM is added to each tube containing the lipid–DNA complexes, and then this solution is added to the cells. (6) After a 5-h incubation at 27°, the transfection mixtures are replaced with 2 ml of Sf-900 II SFM containing antibiotics, and the cells are incubated for 72 h at 27°. At this point, virus is harvested with the culture medium. We typically perform two extra rounds of virus amplification, by using a 1/100 dilution of the medium containing the virus. We dilute into 6 ml (round 1) and 20 ml (round 2) of Sf-900 II SFM containing penicillin/streptomycin in a 50-ml flask with 1×10^6 SF9 cells per milliliter. In each round, the cells are incubated for 72 h at 27°. A successful infection results in about 50% cell death, while the remaining cells are larger and contain granules easily observed by phase-contrast light microscopy. We normally verify protein expression levels at the end of the second round of virus amplification in SF9 cells using different dilutions of the virus stock (from 1/10 to 1/200 for 72 h at 27°; 106 SF9 cells seeded in a six-well plate containing 2 ml of Sf-900 II SFM supplemented with penicillin/streptomycin). We store the virus for up to 6 months in the dark at 4°.

For production of full-length dynamin, we infect 1 liter of Sf-900 II SFM media containing 1×10^6 SF9 cells per milliliter with a 1/1000 dilution of the amplified baculovirus stock. The cells are grown in a 2-liter spinner flask (80 rpm) for 72 h at 27°. Cells are then harvested by centrifugation (4000 rpm, 20 min, 4°) and stored at −20° until protein purification.

We constructed 6-His N-terminal–tagged human dynamin2ΔPRD (lacking the C-terminal, proline-rich domain) in a PET28a bacteria expression vector (Stratagene, La Jolla, CA). Using a Quickchange PCR mutagenesis protocol (Stratagene, La Jolla, CA), we introduced a stop codon at amino acid 747 (Warnock *et al.*, 1997). This recombinant protein is produced in *E. coli* BL21(DE3). In 2 to 4 liters of Luria–Bertani (LB) media, the bacteria are grown at 37° to an OD640, cooled to 18° in a water bath, subsequently induced with 0.25 mM IPTG, and further incubated overnight at 18°.

8.2. Protein purification and storage

N-terminal, His-tagged, full-length human dynamin1 and the truncated N-terminal, His-tagged human dynamin2 are purified using Ni-NTA and hydroxyapatite chromatography. All purification steps are carried out at 4°. Insect or bacteria cells are resuspended in 25 ml HCB250 containing protease inhibitor cocktail tablets (complete without EDTA, Boehringer Manheim, Indianapolis, IN) and cells disrupted by shearing using a laboratory model microfluidizer (Microfluidics, Newton MA). The lysates are then cleared by centrifugation for 1 h at 40,000 rpm (Ti45 rotor, Beckman ultracentrifuge). The supernatants are added to a 2-ml 50% slurry of TALON beads (Clontech, Mountain View, CA). After gentle stirring for 4 h, the beads are loaded into a column and washed with 10 ml of HCB250 supplemented with 20 mM imidazole. Elution of the recombinant proteins (by gravity) then follows upon addition of 0.5 ml HCB250 containing 250 mM imidazole. The fractions containing dynamin are identified by 10% SDS-PAGE and Coomassie blue-staining. They are pooled and supplemented with a calpain inhibitor at 1 μM (Calbiochem, La Jolla, CA) and 5 mM CaCl$_2$. The inhibitor is essential at this point because a metalloprotease, probably activated by Ca^{2+}, seems to be activated during the next fractionation step. The sample is then loaded onto a 2-ml hydroxyapatite column (Bio-Scale^{TM} CHT2-I, 7 $\times$ 52 mm, Biorad, Hercules, CA) pre-equilibrated with 50 mM of K-PO$_4$ buffer. Bound proteins are eluted using a linear gradient from 50 to 500 mM of K-PO$_4$. The peak-containing fractions elute between 300 and 350 mM K-PO$_4$. Purified dynamin1 and dynamin2ΔPRD are stored at $\sim$50 μM in 20% glycerol at $-80°$ until use. Routinely we obtained about 2 mg/l of protein, of which the full-length constructs represent at least 90% of the total amount of protein. Prior to use, frozen aliquots (50 μl) are thawed rapidly to room temperature and centrifuged for 5 min at 14,000 rpm in a microfuge to remove any aggregated proteins.

9. BUFFERS AND REAGENTS

Acid wash buffer: glycine 0.1 M, pH 2.5, and 150 mM NaCl

Penicillin/streptomycin solution: 1$\times$ corresponds to 100 units/ml of penicillin and 100 mg/ml of streptomycin

HEPES column buffer (HCB): 20 mM HEPES, pH 7, 2 mM EGTA, 1 mM MgCl$_2$, and 1 mM dithiothreitol (DTT)

HCB250: HCB supplemented with 250 mM NaCl

K-PO$_4$ hydroxyapatite buffer: 50 or 500 mM, pH 7.2, 1 mM DTT, and 1 μM calpain inhibitor

GTPase buffer: 10 mM Tris, pH 7.2, 2 mM MgCl$_2$, and 20 μM GTPγP32 (2000 dpm/pmol)

Acid-washed charcoal: 10% [w/v] activated charcoal (Sigma, St. Louis, MO) in an acidic solution of 2% (v/v) formic acid/8% (v/v) acetic acid

Malachite green solution: 0.324 mM malachite green, 0.0426% Triton X-100, 16.61 mM ammonium molybdate, and 1.246 M H$_2$SO$_4$. This solution is prepared fresh and used 2 h after; at this point the color of the solution becomes stable.

9.1. Colorimetric assay used during the screen for inhibitors of the stimulated GTPase activity of dynamin

This assay is based on the change of spectral characteristics of malachite green in the presence of free PO_4 ions (Cogan et al., 1999; Maehama et al., 2000). Independent of our work, a similar colorimetric assay was recently developed for studies with dynamin (Leonard et al., 2005). The screen is initiated by mixing 100 nM of purified full-length human dynamin1 with 2 μM GST-Grb2 in a total volume of 30 μl containing 50 mM Tris, pH 7.5, 3 mM MgCl$_2$, 100 mM KCl, and 0.2 mM EGTA. Addition of GST-Grb2 (containing its two SH3 domains) stimulates full-length dynamin GTPase activity (Barylko et al., 1998). We dispense this mixture into 384-well plates containing an optically clear bottom (Krackeler Scientific, Inc., Albany). A fresh malachite green solution is prepared on the day of use. After standing for ~2 h at room temperature, the color of the reagent changes from dark brown to golden yellow and is ready for use. On every plate, we use four columns as controls (DMSO without dynamin, GST-Grb2, or GTP). Approximately 100 nl of each compound (~10 mM dissolved in DMSO) is robotically pin-transferred to the assay plate, providing a final compound concentration range of 20 and 50 μM. The reaction is started by adding ~300 nl GTP (final concentration of 200 μM). After 30 min of incubation at room temperature, the reaction was terminated by the addition of 50 μl of malachite green solution. After 15 min at room temperature, the plates are placed on a plate reader (Perkin Elmer, Shelton, CT) and the absorbance is measured at 650 nm. We display and analyze the resulting data in Microsoft Office Excel (Microsoft, Redmond, WA) with the aid of a macro subroutine that facilitates the rapid identification of inhibitors in a given 384-well plate.

9.2. Radioactive assay for the GTPase activity of dynamin

This assay allows precise quantification of the dynasore activity and is a minor modification of the charcoal-based procedure described by Liu and colleagues (1996). We measure the GTPase activity of full-length dynamin1 or dynamin2 deleted of its PRD domain (dynamin2ΔPRD) at 0.2 μM, a concentration that minimizes spontaneous polymerization (Warnock et al., 1997). The proteins are incubated at ambient temperature (~22°) in a final volume of 100 μl GTPase buffer for up to 30 min using solutions containing 20 or 150 mM NaCl for the low- or high-salt conditions, respectively. GTPase activity is terminated by transferring 10 μl of the reaction mixture into 500 μl of cold acid-washed charcoal (kept in ice in a 1.5-ml eppendorf tube), followed by centrifugation for 10 min (14,000 rpm in a microfuge at 4°). The amount of P32 released by hydrolysis, a measurement of GTP hydrolysis, is determined with a β-counter using 250 μl of the supernatant.

To identify dynasore as a noncompetitive inhibitor, we compared the Km and kcat values of the dynamin with the protein as an enzyme and GTP as a substrate. As shown in Fig. 6.4, the plot of the initial speed of the reaction (Vi) versus the concentration of GTP provides an estimate for Vmax (plateau) and Km (S concentration when Vi is half the Vmax). The Eadie-Hofstee linear transformation (V against V/[s]) can also be used for a more accurate representation; in this case the slope corresponds to −Km and the intercept on the x axis to V/Km.

9.3. Endocytic assay

Dynasore inhibits endocytosis in HeLa cells, human U373-MG astrocytes, COS-1 and BSC-1 cells (monkey cell lines), and mouse hippocampus neurons. In general, dynasore inhibition of endocytosis is measured at a cell density of 40 to 70%. Denser cultures are more resistant to the effects of dynasore.

One convenient way to detect the inhibitory effects of dynasore on endocytosis is to visually monitor its perturbation of the receptor-mediated uptake of transferrin. Transferrin receptors are constitutively internalized by a process that requires the formation of clathrin-coated pits and vesicles and is dynamin dependent (Ehrlich *et al.*, 2004; Hanover *et al.*, 1984). We normally follow the uptake of fluorescently tagged transferrin (e.g., Alexa-568 transferrin, Molecular Probes) at 37°. Before addition of transferrin, the

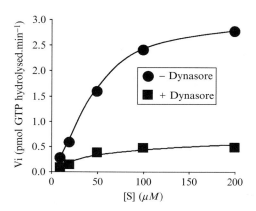

Figure 6.4 Determination of kinetics parameters for the effects of dynasore on the GTPase activity of dynamin. Effect of 40 μM dynasore on the rate of GTPase hydrolysis of 0.2 mM Dyn2ΔPRD determined at ambient temperature. The control experiment was done in the presence of 1% DMSO (vehicle) Initial rates of GTP hydrolysis were determined for different concentrations of GTP (S).

cells (between 40 and 70% confluency and grown on a coverslip) are incubated for 30 min at 37° with 80 μM dynasore or 0.2% DMSO only (vehicle control) in DMEM. In a pulse–chase format, we first allow binding of transferrin to its receptor at the surface of the cells after transfer at room temperature and incubation for 2 min (this step dramatically slows down endocytosis). After three washes with media (e.g., DMEM ± dynasore), cells are transferred to 37° and endocytosis proceeds. The receptor-bound Alexa-568 transferrin is internalized and transported first to peripherally located early endosomes (5 min) and then to the more perinuclear recycling/late endosomes (15 min). At different times, cells can be cooled to 4° (by addition of ice-chilled medium), followed by an acid wash (three consecutive 2-min washes each using 2 ml of acid wash buffer at room temperature under gentle agitation) to remove transferrin still bound at the cell surface. As the final step, the cells are fixed by incubation with a solution of PBS containing 4% PFA for 30 min at room temperature. Coverslips are then mounted on a glass slide, sealed with nail polish and are ready for fluorescence microscopy.

We acquire images using a spinning disk confocal head (Perkin Elmer) attached to an inverted microscope (200M, Zeiss Co.) under control of Slide Book 4 (Intelligent Imaging Innovations, Inc.) as described (Ehrlich *et al.*, 2004). For each cell, we acquired a series of confocal sections imaged every 0.3 μm. The integrated projected fluorescence of 80 to 100 cells is analyzed for each condition (dynasore concentration or time kinetics). To normalize between images, we set the fluorescence intensity of the vehicle control at 100%.

In the above experiment, we observed a strong block in the traffic and accumulation of transferrin with 80 μM dynasore. This inhibition is dose dependent with an IC50 ($\sim$15 μM). The decrease in transferrin uptake is not due to a decrease in the number of transferrin receptors at the cell surface or to a decrease in the association of transferrin with its receptor, as the amount of surface-bound transferrin is the same in cells kept for 30 min at 4° in the presence of dynasore or vehicle control.

Other assays, which are beyond the scope of this paper, are useful for following the effect of dynasore on the endocytosis of other ligands such as LDL, viruses, and bacteria.

ACKNOWLEDGMENT

We thank Matthew D. Shair for use of his laboratory facilities to carry out the synthesis of dynasore. We also thank members of the Kirchhausen lab who participated in the discovery and characterization of dynasore activity, including Chris Brunner, Marcelo Erlich, Ramiro Massol, Werner Boll, and Emmanuel Boucrot. We acknowledge support from the National Institutes of Health (grants GM GM62566, GM03548, and GM075252 to T. K.).

REFERENCES

Barylko, B., Binns, D., Lin, K. M., Atkinson, M. A. L., Jameson, D. M., Yin, H. L., and Albanesi, J. P. (1998). Synergistic activation of dynamin gtpase by grb2 and phosphoinositides. *J. Biol. Chem.* **273,** 3791–3797.

Boucrot, E., Saffarian, S., Massol, R., Kirchhausen, T., and Ehrlich, M. (2006). Role of lipids and actin in the formation of clathrin-coated pits. *Exp. Cell Res.* **312,** 4036–4048 (Epub 2006 Sep 4030.).

Cao, H., Thompson, H. M., Krueger, E. W., and McNiven, M. A. (2000). Disruption of Golgi structure and function in mammalian cells expressing a mutant dynamin. *J. Cell Sci.* **113,** 1993–2002.

Cerny, J., Feng, Y., Yu, A., Miyake, K., Borgonovo, B., Klumperman, J., Meldolesi, J., McNeil, P. L., and Kirchhausen, T. (2004). The small chemical vacuolin-1 inhibits $Ca^{(2+)}$-dependent lysosomal exocytosis but not cell resealing. *EMBO Rep.* **5,** 883–888.

Cogan, E. B., Birrell, G. B., and Griffith, O. H. (1999). A robotics-based automated assay for inorganic and organic phosphates. *Anal. Biochem.* **271,** 29–35.

Cole, N. B., Smith, C. L., Sciaky, N., Terasaki, M., Edidin, M., and Lippincott-Schwartz, J. (1996). Diffusional mobility of Golgi proteins in membranes of living cells. *Science* **273,** 797–800.

Corda, D., Hidalgo Carcedo, C., Bonazzi, M., Luini, A., and Spano, S. (2002). Molecular aspects of membrane fission in the secretory pathway. *Cell. Mol. Life Sci.* **59,** 1819–1832.

Damke, H., Baba, T., Vanderbliek, A. M., and Schmid, S. L. (1995). Clathrin-independent pinocytosis is induced in cells overexpressing a temperature-sensitive mutant of dynamin. *J. Cell Biol.* **131,** 69–80.

Damke, H., Muhlberg, A. B., Sever, S., Sholly, S., Warnock, D. E., and Schmid, S. L. (2001). Expression, purification, and functional assays for self association of dynamin-1. *Methods Enzymol.* **329,** 447–457.

Ehrlich, M., Boll, W., Van Oijen, A., Hariharan, R., Chandran, K., Nibert, M. L., and Kirchhausen, T. (2004). Endocytosis by random initiation and stabilization of clathrin-coated pits. *Cell* **118,** 591–605.

Feng, Y., Yu, S., Lasell, T. K. R., Jadhav, A. P., Macia, E., Chardin, P., Melancon, P., Roth, M., Mitchison, T., and Kirchhausen, T. (2003). Exo1: A new chemical inhibitor of the exocytic pathway. *Proc. Natl. Acad. Sci. USA* **100,** 6469–6474.

Feng, Y., Jadhav, A. P., Rodighiero, C., Fujinaga, Y., Kirchhausen, T., and Lencer, W. I. (2004). Retrograde transport of cholera toxin from the plasma membrane to the endoplasmic reticulum requires the trans-Golgi network but not the Golgi apparatus in Exo2-treated cells. *EMBO Rep.* **5,** 596–601.

Guo, Q., Vasile, E., and Krieger, M. (1994). Disruptions in Golgi structure and membrane traffic in a conditional lethal mammalian cell mutant are corrected by epsilon-COP. *J. Cell Biol.* **125,** 1213–1224.

Gura, T. (2000). A chemistry set for life. *Nature* **407,** 282–284.

Hanover, J. A., Willingham, M. C., and Pastan, I. (1984). Kinetics of transit of transferrin and epidermal growth factor through clathrin-coated membranes. *Cell* **39,** 283–293.

Hill, T., Odell, L. R., Edwards, J. K., Graham, M. E., McGeachie, A. B., Rusak, J., Quan, A., Abagyan, R., Scott, J. L., Robinson, P. J., and McCluskey, A. (2005). Small molecule inhibitors of dynamin I GTPase activity: development of dimeric tyrphostins. *J. Med. Chem.* **48,** 7781–7788.

Kelly, R. B. (1999). New twists for dynamin. *Nat. Cell Biol.* **1,** E8–E9.

Kessels, M. M., Engqvist-Goldstein, A. E. Y., Drubin, D. G., and Qualmann, B. (2001). Mammalian Abp1, a signal-responsive F-actin-binding protein, links the actin cytoskeleton to endocytosis via the GTPase dynamin. *J. Cell Biol.* **153,** 351–366.

Kirchhausen, T. (1999). Cell biology—Boa constrictor or rattlesnake? *Nature* **398,** 470–471.

Koch, A., Yoon, Y., Bonekamp, N. A., McNiven, M. A., and Schrader, M. (2005). A role for Fis1 in both mitochondrial and peroxisomal fission in mammalian cells. *Mol. Biol. Cell* **16,** 5077–5086. (Epub 2005 Aug 5017.)

Kundra, R., and Kornfeld, S. (1998). Wortmannin retards the movement of the mannose 6-phosphate/insulin-like growth factor II receptor and its ligand out of endosomes. *J. Biol. Chem.* **273,** 3848–3853.

Lee, E., and De Camilli, P. (2002). Dynamin at actin tails. *Proc. Natl. Acad. Sci. USA* **99,** 161–166.

Leonard, M., Song, B. D., Ramachandran, R., and Schmid, S. L. (2005). Robust colorimetric assays for dynamin's basal and stimulated GTPase activities. *Methods Enzymol.* **404,** 490–503.

Ling, A., Hong, Y., Gonzalez, J., Gregor, V., Polinsky, A., Kuki, A., Shi, S., Teston, K., Murphy, D., Porter, J., Kiel, D., Lakis, J., *et al.* (2001). Identification of alkylidene hydrazides as glucagon receptor antagonists. *J. Med. Chem.* **44,** 3141–3149.

Liu, J. P., Zhang, Q. X., Baldwin, G., and Robinson, P. J. (1996). Calcium binds dynamin I and inhibits its GTPase activity. *J. Neurochem.* **66,** 2074–2081.

Macia, E., Ehrlich, M., Massol, R., Boucrot, E., Brunner, C., and Kirchhausen, T. (2006). Dynasore, a cell-permeable inhibitor of dynamin. *Dev. Cell* **10,** 839–850.

Maehama, T., Taylor, G. S., Slama, J. T., and Dixon, J. E. (2000). A sensitive assay for phosphoinositide phosphatases. *Anal. Biochem.* **279,** 248–250.

Marks, B., Stowell, M. H. B., Vallis, Y., Mills, I. G., Gibson, A., Hopkins, C. R., and McMahon, H. T. (2001). GTPase activity of dynamin and resulting conformation change are essential for endocytosis. *Nature* **410,** 231–235.

McNiven, M. A. (1998). Dynamin: a molecular motor with pinchase action. *Cell* **94,** 151–154.

McNiven, M. A., Kim, L., Krueger, E. W., Orth, J. D., Cao, H., and Wong, T. W. (2000). Regulated interactions between dynamin and the actin-binding protein cortactin modulate cell shape. *J. Cell Biol.* **151,** 187–198.

Newton, A. J., Kirchhausen, T., and Murthy, V. N. (2006). Inhibition of dynamin completely blocks compensatory synaptic vesicle endocytosis. *Proc. Natl. Acad. Sci. USA* **103,** 17955–17960.

Nichols, B. (2003). Caveosomes and endocytosis of lipid rafts. *J. Cell Sci.* **116,** 4707–4714.

Nieland, T. J. F., Penman, M., Dori, L., Krieger, M., and Kirchhausen, T. (2002). Discovery of chemical inhibitors of the selective transfer of lipids mediated by the HDL receptor SR-BI. *Proc. Natl. Acad. Sci. USA* **99,** 15422–15427.

Ochoa, G. C., Slepnev, V. I., Neff, L., Ringstad, N., Takei, K., Daniell, L., Kim, W., Cao, H., McNiven, M., Baron, R., and De Camilli, P. (2000). A functional link between dynamin and the actin cytoskeleton at podosomes. *J. Cell Biol.* **150,** 377–389.

Orth, J. D., Krueger, E. W., Cao, H., and McNiven, M. A. (2002). The large GTPase dynamin regulates actin comet formation and movement in living cells. *Proc. Natl. Acad. Sci. USA* **99,** 167–172.

Pelish, H. E., Westwood, N. J., Feng, Y., Kirchhausen, T., and Shair, M. D. (2001). Use of biomimetic diversity-oriented synthesis to discover galanthamine-like molecules with biological properties beyond those of the natural product. *J. Am. Chem. Soc.* **123,** 6740–6741.

Pelish, H. E., Peterson, J. R., Salvarezza, S. B., Rodriguez-Boulan, E., Chen, J. L., Stamnes, M., Macia, E., Feng, Y., Shair, M. D., and Kirchhausen, T. (2006). Secramine inhibits Cdc42-dependent functions in cells and Cdc42 activation *in vitro*. *Nat. Chem. Biol.* **2,** 39–46.

Pelletier, L., Jokitalo, E., and Warren, G. (2000). The effect of Golgi depletion on exocytic transport. *Nat. Cell Biol.* **2,** 840–846.

Pless, D. D., and Wellner, R. B. (1996). In vitro fusion of endocytic vesicles: effects of reagents that alter endosomal pH. *J. Cell. Biochem.* **62,** 27–39.

Praefcke, G. J., and McMahon, H. T. (2004). The dynamin superfamily: universal membrane tubulation and fission molecules? *Nat. Rev. Mol. Cell Biol.* **5,** 133–147.

Qualmann, B., Roos, J., DiGregorio, P. J., and Kelly, R. B. (1999). Syndapin I, a synaptic dynamin-binding protein that associates with the neural Wiskott–Aldrich syndrome protein. *Mol. Biol. Cell* **10,** 501–513.

Schrader, M. (2006). Shared components of mitochondrial and peroxisomal division. *Biochim. Biophys. Acta* **1763,** 531–541. (Epub 2006 Feb 2002.)

Sever, S., Muhlberg, A. B., and Schmid, S. L. (1999). Impairment of dynamin's GAP domain stimulates receptor-mediated endocytosis. *Nature* **398,** 481–486.

Sever, S., Damke, H., and Schmid, S. L. (2000a). Dynamin:GTP controls the formation of constricted coated pits, the rate limiting step in clathrin-mediated endocytosis. *J. Cell Biol.* **150,** 1137–1148.

Sever, S., Damke, H., and Schmmid, S. L. (2000b). Garrotes, springs, ratches and whips: putting dynamin models to test. *Traffic* **1,** 385–392.

Spiro, D. J., Boll, W., Kirchhausen, T., and Wessling-Resnick, M. (1996). Wortmannin alters the transferrin receptor endocytic pathway *in vivo* and *in vitro*. *Mol. Biol. Cell* **7,** 355–367.

Stowell, M. H. B., Marks, B., Wigge, P., and McMahon, H. T. (1999). Nucleotide-dependent conformational changes in dynamin: Evidence for a mechanochemical molecular spring. *Nat. Cell Biol.* **1,** 27–32.

Sweitzer, S. M., and Hinshaw, J. E. (1998). Dynamin undergoes a GTP-dependent conformational change causing vesiculation. *Cell* **93,** 1021–1029.

Takei, K., Slepnev, V. I., Haucke, V., and de Camilli, P. (1999). Functional partnership between amphiphysin and dynamin in clathrin-mediated endocytosis. *Nat. Cell. Biol.* **1,** 33–39.

Takei, K., Yoshida, Y., and Yamada, H. (2005). Regulatory mechanisms of dynamin-dependent endocytosis. *J. Biochem. (Tokyo)* **137,** 243–247.

Takizawa, P. A., Yucel, J. K., Veit, B., Faulkner, D. J., Deerinck, T., Soto, G., Ellisman, M., and Malhotra, V. (1993). Complete vesiculation of Golgi membranes and inhibition of protein transport by a novel sea sponge metabolite, ilimaquinone. *Cell* **73,** 1079–1090.

Thompson, H. M., and McNiven, M. A. (2001). Dynamin: switch or pinchase? *Curr. Biol.* **11,** R850.

Torre, E., McNiven, M. A., and Urrutia, R. (1994). Dynamin 1 antisense oligonucleotide treatment prevents neurite formation in cultured hippocampal neurons. *J. Biol. Chem.* **269,** 32411–32417.

Ward, T. H., Polishchuk, R. S., Caplan, S., Hirschberg, K., and Lippincott-Schwartz, J. (2001). Maintenance of Golgi structure and function depends on the integrity of ER export. *J. Cell Biol.* **155,** 557–570. (Epub 2001 Nov 2012.)

Warnock, D. E., Baba, T., and Schmid, S. L. (1997). Ubiquitously expressed dynamin-II has a higher intrinsic GTPase activity and a greater propensity for self-assembly than neuronal dynamin-I. *Mol. Biol. Cell* **8,** 2553–2562.

Wiejak, J., and Wyroba, E. (2002). Dynamin: characteristics, mechanism of action and function. *Cell. Mol. Biol. Lett.* **7,** 1073–1080.

Witke, W., Podtelejnikov, A. V., Dinardo, A., Sutherland, J. D., Gurniak, C. B., Dotti, C., and Mann, M. (1998). In mouse brain profilin I and profilin II associate with regulators of the endocytic pathway and actin assembly. *EMBO J.* **17,** 967–976.

Yang, W. N., and Cerione, R. A. (1999). Endocytosis: Is dynamin a 'blue collar' or 'white collar' worker? *Curr. Biol.* **9,** R511–R514.

Zhang, P. J., and Hinshaw, J. E. (2001). Three-dimensional reconstruction of dynamin in the constricted state. *Nat. Cell Biol.* **3,** 922–926.

CHAPTER SEVEN

Identification and Verification of Sro7p as an Effector of the Sec4p Rab GTPase

Bianka L. Grosshans* *and* Peter Novick[†]

Contents

Abstract

Effectors are operationally defined as proteins that recognize a specific GTPase preferentially in its GTP-bound conformation. Here we present the use of affinity chromatography to identify potential effectors of Sec4p, the Rab GTPase that controls the final stage of the yeast secretory pathway. We describe the

* Novartis Institutes for Biomedical Research, Basel, Switzerland
† Department of Cell Biology, Yale University School of Medicine, New Haven, Connecticut

Methods in Enzymology, Volume 438
ISSN 0076-6879, DOI: 10.1016/S0076-6879(07)38007-5

preparation of the Rab protein affinity matrix and the yeast lysate used in the purification. We also describe the methods used to identify and verify one candidate, Sro7p, as a bona fide Sec4p effector. This includes tests of the specificity and efficiency of binding both *in vitro* and *in vivo*.

1. INTRODUCTION

Rab GTPases control many different aspects of membrane traffic including vesicle budding, cargo selection, vesicle delivery, vesicle tethering, vesicle docking, and fusion of the vesicle and target membranes. They do so by interacting with a variety of structurally and functionally distinct effector molecules, thereby controlling their cellular localization or activity (Grosshans *et al.*, 2006b). In general, effectors preferentially bind to the GTP-bound form of a specific Rab, and this property can be used to operationally define proteins as Rab effectors, at least until their role in membrane traffic can be more thoroughly analyzed (Pfeffer, 2005). A number of different approaches have been used to identify Rab effectors. Some effectors have been found through genetic screens, while others were found by directly selecting for Rab interactors using either the yeast two-hybrid system or affinity chromatography with immobilized, GTP-bound Rab proteins (Grosshans *et al.*, 2006b).

Our focus has been on the final stage of the yeast exocytic pathway. Sec4p is a Rab GTPase found highly concentrated on secretory vesicles destined for fusion with the plasma membrane (Goud *et al.*, 1988). Sec4p has been implicated in three different aspects of vesicular traffic. First, Sec4 is thought to recruit the type V myosin motor Myo2p to secretory vesicles to promote their delivery along polarized actin cables (Pruyne *et al.*, 2004; Walch-Solimena *et al.*, 1997). However, the details of the molecular linkage of Sec4-GTP to Myo2p have not been defined. In addition, our studies have demonstrated that Sec4-GTP directly binds the Sec15p subunit of the exocyst complex (Guo *et al.*, 1999). This interaction is necessary for exocyst assembly and for tethering secretory vesicles to exocytic sites. Finally, genetic experiments have suggested that Sec4p might also regulate the membrane fusion event; Sec9p, one of the exocytic SNAREs (soluble N-ethylmaleimide sensitive attachment protein receptors), was identified as a high copy number suppressor of a mutation in the Sec4p effector domain (Brennwald *et al.*, 1994). However, neither Sec9p, nor either of the other two exocytic SNAREs were found to bind directly to Sec4-GTP (Grote and Novick, 1999). Together these findings prompted us to explore additional approaches to identify effector molecules that might link Sec4p to the Myo2p motor, or to components of the exocytic SNARE complex, or to play unanticipated roles in membrane traffic. We chose to use affinity chromatography to identify novel Sec4p-binding proteins from a total yeast lysate. Here we present the methods used to prepare the affinity resin and yeast extract needed for the affinity chromatography, as well as the approaches used in isolating and identifying candidate Sec4 effector molecules and then verifying their nucleotide-dependent interaction with Sec4p. We will focus on one particular Sec4p effector, Sro7p, as its role in membrane traffic is well documented (Lehman *et al.*, 1999) and, most likely, evolutionarily conserved (Gangar *et al.*, 2005). As we are still analyzing other possible effectors with regards to their interactions with Sec4-GTP and their roles in membrane traffic, we will not discuss them here.

2. EXPERIMENTAL APPROACH TO THE IDENTIFICATION OF SEC4P EFFECTORS

Sec4p effectors, by definition, must bind preferentially to the GTP-bound form of Sec4p. To identify these proteins, we adapted an affinity-chromatography approach from a published protocol previously used to identify Rab5 effectors from brain lysate (Christoforidis and Zerial, 2000). By this approach, parallel purifications are carried out using beads bearing the Rab in its GTP-bound and GDP-bound conformations so that the nucleotide specificity of binding can be assessed by comparison of the eluted proteins. To generate these two forms of Sec4p, EDTA is first used to strip the Mg^{2+} ion that stabilizes the bound nucleotide. After several washes to remove the released nucleotide, Mg^{2+} is added back to the Sec4p beads followed by the addition of the desired nucleotide. Sec4p has a high GDP-off rate (Kabcenell *et al.*, 1990) relative to other Rab proteins, making prolonged treatment with EDTA unnecessary. As nucleotide-free Sec4p is prone to denaturation, this stripping step was shortened to two washes with EDTA, followed by a wash with nucleotide-binding buffer (containing Mg^{2+}) to prepare Sec4p (and a control Rab, Ypt1p) for the subsequent loading of the desired nucleotide. To avoid hydrolysis during the affinity purification procedure, Sec4p was loaded with the nonhydrolyzable analog GTPγS rather than GTP. High salt and EDTA were then used to elute proteins bounds to the Sec4p beads. To confirm that a putative Rab effector is specific for Sec4p, we used the GTP- and GDP-bound forms of a related Rab GTPase as a negative control. For this, we chose Ypt1p, the Rab GTPase required for earlier stages of membrane traffic in yeast (Jedd *et al.*, 1995).

Here, we will describe the purification of recombinant Sec4p and Ypt1p fused to glutathione–S–transferase (GST) from bacteria. We will also describe the loading of these proteins with the required nucleotide or the preparation of nucleotide-free proteins. Moreover, we will further explain how we confirmed the specificity of one of the identified Sec4p effectors, Sro7p.

3. PURIFICATION OF GST-SEC4, GST-YPT1, AND GST, AND NUCLEOTIDE EXCHANGE

3.1. Buffers

Homogenization buffer (HB): PBS, 0.5% Tween 20, 5 mM MgCl$_2$, 5 mM β-mercaptoethanol, 200 μM GDP, 1 mM PMSF, 6 μg/ml chymostatin, 1 μg/ml pepstatin A, 0.5 μg/ml leupeptin, 10 μg/ml antipain, 2 μg/ml aprotinin, 5 μg/ml DNase, 5 μg/ml RNase (to lower viscosity following cell disruption)

Nucleotide exchange (NE) buffer: 20 mM HEPES, pH 7.2, 100 mM KCl, 10 mM EDTA, 5 mM MgCl$_2$, 1 mM DTT

Nucleotide-binding buffer (NB): 20 mM HEPES, pH 7.2, 100 mM KCl, 5 mM MgCl$_2$, 1 mM DTT

3.2. Procedure

In general, GST-fusion proteins were purified from BL21 *Escherichia coli* strains (Novagen) as described previously (Idrissi *et al.*, 2002; Wang *et al.*, 2000). The *SEC4* ORF was amplified by PCR from genomic DNA of the wildtype yeast strain NY 1210 (Mata *ura3-52 leu2-3,112 his3-Δ200 GAL+*), and subsequently inserted into plasmid pGEX5X-1 (Amersham Pharmacia) to create the GST-*SEC4* expression vector (NRB 1245) (Grosshans *et al.*, 2006a). The plasmid carrying GST-*YPT1* (SFNB 455) has been described before (Wang *et al.*, 2000).

Plasmids NRB 1245 (pGEX5X-1-*SEC4*), SFNB 455 (GST-*YPT1*), and pGEX5X-1 (GST) were transformed into the *Escherichia coli* strain BL21. Transformants were inoculated into 50 ml Luria-Bertani (LB) medium containing 50 mg/liter of ampicillin. This preculture was grown overnight at 37 °C. The next morning, bacteria were inoculated into fresh LB$_{AMP}$ medium at a dilution of 1:100. Per assay (i.e., for each nucleotide-bound or -free state), 600 ml of GST-Sec4, 700 ml of GST-Ypt1 or 150 ml of GST culture were required. *E. coli* cultures expressing GST-Sec4 and GST were further incubated at 37 °C until the cultures reached an OD$_{600}$ of 0.6 to 0.8. Production of the fusion proteins was then induced with the addition of 0.1 mM isopropyl-β-D-thiogalactopyranoside (IPTG) for 2 h at 37 °C. In contrast, GST-Ypt1-expressing *E. coli* were grown at 24 °C until the cultures reached an OD$_{600}$ of 0.7 to 0.8, and then protein production was induced overnight (about 18 h) at 24 °C. Cells were pelleted by centrifugation at 5000 rpm (about 3000×g) for 10 min in a Beckman centrifuge and stored at −20 °C until use.

To purify the fusion proteins, *E. coli* cell pellets were resuspended in 15 ml of HB per assay (i.e., for pellets of 600 ml GST-Sec4, 700 ml

GST–Ypt1 or 150 ml GST culture). Cells were subsequently lysed by sonication with 10 repetitions at maximal energy, each 20 s in duration with 10 s on ice between bursts. Cellular debris was removed by centrifugation at $30,000 \times g$ in a Ti70 rotor (Beckman ultracentrifuge) for 20 min at 4 °C. Subsequently, 200 μl of a 50% slurry of glutathione Sepharose beads (Pharmacia; prewashed three times with HB buffer and subsequently adjusted to 50% with HB) were added to the supernatants and incubated for 2 h at 4 °C on a nutator. After binding, the mixture was loaded into Econocolumns (Biorad) and allowed to drain by gravity. Beads were then washed twice with 10 ml of HB buffer and once with 10 ml HB containing 300 mM KCl to strip the beads of nonspecifically bound proteins. Detergent was finally removed by washing with 10 to 20 ml of HB without Tween 20. In order to prepare the Rab proteins for nucleotide exchange, beads were then washed twice with 10 ml of nucleotide exchange buffer (NE) (10 mM EDTA to strip Mg^{2+} and nucleotide from the GTPases) containing 10 μM of either GTPγS or GDP or no nucleotide and once with nucleotide binding buffer (NB) containing either 100 μM GTPγS or GDP or no nucleotide according to subsequent nucleotide loading.

For nucleotide loading, the GST–Rab-containing beads were incubated with 3 mM GTPγS or GDP for 2 h at 37 °C in 1 ml of NB buffer. To obtain the nucleotide-free form, GST–Sec4 was incubated in NE. GST beads were mock-treated in NB without nucleotide.

The procedure described above results in the production of approximately 750 μg of nucleotide-loaded or -free GST–Rab fusions or GST alone bound to glutathione beads as determined by SDS page gel electrophoreses and comparative Coomassie blue staining (BSA as standard).

4. PREPARATION OF THE YEAST EXTRACT FOR SEC4P AFFINITY CHROMATOGRAPHY

4.1. Buffer

NB: 20 mM HEPES, pH 7.2, 100 mM KCl, 5 mM MgCl$_2$, 1 mM DTT

4.2. Procedure

In order to obtain the necessary amount of yeast cells, wildtype yeast strain NY 1210 (Mata $ura3$-52 $leu2$-3,112 $his3$-$\Delta200$ GAL+) was grown overnight in yeast peptone dextrose (YPD) complete medium in 1- or 2-liter cultures (3- or 5-liter flasks) to an OD_{600} of about 1.5. Cells were subsequently pelleted by centrifugation at 3000 rpm (about $1000 \times g$) for 10 min in a Beckman centrifuge. Pellets were resuspended in NB without DTT, and pooled and pelleted again by centrifugation. Supernatants were removed and pellets were kept at $-20\,°C$ until use.

To obtain the yeast extract, 70 ml of yeast pellet (corresponding to about 27 liters of yeast culture) were thawed, completely resuspended in 70 ml of NB buffer containing protease inhibitors (1 mM PMSF, 5 μg/ml pepstatin A, and Complete protease inhibitor cocktail EDTA-free [Roche]). Cells were lysed by homogenization in a microfluidizer (Microfluidics Corporation). About 70 ml of NB was added to increase the disruption of the cells after the first passage through the microfluidizer. After 3 more passages, triton X-100 was added to a final concentration of 1% (from a 10% solution). The mixture was incubated for 15 min on ice and, subsequently, unbroken cells and cell debris were eliminated by centrifugation at $10,000\times g$ for 25 min at $4\,°C$ in a Beckman centrifuge. In order to remove nucleotides, which could interfere with nucleotide loading of the GST–Rab fusions at later steps, the supernatant was dialyzed against 3×8 liters of NB overnight. Subsequently, the extract was centrifuged once more at $10,000\times g$ for 25 min at $4\,°C$ to remove protein aggregates that might have formed during dialysis. The concentration of the extract was finally adjusted to 40 mg/ml (determined by a Bradford test with a BSA standard), which resulted in about 150 ml of wildtype yeast extract.

5. Sec4p Affinity Chromatography

5.1. Buffer

NB: 20 mM HEPES, pH 7.2, 100 mM KCl, 5 mM MgCl$_2$, 1 mM DTT
Elution buffer (EB): 20 mM HEPES pH 7.2, 1.5 M NaCl, 20 mM EDTA,
    1 mM DTT

5.2. Procedure

One hundred microliters of packed glutathione beads bound to nucleotide-loaded or nucleotide-free GST–Rab fusions or GST alone (about 750 μg of protein) were incubated with 25 ml of the wildtype yeast extract (at 40 mg/ml) described above. In the case of GST-Sec4 and GST-Ypt1, the extract was supplemented with either 100 μM of GTPγS or GDP. Additionally, GST-Sec4 was also incubated with extract supplemented with 10 mM EDTA to obtain proteins with affinity for the nucleotide-free form of Sec4p. Binding was allowed to proceed for 2 h at $4\,°C$ on a nutator. Beads were then washed twice with NB containing the appropriate nucleotide or 10 mM EDTA. Subsequently, bound proteins were sequentially eluted with 200 μl and then 100 μl of EB supplemented with 5 mM of the opposing nucleotide. Fifteen microliters of the eluted proteins were

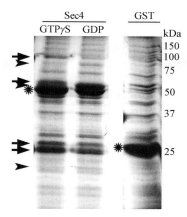

Figure 7.1 Affinity purification of Sec4p-binding proteins. Glutathione beads coated with GST-Sec4p (Sec4) loaded with either GTPγS or GDP were incubated with a wildtype yeast cell extract (strain NY 1210). Copurifying proteins were subsequently eluted, subjected to SDS page electrophoresis and silver stained. GST only was used as a control (right panel). Proteins that exhibit preferential binding to GTPγS-Sec4p are indicated with arrows; arrowheads indicate proteins that preferentially bind to GDP-Sec4p; and asterisks indicate the GST-Sec4p and GST proteins. (From Grosshans, B. L., Andreeva, A., Gangar, A., Niessen, S., Yates, J. R., 3rd, Brennwald, P., and Novick, P. (2006a). The yeast lgl family member Sro7p is an effector of the secretory Rab GTPase Sec4p. *J. Cell Biol.* **172**, 55–66, with permission.)

subjected to SDS page analysis (Fig. 7.1; data not shown) and the rest was TCA precipitated. For this, one-third volume of 100% TCA was added, and the reaction was mixed and incubated on ice overnight. Subsequently, proteins were pelleted by centrifugation at 20,000×*g* for 30 min at 4°C in a Ti70 rotor (Beckman ultracentrifuge). The protein pellet was washed twice with 500 μl of cold acetone and dried.

Proteins bound to GST–Sec4 in its different nucleotide-bound states were identified by mass spectrometry, in collaboration with the laboratory of John Yates (Scripps Research Institute, La Jolla, CA). We excluded from our list of candidate Sec4p effectors those proteins that appeared in all three samples (GTPγS, GDP, and nucleotide-free). Additionally, well-known background proteins were excluded (Ho *et al.*, 2002). We considered proteins to be possible Sec4p effectors if they exhibited at least threefold-higher sequence coverage (percentage of residues in protein sequence that are represented by at least one peptide) in the GTPγS-sample relative to the two control samples. One of the candidate effectors, Sro7p, displayed a sequence coverage of 9.9% in the GTPγS-GST-Sec4 sample, while its coverage in GDP-GST-Sec4 was only 1.9% and it was not detected in the nucleotide-free GST-Sec4 preparation (Table 7.1). In contrast, GDP dissociation inhibitor (GDI), which is known to bind to the GDP-bound form of Rabs, has a sequence coverage of 55.9%, 76.1%, and 73.4% in the samples of the GTPγS-, GDP-bound, and nucleotide-free forms of Sec4p, respectively. The unexpected high amount of GDI found in the GTPγS-Sec4p pull-down might reflect incomplete loading of Sec4p (GTPγS binding as determined with radioactive nucleotide never exceeded about 40% under all temperatures tested; data not shown). For comparison, the sequence of Sec4p itself was covered with 64.7%, 75.8% and 73.5%, respectively (see Table 7.1).

6. CONFIRMATION OF SRO7P AS A SEC4P EFFECTOR

6.1. Buffers

HB: PBS, 0.5% Tween 20, 5 mM MgCl$_2$, 5 mM β-mercaptoethanol, 200 μM GDP, 1 mM PMSF, 6 μg/ml chymostatin, 1 μg/ml pepstatin A, 0.5 μg/ml leupeptin, 10 μg/ml antipain, 2 μg/ml aprotinin, 5 μg/ml DNase, 5 μg/ml RNase (to avoid viscosity problems)

NE buffer: 20 mM HEPES, pH 7.2, 100 mM KCl, 10 mM EDTA, 5 mM MgCl$_2$, 1 mM DTT

NB: 20 mM HEPES, pH 7.2, 100 mM KCl, 5 mM MgCl$_2$, 1 mM DTT

6.2. Procedure

To confirm Sro7p as a Sec4p effector, the affinity-chromatography described above was repeated on a smaller scale with an extract from yeast cells expressing an integrated, HA$_3$-tagged allele of Sro7p. NY 2587 (Mata *SRO-HA$_3$*::kanMX *ura3-52 leu2-3,112 his3-Δ200 GAL+*) was created by homologous recombination using a PCR-based strategy (Longtine *et al.*, 1998). Expression of the Sro7-HA$_3$p fusion protein was checked by Western blot, and its functionality was demonstrated by its ability to suppress the cold sensitivity of an *sro7Δ sro77Δ* double-mutant and the

Table 7.1   Mass spectrometry results for Sro7p, GDI, and Sec4p

Protein	GTPγS-Sec4p[a]	GDP-Sec4p[a]	Nucleotide-free Sec4p[a]
Sec4p	64.7	75.8	73.5
Sro7p	9.9	1.9	—
GDI	55.9	76.1	73.4

[a] Percentage of protein sequence represented by at least one peptide.
From Grosshans, B. L., Andreeva, A., Gangar, A., Niessen, S., Yates, J. R., 3rd, Brennwald, P., and Novick, P. (2006). The yeast lgl family member Sro7p is an effector of the secretory Rab GTPase Sec4p. *J. Cell Biol.* **172,** 55–66.

salt-sensitivity of an *sro7Δ* mutant strain (Grosshans *et al.*, 2006a and data not shown).

Cell pellets derived from 300 ml of GST-Sec4-, 360 ml of GST-Ypt1-, and 20 ml of GST-expressing BL21 *E. coli* cultures (see section 3) were homogenized by sonication in either 30 ml or 15 ml of HB. Cells were sonicated 10 times for 20 s with 10 s on ice between bursts. Cell debris was pelleted by centrifugation at 30,000g for 20 min in a Beckman ultracentrifuge (Ti70 rotor). The supernatants of GST-Sec4- or GST-Ypt1-cultures were divided into three aliquots of 10 ml. Thirty microliters of a 50% suspension of glutathione Sepharose beads (Pharmacia; washed three times with HB and subsequently adjusted to 50% with HB) were then added to each of the 10 ml supernatants and all of the GST supernatant and incubated for 2.5 h at 4 °C on a nutator. The mixture was poured into Econocolumns (Biorad) and washes were performed as described in section 3. For nucleotide loading, 5 μl (50% suspension) of these beads in NB were incubated for 2 h at 37 °C in a total 20 μl NB with either 4 mM GTPγS or GDP or 10 mM EDTA to obtain the nucleotide-free form.

This procedure leads to approximately 15 μg of nucleotide-loaded or nucleotide-free proteins bound to glutathione beads as determined by comparative Coomassie staining (BSA as standard).

In parallel, 2 liters of strain NY 2587 were grown to an OD$_{600}$ of about 1. Cells were pelleted by centrifugation in a Beckman centrifuge at 3000 rpm (about 1000×g) for 5 min and the pellet was kept at −20 °C until use. To obtain the yeast extract, cell pellets (about 5 ml) were completely resuspended in 6 ml of NB buffer. The cell suspension was then subjected to French press lysis (Sim-Amico Spectronic Instruments) at 10,000 psi. The procedure was repeated three times. Cell lysis was between 50 to 80% under these conditions as observed by microscopy. Tween 20 (10% solution) was then added to a final concentration of 0.5%, the extract left on ice for 15 min, and, subsequently, the cell debris removed by centrifugation at 10,000×g for 20 min in a Beckman centrifuge. The supernatant of this centrifugation step was dialyzed against three times 1 liter of NB (no nucleotide) overnight and the extract was then centrifuged again at 10,000×g for 20 min to eliminate protein aggregates. Finally, the protein concentration of the extract was adjusted to 20 mg/ml (as determined by a Bradford test with a BSA standard).

Subsequently, 1 ml of this yeast extract was added to 5 μl of a 50% suspension of glutathione Sepharose beads carrying the nucleotide-loaded GST-Sec4, GST-Ypt1, or GST alone. Next, 100 μM of GTPγS, or GDP, or 10 mM EDTA were added into the appropriate tubes, and the reactions were incubated for 2 h at 4 °C on a nutator. The beads were then pelleted by centrifugation at 1000 rpm for 2 min and washed three times with the appropriate NB buffer (plus nucleotide or EDTA). Finally, beads were boiled in SDS PAGE sample buffer and bound proteins were analyzed by

Western blot. Sro7-HA$_3$p was detected with a rat monoclonal α-HA antibody (3F10, Roche). As controls for nucleotide specificity, the Sec4p effector, Sec15p (Guo *et al.*, 1999), and the Sec4p guanine nucleotide exchange factor, Sec2p (Walch-Solimena *et al.*, 1997), were detected with protein-specific rabbit polyclonal antibodies (P. Novick collection). To compare the levels of the different fusion proteins in the reaction, either a goat α-GST antibody (Sigma) was used to detect the GST moiety or the proteins were visualized by Ponceau staining.

Under these conditions, we found that approximately 2 to 4% of the total pool of Sro7-HA$_3$p bound to GTPγS-Sec4p (Fig. 7.2). The nucleotide specificity of the assay was confirmed by the finding that Sec15p also bound specifically to the GTPγS-bound form of Sec4p, and that Sec2p preferentially bound to the nucleotide-free form of this Rab protein. Interestingly, the amount of available, activated Sec4p seems to be limiting in this assay since the same amount of Sro7-HA$_3$p was found to bind to GTPγS-Sec4p when only half the amount of extract was used (data not shown).

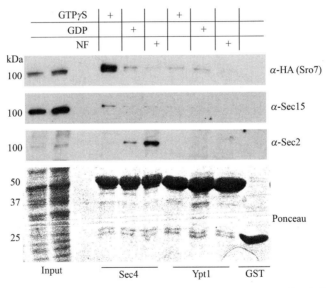

Figure 7.2 Sro7-HA$_3$p binds preferentially to GTPγS-Sec4p. GST-Sec4 (Sec4), GST-Ypt1 (Ypt1) or GST (GST) immobilized on glutathione beads were incubated with an extract of a *SRO7HA$_3$* strain and copurifying proteins were subjected to Western Blot (also see below; could not be labelled there!) analysis using the indicated antibodies. GST-Sec4p and GST-Ypt1p were either stripped of nucleotide (NF) or loaded with GTPγS (GTP) or GDP (GDP). Ponceau staining of the western blot is shown as loading control. The input lanes (Input) represent 0.2 and 0.3%, respectively. (From Grosshans, B. L., Andreeva, A., Gangar, A., Niessen, S., Yates, J. R., 3rd, Brennwald, P., and Novick, P. (2006a). The yeast lgl family member Sro7p is an effector of the secretory Rab GTPase Sec4p. *J. Cell Biol.* **172**, 55–66, with permission.)

7. CO-IMMUNOPRECIPITATION OF SRO7P WITH SEC4P

7.1. Buffers

Immunoprecipitation (IP) buffer: 20 mM HEPES, pH 7.2, 150 mM KCl, 1 mM DTT, 6 mM MgCl$_2$, 1 mM EDTA

7.2. Procedure

While the studies described above indicated that Sro7p is capable of binding to Sec4–GTP *in vitro*, it is important to ask if Sro7p is normally bound to Sec4p *in vivo*. Sro7p is not an abundant protein in yeast (Ghaemmaghami *et al.*, 2003; Huh *et al.*, 2003). Therefore, it was overexpressed, in combination with Sec4p, to evaluate binding of Sro7p to Sec4p *in vivo* by co-immunoprecipitation. As specificity controls, the related Rab GTPase Ypt1p and empty vectors (NRB 529 and 530; Gal1 promoter, *ADH1* terminator, *LEU2* or *URA3* marker) were used. To obtain a galactose-inducible Sro7p construct (plasmid NRB 1246), the *SRO7* ORF was amplified by PCR from the wildtype yeast strain NY 1210 and inserted into plasmid NRB 530 (GAL1 promoter, *ADH1* terminator, *URA3* marker) behind the GAL1 promoter. The plasmids carrying HA-*SEC4* (NRB 833) and HA-*YPT1* (NRB 829) behind the *GAL1* promoter have been described previously (Grote and Novick, 1999). Suitable plasmid combinations were subsequently transformed into NY 1210 to obtain yeast strains NY 2592 through 2595. Overexpression of HA-Sec4p, HA-Ypt1p, and Sro7p was confirmed by Western Blot with either a rat monoclonal α-HA antibody (3F10, Roche) or an Sro7p-specific rabbit polyclonal antibody (Lehman *et al.*, 1999; data not shown).

To overexpress the proteins, yeast strains NY 2592 through NY 2595 were grown overnight at 25 °C in YP media containing raffinose as carbon source to an OD$_{600}$ of about 0.4. Production of HA–Sec4p, HA–Ypt1p and Sro7p was then induced by addition of galactose to a final concentration of 2%. After 45 min incubation, cells were pelleted at 3000 rpm (about 1000×g) for 5 min in a Beckman centrifuge, resuspended in IP buffer and transferred to screw capped tubes. Cells were pelleted again and 2 g of zirconia-silica beads were added to the yeast pellets. The screw-capped tubes were then filled with ice-cold IP buffer containing 1 mM GTP, 1% NP-40 and protease inhibitors (1 mM PMSF, 2 μg/ml pepstatinA, 2 μg/ml chymostatin, 2 μg/ml aprotinin, 2 μg/ml leupeptin, 2 μg/ml antipain). Subsequently, the cells were lysed by homogenization in a mini–Bead Beater (Biospec Products) at full power for 4.5 min. Unbroken cells and cell debris were eliminated by centrifugation at 10,000×g for 20 min at 4 °C in a Microtube centrifuge. The concentration of the yeast extracts was then measured by a Bradford assay with BSA as standard and finally adjusted to 1 mg/ml.

To reduce nonspecific binding in the subsequent immunoprecipitation, 30 μl of a 50% suspension of Protein G Sepharose beads (pre-equilibrated in IP buffer) were added to 500 μl of extract and the reaction was incubated for 90 min at 4 °C. Beads were pelleted, the supernatants transferred to new tubes and incubated overnight at 4 °C with 7.5 μl rat α-HA antibody (3F10, Roche). Subsequently, 6 μl of a 50% suspension of Protein G Sepharose beads (pre-equilibrated with IP buffer) were added and incubated for another 30 min. The beads were pelleted, washed 3 times with IP buffer and the bound proteins were subjected to Western blot analysis. Precipitated HA-tagged Rab GTPases were detected with a mouse monoclonal α-HA antibody (clone 16B12, Convance) and co-precipitating Sro7p was detected with an Sro7p-specific rabbit polyclonal antibody (Lehman *et al.*, 1999).

Using this protocol, we found that approximately 2% of the total pool of Sro7p co-immunoprecipitated with HA-Sec4p (after normalizing for the efficiency of Sec4p precipitation) (Fig. 7.3). This relatively low amount might reflect the predominantly inactive (i.e., GDP-bound) state of Sec4p in a yeast lysate. The specificity of this interaction is illustrated by the finding

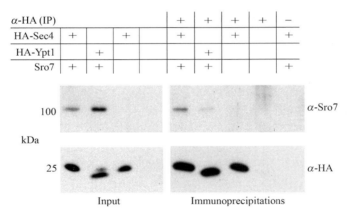

α-HA (IP)					+	+	+	+	−
HA-Sec4	+		+		+		+		+
HA-Ypt1		+				+			
Sro7	+	+			+	+			+

Figure 7.3 Sro7p co-immunoprecipitates with Sec4p. Yeast cells overexpressing HA-Sec4p and Sro7p, HA-Ypt1p and Sro7p, HA-Sec4p alone, or none of these proteins were lysed, and HA-Sec4p and HA-Ypt1p were immunoprecipitated using a rat α-HA antibody (α-HA [IP]). Proteins were subsequently subjected to Western Blot analysis and co-immunoprecipitating Sro7p was detected using an Sro7p-specific antibody (α-Sro7). A mouse α-HA antibody (α-HA) was used to detect the precipitated HA-Sec4p and HA-Ypt1p. The specificity of the observed co-immunopecipitation was shown by omitting the antibody from the reaction (α-HA [IP], lane −). The amount of HA-Sec4p and HA-Ypt1p shown represents 5% of the immunoprecipitation. The inputs represent 0.15% of total protein (Input). (From Grosshans, B. L., Andreeva, A., Gangar, A., Niessen, S., Yates, J. R., 3rd, Brennwald, P., and Novick, P. (2006a). The yeast lgl family member Sro7p is an effector of the secretory Rab GTPase Sec4p. *J. Cell Biol.* **172**, 55–66.)

that only a background amount of Sro7p co-immunoprecipitates with HA-Ypt1p. Moreover, omitting the antibody from the Sec4p/Sro7p immunoprecipitation and overexpressing Sro7p alone did not lead to an Sro7p-signal in the Western Blot.

Taken together, our *in vivo* data support our findings *in vitro* and together with direct binding data (Grosshans *et al.*, 2006a) establish Sro7p as an effector of Sec4p.

REFERENCES

Brennwald, P., Kearns, B., Champion, K., Keranen, S., Bankaitis, V., and Novick, P. (1994). Sec9 is a SNAP-25-like component of a yeast SNARE complex that may be the effector of Sec4 function in exocytosis. *Cell* **79**, 245–258.

Christoforidis, S., and Zerial, M. (2000). Purification and identification of novel Rab effectors using affinity chromatography. *Methods* **20**, 403–410.

Gangar, A., Rossi, G., Andreeva, A., Hales, R., and Brennwald, P. (2005). Structurally conserved interaction of Lgl family with SNAREs is critical to their cellular function. *Curr. Biol.* **15**, 1136–1142.

Ghaemmaghami, S., Huh, W. K., Bower, K., Howson, R. W., Belle, A., Dephoure, N., O'Shea, E. K., and Weissman, J. S. (2003). Global analysis of protein expression in yeast. *Nature* **425**, 737–741.

Goud, B., Salminen, A., Walworth, N. C., and Novick, P. J. (1988). A GTP-binding protein required for secretion rapidly associates with secretory vesicles and the plasma membrane in yeast. *Cell* **53**, 753–768.

Grosshans, B. L., Andreeva, A., Gangar, A., Niessen, S., Yates, J. R., 3rd, Brennwald, P., and Novick, P. (2006a). The yeast lgl family member Sro7p is an effector of the secretory Rab GTPase Sec4p. *J. Cell Biol.* **172**, 55–66.

Grosshans, B. L., Ortiz, D., and Novick, P. (2006b). Rabs and their effectors: achieving specificity in membrane traffic. *Proc. Natl. Acad. Sci. U. S. A.* **103**, 11821–11827.

Grote, E., and Novick, P. J. (1999). Promiscuity in Rab-SNARE interactions. *Mol. Biol. Cell* **10**, 4149–41461.

Guo, W., Roth, D., Walch-Solimena, C., and Novick, P. (1999). The exocyst is an effector for Sec4p, targeting secretory vesicles to sites of exocytosis. *EMBO J.* **18**, 1071–1080.

Ho, Y., Gruhler, A., Heilbut, A., Bader, G. D., Moore, L., Adams, S. L., Millar, A., Taylor, P., Bennett, K., Boutilier, K., Yang, L., Wolting, C., *et al.* (2002). Systematic identification of protein complexes in Saccharomyces cerevisiae by mass spectrometry. *Nature* **415**, 180–183.

Huh, W. K., Falvo, J. V., Gerke, L. C., Carroll, A. S., Howson, R. W., Weissman, J. S., and O'Shea, E. K. (2003). Global analysis of protein localization in budding yeast. *Nature* **425**, 686–691.

Idrissi, F. Z., Wolf, B. L., and Geli, M. I. (2002). Cofilin, but not profilin, is required for myosin-I-induced actin polymerization and the endocytic uptake in yeast. *Mol. Biol. Cell* **13**, 4074–4087.

Jedd, G., Richardson, C., Litt, R., and Segev, N. (1995). The Ypt1 GTPase is essential for the first two steps of the yeast secretory pathway. *J. Cell Biol.* **131**, 583–590.

Kabcenell, A. K., Goud, B., Northup, J. K., and Novick, P. J. (1990). Binding and hydrolysis of guanine nucleotides by Sec4p, a yeast protein involved in the regulation of vesicular traffic. *J. Biol. Chem.* **265**, 9366–9372.

Lehman, K., Rossi, G., Adamo, J. E., and Brennwald, P. (1999). Yeast homologues of tomosyn and lethal giant larvae function in exocytosis and are associated with the plasma membrane SNARE, Sec9. *J. Cell Biol.* **146,** 125–140.

Longtine, M. S., McKenzie, A., 3rd, Demarini, D. J., Shah, N. G., Wach, A., Brachat, A., Philippsen, P., and Pringle, J. R. (1998). Additional modules for versatile and economical PCR-based gene deletion and modification in *Saccharomyces cerevisiae. Yeast* **14,** 953–961.

Pfeffer, S. R. (2005). Structural clues to Rab GTPase functional diversity. *J. Biol. Chem.* **280,** 15485–15488.

Pruyne, D., Legesse-Miller, A., Gao, L., Dong, Y., and Bretscher, A. (2004). Mechanisms of polarized growth and organelle segregation in yeast. *Annu. Rev. Cell Dev. Biol.* **20,** 559–591.

Walch-Solimena, C., Collins, R. N., and Novick, P. J. (1997). Sec2p mediates nucleotide exchange on Sec4p and is involved in polarized delivery of post-Golgi vesicles. *J. Cell Biol.* **137,** 1495–1509.

Wang, W., Sacher, M., and Ferro-Novick, S. (2000). TRAPP stimulates guanine nucleotide exchange on Ypt1p. *J. Cell Biol.* **151,** 289–296.

CHAPTER EIGHT

CHARACTERIZATION OF RAB18, A LIPID DROPLET–ASSOCIATED SMALL GTPASE

Sally Martin *and* Robert G. Parton

Contents

Institute for Molecular Bioscience and Centre for Microscopy and Microanalysis, University of Queensland, Brisbane, Queensland, Australia

Methods in Enzymology, Volume 438
ISSN 0076-6879, DOI: 10.1016/S0076-6879(07)38008-7

Abstract

Lipid droplets are the major intracellular store of lipids in eukaryotic cells. Understanding lipid storage and regulated mobilization of lipids from lipid droplets is essential for understanding the syndromes and diseases associated with excess lipid accumulation. Lipid droplets have been traditionally considered relatively inert structures. However, in recent years it has become apparent that lipid droplets are highly dynamic regulated organelles, which show complex interactions with other cellular compartments. The cellular components involved in regulation of lipid accumulation and release from lipid droplets, and in mediating the complex interactions with other organelles, are only now starting to be unraveled. A particularly important family of proteins in this respect is the Rab GTPases, crucial regulators of membrane traffic. Here we describe the techniques that we used to characterize the regulated association of Rab18 with the surface of lipid droplets. Rab18 provides an excellent marker to follow the dynamics of lipid droplets in living cells. In addition, the study of Rab18 provides insights into the mechanisms involved in the release of lipids from lipid droplets in adipocytes. In 3T3-L1 adipocytes, stimulation of lipolysis increases the association of Rab18 with lipid droplets, suggesting that recruitment of Rab18 is regulated by the metabolic state of individual lipid droplets. The study of Rab18 and its interacting proteins will provide new insights into the complex regulatory mechanisms involved in lipid storage and release.

1. INTRODUCTION

Obesity is a major health concern in the modern world. Regulation of adipose tissue lipid levels occurs through a balance between fat synthesis (lipogenesis) and fat breakdown (lipolysis). Disturbances in the lipolytic pathways in adipose tissue and skeletal muscle are important factors in the increased triglyceride storage contributing to obesity, insulin resistance, and type II diabetes. Cells store excess fatty acids and cholesterol in the form of neutral triglycerides and cholesteryl esters in lipid droplets (LDs) (Murphy, 2001). LDs play a crucial role in maintaining the cellular levels of lipids by regulating the interplay among lipid storage, hydrolysis, and trafficking (Martin and Parton, 2005). The LD is composed of a core of neutral lipids surrounded by a monolayer of phospholipids, cholesterol, and proteins. In the current model of LD formation, neutral lipids are synthesized from fatty acids and cholesterol by enzymes present in the ER and deposited in the hydrophobic domain between the phospholipid leaflets of the ER membrane. This mechanism of formation gives rise to a unique cytosolic organelle surrounded by a monolayer of phospholipids derived from the cytoplasmic leaflet of the ER membrane (Tauchi–Sato et al., 2002).

Virtually all cells have the capacity to form LDs, although adipocytes are the major fat storage cell. Stored neutral lipids can be used for energy metabolism (β-oxidation), phospholipid and lipoprotein synthesis, and steroidogenesis. Until recently, LDs were considered inert storage structures. However, it is now apparent that LDs are dynamic motile organelles that move on microtubules and show multiple interactions with other compartments (reviewed in Martin and Parton, 2005; 2006). Furthermore, in adipocytes there is increasing evidence that LDs undergo morphological rearrangement during lipolysis (Marcinkiewicz et al., 2006) and exist in differently regulated states (Moore et al., 2005). This has generated great interest in the LD as a *bona fide* cellular organelle. These observations indicate that LDs are able to interact, possibly directly, with other organelles, and have the potential capacity to form complex interaction networks similar to other, better-understood intracellular compartments. However, the structural constraints on the LD, including the highly hydrophobic internal environment and the presence of a single limiting phospholipid monolayer rather than the typical bilayer, suggest the involvement of novel, potentially unique, molecular machinery and mechanisms. In recent studies we, and others, have identified the Rab18 small GTPase in LD preparations from a number of cell types (Martin et al., 2005; Ozeki et al., 2005). Rab18 is present on a subset of LDs in most cell types examined, and shows a regulated association with LDs in 3T3-L1 adipocytes, a model system for studying the hormonal regulation of lipolysis (neutral lipid hydrolysis) and lipogenesis (neutral lipid synthesis) (Brasaemle et al., 2000). In this methods paper, we detail the experimental techniques that have been used to analyze LDs and LD-associated proteins, with particular reference to the analysis of Rab18.

2. Cell Systems

Studies of Rab18 LD association can be undertaken in nonadipocyte cell lines where the formation and catabolism of LDs can be regulated by altering the concentration of free fatty acids in the medium. This system is particularly useful for real-time video microscopy and electron microscopy, as well as to study general aspects of LD formation. Rab18 function can also be investigated in adipocyte cell lines and primary adipocytes, where the specific regulation of LD metabolism by catecholamines and insulin, as well as other growth factors, can be investigated.

2.1. Manipulation of LDs in nonadipocyte cell lines

Lipid droplets are present in most cultured cells, although the number, size, and distribution are highly variable. In all cell types we have examined to date, lipid droplets can be readily manipulated by culturing in medium containing elevated free fatty-acid concentrations. In order to deliver free fatty acids to the cell it is helpful to conjugate them to a carrier protein, usually bovine serum albumin (BSA). Oleic acid can be either conjugated directly to BSA before adding to the medium, or added directly to serum-containing medium. For direct conjugation, oleic acid is coupled to fatty acid–free BSA at a molar ratio of 6:1 (oleic acid:albumin molecules) and added to the culture medium to give a final concentration of 50 to 350 μM oleic acid, adapted from Brasaemle et al. (1997). Addition of this conjugate to the cell medium induces the rapid generation of LDs. To generate oleic acid:BSA, oleic acid is melted at 37 °C and 50 mg weighed into a sterile tube. The volume is adjusted to 460 μl, with warm (37 °C) serum-free medium and 40 μl sterile, filtered (0.22 μm filter), 10% fatty acid–free BSA (Sigma/ Calbiochem) added to the tube and mixed well. The mixture is incubated at 37 °C for 30 min with gentle shaking. The medium should change color to orange/yellow and become cloudy. The stock solution (350 mM) can be stored for several weeks at 4 °C as a concentrated stock or diluted directly into growth medium. Prior to dilution of the concentrated stock in the medium it is necessary to warm both to 37 °C and mix them well.

Oleic acid can also be delivered to cells after solubilizing directly in ethanol and diluting in the medium. The mechanism of delivery is less well characterized in this case than following conjugation to BSA, but is likely to involve in situ conjugation of dissolved oleic acid with serum albumin. Unlike LD formation from BSA–oleic acid conjugates, where LD formation is very rapidly induced and can be detected morphologically within 10 min (350 μM oleate), LD formation following direct addition of solubilized oleic acid to the medium is a slower process. However, over longer periods of time it is as effective. Again, medium can be made up in advance and maintained at 37 °C to facilitate conjugation. Weigh oleic acid into a glass vial and dissolve in 100% ethanol to give a concentrated solution of 50 to 100 mM. The concentrated stock solution can be stored at −20 °C and added directly to the tissue culture medium to a final concentration of 20 to 100 μM. Ensure that the final concentration of ethanol in the medium does not exceed 0.1%.

In order to follow LD catabolism in nonadipocyte cell lines, it is possible to grow the cells in delipidated serum for 24 to 48 h. Over this time, pre-existing LDs are catabolized and the components either secreted or metabolized. Lipoprotein-depleted fetal calf serum containing less than 5% normal lipoprotein can be purchased (i.e., Sigma catalog # S5394) or prepared by ultracentrifugation (Goldstein et al., 1983), or delipidated serum generated by stripping the lipids from the medium using alcohols (Cham and Knowles, 1976; Rothblat et al., 1976). In each case, the lipid-depleted serum has to be dialyzed extensively, and it is important to consider that under these conditions there will also be loss of growth factors and small molecules from the treated serum.

2.2. Differentiation of 3T3-L1 adipocytes and manipulation of LDs

The most commonly used and well-characterized model cell line to study regulated lipolysis is the murine embryonic 3T3-L1 line (ATCC CL-173). Fibroblast 3T3-L1 cells can be differentiated to generate 3T3-L1 adipocytes that respond in a similar manner to primary adipocytes in terms of insulin and adrenergic receptor modification of lipolysis. The differentiation of 3T3-L1 adipocytes has been extensively described elsewhere, and only a brief description will be given here. For more detailed methods, see Frost and Lane (1985) and Green and Kehinde (1975). 3T3-L1 fibroblasts are maintained and passaged in DMEM (high glucose) containing 2 mM L-glutamine, and either 10% newborn calf serum or 10% fetal calf serum (FCS). It is very important to the subsequent differentiation process that the cells are not permitted to reach confluence during maintenance. After plating cells for experimental purposes, the differentiation process is initiated 24-h postconfluence by replacing the cell medium with DMEM (high glucose) containing 10% FCS, 2 mM L-glutamine, 0.5 mM isobutyl methylxanthine, 0.25 μM dexamethasone, 1 μg/ml insulin, and 100 ng/ml biotin. After 48 to 72 h, the differentiation medium is carefully removed and replaced with DMEM containing 10% FCS, 2 mM L-glutamine, and 1 μg/ml insulin. Note that at this point in the differentiation process the cells are very loosely attached to the plates and are easily detached during pipetting. After a further 48 to 72 h, remove the insulin-containing medium and replace with DMEM containing 2 mM L-glutamine and 10% FCS. Cells can be maintained in this medium, feeding every 2 and 3 days, and are normally used 8 to 12 days after the initiation of differentiation. The differentiation process can be easily followed using bright-field microscopy or the LDs detected by staining (see section 6) with neutral lipid dyes (Fig. 8.1). It is important to note that some batches of FCS are better able to support differentiation than others. Batch testing FCS is recommended to ensure that the optimal differentiation is achieved.

Lipolysis can be activated in differentiated 3T3-L1 adipocytes by treatment with β-adrenergic agonists (10 μM isoproterenol) or elevation of cAMP (20 μM forskolin) (Brasaemle et al., 2000; Martin et al., 2005). Alternatively, lipolysis can be inhibited by insulin (100 ng/ml) (Zhang et al., 2005), or the activity of isoproterenol blocked by incubation with the β-blocker propranolol (200 μM) (Martin et al., 2005). These modulations can be triggered simply by addition of the reagent to the medium for 10 to 30 min.

2.3. Isolation and culture of primary mouse adipocytes

Primary adipocytes can be readily isolated from adipose tissue by collagenase digestion and lipolysis modulated as in 3T3-L1 adipocytes (Rodbell, 1964). To isolate primary adipocytes, male mice are sacrificed by cervical

Fibroblasts (Day -2)   Day 0   Day +4   Day +8

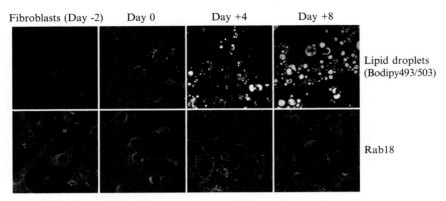

Lipid droplets
(Bodipy493/503)

Rab18

Figure 8.1 Differentiation of 3T3-L1 adipocytes. 3T3-L1 fibroblasts are maintained in culture as subconfluent monolayers (Day -2). To initiate differentiation cells are grown to confluence (Day 0) prior to addition of the differentiation medium (insulin, dexamethasone, IBMX, and biotin). The differentiation of fibroblasts into adipocytes can be followed using Bodipy493/503 to mark LDs. Differentiation is complete from around Day +8, Rab18 labeling develops from a diffuse staining pattern in fibroblasts to a highly punctate pattern around LDs in adipocytes.

dislocation and the epididymal and/or perirenal adipose tissue removed. Tissue is washed in warm, modified Krebs Ringer Phosphate, pH7.4 (KRP) (12.5 mM HEPES, 120 mM NaCl, 6.0 mM KCl, 1.2 mM MgSO$_4$, 1.0 mM CaCl$_2$, 0.4 mM NaH$_2$PO$_4$, 0.6 mM Na$_2$HPO$_4$) containing 0.2% BSA, blotted dry and weighed. After cutting into small (~1 mm^3) pieces, the tissue is resuspended in three times volume/weight KRP containing 4 mg/ml collagenase I (Worthington Biochemical Corp), and incubated at 37 °C with gentle shaking for 30 to 45 min until the tissue is digested. Lumps of undigested material are removed by filtering through a single layer of muslin cloth (or bridal veil), and the cells washed gently twice with KRP containing 0.2% BSA by centrifugation at low speed (1000 rpm, bench-top centrifuge). At this point, the adipocytes will float to the top of the tube while contaminating cell types will pellet and can be removed by aspiration. After washing, the cells are resuspended in KRP containing 0.2% BSA and incubated with gentle shaking for 1 h to equilibrate prior to experimentation.

3. GENERATION OF RAB18 CONSTRUCTS AND POINT MUTATIONS

The high degree of sequence conservation of Rab proteins means that it is possible to predict sites of catalytic activity and nucleotide binding (Seabra and Wasmeier, 2004; Zerial and McBride, 2001). A comparison of the sequence of Rab18 with other characterized Rabs allowed us to design point mutations in Rab18 predicted to generate constitutively active or dominant-negative constructs (Fig. 8.2) (see also Ozeki *et al.*, 2005). To generate GFP-Rab18, the mRab18 ORF was excised from myc-Rab18 using NdeI and BamHI while removing the myc tag, and ligated into pSL1180 (Amersham/Pharmacia). The ORF was subsequently excised with BamHI and PstI and ligated into pEGFP-C1 (Clontech), resulting in an N-terminal GFP tag (Martin *et al.*, 2005). Site-directed mutagenesis of GFP-Rab18 was used to produce point mutations in Rab18 and was carried out essentially according to instructions in the QuickChange site-directed mutagenesis kit (Stratagene). GFP-Rab18S22N was generated using the

```
Rab5A_Mouse    1  MANRGATRPNGPNTGNKICQFKLVLLGESAVGKSSLVLRFVKGQFHEFQE   50
Rab7_Human     1          MTSRKKVLLKVIILGDSGVGKTSLMNQYVNKKFSNQYK   38
Rab18_Mouse    1          MDEDVLTTLKILIIGESGVGKSSLLLRFTDDTFDPELA   38
Rab18_Human    1          MDEDVLTTLKILIIGESGVGKSSLLLRFTDDTFDPELA   38
                          *....*.* ***.**.. ..      *

Rab5A_Mouse   51  STIGAAFLTQTVCLDDTTVKFEIWDTAGQERYHSLAPMYYRGAQAAIVVY  100
Rab7_Human    39  ATIGADFLTKEVMVDDRLVTMQIWDTAGQERFQSLGVAFYRGADCCVLVF   88
Rab18_Mouse   39  ATIGVDFKVKTISVDGNKAKLAIWDTAGQERFRTLTPSYYRGAQGVILVY   88
Rab18_Human   39  ATIGVDFKVKTISVDGNKAKLAIWDTAGQERFRTLTPSYYRGAQGVILVY   88
                  .*** *   .  .* ********... * .**** ..*.

Rab5A_Mouse  101  DITNEESFARAKNWVKELQRQASPN----IVIALSGNKADLANKRAVDFQ  146
Rab7_Human    89  DVTAPNTFKTLDSWRDEFLIQASPRDPENFPFVVLGNKIDLEN-RQVATK  137
Rab18_Mouse   89  DVTRRDTFVKLDNWLNELETYCTRND---IVNMLVGNKIDKEN-REVDRN  134
Rab18_Human   89  DVTRRDTFVKLDNWLNELETYCTRND---IVNMLVGNKIDKEN-REVDRN  134
                  *.*    .*     *   *  .. . *** *  *  *

Rab5A_Mouse  147  EAQSYAD-DNSLLFMETSAKTSMNVNEIFMAIAKKLPKNEPQNPGANSA-  194
Rab7_Human   138  RAQAWCYSKNNIPYFETSAKEAINVEQAFQTIARNALKQETEVELYNEFP  187
Rab18_Mouse  135  EGLKFAR-KHSMLFIEASAKTCDGVQCAFEELVEKIIQTPGLWESENQN-  182
Rab18_Human  135  EGLKFAR-KHSMLFIEASAKTCDGVQCAFEELVEKIIQTPGLWESENQN-  182
                  . . . . *.*** ..*   *   .        *

Rab5A_Mouse  195  RGRGVDLTEPAQP--ARSQCCSN   215
Rab7_Human   188  EPIKLDKNDRAKAS---AESCSC  207
Rab18_Mouse  183  KGVKLSHREESRGGGACGGYCSVL  206
Rab18_Human  183  KGVKLSHREEGQGGGACGGYCSVL  206
                  . . . **
```

Figure 8.2 Sequence alignment of Rab18 with representative endosomal Rab GTPases Structural components of Rab18 can be identified by similarity to other Rab proteins. The conformational switch regions are highlighted in gray boxes. The C-terminal box highlights the CAAX box of Rab18 (Leung *et al.*, 2007). Horizontal lines above the sequence denote GTP-binding domains, an asterisk below the sequence denotes invariant residues, and a dot below the sequence denotes a conservative amino-acid substitution. Amino acids implicated in GTP-GDP exchange and GTPase activity are identified in boldface type. Point mutations in these residues are predicted to result in a loss of GTPase activity leading to constitutive GTP binding and activation (Q67L), constitutive GDP binding (S22N) or nucleotide deficiency (N122I).

following primers: 5′-GCGAGAGTGGGGTGGGCAAGAACTCACTG
CTCCTGAGGTTCACAG-3′ and 5′-CTGTGAACCTCAGGAGCAG
TGAGTTCTTGCCCACCCCACTCTCGC-3′, and GFP-Rab18Q67L
using: 5′-GCAATATGGGATACAGCCGGTCTAGAGAGGTTCAGA
AC-3′ and 5′-GTTCTGAACCTCTCTAGACCGGCTGTATCCCATA
TTGC-3′. All constructs were sequenced using ABI PRISM BigDye
Terminator v3.1 (Applied Biosystems, Foster City, CA) in the Australian
Genome Research Facility, University of Queensland.

4. EXPRESSION OF HETEROLOGOUS PROTEINS IN 3T3-L1 ADIPOCYTES

Expression of heterologous proteins, such as GFP–Rab18, in differentiated 3T3-L1 adipocytes is difficult to achieve by normal cationic lipid
based reagents. One way to overcome this is to express the protein of
interest at the fibroblast stage prior to differentiation. However, this is not
always successful, as transfected cells often do not differentiate to the same
extent as nontransfected cells. Also, it is necessary to ensure that expression
of mutants does not impact on the differentiation processes. With regard to
siRNA-based analysis, it is doubtful that siRNA would remain active in the
cell long enough to detect an effect in the adipocyte, assuming that it did not
affect the differentiation process. The development of stabilized siRNA
(i.e., Stealth™ RNAi, Invitrogen) will help to alleviate some of these
problems in the future. To express proteins and siRNA in mature 3T3-L1
adipocytes, both electroporation (Li and Kandror, 2005) and viral-mediated
transfection (Carlotti et al., 2004; Liao et al., 2006) have been successfully
employed.

4.1. Electroporation of 3T3-L1 adipocytes

Electroporation of adipocytes was adapted from the method of Li and
Kandror (2005). 3T3-L1 adipocytes are grown and differentiated in
10-cm dishes and used 8 days after the initiation of differentiation. After
washing with phosphate buffered saline (PBS), cells are detached from
the dish using 3 ml TryplExpress© (Gibco), and trypsin is removed by
washing twice with PBS using centrifugation at 1200 rpm for 5 min. The
final cell pellet is resuspended in 0.5 ml PBS and transferred to a 4-mm
GenePulser™ electroporation cuvette (BioRad) containing 100 μg DNA.
Electroporation is performed using the BioRad GenePulser II at 960 μF,
0.16 kV, which should generate a time constant of ~19.0 ms for a successful
electroporation. The cells are subsequently transferred into a 15-ml tube
containing growth media and re-plated onto dishes or coverslips for

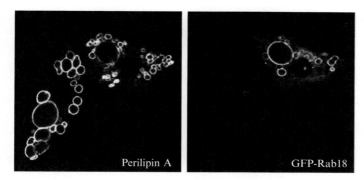

Figure 8.3 Electroporation and expression of GFP-Rab18 in 3T3-L1 adipocytes. GFP-Rab18 was expressed in 3T3-L1 adipocytes by electroporation 8 days after the initiation of differentiation. After 24 h, the cells were fixed and colabeled for perilipin A. GFP-Rab18 was present in a subset of LDs and in the endoplasmic reticulum.

experimentation. Using this technique, GFP-tagged Rab18 has been successfully expressed in 3T3-L1 adipocytes (Fig. 8.3). Due to the high percentage of nonviable cells following electroporation, one 10-cm dish generates only one 6-cm dish of electroporated cells.

4.2. Lentiviral expression of proteins in adipocytes

While electroporation is able to deliver DNA and siRNAs into adipocytes, the efficiency of delivery is low compared with other cell types. Alternative systems that can be used include viral mediated delivery of DNA and shRNA constructs. Previous studies utilizing viral vectors such as adenovirus are relatively inefficient, presumably due to a deficiency in receptors for the virus. More recently lentiviral systems have been successfully employed for the stable expression of transgenic proteins (Carlotti *et al.*, 2004) and shRNA (Liao *et al.*, 2006) in 3T3-L1 adipocytes. Using this technique, a high transfection efficiency and moderate level of expression can be achieved in 3T3-L1 adipocytes with little to no cytotoxicity (Carlotti *et al.*, 2004). Lentiviral vectors consist of a backbone containing the gene of interest. Generation of the viral particles can be achieved by coexpression of the lentiviral vector with helper plasmids (Carlotti *et al.*, 2004), or cotransfection with the ViraPower™ packaging mix into the ViraPower™ 293FT™ producer cell line (Invitrogen). For viral transduction, viral supernatants are added to the cell medium together with 8 μg/ml polybrene overnight. The next day the medium is replaced with fresh medium and the cells maintained in culture for 3 to 6 days prior to experimentation. The appropriate titer of virus expressing GFP-tagged proteins such as GFP-Rab18 can be determined by titration of viral supernatants onto cells of interest and manual scoring of positive (green) cells by fluorescence microscopy.

5. BIOCHEMICAL ISOLATION OF LDS

The high content of neutral lipids in lipid droplets makes them easy to isolate by subcellular fractionation and gradient centrifugation from cells with a large number of LDs, either adipocytes or nonadipocytes treated with oleic acid (one 10-cm dish is sufficient). To increase the yield from cells with only a small number of LDs, a large number of cells are needed. The method of cell disruption used to release LDs varies enormously among laboratories. In our lab, we use sonication or shearing through small-bore needles (25 to 30 g). Alternatively, nitrogen cavitation can be used (Yu et al., 2000). The differences among these methods have not been rigorously examined with respect to yield, purity, or the effects on LD structure.

To isolate LDs from 3T3-L1 adipocytes, we use a modification of the method by Yu et al. (2000). Cells are washed twice using ice-cold dissociation buffer (25 mM Tris-HCl, pH 7.4, 100 mM KCl, 1 mM EDTA, 5 mM EGTA), and subsequently scraped from the dish in 250-μl dissociation buffer containing protease inhibitors (250 μM PMSF, 10 μg/ml aprotinin, 10 μg/ml leupeptin) and phosphatase inhibitors (10 mM NaF, 1 mM NaVO$_3$, 1 mM Na pyrophosphate). After transferring to a microfuge tube on ice, cells are disrupted by sonication using a Branson Sonifier 250 (level 1 constant output), twice for 5 s, cooling on ice between each sonication step. The sonicate is microcentrifuged at 1500×g (4000 rpm), 10 min, 4 °C to pellet cell debris and nuclei. The supernatant (250 μl) is transferred to a fresh eppendorf tube and mixed with an equal volume of 1.08 M sucrose (37.02% w/v). Sucrose is prepared as more than 80% stock solution in UHP water; the concentration is measured using a refractometer, and adjusted to give exactly 80%. The supernatant/sucrose mix is transferred to a 2.2-ml ultracentrifuge tube (Beckman, TLS55) and overlaid sequentially with 700 μl 0.27 M (9.31% w/v) sucrose, 500 μl 0.135 M (4.5% w/v) sucrose and 500 μl Top Solution (25 mM Tris-HCl, pH 7.4, 1 mM EDTA, 1 mM EGTA). Sucrose dilutions are prepared in the dissociation buffer. The gradient is centrifuged at 150,500×g (Beckman, TLS55, 47,000 rpm) for 60 min and 200 μl fractions collect from the top. The LD fraction floats to the very top of the gradient, and care is needed in removing it efficiently. Isolated LDs can be subsequently analyzed using either biochemical measurements (i.e., Western blotting) or fixed directly and analyzed by light or electron microscopy as described in later sections.

6. FLUORESCENCE MICROSCOPY

6.1. LD identification by light microscopy

LDs can be readily detected by light microscopy using either the intrinsic properties of the lipid core in bright field, differential interference contrast (DIC) or phase contrast microscopy or by staining the neutral lipid core using a range of fluorescent and light microscopic stains (Table 8.1). In our experience DIC works well to detect LDs in a number of cell systems and can be coupled to fluorescence microscopy if the microscope system available allows this. Neutral lipid stains also work very well, although with the exception of Bodipy 493/503, they have broad excitation and emission spectra that can complicate detection if the analysis of multiple fluorophores is required. Oil red O is a commonly used LD stain that can be imaged by both bright-field and epifluorescence microscopy, although there is some disagreement in the literature regarding possible effects on LD structure (Andersson *et al.*, 2006; Fukumoto and Fujimoto, 2002; Koopman *et al.*, 2001). Nile red also stains LDs, although this is unstable under intense light and will photo bleach rapidly. Paradoxically, this stain can be used to advantage if a second, more photo-stable fluorophore is also under analysis. In our laboratory, we have made extensive use of the Zeiss LSM510 Meta detector in order to more accurately separate Nile red fluorescence from fluorescent proteins such as GFP. As a final comment on Nile red (and to a lesser extent other neutral lipid stains), the basic property of these molecules is that their fluorescence increases in a hydrophobic environment. Under certain experimental conditions, significant fluorescence can be detected in lysosomes and other organelles. In addition, background fluorescence will increase if true LDs are absent from the cell. New users should use caution in interpreting data obtained using these stains.

6.2. Fixation and indirect immunolabeling of LD proteins

The preservation of LDs for light microscopy can be particularly prone to difficulties. Several studies have detailed appropriate fixation and permeabilization conditions to preserve the LDs in their native state (DiDonato and Brasaemle, 2003; Ohsaki *et al.*, 2005). The use of organic solvents must be avoided as these can dissolve the lipid components and alter the surface of the LDs (DiDonato and Brasaemle, 2003). Fixation by paraformaldehyde (3 to 5%) in PBS followed by permeabilization using either 0.1% saponin (Martin *et al.*, 2005; Ohsaki *et al.*, 2005) or 0.01% digitonin (Ohsaki *et al.*, 2005) permits optimal labeling of LDs while preserving LD structure. If 0.1% Triton X-100 is used as a permeabilization agent, paraformaldehyde should be supplemented with small amounts of glutaraldehyde (0.01 to 0.025%) during fixation (Ohsaki *et al.*, 2005). Even following optimal procedures,

Table 8.1 Neutral lipid cell stains

Stain	Chemical Name/ Synonyms	Solubilisation	Working dilution	Ex/Em	References
Nile Red	9-diethylamino-5H-benzo[α] phenoxazine-5-one	Saturated solution in acetone	0.01–0.1%	410–590/ 500–800 nm	Greenspan and Fowler, (1985) Greenspan, Mayer and Fowler, (1985)
Bodipy493/503© (Invitrogen)	4,4,-difluoro-1,3,5,7,8, -pentamethyl-4-bora-3a,4a-diaza-sindacene	Saturated solution in ethanol	0.2–0.5% on fixed cells 0.025% on live cells	488/503 nm	Gocze and Freeman, (1994)
Oil Red O	Solvent Red27 Sudan Red 5B	60% isopropanol	0.2%	Ex540–580 nm	Andersson et al., (2006)

it is still possible to detect alterations in the surface of the LDs using some proteins. One clear example of this is the morphology of LDs containing GFP-Rab18 in fixed cells compared to live cells (see figure 1 in Martin *et al.*, 2005). Whereas the GFP-Rab18 completely surrounds the LD in the live cell, it shows only partial enwrapping in fixed cells. These alterations are protein specific (i.e., not detected with YFP-Cav3^{DGV}, for example) and should be kept in mind when analyzing LD association.

In order to immunolabel LDs, cells grown on coverslips are briefly washed once in PBS and fixed for a minimum of 30 min in 4% paraformaldehyde in PBS at room temperature. Following fixation, cells are permeabilized with 0.1% saponin in PBS (made up freshly from powder) for 10 min, quenched in 50 mM NH$_4$Cl in PBS for 10 min, and blocked in 0.2% BSA/0.2%, cold-water, fish-skin gelatin in PBS for 10 min (blocking solution). Primary antibodies are applied in blocking solution at an appropriate dilution for 45 min to 1 h at room temperature. Coverslips are subsequently washed three times for 2 min each in PBS and incubated with secondary fluorescently conjugated antibodies in blocking solution for 30 to 45 min. If a neutral lipid stain is to be included in the labeling procedure, this is added with the secondary antibody. Coverslips are again washed three times for 2 min each in PBS, once briefly in UHP water, and mounted in Mowiol. The Mowiol is dried at 37 °C prior to imaging. Cells prepared in this way can be stored at 4 °C for several weeks. Using confocal microscopy, LDs can be imaged in three dimensions (3D) and volume measurements made. For details on the volume quantitation of LDs in 3D, see the work of Bostrom and colleagues (Andersson *et al.*, 2006; Bostrom *et al.*, 2005).

6.3. Immunolabeling of isolated LD fractions

LDs prepared by gradient centrifugation as described above (section 5) can be directly applied to coverslips and labeled for light or fluorescence microscopy. Following removal from the gradient, the LD fraction (or a portion of it) is mixed with an equal volume of 8% PFA in PBS and fixed at room temperature for 30 min. For long-term storage, fixed fractions are maintained at 4 °C. To attach LDs to glass coverslips, the coverslips (12 mm^2) are first coated with 0.05% poly L-lysine for 30 min, washed with UHP water and dried. Drops of the LD fraction (20 μl) are placed onto clean parafilm and the coverslips floated on top for 10 min at room temperature. After washing with PBS, labeling and/or staining can be carried out as described above for whole cells (section 6.2).

6.4. Real-time video microscopy

The formation, motility, and interaction of LDs with other organelles can be detected by expression of GFP-Rab18 (Martin *et al.*, 2005) or YFP-ADRP (Targett-Adams *et al.*, 2003), as well as using neutral lipid stains (i.e., Nile Red [Pol *et al.*, 2004; Valetti *et al.*, 1999], Bodipy 493/503) or DIC. Cells for real-time microscopy are plated onto glass bottomed tissue culture dishes (MatTek Corp.) and transferred into CO_2-independent medium (Invitrogen) supplemented with 0.1% fatty-acid–free BSA (Calbiochem) in the presence or absence of 350 μM oleic acid. Time series are collected at 37 °C using an Axiovert 200 M SP LSM 510 META, confocal, laser scanning microscope equipped with a heated stage and a heated 100× oil immersion objective. As prolonged imaging can damage the cells, real-time data collection is limited to a maximum of 1.5 h, and laser power is kept to a minimum. Time series images are collected using a 488-nm excitation laser line at less than 20% maximum power using the Zeiss LSM510 Meta software for GFP or YFP. Images are converted to 8-bit TIFF files and further analyzed using ImageJ software (National Institutes of Health, Bethesda, MD). Quick-time movies can be assembled using ImageJ 1.33, and still images can be compiled using Adobe Photoshop 7.0.

7. ELECTRON MICROSCOPY

Electron microscopy of LDs has added fundamentally to our understanding of their structure, formation, and interaction with other organelles. Standard resin embedded fixed sections and more recently high pressure frozen, fixed, embedded sections demonstrate the extensive network of LD associated membranes distinct from the LD phospholipid monolayer (Martin and Parton, 2006). Despite advances in light microscopy, electron microscopy remains the only technique with sufficient resolution to unequivocally identify whether putative lipid droplet-associated proteins associate with the lipid droplet core, surface or with lipid droplet-associated membranes. Here we describe the detailed methods used for Rab18 localization on ultrathin sections of cultured cells based on the methods of Tokuyasu (Liou *et al.*, 1996; Tokuyasu, 1980). Solutions needed include 2% methylcellulose and PVP-sucrose.

A 2% solution of methylcellulose (25 centipoise, Sigma M6385) is prepared by stirring the powder into hot water. Transfer onto ice and continue stirring until the solution cools, and then stir at 4 °C overnight. The solution is stored at 4 °C for several days to make sure that it is fully dissolved. Centrifuge at high speed (e.g., 60,000 rpm in a Beckman 60Ti or 70Ti rotor) for 60 min at 4 °C. The supernatant can either be removed from over the pellet, or the tubes can be stored in the refrigerator for 2 to 3 weeks. The methylcellulose should be removed without disturbing the pellet at the bottom of the tube.

For 20 ml of 15% PVP, weigh 3 g of polyvinylpyrollidone (Sigma, MW 10,000) into a 50-ml tube and make a paste with 0.6 ml of 1.1 M Na$_2$CO$_3$.

To this add 17 ml of 2 M sucrose in 0.1 M phosphate buffer, pH 7.4. Mix gently, cover, and leave overnight at room temperature to allow the minute air bubbles in the mixture to escape into the air, leaving behind a clear solution. Store at 4 °C or in aliquots at –20 °C.

7.1. Fixation and cryoprotection

Fixation for immunoelectron microscopy (immunoEM) is performed using either 0.1% glutaraldehyde/4% PFA in PHEM buffer (25 mM HEPES, 10 mM EGTA, 60 mM PIPES, 2 mM MgCl$_2$, pH 6.9) for 1 h at RT or using 8% PFA in PHEM buffer for 24 h. The precise fixative required depends on the accessibility and stability of the epitope and has to be determined for each individual protein and antibody by trial and error. After fixation samples (tissue or adherent cells) are cryoprotected using 2.3- or 2.6-M sucrose in 0.1 M phosphate buffer, pH 7.4, or in PVP-sucrose.

For cryoprotection, adipose tissue pieces can be trimmed into small (~1 mm^3) pyramid-shaped pieces and placed directly in cryoprotectant. Adherent cells are washed with 0.1 M phosphate buffer, pH 7.4, and scraped gently into 1% gelatin in 0.1 M phosphate buffer, pH 7.4. Use a minimum of one 6-cm diameter dish to obtain a good pellet. Scraped cells are transferred to microcentrifuge tubes and pelleted (10,000 rpm, 10 s). The supernatant is removed and the cell pellet resuspended in warm 10% gelatin. Repeat the centrifugation once and cool the final pellet in 10% gelatin prior to trimming into blocks and cryoprotection as above. Leave the tissue or cells in cryoprotectant for a minimum of 2 h at room temperature (non–gelatin-embedded samples) or overnight at 4 °C. Blocks for sectioning are mounted onto the specimen stub, excess sucrose removed with a filter paper and the specimen stub with block quickly immersed in liquid N$_2$. Samples can be stored in liquid N$_2$ indefinitely.

7.2. Cryosectioning

Precool the chamber of an ultracryomicrotome to –80 °C and trim blocks prior to sectioning using either a diamond trimming tool or a glass knife. Using a trimming tool or glass knife, carefully advance the cutting edge to the block face and commence facing the block. This is done by taking off sections (100 to 300 μm thick) from the block surface until it is flat. Sections that are too thick can cause fractures resulting in an unpolished surface or even loss of the block. Once the surface is flat, the sides can be trimmed so as to give the block a square or oblong shape. To commence sectioning allow the chamber and knife to equilibrate to –120 °C. Arrange cut sections with a dog hair (Dalmatian is particularly good) or an eyelash to allow knife to section cleanly and to arrange sections for pick-up. An important modification for studies of LDs in cultured cells was the preparation of relatively thick sections of 100 nm as well as the more usual 60-nm thickness. This modification

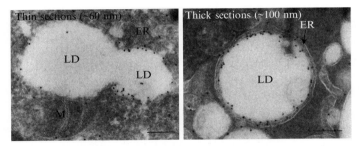

Figure 8.4 ImmunoEM of GFP-Rab18. GFP-Rab18 was expressed in Vero cells for 24 h, cells fixed in 2% PFA/0.2% glutaraldehyde in PHEM buffer, and processed for immunoEM as described in section 7. Cryosections were immunolabeled using anti-GFP antibodies and labeling detected using 10-nm protein A gold. In the thin sections, both labeled and unlabeled LDs are present. However, the discrete boundary of each LD was not preserved resulting in a fused structure. Additionally, the LD surface-associated membranes, including endoplasmic reticulum, are difficult to detect. In thicker sections, the LD structure and the encircling ER are structurally intact although in these sections smaller membrane structures are no longer detected.

provided greatly improved preservation of highly labile lipid droplets in aldehyde-fixed tissues (Fig. 8.4) (see also Martin *et al.*, 2005). Remove sections from the knife using a loop containing a drop of 1:1 mix of methylcellulose/2.3 M sucrose. Allow the drop to thaw in the loop and apply to grid (still attached to support). Label immediately or store at 4 °C for 24 to 48 h.

7.3. Immunolabeling

All procedures are performed on drops on parafilm at room temperature, ensuring that the reverse side of the grid remains dry. Wash the grids twice for 1 min each in PBS, and quench five times, 1 min each, in 50 mM glycine in PBS. Block the grids for 10 min in PBS containing 0.1% BSA and 0.2% fish skin gelatin. Dilute the primary antibody in blocking solution then spin 1 min in a microfuge to remove any aggregates. Incubate each grid on a 5 μl drop for 30 min. Wash with PBS containing 0.1% BSA four times, 5 min each, and incubate each grid on 5-μl Protein A Gold in blocking solution for 20 min. Wash with PBS five times at 5 min each, and fix on 1% glutaraldehyde in PBS for 5 min. Wash with UHP water seven times at 2 min each, and ensure that no PBS is carried along during washing steps by rinsing the forceps, as it is essential to remove salts that will precipitate uranyl.

7.4. Contrasting and drying

To contrast the sections mix 2% methylcellulose with a 4% aqueous uranyl acetate solution to give a final concentration of 0.4% uranyl acetate. Put drops of this solution onto a clean surface (parafilm) on ice.

Take immunolabeled grids from last water wash step above and float the grids, sections down , on the drops for 10 min. Loop off each grid individually from the methylcellulose-uranyl acetate solution using a large wire loop (formed around a 3.5-mm–diameter electric drill bit). Remove excess liquid from the loop by dragging the loop over the surface of a filter paper. Note that the final thickness of the methylcellulose is critical for the production of contrast and fine structure preservation. If this layer is too thin, artifacts can be introduced from air drying, whereas a too thick layer will reduce contrast. Allow the grid to dry in the loop. When the grids are dry ($\sim$10 min) they can be removed carefully from the loops using fine forceps and viewed in a transmission electron microscope.

7.5. EM Examination and evaluation

In addition to determination of the localization of LD proteins in cryosections of adipocyte and nonadipocyte cell lines, stereological techniques can be used to rapidly estimate the volume density (Vv) or volume fraction of an object that is embedded in a larger structure in both cryosections and resin-embedded sections. In this case, the object of interest is the LD, which can be related to the volume of the cytoplasm to measure changes in lipid droplet formation or catabolism. This analysis relies on the fact that the larger the volume of the lipid droplets, the greater the chance that random sections will contain profiles of lipid droplets. In fact, the volume density of the object of interest is obtained simply by measuring the ratio of the total profile area of the object in the sections relative to the total area of the reference space in the sections. Digital capture of images with a CCD camera and immediate "on screen" analysis offers a simple and efficient way of quantitating lipid droplet volume under different experimental conditions. Random sections of plastic embedded cells are analyzed by overlaying a grid over the sections and relating points over the lipid droplets, recognized by their characteristic morphology, to the cytoplasmic volume. This provides an estimate of the volume of the lipid droplets relative to the cytoplasmic volume. Ideally the cytoplasmic volume should be known (or measured by independent techniques) so that the absolute volume of the lipid droplets can be determined from the Vv. Alternatively, the investigator should ascertain by some independent means that the cytoplasmic volume is the same under the different experimental conditions. For further reading on sampling, interpretation, and other stereological techniques, see Griffiths' *Fine Structure Immunocytochemistry* (1993).

To perform a stereological analysis on LDs in resin embedded sections, prepare ultrathin sections and stain with uranyl acetate and lead citrate if required. View grids in the TEM (Jeol 1011 equipped with Morada cooled-CCD camera (Olympus Soft Imaging Solutions Gmbh, Germany) with AnalySIS software (iTEM Soft Imaging Systems, Olympus) at appropriate magnification (e.g., 12,000×). Capture images at random, for example, by moving sample by a set distance in one direction. Grids of different sizes can be preset as macros in AnalySIS. After capturing or freezing each image (or in live mode), overlay the appropriate grid (e.g., 1000 nm × 1000 nm) onto the image. Count points lying over the cytoplasm and those over LD profiles. If necessary, a double-lattice grid of two different grid distances can be used so that larger areas can be sampled more efficiently (in this case, the points over the larger reference space would be multiplied by the difference in area of the two grid sizes). Determine Vv by measuring the ratio of points over the lipid droplets to points over the cytosol. This can be expressed as a percentage of cytoplasmic volume occupied by lipid droplets. At least three independent experiments should be quantitated in this fashion to obtain a mean and standard error.

7.6. ImmunoEM of isolated LD fractions

Whole mounts of isolated LDs can be used to determine purity of LD preps and to label LD proteins that are difficult to label in cell sections. Fixation can be performed on all or part of the LD fraction following removal from the gradient by mixing with an equal volume of 8% PFA in PBS at room temperature for 30 min (this can also be stored for several weeks at 4 °C). Poly-L-Lysine, carbon- and formvar-coated copper grids (100 mesh hexagonal) can then be floated on small drops (5 μl) of the fixed fraction for 10 min. Alternatively, poly-L-Lysine, carbon-coated, formvar-coated copper grids (100-mesh hexagonal) can be floated on 5-μl drops of the LD fraction directly from the gradient for 5 min, followed by floatation on drops of 4% paraformaldehyde for 30 min. In either case, grids containing LDs are subsequently washed three times with 0.15 M glycine in PBS (2 min each) then labeled as described above for cell and tissue sections prior to contrasting and drying in methylcellulose/uranylacetate. It should be noted that this infrequently used method is extremely simple, does not require sectioning, and can be used to check any fractions for purity as long as the antigens to be detected are cytoplasmically exposed.

8. SUMMARY

The mechanisms by which lipids are moved around cells, stored, and released in response to physiological needs are relatively poorly characterized in comparison to protein trafficking pathways. This is reflected in our fairly superficial understanding of the lipid droplet, one of the least characterized cellular organelles. Yet an understanding of lipid droplets is vital for numerous diseases state and health conditions. The surge of interest in lipid droplets in recent years has provided rapid insights into the protein constituents of the lipid droplet and how this is modulated under different conditions. The regulated association of Rab18 with the lipid droplet surface indicates an important role of this small GTPase in lipid mobilization during lipolysis, while the fairly ubiquitous expression of Rab18 in various tissues indicates that the function of Rab18 is not restricted to lipid storage tissues. Together with existing knowledge of the mechanism of Rab protein function in other systems, the methods described here should help to improve our understanding of Rab18 function and lipid droplet cell biology, in health and in disease.

ACKNOWLEDGMENTS

The authors acknowledge the support of the National Health and Medical Research Council of Australia, the Diabetes Australia Research Trust, and the National Institutes of Health, United States. We are grateful to Cynthia Corley-Mastick, Lars Kuerschner, and Samantha Murphy for invaluable discussions during the preparation of this manuscript. Confocal microscopy was performed at the ACRF/IMB Dynamic Imaging Facility for Cancer Biology, established with funding from the Australian Cancer Research Foundation. The Institute for Molecular Bioscience is a Special Research Centre of the Australian Research Council.

REFERENCES

Andersson, L., Bostrom, P., Ericson, J., Rutberg, M., Magnusson, B., Marchesan, D., Ruiz, M., Asp, L., Huang, P., Frohman, M. A., Boren, J., and Olofsson, S. O. (2006). PLD1 and ERK2 regulate cytosolic lipid droplet formation. *J. Cell Sci.* **119,** 2246–2257.

Bostrom, P., Rutberg, M., Ericsson, J., Holmdahl, P., Andersson, L., Frohman, M. A., Boren, J., and Olofsson, S. O. (2005). Cytosolic lipid droplets increase in size by microtubule-dependent complex formation. *Arterioscler. Thromb. Vasc. Biol.* **25,** 1945–1951.

Brasaemle, D. L., Barber, T., Kimmel, A. R., and Londos, C. (1997). Post-translational regulation of perilipin expression. Stabilization by stored intracellular neutral lipids. *J. Biol. Chem.* **272,** 9378–9387.

Brasaemle, D. L., Levin, D. M., Adler-Wailes, D. C., and Londos, C. (2000). The lipolytic stimulation of 3T3-L1 adipocytes promotes the translocation of hormone-sensitive lipase to the surfaces of lipid storage droplets. *Biochim. Biophys. Acta* **1483,** 251–262.

Carlotti, F., Bazuine, M., Kekarainen, T., Seppen, J., Pognonec, P., Maassen, J. A., and Hoeben, R. C. (2004). Lentiviral vectors efficiently transduce quiescent mature 3T3-L1 adipocytes. *Mol. Ther.* **9,** 209–217.

Cham, B. E., and Knowles, B. R. (1976). A solvent system for delipidation of plasma or serum without protein precipitation. *J. Lipid Res.* **17,** 176–181.

DiDonato, D., and Brasaemle, D. L. (2003). Fixation methods for the study of lipid droplets by immunofluorescence microscopy. *J. Histochem. Cytochem.* **51,** 773–780.

Frost, S. C., and Lane, M. D. (1985). Evidence for the involvement of vicinal sulfhydryl groups in insulin-activated hexose transport by 3T3-L1 adipocytes. *J. Biol. Chem.* **260,** 2646–2652.

Fukumoto, S., and Fujimoto, T. (2002). Deformation of lipid droplets in fixed samples. *Histochem. Cell Biol.* **118**, 423–428.

Goldstein, J. L., Basu, S. K., and Brown, M. S. (1983). Receptor-mediated endocytosis of low-density lipoprotein in cultured cells. *Methods Enzymol.* **98**, 241–260.

Green, H., and Kehinde, O. (1975). An established preadipose cell line and its differentiation in culture. II. Factors affecting the adipose conversion. *Cell* **5**, 19–27.

Griffiths, G. (1993). In "Fine structure immunocytochemistry." Berlin: Springer-Verlag, Berlin.

Koopman, R., Schaart, G., and Hesselink, M. K. (2001). Optimisation of oil red O staining permits combination with immunofluorescence and automated quantification of lipids. *Histochem. Cell Biol.* **116**, 63–68.

Leung, K. F., Baron, R., Ali, B. R., Magee, A. I., and Seabra, M. C. (2007). Rab GTPases containing a CAAX motif are processed post-geranylgeranylation by proteolysis and methylation. *J. Biol. Chem.* **282**, 1487–1497.

Li, L. V., and Kandror, K. V. (2005). Golgi-localized, gamma-ear-containing, Arf-binding protein adaptors mediate insulin-responsive trafficking of glucose transporter 4 in 3T3-L1 adipocytes. *Mol. Endocrinol.* **19**, 2145–2153.

Liao, W., Nguyen, M. T., Imamura, T., Singer, O., Verma, I. M., and Olefsky, J. M. (2006). Lentiviral short hairpin ribonucleic acid-mediated knockdown of GLUT4 in 3T3-L1 adipocytes. *Endocrinology* **147**, 2245–2252.

Liou, W., Geuze, H. J., and Slot, J. W. (1996). Improving structural integrity of cryosections for immunogold labeling. *Histochem. Cell Biol.* **106**, 41–58.

Marcinkiewicz, A., Gauthier, D., Garcia, A., and Brasaemle, D. L. (2006). The phosphorylation of serine 492 of perilipin a directs lipid droplet fragmentation and dispersion. *J. Biol. Chem.* **281**, 11901–11909.

Martin, S., Driessen, K., Nixon, S. J., Zerial, M., and Parton, R. G. (2005). Regulated localization of Rab18 to lipid droplets: effects of lipolytic stimulation and inhibition of lipid droplet catabolism. *J. Biol. Chem.* **280**, 42325–42335.

Martin, S., and Parton, R. G. (2005). Caveolin, cholesterol, and lipid bodies. *Semin. Cell Dev. Biol.* **16**, 163–174.

Martin, S., and Parton, R. G. (2006). Lipid droplets: a unified view of a dynamic organelle. *Nat. Rev. Mol. Cell Biol.* **7**, 373–378.

Moore, H. P., Silver, R. B., Mottillo, E. P., Bernlohr, D. A., and Granneman, J. G. (2005). Perilipin targets a novel pool of lipid droplets for lipolytic attack by hormone-sensitive lipase. *J. Biol. Chem.* **280**, 43109–43120.

Murphy, D. J. (2001). The biogenesis and functions of lipid bodies in animals, plants and microorganisms. *Prog. Lipid Res.* **40**, 325–438.

Ohsaki, Y., Maeda, T., and Fujimoto, T. (2005). Fixation and permeabilization protocol is critical for the immunolabeling of lipid droplet proteins. *Histochem. Cell Biol.* **124**, 445–452.

Ozeki, S., Cheng, J., Tauchi-Sato, K., Hatano, N., Taniguchi, H., and Fujimoto, T. (2005). Rab18 localizes to lipid droplets and induces their close apposition to the endoplasmic reticulum-derived membrane. *J. Cell Sci.* **118**, 2601–2611.

Pol, A., Martin, S., Fernandez, M. A., Ferguson, C., Carozzi, A., Luetterforst, R., Enrich, C., and Parton, R. G. (2004). Dynamic and regulated association of caveolin with lipid bodies: modulation of lipid body motility and function by a dominant negative mutant. *Mol. Biol. Cell* **15**, 99–110.

Rodbell, M. (1964). Metabolism of Isolated Fat Cells. I. Effects of hormones on glucose metabolism and lipolysis. *J. Biol. Chem.* **239**, 375–380.

Rothblat, G. H., Arbogast, L. Y., Ouellette, L., and Howard, B. V. (1976). Preparation of delipidized serum protein for use in cell culture systems. *In Vitro* **12**, 554–557.

Seabra, M. C., and Wasmeier, C. (2004). Controlling the location and activation of Rab GTPases. *Curr. Opin. Cell Biol.* **16,** 451–457.

Targett-Adams, P., Chambers, D., Gledhill, S., Hope, R. G., Coy, J. F., Girod, A., and McLauchlan, J. (2003). Live cell analysis and targeting of the lipid droplet-binding adipocyte differentiation-related protein. *J. Biol. Chem.* **278,** 15998–16007.

Tauchi-Sato, K., Ozeki, S., Houjou, T., Taguchi, R., and Fujimoto, T. (2002). The surface of lipid droplets is a phospholipid monolayer with a unique fatty acid composition. *J. Biol. Chem.* **277,** 44507–44512.

Tokuyasu, K. T. (1980). Immunochemistry on ultrathin frozen sections. *Histochem. J.* **12,** 381–403.

Valetti, C., Wetzel, D. M., Schrader, M., Hasbani, M. J., Gill, S. R., Kreis, T. E., and Schroer, T. A. (1999). Role of dynactin in endocytic traffic: effects of dynamitin overexpression and colocalization with CLIP-170. *Mol. Biol. Cell* **10,** 4107–4120.

Yu, W., Cassara, J., and Weller, P. F. (2000). Phosphatidylinositide 3-kinase localizes to cytoplasmic lipid bodies in human polymorphonuclear leukocytes and other myeloid-derived cells. *Blood* **95,** 1078–1085.

Zerial, M., and McBride, H. (2001). Rab proteins as membrane organizers. *Nat. Rev. Mol. Cell Biol.* **2,** 107–117.

Zhang, J., Hupfeld, C. J., Taylor, S. S., Olefsky, J. M., and Tsien, R. Y. (2005). Insulin disrupts beta-adrenergic signalling to protein kinase A in adipocytes. *Nature* **437,** 569–573.

CHAPTER NINE

ANALYSIS ON THE EMERGING ROLE OF RAB3 GTPASE-ACTIVATING PROTEIN IN WARBURG MICRO AND MARTSOLF SYNDROME

Ayuko Sakane,* Jun Miyoshi,[†] Yoshimi Takai,[‡] *and* Takuya Sasaki*

Contents

Abstract

Evidence is accumulating that Rab3A plays a key role in neurotransmitter release and synaptic plasticity. Recently mutations in the catalytic subunit p130 and the noncatalytic subunit p150 of Rab3 GTPase–activating protein were found to cause Warburg Micro syndrome and Martsolf syndrome, respectively, both of which exhibit mental retardation. We have found that loss of p130 in mice results in inhibition of Ca^{2+}-dependent glutamate release from cerebrocortical synaptosomes and alters short-term plasticity in the hippocampal CA1 region, probably through the accumulation of the GTP-bound form of Rab3A. Here, we describe the procedures for the measurement of the GTP-bound pool of Rab3A with pull-down assay using mouse brains and the biochemical method for the measurement of glutamate release from mouse synaptosomes.

* Department of Biochemistry, Institute of Health Biosciences, The University of Tokushima Graduate School, Tokushima, Japan
† Department of Molecular Biology, Osaka Medical Center for Cancer and Cardiovascular Diseases, Osaka, Japan
‡ Division of Molecular and Cellular Biology, Department of Biochemistry and Molecular Biology, Kobe University Graduate School of Medicine/Faculty of Medicine, Kobe, Japan

Methods in Enzymology, Volume 438
ISSN 0076-6879, DOI: 10.1016/S0076-6879(07)38009-9

1. INTRODUCTION

The Rab family small G proteins have been regarded as principal classes of GTPases that play an important role in regulation of intracellular vesicle transport (Takai *et al.*, 2001; Zerial and McBride, 2001). The Rab3 subfamily, which consists of Rab3A, -3B, -3C, and -3D, is associated with secretory granules or vesicles, and plays a crucial role in regulated exocytosis (Geppert and Südhof, 1998; Takai *et al.*, 1996, 2001).

Rab3A is the most abundant isoform in the brain, where it is localized on synaptic vesicles. A detailed electrophysiological analysis of Rab3A-deficient mice revealed that Rab3A is not essential for synaptic transmission, but performs a modulatory function that acts at the Ca^{2+}-triggered fusion step (Südhof, 2004). Rab3A is also required for hippocampal CA3 mossy fiber long-term potentiation (LTP) (Südhof, 2004). A recent study on the Rab3 single, double, triple and quadruple knockout (KO) mice reported that all single and double KO mice are viable and fertile, whereas quadruple KO mice die due to respiratory failure shortly after birth (Schluter *et al.*, 2004). Most triple KO mice are unable to survive whenever Rab3A is one of the three deleted proteins. Furthermore, analysis of transmitter release at embryonic autaptic cultures from the quadruple-deficient hippocampal neurons revealed a decrease of about 30% in evoked transmitter release due to a decrease in release probability, which is not observed in Rab3A single KO mice.

Like other Rab family members, Rab3A cycles between the GDP-bound inactive and GTP-bound active forms (Geppert and Südhof, 1998; Takai *et al.*, 1996, 2001). This cycling is regulated by three types of regulators: Rab GDP dissociation inhibitor (GDI), Rab3 GDP/GTP exchange protein (GEP), and Rab3 GTPase–activating protein (GAP) (Geppert and Südhof, 1998; Takai *et al.*, 1996, 2001). Because the cyclical activation is coupled with membrane association and allows both spatial and temporal control of Rab3A activity, these regulators are thought to be important for the proper functioning of Rab3A in synaptic vesicle transport. The significance of Rab GDIα, a neuron-specific isoform, and Rab3 GEP in synaptic transmission and plasticity has been shown by the knockout studies in mice and identification of mutations in the Rab GDIα that cause human X-linked nonspecific mental retardation (D'Adamo *et al.*, 1998; Ishizaki *et al.*, 2000; Tanaka *et al.*, 2001; Yamaguchi *et al.*, 2002).

Rab3 GAP, which is specific for the Rab3 subfamily, consists of two subunits: the catalytic subunit p130 and the noncatalytic subunit p150 (Fukui *et al.*, 1997; Nagano *et al.*, 1998). Rab3 GAP is ubiquitously expressed and enriched in the synaptic soluble fraction of brain (Oishi *et al.*, 1998). We have generated mice lacking p130 and shown that GTP-Rab3A accumulates in p130-deficient mouse brains (Sakane *et al.*, 2006). We have also shown that loss of p130 in mice results in the inhibition of Ca^{2+}-dependent glutamate release from cerebrocortical synaptosomes and altered short-term plasticity in the hippocampal CA1 region. These results suggest that Rab3 GAP contributes to maintaining a limiting amount of GTP–Rab3A by stimulating GTP hydrolysis during the modulation of synaptic transmission and plasticity.

The significance of Rab3 GAP in neural function has been also shown by the human diseases. Aligianis *et al.* (2005) have reported that Warburg Micro syndrome, caused by mutations of Rab3 GAP p130 gene, is a rare autosomal recessive disorder that results in ocular defects, cerebral malformations, severe mental retardation, and gonadal hormonal dysfunction. More recently, mutations in Rab3 GAP p150 have been detected in the related Martsolf syndrome, which shares clinical features but is a milder disorder (Aligianis *et al.*, 2006). This chapter describes the procedures for the measurement of the GTP-bound pool of Rab3A with pull-down assay from mouse brains. It also shows the method used for the measurement of glutamate release from mouse synaptosomes.

2. MATERIALS

(p-Amidinophenyl)methanesulfonyl fluoride (APMSF), reduced glutathione, $NADP^+$, fura-2-acetoxymethyl ester, L–glutamate dehydrogenase, leupeptin, and isopropyl-β-D-thiogalactopyranoside (IPTG) are from Wako Pure Chemicals Industries(Osaka, Japan). Nonidet P-40 is from Nakalai Tesque (Kyoto, Japan). Dulbecco's modified Eagle's medium (DMEM) and fetal bovine serum (FBS) are from Sigma (St. Louis, MO). Hybridoma cells expressing the anti-Myc mouse monoclonal antibody (9E10) are purchased from American Type Culture Collection (Rockville, MD). Horseradish peroxidase–linked anti-mouse IgG antibody is from Jackson ImmunoResearch Laboratories (West Grove, PA). FuGENE 6 is from Roche (Basel, Switzerland). Glutathione Sepharose 4B and ECL plus reagents are from GE Healthcare Biosciences (Piscataway, NJ). Immobilon-P membrane is from Millipore (Benford, MA). All other chemicals are of reagent grade.

Plasmids for expression of Rab3A are constructed as follows. The cDNA fragments encoding a constitutively active form of bovine Rab3A (Rab3A Q81L) and a dominant negative form (Rab3A T36N) are inserted into pEFBOS-Myc vector to construct pEFBOS-Myc-Rab3AQ81L and pEFBOS-Myc-Rab3AT36N, respectively.

To produce a GST fusion protein containing the Rab3-binding domain of Rim1α (Rim1αN), the cDNA encoding the N-terminus of rat Rim1α

(amino acids 1 to 200) is amplified from rat brain cDNA, and its cDNA fragment is inserted into pGEX-6P (GE Healthcare Biosciences, Piscataway, NJ).

3. METHODS

3.1. Pull-down assay for the measurement of the GTP-bound pool of Rab3A

3.1.1. Purification of GST-Rim1αN
GST-Rim1αN is expressed in the *Escherichia coli* strain DH5α after being induced by IPTG at a final concentration of 0.1 mM and is then purified on a glutathione Sepharose 4B column.

3.1.2. Culture
Baby hamster kidney (BHK) cells are maintained at 37° in a humidified atmosphere of 5% CO_2 and 95% air (v/v) in DMEM containing 10% FBS, 100 U/ml penicillin, and 100 mg/ml streptomycin.

3.1.3. Generation of p130-deficient mice
To disrupt Rab3 GAP p130 in embryonic stem (ES) cells, gene targeting is used to replace the 3′ half of the exon 1, the exon 2, and the 5′ portion of the downstream intron with an MC1–neomycin resistance cassette (Sakane *et al.*, 2006). The targeting vector is electroporated into RW4 ES cells, which are then selected with G418. The genotypes of the G418–resistance colonies are confirmed by Southern hybridization. ES cells are microinjected into E3.5 C57BL/6J blastocysts and transferred to MCH pseudopregnant foster mothers to generate chimeras, which are mated with BDF1 mice for germline transmission. Mice carrying mutant alleles are also backcrossed with C57BL/6 mice. Genotyping is performed by Southern hybridization using 5′ probe and/or 3′ probe and by PCR using primers specific for the neo gene and for the replaced p130 gene.

3.1.4. GST pull-down assay
BHK cells (6×10^5 cells/60-mm dish) are transiently transfected by FuGENE 6 with 5 μg of pEFBOS-Myc-Rab3AQ81L or pEFBOS-Myc-Rab3AT36N. The cells are then incubated at 37° for 48 h. After the incubation, the cells are washed once and scraped from the dishes in phosphate-buffered saline (PBS), followed by centrifugation at $500 \times g$ at 4° for 7 min. The cells are lysed in 300 μl of buffer A (10 mM Tris/HCl at pH 8.0, 150 mM NaCl, 1 mM EDTA, 1%[w/v] Nonidet P-40, 10 μM p-APMSF, and 10 μM leupeptin). The cell lysates are centrifuged at $16,100 \times g$ at 4° for 5 min. An aliquot of the supernatant is saved and used

to verify the total amount of mutant Rab3A expressed in the cells. The rest of the supernatant is mixed with 4 μg GST-Rim1αN attached to glutathione Sepharose 4B beads and incubated at 4° for 1 h. The beads are then washed four times in 500 μl of buffer A and resuspended in SDS sample buffer. Comparable amounts of the proteins that remain associated with the beads are separated by SDS-PAGE and transferred to Immobilon-P membrane, and blocked for 1 h in Tris-buffered saline containing 5% skimmed milk. After incubation with a monoclonal antibody against the Myc epitope tag (9E10) for 1 h and then with the peroxidase-conjugated anti-mouse IgG for 1 h, the blots are developed with ECL plus reagents. In these experiments, GST-Rim1αN interacts with Myc-Rab3AQ81L, but not with Myc-Rab3AT36N. This result indicates that this assay can specifically detect GTP-Rab3A in the cells (Fig. 9.1A).

For the measurement of the GTP-bound pool of Rab3A in the mouse brains, wildtype or Rab3 GAP p130-deficient mouse brains are homogenized in buffer A, followed by centrifugation at $500 \times g$ at 4° for 7 min. The supernatant is immediately mixed with 4 μg GST-Rim1αN attached to glutathione Sepharose 4B beads and incubated at 4° for 1 h. After the incubation, the beads are washed four times in buffer A and resuspended in SDS sample buffer. Comparable amounts of the proteins that remain associated with the beads are separated by SDS-PAGE. The fraction of Rab3 bound to the affinity column is determined by Western blotting using a monoclonal antibody against Rab3A. In these experiments, the GTP-Rab3A level is increased in the p130-deficient brains compared with the wildtype brains. This indicates that p130 functions as a GAP for Rab3A *in vivo* and that GTP-Rab3A accumulates in the p130-deficient mouse brains (Fig. 9.1B).

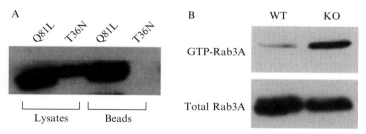

Figure 9.1 Pull-down assay. (A) BHK cells are transfected with pEFBOS-Myc-Rab3AQ81L or -Rab3AT36N. After 48 h, the mutants are isolated using a GST-Rim1αN affinity column and visualized by Western blotting using an anti-Myc antibody. (B) Wildtype or p130-deficient mouse brains are homogenized. GTP-Rab3A is then purified from the homogenates with a GST-Rim1αN affinity column and visualized by Western blotting using an anti-Rab3A antibody. (Adapted from Sakane, A., Manabe, S., Ishizaki, H., Tanaka-Okamoto, M., Kiyokage, E., Toida, K., Yoshida, T., Miyoshi, J., Kamiya, H., Takai, Y., and Sasaki, T. (2006). Rab3 GTPase-activating protein regulates synaptic transmission and plasticity through the inactivation of Rab3. *Proc. Natl. Acad. Sci. USA* **103**, 10029–10034.)

3.2. Assay for glutamate release from synaptosomes

3.2.1. Preparation of synaptosomes

Synaptosomes are prepared as previously described (Dunkley *et al.*, 1986). Cerebral cortices from eight wildtype or eight p130-deficient mice are homogenized in 20 ml of buffer B (320 mM sucrose and 1 mM EDTA at pH 7.4) using a Teflon-glass homogenizer. The homogenate is then centrifuged at 1000×g at 4° for 10 min. The supernatant is collected as the S1 fraction. The resulting pellet is resuspended and homogenized in 20 ml of buffer B using a Teflon-glass homogenizer. The homogenate is diluted with buffer B, followed by centrifugation at 1000×g at 4° for 10 min. The supernatant is collected as the S2 fraction. The S2 fraction is combined with the S1 fraction and centrifuged at 15,000×g at 4° for 30 min. The resulting pellet is suspended in buffer C (320 mM sucrose, 1 mM EDTA at pH 7.4, and 1 mM DTT) and loaded on to a discontinuous Percoll gradient consisting of 23%, 10%, and 3% (v/v) Percoll in buffer C. The gradient is centrifuged at 32,500×g at 4° for 5 min. Synaptosomes enriched under the 10% layer are recovered and diluted with buffer D (140 mM NaCl, 5 mM KCl, 10 mM Hepes/NaOH at pH 7.4, 5 mM NaHCO$_3$, 1.2 mM NaH$_2$PO$_4$, 1 mM MgCl$_2$, and 10 mM glucose) and centrifuged at 15,000×g at 4° for 15 min. The pellet is washed in buffer D and centrifuged at 15,000×g at 4° for 15 min again. This synaptosomal pellet is resuspended with buffer D, and stored on ice before use.

3.3. Measurement of glutamate release from synaptosomes

Glutamate release is assayed by on-line fluorometry as described by Nicholls *et al.* (1987). A solution containing 1.0 mg synaptosomes in 1.5 ml of buffer D is prepared and pre-warmed at 37° for 15 min. A 1.5-ml aliquot is transferred to a stirred cuvette in a JASCO FP-6300 Spectrofluorometer. Either 1.3 mM CaCl$_2$ or 0.5 mM EGTA is added together with 1 mM NADP$^+$ to the cuvette and incubated under stirring at 37° for 2 min, followed by addition of 50 U L-glutamate dehydrogenase. After further incubation at 37° for 3 min, 30 mM KCl is added, and the incubation is extended for an additional 5 min. Generation of NADPH is measured at excitation and emission wavelengths of 340 and 460 nm, respectively (Fig. 9.2A). Internal standards are made by adding 10 nmol of L-glutamate at the end of each experiment.

3.4. Measurement of cytoplasmic [Ca^{2+}] in synaptosomes

Synaptosomes are incubated in buffer D containing additionally 5 μM fura-2-acetoxymethyl ester and 0.1 mM CaCl$_2$ at 37° for 30 min as previously described (Kauppinen *et al.*, 1988). After fura-2 loading, synaptosomes

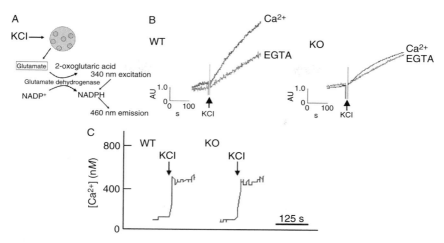

Figure 9.2 Glutamate release from the p130-deficient synaptosomes. (A) Schematic representation of the measurement of glutamate release from synaptosomes. Glutamate released from the synaptosomes by KCl-induced depolarization is converted to 2-oxoglutaric acid by glutamate dehydrogenase. Generation of NADPH in this reaction is spectrofluorometrically measured at excitation and emission wavelengths of 340 and 460 nm, respectively. (B) Glutamate release from the wildtype (WT) or p130-deficient (KO) synaptosomes is spectrofluorometrically recorded. Incubations are performed in the presence of 1.3 mM CaCl$_2$ (Ca^{2+}) or 0.5 mM EGTA (EGTA). (C) Ca^{2+} influx into the wildtype (WT) or p130-deficient (KO) synaptosomes is evoked by KCl-induced depolarization and is monitored through intrasynaptosomal fluorescence of fura-2. Intrasynaptosomal [Ca^{2+}] is elevated in the p130-deficient synaptosomes to a similar level as observed in the wildtype synaptosomes. (Adapted from Sakane, A., Manabe, S., Ishizaki, H., Tanaka-Okamoto, M., Kiyokage, E., Toida, K., Yoshida, T., Miyoshi, J., Kamiya, H., Takai, Y., and Sasaki, T. (2006). Rab3 GTPase-activating protein regulates synaptic transmission and plasticity through the inactivation of Rab3. *Proc. Natl. Acad. Sci. USA.* **103**, 10029–10034.)

are centrifuged at 10,000×g for 5 min, resuspended in buffer D and transferred to a stirred cuvette in a JASCO FP-6300 Spectrofluorometer. Add 1.3 mM CaCl$_2$ to the cuvette and after further incubation at 37° for 5 min, add 30 mM KCl. Emission of 505 nm is determined in response to alternate excitation at 340 and 380 nm at intervals of 2 s. Cytoplasmic free Ca^{2+} concentration is calculated as described (Grynkiewicz *et al.*, 1985).

In these experiments, KCl-induced depolarization leads to Ca^{2+}-dependent glutamate release from synaptosomes of the wildtype mice, whereas the p130-deficient synaptosomes show weak Ca^{2+}-dependent release (see Fig. 9.2B). However, cytoplasmic free Ca^{2+} concentration is elevated to a similar extent in the wildtype and p130-deficient synaptosomes (see Fig. 9.2C).

4. COMMENTS

For the measurement of the GTP-bound pool of Rab proteins in the cell lysates or tissue homogenates, it is the most important to select the detergent used in the pull-down assay. For this purpose, the pull-down assay using the constitutively active and dominant negative forms of Rab proteins are necessary to determine which detergent is most appropriate for the assay.

REFERENCES

Aligianis, I. A., Johnson, C. A., Gissen, P., Chen, D., Hampshire, D., Hoffmann, K., Maina, E. N., Morgan, N. V., Tee, L., Morton, J., Ainsworth, J. R., Horn, D., *et al.* (2005). Mutations of the catalytic subunit of RAB3GAP cause Warburg Micro syndrome. *Nat. Genet.* **37,** 221–223.

Aligianis, I. A., Morgan, N. V., Mione, M., Johnson, C. A., Rosser, E., Hennekam, R. C., Adams, G., Trembath, R. C., Pilz, D. T., Stoodley, N., Moore, A. T., Wilson, S., *et al.* (2006). Mutation in Rab3 GTPase-activating protein (RAB3GAP) noncatalytic subunit in a kindred with Martsolf syndrome. *Am. J. Hum. Genet.* **78,** 702–707.

D'Adamo, P., Menegon, A., Lo Nigro, C., Grasso, M., Gulisano, M., Tamanini, F., Bienvenu, T., Gedeon, A. K., Oostra, B., Wu, S. K., Tandon, A., Valtorta, F., *et al.* (1998). Mutations in GDI1 are responsible for X-linked non-specific mental retardation. *Nat. Genet.* **19,** 134–139.

Dunkley, P. R., Jarvie, P. E., Heath, J. W., Kidd, G. J., and Rostas, J. A. (1986). A rapid method for isolation of synaptosomes on Percoll gradients. *Brain Res.* **372,** 115–129.

Fukui, K., Sasaki, T., Imazumi, K., Matsuura, Y., Nakanishi, H., and Takai, Y. (1997). Isolation and characterization of a GTPase activating protein specific for the Rab3 subfamily of small G proteins. *J. Biol. Chem.* **272,** 4655–4658.

Geppert, M., and Südhof, T. C. (1998). RAB3 and synaptotagmin: the yin and yang of synaptic membrane fusion. *Annu. Rev. Neurosci.* **21,** 75–95.

Grynkiewicz, G., Poenie, M., and Tsien, R. Y. (1985). A new generation of Ca^{2+} indicators with greatly improved fluorescence properties. *J. Biol. Chem.* **260,** 3440–3450.

Ishizaki, H., Miyoshi, J., Kamiya, H., Togawa, A., Tanaka, M., Sasaki, T., Endo, K., Mizoguchi, A., Ozawa, S., and Takai, Y. (2000). Role of rab GDP dissociation inhibitor alpha in regulating plasticity of hippocampal neurotransmission. *Proc. Natl. Acad. Sci. USA* **97,** 11587–11592.

Kauppinen, R. A., McMahon, H. T., and Nicholls, D. G. (1988). Ca^{2+}-dependent and Ca^{2+}-independent glutamate release, energy status and cytosolic free Ca^{2+} concentration in isolated nerve terminals following metabolic inhibition: Possible relevance to hypoglycaemia and anoxia. *Neuroscience* **27,** 175–182.

Nagano, F., Sasaki, T., Fukui, K., Asakura, T., Imazumi, K., and Takai, Y. (1998). Molecular cloning and characterization of the noncatalytic subunit of the Rab3 subfamily-specific GTPase-activating protein. *J. Biol. Chem.* **273,** 24781–24785.

Nicholls, D. G., Sihra, T. S., and Sanchez-Prieto, J. (1987). Calcium-dependent and -independent release of glutamate from synaptosomes monitored by continuous fluorometry. *J. Neurochem.* **49,** 50–57.

Oishi, H., Sasaki, T., Nagano, F., Ikeda, W., Ohya, T., Wada, M., Ide, N., Nakanishi, H., and Takai, Y. (1998). Localization of the Rab3 small G protein regulators in nerve terminals and their involvement in Ca^{2+}-dependent exocytosis. *J. Biol. Chem.* **273**, 34580–34585.

Sakane, A., Manabe, S., Ishizaki, H., Tanaka-Okamoto, M., Kiyokage, E., Toida, K., Yoshida, T., Miyoshi, J., Kamiya, H., Takai, Y., and Sasaki, T. (2006). Rab3 GTPase-activating protein regulates synaptic transmission and plasticity through the inactivation of Rab3. *Proc. Natl. Acad. Sci. USA* **103**, 10029–10034.

Schluter, O. M., Schmitz, F., Jahn, R., Rosenmund, C., and Südhof, T. C. (2004). A complete genetic analysis of neuronal Rab3 function. *J. Neurosci.* **24**, 6629–6637.

Südhof, T. C. (2004). The synaptic vesicle cycle. *Annu. Rev. Neurosci.* **27**, 509–547.

Takai, Y., Sasaki, T., Shirataki, H., and Nakanishi, H. (1996). Rab3A small GTP-binding protein in Ca^{2+}-dependent exocytosis. *Genes Cells* **1**, 615–632.

Takai, Y., Sasaki, T., and Matozaki, T. (2001). Small GTP-binding proteins. *Physiol. Rev.* **81**, 153–208.

Tanaka, M., Miyoshi, J., Ishizaki, H., Togawa, A., Ohnishi, K., Endo, K., Matsubara, K., Mizoguchi, A., Nagano, T., Sato, M., Sasaki, T., and Takai, Y. (2001). Role of Rab3 GDP/GTP exchange protein in synaptic vesicle trafficking at the mouse neuromuscular junction. *Mol. Biol. Cell* **12**, 1421–1430.

Yamaguchi, K., Tanaka, M., Mizoguchi, A., Hirata, Y., Ishizaki, H., Kaneko, K., Miyoshi, J., and Takai, Y. (2002). A GDP/GTP exchange protein for the Rab3 small G protein family up-regulates a postdocking step of synaptic exocytosis in central synapses. *Proc. Natl. Acad. Sci. USA* **99**, 14536–14541.

Zerial, M., and McBride, H. (2001). Rab proteins as membrane organizers. *Nat. Rev. Mol. Cell Biol.* **2**, 107–117.

CHAPTER TEN

IDENTIFICATION AND CHARACTERIZATION OF JRAB/MICAL-L2, A JUNCTIONAL RAB13-BINDING PROTEIN

Noriyuki Nishimura *and* Takuya Sasaki

Contents

Abstract

The Rab family small G proteins are localized to distinct subsets of intracellular membranes and play a key role in membrane traffic through the interaction with their specific effector protein(s). Rab13 is identified as a plaque protein at tight junctions (TJs) and has been shown to regulate the assembly of functional TJs in epithelial cells. We have demonstrated that Rab13 mediates the endocytic recycling of integral TJ protein occludin, and identified a junctional Rab13-binding protein (JRAB)/molecule interacting with CasL-like 2 (MICAL-L2) as a Rab13 effector protein using a yeast two-hybrid system. JRAB/MICAL-L2 has a calponin-homology domain in the N-terminus, a LIM domain in the middle, and a coiled-coil domain at the C-terminus, and specifically binds to the GTP-bound form of Rab13 via its C-terminus. It is localized to TJs in epithelial cells and distributed along stress fibers in fibroblasts. In epithelial cells, JRAB/MICAL-L2 as well as Rab13 mediates the endocytic recycling of occludin, but not

Department of Biochemistry, Institute of Health Biosciences, The University of Tokushima Graduate School, Tokushima, Japan

Methods in Enzymology, Volume 438
ISSN 0076-6879, DOI: 10.1016/S0076-6879(07)38010-5

transferrin receptor, and the formation of functional TJs. This chapter describes the procedures for the isolation of JRAB/MICAL-L2 and the analysis of its functions.

1. INTRODUCTION

The Rab family small G proteins play a key role in membrane traffic (Pfeffer and Aivazian, 2004; Takai *et al.*, 2001; Zerial and McBride, 2001). At least 63 different Rab family members that reside in distinct subsets of intracellular membranes are identified in mammalian cells. As a molecular switch, Rab proteins exist as inactive GDP-bound or active GTP-bound forms. GTP-bound Rab proteins interact with their specific effector protein(s) that mediate at least one element of their downstream effects.

Tight junctions (TJs) are continuous, circumferential belt–like structures located at the apical end of the intercellular space. TJs delineate the boundaries between the apical and basolateral domains of the plasma membrane (PM) of epithelial cells, and regulate the solute flux through the paracellular space (Anderson *et al.*, 2004; Schneeberger and Lynch, 2004; Tsukita *et al.*, 2001). TJs consist of integral TJ proteins and TJ plaque proteins. Integral TJ proteins including occludin and claudins mediate cell–cell adhesion and create physical intercellular barrier. Although the precise function of occludin remains to be determined, claudins are primary structural components of TJ strands. TJ plaque proteins cluster integral TJ proteins and form an organizing platform for a variety of scaffolding, signaling, and membrane traffic proteins. In addition to specific TJ proteins, a circumferential actin belt that encircles each epithelial cell at the level of TJs is critical for formation and maintenance of TJs (Bershadsky, 2004; Vasioukhin and Fuchs, 2001). Among TJ plaque proteins, zonula occludens (ZO) family proteins, cingulin, and JACOP are shown to bind to F-actin and are implicated in linking TJs to a circumferential actin belt (D'Atri and Citi, 2001; Ohnishi *et al.*, 2004; Wittchen *et al.*, 1999).

The high degree of TJ integrity observed in normal epithelial cells is often lost in cancer cells. Deregulation of specific TJ proteins is shown to correlate with invasiveness and metastatic potential of cancers. Occludin downregulation is often detected in gastrointestinal and endometrial cancers (Kimura *et al.*, 1997; Tobioka *et al.*, 2004). Similarly, claudin-1 and claudin-7 are reduced in breast cancers (Kominsky *et al.*, 2003; Kramer *et al.*, 2000). Furthermore, expression of exogenous occludin rescues Raf1-transformed cells from their tumorigenic phenotype (Li and Mrsny, 2000). Expression of claudin-4 in pancreatic cancer cells also reduces the invasiveness of these cells (Michl *et al.*, 2003). These observations suggest a tumor-suppressive role of occludin and claudins. On the other hand, claudin-3 and claudin-4 are upregulated in ovarian and prostate cancers, indicating a tumor-promoting role of claudins (Hough *et al.*, 2000; Long *et al.*, 2001). The molecular mechanisms underlying these tissue- and/or cellular context–dependent functions of TJ proteins remain to be determined.

Two Rab family members, Rab3B and Rab13, are identified as TJ plaque proteins in epithelial cells and Rab13 is implicated in the assembly of functional TJs and neuronal plasticity (Di Giovanni *et al.*, 2005; Marzesco *et al.*, 2002; Weber *et al.*, 1994; Zahraoui *et al.*, 1994). We have found that Rab13 regulates the endocytic recycling of occludin and identified a junctional Rab13-binding protein (JRAB)/molecule interacting with CasL-like 2 (MICAL-L2) as a Rab13 effector protein using a yeast two-hybrid system. JRAB/MICAL-L2 is localized to TJs in epithelial MTD-1A cells and distributed along stress fibers in fibroblastic NIH3T3 cells. JRAB/MICAL-L2 as well as Rab13 mediates the endocytic recycling of occludin, but not transferrin receptor (TfR), and the formation of functional TJs in epithelial cells (Morimoto *et al.*, 2005; Terai *et al.*, 2006). This chapter describes the procedures for the isolation of JRAB/MICAL-L2 and the analysis of its properties and function.

2. MATERIALS

Rat anti-occludin (MOC37) antibody is a kind gift from S. Tsukita (Kyoto University, Kyoto, Japan). Rabbit anti-ZO-1 and mouse anti-TfR antibodies are purchased from Zymed (San Francisco, CA), rabbit anti-GFP from Molecular Probes (Eugene, OR), mouse anti-Myc (9E10) from ATCC (Manassas, VA), and mouse anti-HA (12CA5) and rat anti-HA (3F10) from Roche (Mannheim, Germany). Alexa 488- and Alexa 594-conjugated secondary antibodies are from Molecular Probes (Eugene, OR). Horseradish peroxidase-coupled secondary antibody is from Jackson ImmunoResearch Laboratories (West Grove, PA). Rhodamine-phalloidine is from Molecular Probes (Eugene, OR). 4- and 40-kDa FITC-dextran are from Sigma (St. Louis, MO). Formaldehyde is from Polyscience (Warrington, PA). Block Ace is from Dainippon Pharmaceutical (Osaka, Japan). (4-Amidinophenyl) methanesulfonyl fluoride (APMSF), 3-(3-cholamidepropyl) dimethylammonio-1-propanesulphonate (CHAPS), and iodoacetamide are from Wako Pure Chemicals (Osaka, Japan). 3-Amino-1, 2, 4-triazole (3-AT) and 2-mercaptoethanesulfonic acid (MESNA) are from Sigma (St. Louis, MO). GTPγS is from Roche (Mannheim, Germany). Sulfo-NHS-SS-biotin and NeutrAvidin beads are from Pierce (Rockford, IL). Glutathione-Sepharose beads, protein G-Sepharose beads, HiTrap NHS-activated column, and ECL-Plus kit are from GE Healthcare (Piscataway, NJ). Lipofectamine 2000 is from Invitrogen (Carlsbad, CA). MBS Mammalian Transfection kit is from

Stratagene (La Jolla, CA). Transwell filter (polycarbonate membranes with 12-mm diameter and 0.4-μm pore size) is from Corning (Corning, NY). Dulbecco's modified Eagle's medium (DMEM) and fetal bovine serum (FBS) are from Sigma (St. Louis, MO).

Plasmids for expression of Rab proteins (Rab1A, Rab5A, and Rab13) are constructed as follows. Rab1A, Rab5A, and Rab13 cDNAs are isolated by RT-PCR from MDCK, MDCK, and Caco2 cells, respectively. Rab13 T22N, Rab13 Q67L, and Rab13 N121I mutants are constructed using the Quick Change Mutagenesis kit (Stratagene, La Jolla, CA). Rab1A, Rab5A, Rab13, Rab13 T22N, Rab13 Q67L, and Rab13 N121I are cloned into the mammalian expression vector pCI-neo-HA to express N-terminal HA-tagged proteins and the yeast two-hybrid bait vector pGBDU-C1, respectively. The mouse 11-day embryo cDNA library in the yeast two-hybrid prey vector pACT2 is obtained from Clontech (Palo Alto, CA).

Plasmids for expression of JRAB/MICAL-L2 are constructed as follows. A full-length cDNA of JRAB/MICAL-L2 (JRAB/MICAL-L2-F, amino acids 1 to 1,009) is obtained by RT-PCR from mouse brain cDNA. JRAB/MICAL-L2-N (amino acids 1 to 805) and JRAB/MICAL-L2-C (amino acids 806 to 1,009) cDNAs are generated via PCR using JRAB/MICAL-L2-F cDNA as a template (Fig.10.1A). JRAB/MICAL-L2-F, JRAB/MICAL-L2-N, and JRAB/MICAL-L2-C are cloned into the mammalian expression vector pCI-neo-Myc to express N-terminal Myc-tagged proteins.

Rat anti-JRAB/MICAL-L2 polyclonal antibody is generated as follows. JRAB/MICAL-L2-C (amino acids 806 to 1009) is cloned into pGEX-6P1 vector to express N-terminal GST-tagged JRAB/MICAL-L2-C protein. GST and GST-JRAB/MICAL-L2-C fusion proteins are produced in *Escherichia coli* DH5α strain and purified using glutathione-Sepharose beads according to the manufacturer's instructions. Two milligrams of GST or GST-JRAB/MICAL-L2-C protein are immobilized on HiTrap NHS-activated columns according to the manufacturer's instructions. Two female Wistar rats are immunized with 100 μg of GST-JRAB/MICAL-L2-C protein twice at 4-week intervals, after which whole blood is collected. Crude immunoglobulin fractions are prepared by ammonium sulfate precipitation and passed through a GST-immobilized column to remove an anti-GST antibody. The anti-JRAB/MICAL-L2 polyclonal antibody is further purified on a GST-JRAB/MICAL-L2-C-immobilized column according to the manufacturer's instructions.

Recombinant adenovirus for expression of GFP, Rab13 Q67L, JRAB/MICAL-L2-F, JRAB/MICAL-L2-N, and JRAB/MICAL-L2-C proteins are constructed as follows. Recombinant adenovirus (Ad-EGFP, Ad-EGFP-Rab13 Q67L, Ad-Myc-JRAB/MICAL-L2-F, Ad-Myc-JRAB/MICAL-L2-N, and Ad-Myc-JRAB/MICAL-L2-C) is constructed using the

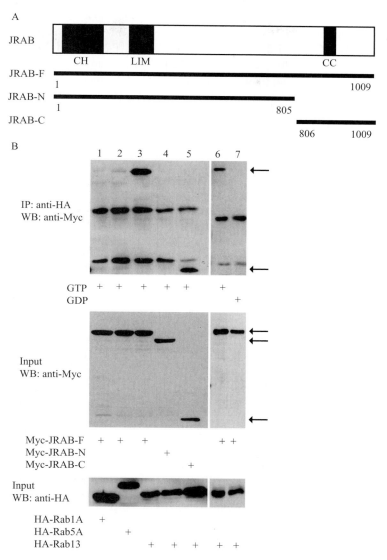

Figure 10.1 Co-immunoprecipitation of JRAB/MICAL-L2 with Rab13. (A) Structures of the full length and various fragments of JRAB/MICAL-L2. CH, calponin homology domain; LIM, LIM domain; CC, coiled-coil domain. (B) BHK cells expressing Myc-JRAB/MICAL-L2-F, Myc-JRAB/MICAL-L2-N, or Myc-JRAB/MICAL-L2-C together with HA-Rab1A, HA-Rab5A, or HA-Rab13 are subjected to co-immunoprecipitation assay. The arrows indicate JRAB/MICAL-L2-F, JRAB/MICAL-L2-N, and JRAB/MICAL-L2-C. IP, immunoprecipitation. WB, Western blot. Lanes 1 to 3, IP of Rab1A/5A/13 (GTP) with JRAB-F. Lanes 3 to 5, IP of Rab13 (GTP)

Transpose-Ad Adenoviral Vector System (Qbiogene, Carlsbad, CA) according to the manufacturer's instruction. Briefly, EGFP, EGFP-Rab13 Q67L, Myc-JRAB/MICAL-L2-F, Myc-JRAB/MICAL-L2-N, and Myc-JRAB/MICAL-L2-C cDNAs are cloned into a pCR259 transfer vector. Then, a recombinant adenoviral plasmid is generated by Tn7-mediated transposition in *Escherichia coli*. The resulting plasmid is linearized by *PacI*-digestion and transfected into QBI-HEK293 cells using the MBS Mammalian Transfection kit. After 24 h, the cells are split into a 96-well plate and incubated at 37° for 10 to 14 days. Screening of recombinant adenovirus is performed by PCR and Western blot analysis. Recombinant adenovirus is amplified in QBI-HEK293 cells and its titer is determined by MOI test.

3. METHODS

3.1. Identification of JRAB/MICAL-L2 with yeast two-hybrid screening

The yeast strain PJ69-4A (*MATa trp1-901 leu2-3112 ura3-52 his3-200 gal4Δgal80Δ GAL2-ADE2 LYS2::GAL1-HIS3 met2::GAL7-lacZ*) is sequentially transformed with pGBDU-Rab13 Q67L and the mouse 11-day embryo cDNA library. Transformants (2.4×10^6) are grown on a SC-Ura-Leu-His plate containing 3 mM 3-AT at 30° for 10 days. The colonies grown on a SC-Ura-Leu-His plate are further screened by the growth on a SC-Ura-Leu-Ade plate and the β-galactosidase activity as described previously (James *et al.*, 1996). The library plasmids are isolated from the colonies and sorted into groups according to the *Hae III*–digested pattern of their PCR-amplified inserts. A representative plasmid from each group is cotransformed with pGBDU-Rab13 Q67L into PJ69-4A, and only plasmids showing interaction with Rab13 Q67L are subjected to DNA sequence analysis. Three clones encode JRAB/MICAL-L2 and a representative clone is cotransformed into PJ69-4A with Rab1A, Rab5A, Rab13, Rab13 T22N, Rab13 Q67L, or Rab13 N121I to further analyze the specificity of its interaction with Rab13. This clone specifically interacts with Rab13 and Rab13 Q67L, but not with Rab1A, Rab5A, Rab13 T22N, and Rab13 N121I. These results suggest that JRAB/MICAL-L2 is a Rab13 effector protein.

with JRAB-F/N/C. Lanes 6 and 7, IP of Rab13 (GTP/GDP) with JRAB-F. (Adapted from Terai, T., Nishimura, N., Kanda, I., Yasui, N., and Sasaki, T. (2006). JRAB/MICAL-L2 is a junctional Rab13-binding protein mediating the endocytic recycling of occludin. *Mol. Biol. Cell* **17**, 2465–2475.)

3.2. Co-immunoprecipitation of JRAB/MICAL-L2 with Rab13

BHK cells are plated and cultured in DMEM with 10% FBS at 37° (5% CO_2 and 95% air) for 16 to 24 h and transfected with pCI-neo-HA-Rab and pCI-neo-Myc-JRAB/MICAL-L2 using Lipofectamine 2000 transfection reagent according to manufacture's instruction. At 36 to 48 h after the transfection, cells are lysed with 25 mM Tris/HCl (pH 7.5) containing 0.5% CHAPS, 125 mM NaCl, 1 mM MgCl$_2$, 10 μg/ml APMSF, and 100 μM GTPγS/GDP at 4° for 15 min. The lysates are centrifuged at 15,000×g at 4° for 30 min, and the supernatants are immunoprecipitated with anti-HA antibody bound to protein G-Sepharose beads at 4° for 2 h. The beads are washed three times with 25 mM Tris/HCl (pH 7.5) containing 0.1% CHAPS, 300 mM NaCl, 1 mM MgCl$_2$, and 10 μM GTPγS/GDP, and eluted with sodium dodecyl sulfate (SDS) sample buffer. The samples are subjected to SDS-polyacrylamide gel electrophoresis (PAGE) and transferred to a polyvinylidene difluoride (PVDF) membrane. After blocking with Block Ace, the blot is sequentially incubated with anti-HA or anti-Myc antibody and horseradish peroxidase-coupled secondary antibody. The immunoreactive proteins are detected with an ECL-Plus kit. Quantitation is performed on scans of autoradiograph films with nonsaturated signals using Image J 1.36 program (http://rsb.info.nih.gov/ij/).

JRAB/MICAL-L2 has a calponin-homology (CH) domain in its N-terminal region (amino acids 3 to 102), a LIM domain in its middle region (amino acids 187 to 241), and a coiled-coil (CC) domain at the C-terminus (amino acids 842 to 880) (see Fig.10.1A). When JRAB/MICAL-L2-F is divided into JRAB/MICAL-L2-N (N-terminal region containing LIM and CH domains, amino acids 1 to 805) and JRAB/MICAL-L2-C (C-terminal region containing CC domain, amino acids 806 to 1,009) (see Fig.10.1A), Myc-JRAB/MICAL-L2-F and Myc-JRAB/MICAL-L2-C, but not Myc-JRAB/MICAL-L2-N, can bind to HA-Rab13 (see Fig.10.1B, lanes 3, 4, and 5). Myc-JRAB/MICAL-L2 specifically interacts with HA-Rab13, but not with HA-Rab1A and HA-Rab5A (see Fig.10.1B, lanes 1, 2, and 3). Furthermore, Myc-JRAB/MICAL-L2 is co-immunoprecipitated with HA-Rab13 in the presence of GTPγS, but not GDP, in BHK cells (see Fig.10.1B, lanes 6 and 7). These results indicate that JRAB/MICAL-L2 specifically interacts with the GTP-bound form of Rab13 through its C-terminus.

3.3. Intracellular localization of JRAB/MICAL-L2

MTD-1A cells are grown on Transwell filters for 3 to 4 days and fixed with 10% trichloroacetic acid in PBS on ice for 15 min. After permeabilization with 0.2% Triton X-100 in PBS for 15 min, cells are blocked with 5% goat serum in PBS for 60 min. Then cells are sequentially incubated with anti-JRAB/MICAL-L2 and anti-ZO-1 antibodies for 60 min, and with Alexa 488- and Alexa 594-conjugated secondary antibodies (Molecular Probes, Eugene, OR) at room temperature for 60 min. Fluorescent images are acquired using a Radiance 2000 confocal laser-scanning microscope (Bio-Rad, Hercules, CA). JRAB/MICAL-L2 is detected both in the cell–cell contacts, where it is mostly colocalized with ZO-1, and in the cytosol of polarized MTD-1A cells. A vertical section of a confocal image revealed that JRAB/MICAL-L2 as well as ZO-1 is present at apical side of cell–cell contacts, indicating that a considerable part of JRAB/MICAL-L2 protein is localized to TJ (Fig.10.2A).

NIH3T3 cells are grown on glass coverslips for 1 day and fixed with 1% formaldehyde in PBS at room temperature for 15 min. Cells are then

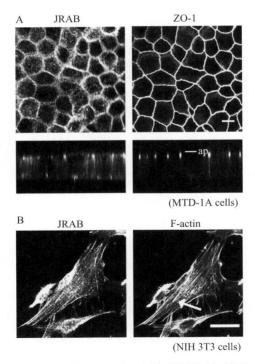

Figure 10.2 Intracellular localization of JRAB/MICAL-L2. MTD-1A cells grown on Transwell filters (A) and NIH3T3 cells grown on glass coverslips (B) are double-labeled with anti-JRAB/MICAL-L2 and anti-ZO-1 antibodies (A) and with anti-JRAB/MICAL-L2 antibody and rhodamine-phalloidine (B), respectively, and analyzed by confocal laser scanning microscopy. Vertical sectional images are generated and shown in the lower panels (A). ap, the level of apical membranes. Arrow, stress fibers. Bar, 10 μm. (Adapted from Terai, T., Nishimura, N., Kanda, I., Yasui, N., and Sasaki, T. (2006). JRAB/MICAL-L2 is a junctional Rab13-binding protein mediating the endocytic recycling of occludin. *Mol. Biol. Cell* **17**, 2465–2475.)

processed for immunofluorescence microscopy labeling with anti-JRAB/ MICAL-L2 and Alexa 488–conjugated secondary antibodies together with rhodamine-phalloidine. JRAB/MICAL-L2 is distributed along stress fibers in addition to the PM of NIH3T3 cells, suggesting the association of JRAB/ MICAL-L2 with actin cytoskeleton (Fig. 10.2B).

3.4. Involvement of JRAB/MICAL-L2 in the Formation of Functional TJ

Recycling assay MTD-1A cells are plated and cultured in DMEM with 10% FBS at $37°$ (5% CO_2 and 95% air) for 24 to 48 h and infected with Ad-EGFP, Ad-EGFP-Rab13 Q67L, Ad-Myc-JRAB/MICAL-L2-F, Ad-Myc-JRAB/MICAL-L2-N, or Ad-Myc-JRAB/MICAL-L2-C. At 24–36 h after the infection, cell-surface proteins are biotinylated with 0.5 mg/ml sulfo-NHS-SS-biotin in PBS containing 0.9 mM $CaCl_2$ and 0.33 mM $MgCl_2$ (PBS/CM) at $4°$ for 30 min, quenched with 50 mM NH_4Cl in PBS/CM at $4°$ for 15 min, and incubated at $37°$ for 15 min to allow endocytosis. The remaining biotin on the cell-surface is stripped with 50 mM MESNA in 100 mM Tris/HCl (pH 8.6) containing 100 mM NaCl and 2.5 mM $CaCl_2$ at $4°$ for 30 min, and quenched with 5 mg/ml iodoa-cetamide in PBS/CM at $4°$ for 15 min. Cells are again incubated at $37°$ for 15 min to allow recycling of endocytosed cargo proteins back to the cell surface. Subsequently, the newly appeared cell-surface biotin is again stripped as described above. Cells are lysed with 50 mM Tris/HCl (pH 8.0) containing 1.25% Triton X-100, 0.25% SDS, 150 mM NaCl, 5 mM EDTA, and 10 μg/ml APMSF on ice for 15 min. The lysates are centri-fuged at 15,000×g at $4°$ for 15 min, and the supernatants are incubated with NeutrAvidin beads at $4°$ for 2 h. The beads are washed three times with 50 mM Tris-HCl (pH 8.0) containing 0.5% Triton X-100, 0.1% SDS, 150 mM NaCl, and 5 mM EDTA, and eluted with SDS sample buffer. The samples are subjected to SDS-PAGE and transferred to a PVDF membrane. After blocking with 100% Block Ace, the blot is sequentially incubated with anti-occludin or anti-TfR and horseradish peroxidase-coupled secondary antibodies. The immunoreactive proteins are detected as described above.

While the recycling of occludin in MTD-1A cells infected with Ad-Myc-JRAB/MICAL-L2-F or Ad-Myc-JRAB/MICAL-L2-C is com-parable to that in Ad-EGFP–infected MTD-1A cells, it is inhibited in Ad-Myc-JRAB/MICAL-L2-N–infected cells as potently as in Ad-EGFP-Rab13 Q67L–infected cells (Fig.10.3A). The recycling of TfR in MTD-1A cells is not affected by the expression of Myc-JRAB/MICAL-L2-F, Myc-JRAB/MICAL-L2-N, Myc-JRAB/MICAL-L2-C, or EGFP-Rab13 Q67L (Fig.10.3A). These results suggest that JRAB/MICAL-L2

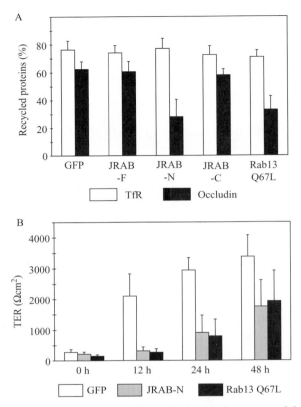

Figure 10.3 Involvement of JRAB/MICAL-L2 in the formation of functional TJs. (A) MTD-1A cells expressing GFP, Myc-JRAB/MICAL-L2-F, Myc-JRAB/MICAL-L2-N, Myc-JRAB/MICAL-L2-C, or GFP-Rab13 Q67L are subjected to recycling assay. Recycled Tf R and occludin represent the percentage of endocytosed Tf R and occludin, respectively, and are expressed as the mean and SEM of three independent experiments. (B) MTD-1A cells expressing GFP, Myc-JRAB/MICAL-L2-N, or GFP-Rab13 Q67L are subjected to Ca^{2+}-switch assay. TER is measured at 0, 12, 24, and 48 h after Ca^{2+}-restoration and presented as the mean and SEM of three independent experiments performed in duplicate. (Adapted from Terai, T., Nishimura, N., Kanda, I., Yasui, N., and Sasaki, T. (2006). JRAB/MICAL-L2 is a junctional Rab13-binding protein mediating the endocytic recycling of occludin. *Mol. Biol. Cell* **17**, 2465–2475.)

and Rab13 specifically regulate the endocytic recycling of occludin in MTD-1A cells.

Transepithelial electrical resistance (TER) MTD-1A cells are plated onto Transwell filters as an instant confluent monolayer and grown in DMEM with 10% FBS (normal calcium medium; NCM) for 3 to 4 days. Cells are then infected with Ad-EGFP, Ad-Myc-JRAB/MICAL-L2-N, or

Ad-EGFP-Rab13 Q67L for 24 h, and sequentially incubated in Ca^{2+}-free minimal essential medium without FBS (low calcium medium; LCM) for 1 h, in LCM with 20 mM EGTA for 10 min to remove extracellular Ca^{2+}, and in NCM to restore it. TER is measured directly in culture media at 0, 12, 24, and 48 h after restoring Ca^{2+} using a Millicell-ERS epithelial volt-ohm-meter (Millipore, Billerica, MA). TER values are calculated by subtracting the background TER of blank filters and by multiplying the surface area of the filter. MTD-1A cells expressing GFP, JRAB/MICAL-L2-N, or Rab13 Q67L all go from the same maximal TER ($>3,000 \ \Omega cm^2$) to the same minimal TER (10 to 20 Ωcm^2) following incubation in Ca^{2+}-chelated medium. Subsequent incubation in physiological Ca^{2+} medium triggers a time-dependent increase in TER. MTD-1A cells expressing JRAB/MICAL-L2-N as well as Rab13 Q67L exhibit a significantly delayed increase in TER compared to GFP-expressing MTD-1A cells (Marzesco *et al.*, 2002; Terai *et al.*, 2006) (see Fig.10.3B).

4. COMMENTS

The cell-surface biotinylation is proven to be a powerful method to examine the intracellular transport of transmembrane proteins to and from the cell surface. To achieve the selective labeling and removal of only the cell-surface proteins, the cleavable, water-soluble and membrane-impermeable biotin analog, sulfo-NHS-SS-biotin, is used. Endocytosis is detected by the accumulation of the sulfo-NHS-SS-biotin–labeled cargo proteins within the cells, which are protected from reduction by MESNA. Recycling is detected by the disappearance of the endocytosed sulfo-NHS-SS-biotin–labeled cargo proteins on the cell surface, which are subjected to reduction by MESNA. Using this method, we have successfully demonstrated that occludin is endocytosed and recycled back to the cell surface in a Rab13- and JRAB/MICAL-L2–dependent manner (Morimoto *et al.*, 2005; Terai *et al.*, 2006).

REFERENCES

Anderson, J. M., Van Itallie, C. M., and Fanning, A. S. (2004). Setting up a selective barrier at the apical junction complex. *Curr. Opin. Cell Biol.* **16,** 140–145.

Bershadsky, A. (2004). Magic touch: How does cell–cell adhesion trigger actin assembly? *Trends Cell Biol.* **14,** 589–593.

D'Atri, F., and Citi, S. (2001). Cingulin interacts with F-actin *in vitro*. *FEBS Lett.* **507,** 21–24.

Di Giovanni, S., De Biase, A., Yakovlev, A., Finn, T., Beers, J., Hoffman, E. P., and Faden, A. I. (2005). *In vivo* and *in vitro* characterization of novel neuronal plasticity factors identified following spinal cord injury. *J. Biol. Chem.* **280,** 2084–2091.

Hough, C. D., Sherman-Baust, C. A., Pizer, E. S., Montz, F. J., Im, D. D., Rosenshein, N. B., Cho, K. R., Riggins, G. J., and Morin, P. J. (2000). Large-scale serial analysis of gene expression reveals genes differentially expressed in ovarian cancer. *Cancer Res.* **60,** 6281–6287.

James, P., Halladay, J., and Craig, E. A. (1996). Genomic libraries and a host strain designed for highly efficient two-hybrid selection in yeast. *Genetics* **144,** 1425–1436.

Kimura, Y., Shiozaki, H., Hirao, M., Maeno, Y., Doki, Y., Inoue, M., Monden, T., Ando-Akatsuka, Y., Furuse, M., Tsukita, S., and Monden, M. (1997). Expression of occludin, tight-junction-associated protein, in human digestive tract. *Am. J. Pathol.* **151,** 45–54.

Kominsky, S. L., Argani, P., Korz, D., Evron, E., Raman, V., Garrett, E., Rein, A., Sauter, G., Kallioniemi, O., and Sukumar, S. (2003). Loss of the tight junction protein claudin-7 correlates with histological grade in both ductal carcinoma in situ and invasive ductal carcinoma of the breast. *Oncogene* **22,** 2021–2033.

Kramer, F., White, K., Kubbies, M., Swisshelm, K., and Weber, B. H. (2000). Genomic organization of claudin-1 and its assessment in hereditary and sporadic breast cancer. *Hum. Genet.* **107,** 249–256.

Li, D., and Mrsny, R. J. (2000). Oncogenic Raf-1 disrupts epithelial tight junctions via downregulation of occludin. *J. Cell Biol.* **148,** 791–800.

Long, H., Crean, C. D., Lee, W. H., Cummings, O. W., and Gabig, T. G. (2001). Expression of Clostridium perfringens enterotoxin receptors claudin-3 and claudin-4 in prostate cancer epithelium. *Cancer Res.* **61,** 7878–7881.

Marzesco, A. M., Dunia, I., Pandjaitan, R., Recouvreur, M., Dauzonne, D., Benedetti, E. L., Louvard, D., and Zahraoui, A. (2002). The small GTPase Rab13 regulates assembly of functional tight junctions in epithelial cells. *Mol. Biol. Cell* **13,** 1819–1831.

Michl, P., Barth, C., Buchholz, M., Lerch, M. M., Rolke, M., Holzmann, K.-H., Menke, A., Fensterer, H., Giehl, K., Löhr, M., Leder, G., Iwamura, T., *et al.* (2003). Claudin-4 expression decreases invasiveness and metastatic potential of pancreatic cancer. *Cancer Res.* **63,** 6265–6271.

Morimoto, S., Nishimura, N., Terai, T., Manabe, S., Yamamoto, Y., Shinahara, W., Miyake, H., Tashiro, S., Shimada, M., and Sasaki, T. (2005). Rab13 mediates the continuous endocytic recycling of occludin to the cell surface. *J. Biol. Chem.* **280,** 2220–2228.

Ohnishi, H., Nakahara, T., Furuse, K., Sasaki, H., Tsukita, S., and Furuse, M. (2004). JACOP, a novel plaque protein localizing at the apical junctional complex with sequence similarity to cingulin. *J. Biol. Chem.* **279,** 46014–46022.

Pfeffer, S. R., and Aivazian, D. (2004). Targeting Rab GTPases to distinct membrane compartments. *Nat. Rev. Mol. Cell Biol.* **5,** 886–896.

Schneeberger, E. E., and Lynch, R. D. (2004). The tight junction: A multifunctional complex. *Am. J. Physiol. Cell Physiol.* **286,** C1213–C1228.

Takai, Y., Sasaki, T., and Matozaki, T. (2001). Small GTP-binding proteins. *Physiol. Rev.* **81,** 153–208.

Terai, T., Nishimura, N., Kanda, I., Yasui, N., and Sasaki, T. (2006). JRAB/MICAL-L2 is a junctional Rab13-binding protein mediating the endocytic recycling of occludin. *Mol. Biol. Cell* **17,** 2465–2475.

Tobioka, H., Isomura, H., Kokai, Y., Tokunaga, Y., Yamaguchi, J., and Sawada, N. (2004). Occludin expression decreases with the progression of human endometrial carcinoma. *Hum. Pathol.* **35,** 159–164.

Tsukita, S., Furuse, M., and Itoh, M. (2001). Multifunctional strands in tight junctions. *Nat. Rev. Mol. Cell Biol.* **2,** 285–293.

Vasioukhin, V., and Fuchs, E. (2001). Actin dynamics and cell–cell adhesion in epithelia. *Curr. Opin. Cell Biol.* **13,** 76–84.

Weber, E., Berta, G., Tousson, A., St John, P., Green, M. W., Gopalokrishnan, U., Jilling, T., Sorscher, E. J., Elton, T. S., Abrahamson, D. R., and Kirk, K. L. (1994). Expression and polarized targeting of a rab3 isoform in epithelial cells. *J. Cell Biol.* **125,** 583–594.

Wittchen, E. S., Haskins, J., and Stevenson, B. R. (1999). Protein interactions at the tight junction. Actin has multiple binding partners, and ZO-1 forms independent complexes with ZO-2 and ZO-3. *J. Biol. Chem.* **274,** 35179–35185.

Zahraoui, A., Joberty, G., Arpin, M., Fontaine, J. J., Hellio, R., Tavitian, A., and Louvard, D. (1994). A small rab GTPase is distributed in cytoplasmic vesicles in non polarized cells but colocalizes with the tight junction marker ZO-1 in polarized epithelial cells. *J. Cell Biol.* **124,** 101–115.

Zerial, M., and McBride, H. (2001). Rab proteins as membrane organizers. *Nat. Rev. Mol. Cell Biol.* **2,** 107–117.

CHAPTER ELEVEN

A SMAP Gene Family Encoding ARF GTPase-Activating Proteins and Its Implication in Membrane Trafficking

Kenji Tanabe,* Shunsuke Kon,* Nobuyuki Ichijo, Tomo Funaki, Waka Natsume, Toshio Watanabe, *and* Masanobu Satake

Contents

Abstract

SMAP1 and SMAP2 proteins constitute a subfamily of the Arf-specific GTPase-activating proteins. Both SMAP proteins bind to clathrin heavy chains and are involved in the trafficking of clathrin-coated vesicles. In cells, SMAP1 regulates Arf6-dependent endocytosis of transferrin receptors from the coated pits of the

Department of Molecular Immunology, Institute of Development, Aging and Cancer, Tohoku University, Sendai, Japan
* These two authors contributed equally to this chapter

Methods in Enzymology, Volume 438
ISSN 0076-6879, DOI: 10.1016/S0076-6879(07)38011-7

plasma membrane, whereas SMAP2 regulates Arf1-dependent retrograde transport of TGN38 from the early endosome to the trans-Golgi network. The common and distinct features of SMAP1 and SMAP2 activity provide a valuable opportunity to examine the differential regulation of membrane trafficking by these two proteins. In this chapter, we describe several basic experimental procedures that have been used to study the regulation of membrane trafficking using SMAP proteins, including a GAP assay as well as procedures to study the transport of transferrin receptors and TGN38. In addition, a yeast two-hybrid system is described because of its utility in identifying novel molecules that interact with SMAP.

1. INTRODUCTION

1.1. Arf and ArfGAP

Arfs and their related Arf-like proteins are essential regulators of membrane trafficking. Arfs are low molecular mass GDP/GTP-binding GTPases that cycle between GDP-bound inactive and GTP-bound active forms. The interaction of Arf with an Arf-specific guanine nucleotide-exchanging factor (GEF) replaces GDP with GTP, which converts the Arf to an active state. Interaction of Arf with an Arf-specific GTPase-activating protein (GAP) triggers the GTPase activity of Arf, resulting in an inactive state. In mammals, six Arf homologs (Arf1-6) and numerous other Arf-like proteins have been identified. At least 18 distinct genes have been reported to code for ArfGAPs in mammals. The ArfGAPs can be categorized into seven subfamilies based on their structural features (see Fig. 1 in Randazzo and Hirsh, 2004, and Fig. 3 in Tanabe *et al.*, 2006). They include ArfGAP1/3, GIT1/2, ASAP/AMAP1–3, ACAP1–3, ARAP1–3, and AGAP1–3. The specific combination of an Arf isoform and ArfGAP sub-type in cells appears to be an important determinant of vesicle formation and budding from different parts of cellular membranes. The SMAP1 and SMAP2 genes that we have recently identified have been shown to constitute a small subfamily of the ARFGAPs (Tanabe *et al.*, 2005, 2006; Natsume *et al.*, 2006).

1.2. Common features of SMAP1 and SMAP2

The amino acid (aa) sequences of the SMAP proteins are shown in Fig. 11.1. Murine SMAP1 and SMAP2 are composed of 440 and 428 aa residues, respectively, and share 50% overall sequence homology. There are four distinct regions of the proteins that show a marked degree of

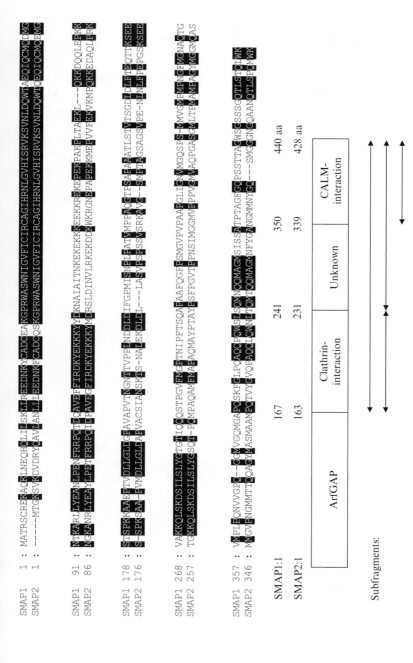

Figure 11.1 An upper panel shows the aa sequence of SMAP1 and SMAP2 proteins in which identical residues are indicated by white letters in a black background. The SMAP1 and SMAP2 proteins are constituted by 440 and 428 aa residues, respectively. A lower panel shows the domain architecture found in the SMAP protein. Four subdomains are found within the protein. They are an ArfGAP domain, a clathrin-interacting domain, a domain whose function is yet unknown, and a CALM-interacting domain. The boundaries between each domain are indicated by the aa number. Bars with bidirectional arrowheads indicate the fragments that were used as bait in the yeast two-hybrid assays.

conservation between SMAP1 and SMAP2. Interestingly, each region appears to correspond to a functionally important domain of the protein.

The amino-terminal region of the SMAP proteins is an ArfGAP domain that shows 85% aa identity between SMAP1 and SMAP2. Indeed, both SMAP1 and SMAP2 proteins exhibit GAP activity on a GTP-loaded Arf protein *in vitro*, but neither SMAP1 nor SMAP2 show preference for either Arf1 or Arf6 as substrates. The second-most amino-terminal region and the carboxy-terminal region of each SMAP protein interact with clathrin heavy chains and the clathrin assembly protein CALM, respectively (note that, to our knowledge, only SMAP2 binding to CALM has been demonstrated experimentally). The precise mechanism for how these two interacting domains contribute to the vesicle formation is not known; however, the presence of clathrin- and CALM-interacting domains is relevant to our observations that both SMAP proteins are involved in the transport of clathrin-coated vesicles. As for the clathrin-binding domains, aa sequences LLGLD (aa 192–196) of SMAP1 and LLGLD (aa 187–191) and DLL (aa 212–214) of SMAP2 are thought to be responsible for the interaction with clathrin. A consensus motif that is essential for the CALM interaction has not yet been deduced, although the carboxy-terminal region contains a unique abundance of Gln, Gly, Met, and Pro residues. There is one region between the clathrin- and CALM-binding domains whose function is not known. This region shows 69% homology between SMAP1 and SMAP2 and does not appear to have a homologous sequence in the protein database. It is possible that this unknown region might have a novel function in vesicle formation.

1.3. Distinct features of SMAP1 and SMAP2

In addition to the similarities described above, SMAP1 and SMAP2 also exhibit features that distinguish the two proteins from each other. The first difference is their subcellular localization. SMAP1 is diffusely localized in the cytoplasm with high concentrations in the juxtamembranous region, whereas SMAP2 is localized as punctuates on the early endosome as well as within the trans-Golgi network (TGN). The second difference is the clathrin-adaptor molecules that these proteins utilize. SMAP1 colocalizes with AP-2, whereas SMAP2 colocalizes with AP-1 and EpsinR. The third difference is the pathway of membrane trafficking with which SMAP1 and SMAP2 interact. Overexpression of SMAP1 abrogates the internalization of transferrin receptors from the plasma membrane of HeLa cells, whereas overexpression of SMAP2 delays the transport of TGN38 from the early endosome to the TGN in Cos-7 cells. The fourth difference is the sensitivity of SMAP1- and SMAP2-transfected cells to brefeldin A, an inhibitor of Arf1-GEFs, but not Arf6-GEFs. Cells overexpressing GAP-negative mutant of SMAP1 are sensitive to brefeldin A, whereas those overexpressing GAP-negative mutant of SMAP2 are resistant to the Arf1-GEF inhibitor. This indicates that active form of Arf1 accumulated in cells overexpressing GAP-negative mutant of SMAP2. Based on these differences, we have proposed that SMAP1 regulates Arf6-dependent endocytosis of transferrin receptors from coated pits of the plasma membrane, whereas SMAP2 regulates Arf1-dependent retrograde transport of TGN38 from the early endosome to the TGN.

1.4. Issues and experimental procedures to study SMAP functions

Both the common and distinct features of SMAP1 and SMAP2 raise several interesting issues and challenges in studying SMAP function. These issues include: (1) the specificity of GAP activity of the two SMAPs on Arf1 and Arf6 (both SMAP proteins act nonpreferentially on Arf1 and Arf6 *in vitro*, whereas SMAP1 and SMAP2 appear to act preferentially on Arf6 and Arf1, respectively, *in vivo*); (2) the differential subcellular localization of SMAP1 and SMAP2 (which might help explain specificity issue); (3) the mechanisms responsible for how interactions with clathrin and/or CALM contributes to vesicle formation; and (4) the potential role of a yet unknown, but conserved region of the two proteins.

In this chapter, we describe several basic experimental procedures that are necessary to pursue the issues described above. Elucidating these issues is considered to be pivotal not only for studying SMAP, but also for comparing the SMAP proteins with other ArfGAPs. SMAP appears to stand as a unique subfamily among the ArfGAPs because SMAP possesses clathrin- and CALM-binding regions, and is involved in a critical trafficking pathway from the plasma membrane to the TGN via early endosomes. The experimental procedures described below include a GAP assay as well as examining the transport of transferrin receptors, E-cadherin, and TGN38. E-cadherin is mentioned because SMAP1 plays a significant role in E-cadherin internalization (Kon *et al.*, 2008). In addition, we describe a yeast two-hybrid screen that we have found useful in detecting the interaction between SMAP2 and CALM and that we expect to be useful in identifying additional SMAP-interacting molecules.

At the end of chapter, we mention gene targeting of SMAP in mice and summarize the current knowledge regarding the human SMAP genes. These two aspects are still premature, but are certainly important for the roles of SMAP in human disease.

2. Experimental Procedures and Results

2.1. Preparation of recombinant proteins

Preparation of SMAP proteins is necessary to explore their GAP activity. We have used an ArfGAP domain (aa 1–255 of SMAP1 and aa 1–163 of SMAP2, respectively) for bacterial expression because expression of the intact form of SMAP1 or SMAP2 in *Escherichia coli* resulted in an insoluble or highly toxic protein in these cells. Each recombinant protein was tagged with glutathione S-transferase (GST) at the amino-terminus and with 6′His at the carboxy-terminus. The recombinant cDNAs were inserted into a pGEX5X-3 vector (Amersham Pharmacia Biotech, Piscataway, NJ).

1. Transform the *E. coli* strain BL21 with the expression plasmid, and culture it overnight at 37° in 10 ml Luria–Bertani (LB) medium containing ampicillin.
2. Transfer the above culture into 250 ml fresh LB medium containing ampicillin and incubate at 37° for 3 h.
3. Add isopropyl-β-D-thiogalactoside (IPTG) (0.4 mM at final concentration) to induce protein expression.
3. After 1 h at 37°, harvest the cells by centrifugation at 5000 rpm for 15 min at 4°.
4. Purify the GST–fusion protein using a B-PER GST-spin purification kit (Pierce Chemical, Rockford, IL). Suspend the harvested cells in 10 ml B-PER supplemented with 5 mM MgCl$_2$, 5 μg/ml DNase I, and 200 μg/ml lysozyme, and gently shake the mixture for 10 min at room temperature (RT).
5. Remove debris and inclusion bodies by centrifugation at 15 000 rpm for 15 min at 4°.
6. Add 1 ml immobilized glutathione beads (50% slurry) to the supernatant and gently shake the mixture for 10 min at RT followed by centrifugation at 5000 rpm for 2 min.
7. Wash the beads twice with wash buffer, suspend them in 0.25 ml wash buffer, and transfer them to a microfilter spin column.
8. Follow the next procedure according to manufacturer's instructions. To remove the GST tag, digest the fusion protein (<300 mg/ml) with 4 U/ml Factor Xa (Novagen, Madison, WI) for 16 h at 37°. Purify the GST-free protein using a 6×His spin purification kit (Pierce Chemical). Measure the protein concentration using a BCA protein assay kit (Pierce Chemical). Purity of the recovered protein should be ∼95% as determined by SDS–PAGE.

2.2. *In vitro* GAP assay

GAP activity of SMAP proteins was measured as described in the literature (Randazzo *et al.*, 1994; Randazzo, 1997).

1. Incubate recombinant non-myristoylated Arf1 or Arf6 proteins with $[\alpha^{32}P]$ GTP in GTP loading buffer for 60 min at 30°. The buffer is 25 mM HEPES, pH 7.4, 100 mM NaCl, 25 mM KCl, 2.5 mM MgCl$_2$, 1 mM EDTA, 0.1% (v/v) Triton X-100, 1.25 U/ml pyruvate kinase, 3 mM phospho(enol)pyruvate, 1 mM DTT, and 9.25 MBq $[\alpha^{32}P]$GTP.
2. Initiate GTP hydrolysis at 30° by adding SMAP1 or SMAP2 protein in GTP-hydrolysis buffer (25 mM HEPES, pH 7.4, 2.5 mM MgCl$_2$, 100 mM NaCl, 1 mM GTP, 0.1% [v/v] Triton X-100, and 1 mM DTT).
3. Terminate the reaction by adding a 50-fold volume of wash buffer (20 mM Tris-HCl, pH 8.0, 100 mM NaCl, 5 mM MgCl$_2$, and 2 mM DTT).
4. Remove unbound nucleotides by passing the reaction mixture through a 0.45 μm nitrocellulose filter (Millipore, Bedford, MA) and washing the filter with a 10-fold volume of wash buffer.
5. Elute GTP and GDP from the Arf protein with 1 ml 2 M formic acid.
6. Separate GDP from GTP by thin layer chromatography, developed in 1 M LiCl/1 M formic acid.
7. Dry and expose the plate to a phosphorimager screen. Detect and quantify radioactivity using Bio Imaging Analyzer (Fuji Photo Film Co., Ltd., Tokyo, Japan).

2.3. Expression plasmids for cell culture

Epitope-tagged SMAP1 and SMAP2 cDNAs were used to transfect cultured cells. The GenBank accession numbers of cDNAs are BC006946 for murine SMAP1 and BC052413 for murine SMAP2. An epitope derived from influenza hemagglutinin (HA) was tagged at the amino-terminus of each protein and the tagged cDNA was inserted into a pcDNA3 vector (Invitrogen, Carlsbad, CA). In some cases, Myc-tagged SMAP was also used. SMAP1R61Q and SMAP2R56Q were used to generate GAP-negative mutants because these mutations have been shown to eliminate GAP activity. Mutations were also introduced into the clathrin-binding motifs of SMAP. SMAP1 residues LLGLD (aa 192–196) and SMAP2 residues LLGLD (aa 187–191) and DLL (aa 212–214) were replaced with alanine. These mutants were unable to interact with clathrin heavy chains and were designated clathrin box mutants (CBm) of SMAP.

2.4. Internalization of transferrin

To examine the effect of HA–SMAP1 overexpression on transferrin internalization, we observed transfected HeLa cells using immunofluorescence microscopy. Representative results from these studies are shown in Fig. 11.2. The HA–SMAP1– and HA–SMAP1R61Q-transduced HeLa cells did not incorporate transferrin into the cytoplasm (indicated by arrows), whereas the untransduced cells did.

1. Culture HeLa cells in DMEM supplemented with 10% (v/v) FBS on a six-well plate coverslip.
2. Transfect cells with a SMAP1-expression vector using Effectene (QIAGEN, Chatsworth, CA) according to the manufacturer's instructions.

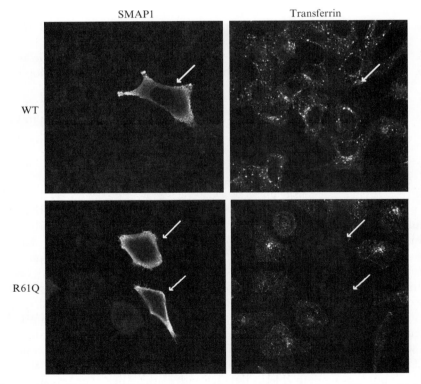

Figure 11.2   Effect of SMAP1 overexpression on transferrin internalization. HeLa cells were transiently transfected by HA-SMAP1 (wildtype) or HA-SMAP1R61Q (a GAP-negative mutant), incubated with dye-conjugated transferrin, and processed for immunofluorescent analyses. Note that the untransfected cells incorporated transferrin into the cytoplasm, whereas the SMAP1-transduced cells (indicated by arrows) did not.

3. At 24 h after transfection, incubate the cells with 25 μg/ml Alexa488-conjugated human transferrin (Molecular Probe, Eugene, OR) at 37° for 15 min. After incubation, wash the cells with PBS.

4. Fix the cells with 3.7% (v/v) formaldehyde in PBS for 15 min at RT, and wash three times with PBS.

5. Immerse the cells in 1% (w/v) bovine serum albumin (BSA) and 0.1% (v/v) Triton X-100 in PBS for 15 min at RT. This step represents the blocking reaction.

6. Incubate the cells with rat anti-HA monoclonal antibody (3F10, Roche Diagnostics, Indianapolis, IN) and then with Cy3-conjugated goat anti-rat IgG antibody (Chemicon, Temecula, CA) at RT for 45 min each. Anti-HA and anti-rat IgG antibodies are diluted 100- and 200-fold, respectively, in PBS containing 1% (w/v) BSA. Wash the cells with PBS after each antibody incubation.

7. Mount the coverslips on glass slides using Vectashield (Vector Laboratories, Burlingame, CA) and observe the cells using a confocal microscope (LSM-410, Zeiss, Thornwood, NY).

9. Quantify the number of transferrin-internalized cells among the SMAP1-positive population.

2.5. Internalization of E-cadherin

To examine the effect of HA-SMAP1 overexpression on the internalization of E-cadherin, we observed transfected Mardin-Darby canine kidney (MDCK) epithelial cells using immunofluorescence microscopy.

1. Culture MDCK cells in DMEM supplemented with 10% (v/v) FBS and 50 μg/ml gentamycin on a six-well plate coverslip.

2. Transfect 50 to 70% confluent cells with a SMAP1-expression vector using Lipofectamine 2000 (Invitrogen), according to the manufacturer's instructions. First, add 10 μl Lipofectamine 2000 to 250 μl DMEM, mix well, and allow the mixture to set for 5 min at RT. Second, add 4 μg plasmid DNA in 250 μl DMEM to the mixture and allow this lipid-DNA complex solution to sit for 20 min at RT. Third, add the solution to the cultured cells, adding only a drop at a time.

3. Incubate the cells for 48 h at 37° or until a confluent monolayer is reached.

4. To induce the E-cadherin internalization, incubate cells with 100 nM 12-0-tetradecanoylphorbol-13-acetate (TPA) at 37° for 2 h and then briefly wash the cells with PBS.

5. Fix the cells with 3.7% (w/v) formaldehyde in PBS for 15 min, and then wash them three times with PBS.

6. Immerse the cells in 1% (w/v) BSA and 0.1% (v/v) Triton X-100 in PBS for 15 min at RT. This step represents the blocking reaction.

7. Incubate the cells with the mixture of mouse anti-E-cadherin antibody (Transduction Laboratories, Lexington, KY) and rat anti-HA antibody at RT for 1 h. Each antibody is diluted 100-fold in PBS containing 1% (w/v) BSA. After incubation, wash the cells three times with PBS.

8. Incubate the cells with the mixture of Alexa488-conjugated goat anti-mouse IgG (Molecular Probes) and Cy3-conjugated goat anti-rat IgG at RT for 45 min. Each antibody is diluted 200-fold in PBS containing 1% (w/v) BSA. After incubation, wash the cells three times with PBS.

9. Mount the coverslips on glass slides with Vectashield, and observe the cells using a confocal microscope.

10. Quantify the number of E-cadherin–internalized cells in the SMAP1-positive population.

2.6. Transport of CD25-TGN38

CD25-TGN38 is a chimeric protein in which an extracellular domain of CD25 and a transmembrane and cytoplasmic domain of TGN38 are fused using PCR. To assess the involvement of SMAP2 in early endosome-to-TGN trafficking, we used this chimera, which has been shown to be transported from the plasma membrane to the endosome and subsequently to the TGN (Ghosh *et al.*, 1998). Representative results from these studies are shown in Fig. 11.3. In the cells indicated by arrowheads, TGN38 accumulated on the TGN, whereas in the cells indicated by arrows, TGN38 remained on the early endosome and did not reach the TGN.

1. Culture Cos-7 cells in DMEM supplemented with 10% (v/v) FBS on a coverslip using six-well plates.

2. Cotransfect cells with HA-SMAP2- and CD25-TGN38-expression vectors using Effectene (Qiagen). Mix 0.4 μg of each plasmid DNA with 92.8 μl EC buffer and 6.4 μl of enhancer. Vortex the mixture and allow it to sit at RT for 5 min. Add 8 μl of Effectene to the mixture and allow it to sit at RT for 10 min. Mix the solution with 0.6 ml of DMEM, and add this combined mixture to the cells. Incubate the cells for 24 h.

3. Incubate the cells in 0.1% (w/v) BSA in DMEM at 37° for 30 min.

4. Add 10 μg/ml mouse anti-CD25 antibody (Upstate Biotechnology, Lake Placid, NY) in DMEM to the cells and incubate at 37° for 15 to 45 min. This incubation allows the antibody-conjugated chimeric proteins to be internalized. After incubation, wash the cells three times with PBS.

5. Fix the cells with 3.7% (w/v) formaldehyde in PBS for 15 min, and then briefly wash them with PBS.

6. Permeabilize the cells with 0.1% TritonX-100 (v/v) in PBS containing 1% (w/v) BSA for 15 min.

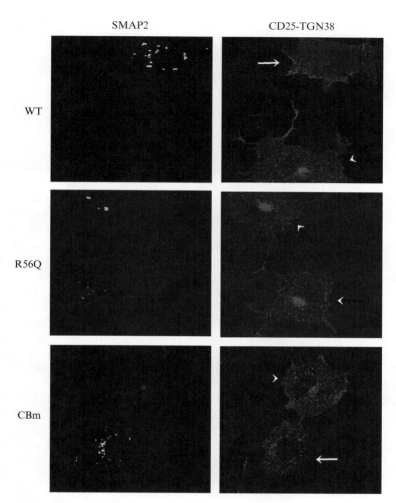

Figure 11.3 Effect of SMAP2 overexpression on the transport of CD25-TGN38. Cos-7 cells were transiently cotransfected with CD25-TGN38, and HA-SMAP2 (wildtype), HA-SMAP2R56Q (a GAP-negative mutant), or HA-SMAP2CBm (a clathrin box mutant). The cells were incubated with anti-CD25 antibody and processed for immuno-fluorescence. Note that TGN38 accumulated on the TGN in the cells indicated by arrow-heads and that the molecule did not reach TGN, but remained on the early endosome, in the cells indicated by arrows.

7. Incubate the cells with rat anti-HA antibody at RT for 1 h. The antibody is diluted 100-fold in PBS containing 1% (w/v) BSA. Wash the cells three times with PBS.
8. Incubate the cells with the mixture of Alexa488-conjugated goat anti-mouse IgG and Cy3-conjugated goat anti-rat IgG at RT for 45 min.

Each antibody is diluted 200-fold as described above. Wash the cells three times with PBS.

9. Mount coverslips on glass slides with Vectashield and observe the cells using a confocal microscope.

10. Quantify the number of CD25-TGN38–internalized cells in the SMAP2-positive population.

2.7. Yeast two-hybrid screening with SMAP as bait

Searching for candidate molecules that interact with SMAP proteins is an important step toward understanding the molecular function of SMAPs. Yeast two-hybrid screening is a convenient and useful method for this purpose. We used the MATCHMAKER Two-Hybrid System 3 (Clontech, Palo Alto, CA) with SMAP as bait and a cDNA library from mouse brain or testes as prey. A candidate region responsible for the interaction would be located downstream of the ArfGAP domain. We initially used aa 167–440 of SMAP1 as bait (see Fig. 11.1). Unexpectedly, a colony harboring only this SMAP1 fragment grew in the selection medium (SD/-Trp, -Leu, -His, -Ade, +5mM 3-amino-1, 2, 4-triazole [3AT]), indicating that the aa 167–440 fragment of SMAP1 fused to a GAL4DNA-binding domain by itself can activate the reporter gene in yeast cells. Therefore, we divided the aa 167–440 region of SMAP1 into several fragments as shown in Fig. 11.1. Using each of the subdivided fragments of SMAP1 as bait, we continued to observe self-activation of the reporter gene. On the other hand, colonies containing each fragment derived from SMAP2 (aa 163–428, aa 163–231, aa 231–428, aa 339–428, respectively) grew in the nonselection medium (SD/-Trp, -Leu) and not in the selection medium. Thus, only fragments of SMAP2 were used as bait for the yeast two-hybrid screening. The following screening protocol is a modification of manufacturer's manual (Yeast Protocol Handbook).

2.8. Library transformation and screening protocols

The pGBKT7-SMAP2 (aa 163–428) plasmid was introduced into AH109, and the transformant was selected on SD/-Trp plates. The resulting strain, AH109/pGBKT7-SMAP2 was used for further study.

1. Inoculate a 15-ml tube containing 2ml SD/-Trp medium with a single colony of AH109/pGBKT7-SMAP2. Incubate at 30° overnight while shaking at 250 rpm.

2. Inoculate a 500-ml flask containing 50 ml SD/-Trp with 2 ml of the culture that was incubated overnight. Incubate this solution at 30°

overnight while shaking at 250 rpm. Absorbance at 600 nm should reach 1.1 to 1.2.

3. Prepare 100 ml YPDA (yeast extract/peptone/dextrose/adenine) in a 500-ml flask. Add the AH109/pGBKT7-SMAP2 from step 2 to reach a final 0.4 OD600. Incubate at 30° for 5.5 h while shaking at 250 rpm. Absorbance should reach 1.3 to 1.5.

4. Divide the culture into two 50-ml tubes and centrifuge at 1800 rpm for 5 min. Decant the supernatant.

5. Suspend each pellet in 40 ml TE (Tris/EDTA buffer). Spin and decant the supernatant.

6. Suspend each pellet in 7.5 ml LiAc/TE (100 mM LiAc, 10 mM Tris-HCl, pH, 7.5, and 1 mM EDTA). Combine the suspension, spin, and decant the supernatant.

7. Suspend the pellet in an appropriate volume of LiAc/TE. If the absorbance of step 3 is 1.5, use 2 ml.

8. Place 15 μg of library plasmid DNA and 100 μg (100 μl of 1 mg/ml) of denatured herring sperm DNA (Clontech) in a 50-ml tube. Add 1 ml of the suspension from step 7 and mix well with a pipette.

9. Add 7 ml 40% (w/v) polyethylene glycol in LiAc/TE and mix well by inversion. Incubate the mixture at 30° for 30 min while shaking at 250 rpm.

10. Add 880 μl DMSO and mix by inversion. Heat shock at 42° for 15 min while swirling constantly. After the heat shock, rapidly chill the tube on ice for 2 min.

11. Centrifuge at 1800 rpm for 5 min.

12. Wash the cells with 10 ml YPDA twice.

13. Suspend the pellet in 35 ml YPDA. Transfer the suspension into a 500-ml flask and incubate it at 30° for 1 h while shaking at 250 rpm.

14. Centrifuge at 1800 rpm for 5 min and then decant the supernatant.

15. Suspend the cell pellet in 40 ml TE. Spin and decant the supernatant.

16. Suspend the cell pellet in 5 ml TE. To measure the transformation efficiency, plate $1/10^4$ of total suspension on SD/-Trp, -Leu medium. We obtained an efficiency of 0.76×10^5 colonies/μg library DNA.

17. Spread 125 μl each of suspension on one 100-mm plate of SD/-Trp, -Leu, -His, -Ade, +2.5 mM 3AT medium. We screened 4.3×10^6 independent colonies.

18. Incubate the plates at 30° for 1 week. Pick up and streak grown colonies on SD/-Trp, -Leu, -His, -Ade, +2.5 mM 3AT/X-β-Gal (20 μg/ml) plates. Select Ade+/His+/Mel+ (X-β-Gal+) colonies for further analyses. We obtained 179 positive colonies by screening a mouse brain library.

2.9. Interaction of SMAP2 with sumoylation factors

Among the cDNA clones isolated using yeast two-hybrid screening, there were Ubc9, protein inactivator of activated STAT proteins 1 (PIAS1), and PIAS3. Ubc9 represents an E2 enzyme involved in a sumoylation pathway, whereas PIAS1 and PIAS3 represent E3 ligases. The cDNAs recovered in pGADT7 corresponded to aa 1–158 of Ubc9 (total 158 aa), aa 1–230 of PIAS1 (total 651 aa), and aa 338–570 of PIAS3 (total 619 aa), respectively.

Due to the fact that several subfragments of SMAP2 can be used as bait in pGBKT7, we examined which region of SMAP2 interacts with Ubc9, PIAS1, and PIAS3. The results obtained for PIAS3 are shown in the upper panel of Fig. 11.4. Each combination of bait and prey (nos. 1 to 10) grew as colonies on the nonselection medium (SD/-Trp, -Leu), whereas only the combination of SMAP2 (aa 163–428) or SMAP2 (aa 339–428) and PIAS3 (Nos. 7 and 10, respectively) grew on the selection medium (SD/-Trp, -Leu, -His, -Ade, +5 mM 3AT). Thus, the region that is responsible for the interaction of SMAP2 with PIAS3 appears to lie in the aa 339–428 region of SMAP2, which corresponds to a CALM-interacting domain. On the other hand, when using PIAS1 or Ubc9 as prey, only the combinations of no. 7 (aa 163–428) and no. 9 (aa 231–428) allowed colonies to grow in the nonselection medium (data not shown). Therefore, we

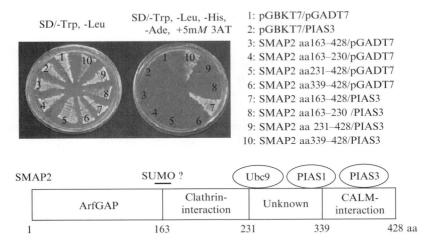

Figure 11.4 Interaction of SMAP2 with sumoylation factors. In the upper panel, SMAP2 was tested for its ability to interact with PIAS3. Yeast cells were cotransformed by the indicated combination of bait (SMAP2 in pGBKT7) and prey (PIAS3 in pGADT7) vectors, and grown on nonselective (SD/-Trp, -Leu) and selective (SD/-Trp, -Leu, -His, -Ade, +5 mM 3AT) media, respectively. In the lower panel, the subdomains of SMAP2 that likely interact with Ubc9, PIAS1, and PIAS3 are indicated. A putative sumoylation site (aa 161–164) in SMAP2 is also indicated.

deduced that the unknown region of SMAP2 (aa 231–339) is likely involved in the interaction of SMAP2 with PIAS1 and Ubc9. It was impossible to directly test the aa 231–339 fragment of SMAP2 using the yeast two-hybrid screen, however, because the aa 231–339 fragment alone activated the reporter. These results from the yeast two-hybrid screening are schematically illustrated in the lower panel of Fig. 11.4.

It is not clear what significance the interactions described above might have on the function of SMAP2 or whether SMAP2 is sumoylated. It must be noted, however, that the SMAP2 protein contains a possible sumoylation site at aa 161–164 (VKMP) as determined by SUMOplot Prediction (score 0.82).

2.10. The human SMAP genes

The human SMAP1 (AY055004) and SMAP2 (BC021133) genes are located at 6q12–q13 and 1p35.3–p34.1, respectively. Loci harboring SMAP1 or SMAP2 do not contain any OMIM information. Single nucleotide polymorphisms (SNP) in the exon sequences of SMAP1 are as follows according to the National Center for Biotechnology Information (NCBI) database: A or T at codon 3 of aa 10 (both for Ala); T or C at codon 2 of aa 185 (T for Val and C for Ala); A or C at codon 2 of aa 367 (A for His and C for Pro); and A or T at codon 3 of aa 387 (both for Ala). An SNP of SMAP2 is G or C at codon 3 of aa 223 (both for Ser). The significance of these SNPs and the relevance of SMAP genes to human diseases are currently unknown.

We have generated SMAP-targeted mice including SMAP1(-/-), SMAP2(-/-), and SMAP1(-/-); SMAP2(-/-). Mice of each genotype have exhibited interesting pathological phenotypes thus far. We anticipate that further study of these genetically modified mice will lead to discovery of the possible involvement of SMAP genes in human diseases.

REFERENCES

Ghosh, R., Mallet, W. G., Soe, T. T., McGraw, T. E., Frederick, R., and Maxfield, F. R. (1998). An endocytosed TGN38 chimeric protein is delivered to the TGN after trafficking through the endocytic recycling compartment in CHO cells. *J. Cell Biol.* **142,** 923–936.

Kon, S., Tanabe, K., Watanabe, T., Sabe, H., and Satake, M. (2008). Clathrin dependent endocytosis of E-cadherin is regulated by the Arf6GAP isoform SMAP1. *Exp. Cell Res.* (in press).

Natsume, W., Tanabe, K., Kon, S., Yoshida, N., Watanabe, T., Torii, T., and Satake, M. (2006). SMAP2, a novel ARF GTPase-activating protein, interacts with clathrin and clathrin assembly protein, and functions on the AP-1-positive early endosome/trans-Golgi-network. *Mol. Biol. Cell* **17,** 2592–2603.

Randazzo, P. A. (1997). Functional interaction of ADP-ribosylation factor 1 with phosphatidylinositol 4,5-bisphosphate. *J. Biol. Chem.* **272,** 7688–7692.

Randazzo, P. A., and Hirsch, D. S. (2004). Arf GAPs: Multifunctional proteins that regulate membrane traffic and actin remodelling. *Cell Signal.* **4,** 401–413.

Randazzo, P. A., Terui, T., Sturch, S., and Kahn, R. A. (1994). The amino terminus of ADP-ribosylation factor (ARF) 1 is essential for interaction with Gs and ARF GTPase-activating protein. *J. Biol. Chem.* **269,** 29490–29494.

Tanabe, K., Kon, S., Natsume, W., Torii, T., Watanabe, T., and Satake, M. (2006). Involvement of a novel ArfGAP protein, SMAP, in membrane trafficking: Implications in cancer cell biology. *Cancer Sci.* **97,** 801–806.

Tanabe, K., Torii, T., Braesch-Andersen, S., Watanabe, T., and Satake, M. (2005). A novel GTPase-activating protein for ARF6 directly interacts with clathrin and regulates the clathrin-dependent endocytosis. *Mol. Biol. Cell* **16,** 1617–1628.

CHAPTER TWELVE

In Vitro and In Vivo Analysis of Neurotrophin-3 Activation of Arf6 and Rac-1

Pedro F. Esteban,* Paola Caprari,* Hye-Young Yoon,[†]
Paul A. Randazzo,[†] *and* Lino Tessarollo*

Contents

Abstract

Arf GTP-binding proteins and Rho-family GTPases play key roles in regulating membrane remodeling and cytoskeletal reorganization involved in cell movement. Several studies have implicated neurotrophins and their receptors as upstream activators of these small GTP-binding proteins, however, the mechanisms and the cell type specificity of this neurotrophin activity are still under investigation. Here we describe the rationale and protocols used for the dissection of an NT3 activated pathway that leads to the specific activation of Arf6 and Rac1.

1. Introduction

Neurotrophins are structurally related growth factors including nerve growth factor (NGF), brain–derived neurotrophic factor (BDNF), neurotrophin-3 and –4/5 (NT3 and NT-4/5) that control the growth and

* Neural Development Group, Mouse Cancer Genetics Program, National Cancer Institute, Frederick, Maryland
† Laboratory of Cellular and Molecular Biology, National Cancer Institute, Bethesda, Maryland

Methods in Enzymology, Volume 438
ISSN 0076-6879, DOI: 10.1016/S0076-6879(07)38012-9

differentiation programs of various cell lineages (Huang and Reichardt, 2001). These activities are mediated by two types of receptors. The Trk tyrosine kinase receptors bind selectively to neurotrophins, with NGF binding to TrkA; BDNF and NT-4/5 binding to TrkB; and NT3 binding to TrkC. Trk receptors mediate most of neurotrophins' trophic actions. The p75 neurotrophin receptor is a member of the TNF receptor family, binds neurotrophins with similar affinity, and has an intracellular domain that lacks intrinsic enzymatic activity, but can initiate neurotrophin mediated cell death (Kaplan and Miller, 2000). Neurotrophin functions have been studied mainly in cell populations of the nervous system, although neurotrophins have been shown to exert biological activities on a variety of cell populations outside the nervous system as well (Tessarollo, 1998).

The best-established roles for neurotrophins relate to their ability to control cell survival in the nervous system (Tessarollo, 1998). This function has been associated with Trk activation of intracellular signaling cascades, including the Ras/ERK protein kinase pathway, the PI3K/Akt kinase pathway and phospholipase C-γ1 (PLC-γ1) (Kaplan and Miller, 2000). The role of p75 is more complicated since it can augment the Trk prosurvival effect when both receptors are coexpressed, and can induce apoptosis when activated alone by a neurotrophin. For its role in survival, p75 activates NF-kB, while for apoptosis induction, it activates the c-jun N-terminal kinase (JNK) pathway (Kaplan and Miller, 2000). In addition to controlling cell survival of various populations of the nervous system, neurotrophins play an important biological role in the formation of neural networks and regulation of synaptic plasticity (McAllister *et al.*, 1999). These processes involve cell membrane trafficking and modifications of the cellular cytoskeleton. However, the molecular basis by which neurotrophins influence these cellular functions is not well understood.

A number of recent studies have shown that small GTP-binding proteins are downstream effectors of neurotrophin signaling (Hempstead, 2005). Small GTP-binding proteins are a large class of proteins that have been well characterized for their role as molecular switches that alter the cytoskeleton to mediate morphological changes and cell migration. In the nervous system, these proteins have been shown to regulate a variety of processes including dendritic development and synaptic activity in neurons (Ashery *et al.*, 1999; Hernandez-Deviez *et al.*, 2002, 2004), migration of Schwann cells (Yamauchi *et al.*, 2003, 2004), and cell death of oligodendrocytes (Harrington *et al.*, 2002).

Small GTP-binding proteins include the Arf GTP-binding proteins and Rho-family GTPases. Arf proteins regulate membrane traffic in various subcellular compartments of the cell as well as assembly of the actin cytoskeleton (Nie and Randazzo, 2006). The most extensively studied Arf family proteins are Arf1 and Arf6. Arf1 is primarily localized to the Golgi, but also functions in the endocytic compartment. Arf6 associates with a tubular-vesicular endocytic compartment and, when activated by specific guanine nucleotide exchange factors (GEFs), the plasma membrane (Nie et al., 2003). Recently, it has been suggested that Arf6 plays an important role in synaptic transmission, most likely by making more vesicles available for fusion at the plasma membrane (Ashery et al., 1999). Moreover, Arf6 and Rac1 have been shown to regulate dendritic arbor development, at least in isolated hippocampal neurons (Hernandez-Deviez et al., 2002).

Several studies have shown that most neurotrophins and their receptors or even other ligand/receptor systems can activate the same GTP-binding protein (e.g., Rac1), leading to different cellular outcomes (Harrington et al., 2002; Hempstead, 2005; Miyamoto et al., 2006). These findings have raised the important questions of which and how many pathways activate a specific GTP-binding protein, and in what hierarchical order they function to determine a specific cellular outcome. While the complexity of this task is obvious, recent work has started to identify some of the pathways leading to the activation of small GTP-binding proteins (Miyamoto et al., 2006; Yamauchi et al., 2003, 2004, 2005a, 2005b). The in vivo integration of how these pathways may work in concert to affect specific cellular phenotypes will be a challenging task. Key tools to address these issues are provided first by in vitro systems that can help validating the interactions of the different component of the pathways. Second, in genetically engineered mouse models certain pathways can be eliminated, proving their physiological relevance on specific cellular changes.

Among neurotrophins, NT3 is the most challenging to study for its ability to activate small GTP-binding proteins. It is the most promiscuous neurotrophin, since in addition to binding to the TrkC receptor and p75, it can also activate TrkA and TrkB in specific cellular contexts. Moreover, the trkC locus encodes a variety of receptors, including truncated isoforms that lack kinase activity in addition to the canonical full-length tyrosine kinase receptor (Tessarollo, 1998). Interestingly, NT3 can activate Rac1 through the TrkC-kinase (TrkC.Kin) receptor, as well as the TrkC.T1 truncated isoform (Esteban et al., 2006; Yamauchi et al., 2005b) (Fig. 12.1). NT3 binding to TrkC.Kin leads to the activation of the GEF Tiam1 (T lymphoma invasion and metastasis) that, in concert with Ras, activates Rac1 and stimulates Schwann cell migration (Yamauchi et al., 2005b). The Ras/Tiam1/Rac1 signaling pathway is also present in neurons (Miyamoto et al., 2006) where the truncated receptor TrkC.T1 can lead to Rac1 activation through an independent signaling pathway including tamalin/ARNO/Arf6/Rac1 (Esteban et al., 2006) (Fig. 12.2). Thus, it is particularly challenging to dissect the significance of how a cell bearing multiple TrkC isoforms with identical affinity for a neurotrophin can activate the same GTP-binding protein, although through different pathways (see Fig. 12.2). One possibility is that the expression of various TrkC receptor isoforms is

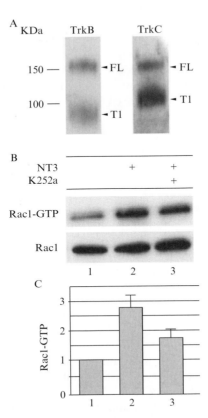

Figure 12.1 NT3 induces Rac1 activation in primary neurons. (A) Western blot analysis of wheat germ agglutinin–sepharose concentrated Trk protein from cultured cortical primary neurons. Blots were analyzed with a TrkB- (left, TrkB out) (Dorsey *et al.*, 2006) or a TrkC-specific (656) antiserum (Esteban *et al.*, 2006). (B) Rac1 activation in primary neurons. Cortical neurons were dissected from C57BL/6 wildtype newborn mice and plated on poly-L-lysine–coated 10-mm petri dishes as described in the Methods section. After 5 days *in vitro*, neurons were starved for 4 h (with unsupplemented DMEM) and stimulated with 100 ng/ml NT3 for 5 min (B lane 2), or pretreated with the kinase inhibitor K252a for 45 min before NT3 treatment (B lane 3). Rac1 activity (B, top panel) to total Rac1 protein normalized to the basal activity (lane 1) is shown. Input for Rac1 represents 5% of the total cell extract used to detect the active Rac1-GTP (top panel). (C) Quantification of Rac1 activation in response to NT3 treatment. Fold difference, mean (± standard error of the mean) relative to untreated cells (lane 1); means (± standard error of the mean) from three independent experiments are shown.

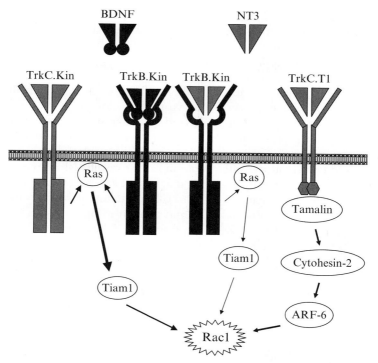

Figure 12.2 NT3 employs different signaling pathways to activate Rac1. Schematic representation of the signaling pathways used by NT3 and BDNF to activate Rac1. NT3 binding to TrkC.Kin induces Rac1 activation by recruiting and activating Ras and subsequently Tiam1. When binding to TrkC.T1, NT3 causes the recruitment of tamalin to the TrkC.T1 cytoplasmic domain. In turn, tamalin activates the guanine nucleotide exchange factors cytohesin-2/ARNO (Kitano *et al.*, 2002), which causes the switch of Arf6 to its active form. Activated Arf6 induces Rac1 activation and is translocated to the membrane where ruffling and actin reorganization take place. Arrow thickness indicates the level of activity of a specific pathway (the thickest, the highest) to underline that the level of Rac1 activation may be lower by TrkB.Kin when binding its nonpreferred ligand NT3 instead of its high-affinity ligand BDNF.

spatiotemporally regulated in different cell compartments. This subcellular localization may allow a particular TrkC isoform to couple with a specific Rac1 GEF. Moreover, a cellular response could be determined not only by Rac1 activation independently, but also by its activation in concert with other small GTPases. This hypothesis would imply that what determines a specific cell phenotype is not the activation of Rac1 alone, but rather the combined activation of two or more GTP-binding proteins (e.g., Cdc42, RhoA, etc.). So, although many studies have identified Rac1 as a specific component of a downstream signaling pathway, it is still to be determined whether other small GTPase are part of the molecular machinery leading to the cellular outcome caused by the extracellular cue (Hempstead, 2005).

Here we describe how we have approached the dissection of a pathway leading to the activation of the small GTP-binding proteins ARF6 and Rac1 by NT3 (Esteban *et al.*, 2006).

Following the identification of the scaffolding protein tamalin as a protein able to bind to a conserved cytoplasmic domain of the TrkC.T1 truncated receptor by yeast-two hybrid, we confirmed this interaction *in vitro* by co-transfection and co-immunoprecipitation assay in HEK293 cells. In fact, this cell line does not express either Trk receptors or p75, which could confound the interpretation of our results. Moreover, it does express cytohesins, a family of GEFs that have been previously shown to interact with tamalin (Kitano *et al.*, 2002). Importantly, *in situ* hybridization indicated that expression of TrkCT1 and tamalin in mouse brain overlaps, suggesting physiological relevance of our finding. A key finding of the study was that NT3 in cultured mouse hippocampal neurons promoted coloca-lization of the endogenous TrkC.T1 and tamalin proteins. However, since mouse hippocampal neurons express both TrkC.Kin and TrkC.T1 receptor isoforms, we had to use hippocampal neurons derived from a TrkC.Kin-deficient mouse (Fig. 12.3). This is important because any change caused by NT3 treatment in neurons would have been difficult to interpret because of the widespread presence of TrkC.Kin isoform (Esteban *et al.*, 2006; Menn *et al.*, 2000). Tamalin forms a complex with the guanine nucleotide exchange factor cytohesin2/ARNO (for the Arf

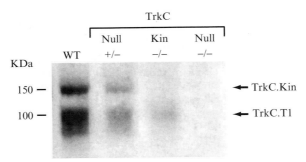

Figure 12.3   Western blot analysis of wheat germ agglutinin–sepharose concentrated TrkC protein in brain from mice carrying different targeted mutations. Wildtype (WT) and trkC +/- (Null) brain samples are from mice at postnatal day 40; trkC kinase -/- (Kin) and trkC -/- (Null) mice are early postnatal lethal, so these samples are from embryonic day 17.5. Note that the reduced level of TrkC.T1 in the trkC kinase -/- brain is due to the general lower level of TrkC receptors during embryonic development. An antiserum directed against the juxtamembrane domain of the trkC receptor (serum 656) was employed for the analysis. The signal detected below the truncated receptor in the first three lanes is caused by protein degradation.

Nucleotide-Binding Site Opener) and, in the HEK293 line with endogenous tamalin that was stably transfected with TrkC.T1, NT3 promoted transloca-tion of Arf6 to actin-rich membrane ruffles. Arf6 translocation was blocked by tamalin dominant/negatives and enhanced by tamalin overexpression. Indeed, overexpression of tamalin in the HEK 293 cells expressing TrkC. T1 caused the formation of cellular protrusion following NT3 treatment. The physiological significance of this cell phenotype is still unknown. Likewise, it is still not known how activation of Arf6 and Rac1 by NT3 affects cell morphology and/or function of neurons, the cell type expressing the highest levels of TrkC and tamalin. The following methodology and reagents should help elucidate the function of this pathway *in vivo*.

2. METHODS

2.1. Mouse strains and cell lines

The optimal way to dissect NT3/TrkC-mediated pathways *in vivo* employs specific mouse mutants that have been previously published, namely, the TrkC.Kin, TrkC.null, and the p75-deficient mouse models (see Fig. 12.3) (Klein *et al.*, 1994; Lee *et al.*, 1992; Tessarollo *et al.*, 1997). The TrkC.Kin and p75 mutant mice are available from Jackson Laboratories (Bar Harbor, ME), whereas the TrkC.null mouse strain is available from our own laboratory. The power of using these models is best illustrated in the paper by Esteban *et al.* (2006), where only by using neurons isolated from the TrkC.Kin mutant mouse was it possible to unequivocally demonstrate that the truncated TrkC receptor can recruit tamalin upon NT3 stimulation.

2.2. Neuron cultures

Hippocampal or cortical neurons are dissected and cultured from E16.5 (embryonic day 16.5) mice. Protocols for embryo dissection and neuronal cultures are widely available and described in detail elsewhere (e.g., *Current Protocols in Neuroscience*; Vicario-Abejon, 1997). However, in our experience we found that an untrained investigator should visit a laboratory where this technique is performed regularly to learn not only the procedure but also how to execute it in a timely manner to minimize cell trauma. Briefly, embryos are removed from the uterus and freed from the yolk sac. Heads removed from the embryos by decapitation are placed on a 35-mm Petri dish with 2 ml of cold Leibovitz's L-15 medium. The brains are freed of the meninges, and the dissected hippocampi or cortices are digested with trypsin, and dissociated by trituration in minimal essential medium (MEM) 10/10 (MEM with Earle's salts/2 mM glutamine/10% [vol/vol] fetal bovine serum/10% [vol/vol] horse serum/penicillin [100 units/ml]/streptomycin [100 units/ml])

(Bambrick *et al.*, 1995). Cells are then plated in serum-free supplement B27 (InVitrogen) at 104 cells/cm^2 on 12-mm glass coverslips for immunostaining and at 5 × 10^5 cells/35-mm dish for Western blots. At 1 day *in vitro*, MEM 10/ 10 is replaced with DMEM supplemented with B27 and 10 μM cytosine β-D-arabinofuranoside to kill proliferating cells, including astrocytes. Glass coverslips are coated with poly-L-lysine (Sigma, St. Louis, MO) and mouse laminin. Plastic dishes are coated with poly-L-lysine alone. Unless otherwise indicated, all cell culture reagents are obtained from Invitrogen (Carlsbad, CA).

Importantly, for TrkC immunostaining experiments, primary neurons need to be cultured for at least 3 weeks such that they produce enough endogenous TrkC receptor that can be detected by the monoclonal B27 antibody (Esteban *et al.*, 2006). (The B27 antibody can be requested from Uri Saragovi, McGill University, Montreal, Canada).

3. CELL CULTURE AND TRANSFECTION

The dissection of the TrkC.T1-activated pathway was performed in the human embryonic kidney 293 (HEK 293) cells because they lack any Trk or p75 neurotrophin receptors. This approach allows detection of cellular responses that are specific to TrkC.T1. HEK 293 cells are maintained in Dulbecco's modified Eagle medium (DMEM) supplemented with 10% FBS, 100 units/ml penicillin and 100 μg/ml streptomycin in a 37°, 5% CO$_2$ incubator. A clone of HEK 293 cells stably expressing TrkC.T1 (HEK293-TrkC.T1) was grown in the same manner, with the addition of 1 ng/ml puromycin.

4. DETERMINATION OF ACTIN-RICH PROTRUSION FORMATION

In some cell types, activation of recombinant Arf6 induces the formation of actin-rich protrusions (Radhakrishna *et al.*, 1996). We have used formation of these protrusions as a surrogate measure of Arf6 activation (Esteban *et al.*, 2006). To assess the effect of TrkC.T1 on the formation of Arf6-dependent actin-rich protrusions, we compare cells expressing TrkC.T1 to control cells lacking the receptor. We transiently express Arf6. The addition of another component of the pathway, tamalin, further drives formation of protrusions making the differences resulting from NT3 treatment more dramatic and easier to interpret. For transfection, control HEK293 and HEK293-TrkC.T1 cells were grown in six-well cell-culture plates (35-mm wells) to 60% confluence. Cells were transfected with mammalian expression vector, pcDNA3.1 (+), containing the open reading frame for FLAG-tagged wildtype human ARF6, and similarly constructed expression vectors for Myc-tagged wildtype tamalin or HA-tagged mutant 1~189 tamalin using the Fugene 6 reagent (Roche) following the manufacturer's instructions. Thirty-two hours after transfection, the cells were re-seeded on coverslips, coated with 15 μg/ml fibronectin (1 h, 37°), in serum-free media (OPTi-MEM, Invitrogen), and incubated for 12 h at 37°.

To determine responses to NT3, the cells were stimulated with 100 ng/ml NT3 for 10 min at room temperature. Immediately following treatment, the cells were fixed in phosphate-buffered saline (PBS) containing 2% formaldehyde (1 ml/well) for 10 min at room temperature. The cells were rinsed three times with 2 ml PBS to remove excess formaldehyde and once in blocking buffer (PBS containing 10% FBS and 0.04% sodium azide). The cells were permeabilized with staining buffer (PBS containing 10% FBS, 0.2% saponin, and 0.04% sodium azide) for 10 min at room temperature, and then incubated with the appropriate primary antibodies for 1 h at room temperature. Antibodies used were polyclonal anti-TrkC 656 (1:100) (a gift of Dr. Pantelis Tsoulfas, Miami University School of Medicine, Miami, FL); polyclonal anti-HA (1:1000) (Abcam, catalog no. ab9110); mouse monoclonal anti-HA (1:1000) (Roche Molecular Biochemicals, catalog no. 1583816); mouse monoclonal M5 anti-FLAG (1:1000) (Sigma, catalog no. F4042); rabbit polyclonal anti-FLAG (1:1000, Sigma, catalog no. F7425); polyclonal anti-Myc (1:100) (Santa Cruz Biotechnology Inc.); and mouse monoclonal anti-Myc (1:1000) (Upstate, clone 9E10, catalog no. 05–419). Unbound antibody was removed with three 10-min washes in blocking buffer. The cells were then incubated with FITC-conjugated goat anti-mouse IgG (Jackson Immuno Research, catalog no. 115-095-003); Alexa Fluor® 633-conjugated goat anti-rabbit secondary antibody (Invitrogen, catalog no. A21072, to detect tamalin and Arf6, respectively); and rhodamine-conjugated phalloidin (Invitrogen, catalog no. R415, to detect polymerized actin) for 1 h at room temperature. After three 10-min washes with blocking buffer, the coverslips were rinsed in PBS and distilled water. Glass coverslips were mounted on slides with DAKO Fluorescent Mounting Medium (DAKO). Fluorescence images were obtained with a Zeiss LSM 510 confocal microscope using a 63×/1.3 NA objective lens and collected at 1.4× zoom using the Zeiss LSM software package.

To interpret the results of these experiments, controls of HEK 293 cells that were not transfected, or transfected with the expression vector for Arf6 alone, for tamalin alone, or Arf6 and tamalin together, should be examined. The same combinations should be examined for the HEK 293 cells expressing TrkC.T1. The amounts of DNA used for the transfections should be optimized to obtain similar levels of protein produced by the different cDNA vectors. Because of the cDNA size difference, 2 μg of the Arf6 plasmid and 6 μg of the tamalin plasmid is a reasonable starting point.

The most dramatic differences are seen between HEK 293 and HEK 293-TrkC.T1 cells with both cell lines expressing Arf6, tamalin, and treated with both cell lines expressing Arf6 and tamalin, after treatment with NT3. Similar differences are observed in comparing HEK 293-TrkC.T1 cells expressing Arf6 and tamalin before and after NT3 treatment.

4.1. Rac1 activation

The changes in the actin cytoskeleton and signaling mediated by TrkC.T1 are downstream of Arf6-dependent activation of Rac (Donaldson, 2003; Radhakrishna *et al.*, 1999; Zhang *et al.*, 1998). The roles of various components of the signaling pathway (i.e., TrkC.T1, Arf6, and tamalin) can be determined using the HEK293 cell model as described above. In this case, the parental HEK 293 and the stable cell line HEK 293.TrkC.T1 are transiently transfected with plasmids for tamalin, Arf1T31N, and Arf6T27N. The GTP-bound "activated" form of Rac (or other Rho family members) can be detected by an assay using fragments of Rac or Rho binding proteins.

For detection of Rac1 activity in primary neurons, cells are dissected from mouse cortices or hippocampi as described above and plated on dishes coated with poly-L-lysine.

The assay to determine relative activation of Rac1, Cdc42, or RhoA *in vivo* is based on the preferential binding of Rac1•GTP and Cdc42•GTP, as compared to Rac1•GDP and Cdc42•GDP, to the p21-binding domain (PBD) of p21-activated protein kinase 1 (PAK1), and the preferential binding of RhoA•GTP, relative to RhoA•GDP to the Rho binding domain of rhotekin (Ren *et al.*, 1999). GST fused to PBD or rhotekin is mixed with cell lysates. The fusion proteins are precipitated by binding to glutathione conjugated to agarose beads. Rho, Rac1, or Cdc42 in the precipitate, presumed to be the GTP-bound forms, are detected by immunoblotting. Commercial kits are available that provide as main reagent glutathione-agarose beads loaded with GST fusions of the PBDomain of PAK1 or RBDomain of rhotekin (e.g., Chemicon International, Rac/Cdc42 activation assay kit, catalog no. catalog no. SGT445; Pierce, EZ-Detect Rac1 Activation Kit catalog no. 89856). It is also straightforward to express the fusion proteins in bacteria and adsorb the protein to glutathione agarose beads, which are available from GE-Biosciences.

In a typical experiment in which Rac1•GTP, Cdc42•GTP, or RhoA•GTP levels are determined, cells are grown in 35-mm wells. After treatment with NT3 for 5 min, cells are lysed in 0.5 ml of 50 mM Tris-HCl, pH 7.5, 100 mM NaCl, 2 mM MgCl$_2$, 1% Triton X-100, and protease inhibitors. We have used complete protease inhibitor tablets from Roche (catalog no. 11 836 153 001). The cell lysates are cleared by addition of 20 μl of sepharose CL-4B beads, mixing and separating the beads from the lysates by centrifugation at $16,000 \times g$ for 20 s. Fifty μg of the fusion protein immobilized on 20 μl of glutathione-sepharose CL-4B (or equivalent if using materials from a commercial kit) are added to the cleared lysates, and the mixture is incubated for 1 h at 4°. The beads are collected by centrifugation for 30 s at $13,000 \times g$. The beads are then washed three times with the lysis buffer. The proteins are extracted from the beads with 60 μl SDS-PAGE sample buffer at 95° for 5 min. Proteins are separated by PAGE using a 15% gel, transferred to Immobilon P membrane (Millipore, Bedford, MA), and detected by immunoblotting for the GTP-binding protein. The amount of activated relative to total GTP-binding protein can be determined by comparison of the signal to that obtained by immunoblot analysis of 1% of the total cell lysate.

There are several possible sources of variability in this assay. First is loss of GTP from the GTP-binding protein due to either hydrolysis or dissociation. These sources of loss of signal are minimized by strict temperature control throughout the procedure, maintaining the samples at 4° at all times, and by immediately processing the samples. Particularly when using GST fusion proteins already bound to beads, we have found variability as a consequence of lack of reproducibility in delivery of specific volumes of beads. This variability is minimized by using wide-bore pipette tips or by clipping the ends of the pipette tips to increase the bore. Linearity of signals is difficult to achieve in immunoblots as well. For this reason, we prefer titrating all samples and determining volumes that provide equivalent signals. For example, 5, 15, and 30 μl of beads extract, following extraction from each sample are separated by SDS-PAGE and processed for immunoblotting. If similar signals are obtained for 5 μl of sample A and 15 μl of sample B, we would conclude that sample A had threefold the amount of activated GTP-binding protein as sample B.

ACKNOWLEDGMENT

This research was supported by the Intramural Research Program of the National Institutes of Health, National Cancer Institute, Center for Cancer Research.

REFERENCES

Ashery, U., Koch, H., Scheuss, V., Brose, N., and Rettig, J. (1999). A presynaptic role for the ADP ribosylation factor (ARF)-specific GDP/GTP exchange factor msec7-1. *Proc. Natl. Acad. Sci. USA.* **96,** 1094–1099.

Bambrick, L. L., Yarowsky, P. J., and Krueger, B. K. (1995). Glutamate as a hippocampal neuron survival factor: An inherited defect in the trisomy 16 mouse. *Proc. Natl. Acad. Sci. USA.* **92,** 9692–9696.

Donaldson, J. G. (2003). Multiple roles for Arf6: Sorting, structuring, and signaling at the plasma membrane. *J. Biol. Chem.* **278,** 41573–41576.

Dorsey, S. G., Renn, C. L., Carim-Todd, L., Barrick, C. A., Bambrick, L., Krueger, B. K., Ward, C. W., and Tessarollo, L. (2006). *In vivo* restoration of physiological levels of truncated TrkB. T1 receptor rescues neuronal cell death in a trisomic mouse model. *Neuron* **51,** 21–28.

Esteban, P. F., Yoon, H. Y., Becker, J., Dorsey, S. G., Caprari, P., Palko, M. E., Coppola, V., Saragovi, H. U., Randazzo, P. A., and Tessarollo, L. (2006). A kinase-deficient TrkC receptor isoform activates Arf6-Rac1 signaling through the scaffold protein tamalin. *J. Cell Biol.* **173,** 291–299.

Harrington, A. W., Kim, J. Y., and Yoon, S. O. (2002). Activation of Rac GTPase by p75 is necessary for c-jun N-terminal kinase-mediated apoptosis. *J. Neurosci.* **22,** 156–166.

Hempstead, B. L. (2005). Coupling neurotrophins to cell migration through selective guanine nucleotide exchange factor activation. *Proc. Natl. Acad. Sci. USA.* **102,** 5645–5646.

Hernandez-Deviez, D. J., Casanova, J. E., and Wilson, J. M. (2002). Regulation of dendritic development by the ARF exchange factor ARNO. *Nat. Neurosci.* **5,** 623–624.

Hernandez-Deviez, D. J., Roth, M. G., Casanova, J. E., and Wilson, J. M. (2004). ARNO and ARF6 regulate axonal elongation and branching through downstream activation of phosphatidylinositol 4-phosphate 5-kinase alpha. *Mol. Biol. Cell* **15,** 111–120.

Huang, E. J., and Reichardt, L. F. (2001). Neurotrophins: Roles in neuronal development and function. *Annu. Rev. Neurosci.* **24,** 677–736.

Kaplan, D. R., and Miller, F. D. (2000). Neurotrophin signal transduction in the nervous system. *Curr. Opin. Neurobiol.* **10,** 381–391.

Kitano, J., Kimura, K., Yamazaki, Y., Soda, T., Shigemoto, R., Nakajima, Y., and Nakanishi, S. (2002). Tamalin, a PDZ domain-containing protein, links a protein complex formation of group 1 metabotropic glutamate receptors and the guanine nucleotide exchange factor cytohesins. *J. Neurosci.* **22,** 1280–1289.

Klein, R., Silos-Santiago, I., Smeyne, R. J., Lira, S. A., Brambilla, R., Bryant, S., Zhang, L., Snider, W. D., and Barbacid, M. (1994). Disruption of the neurotrophin-3 receptor gene trkC eliminates Ia muscle afferents and results in abnormal movements. *Nature* **368,** 249–251.

Lee, K. F., Li, E., Huber, L. J., Landis, S. C., Sharpe, A. H., Chao, M. V., and Jaenisch, R. (1992). Targeted mutation of the gene encoding the low affinity NGF receptor p75 leads to deficits in the peripheral sensory nervous system. *Cell* **69,** 737–749.

McAllister, A. K., Katz, L. C., and Lo, D. C. (1999). Neurotrophins and synaptic plasticity. *Annu. Rev. Neurosci.* **22,** 295–318.

Menn, B., Timsit, S., Represa, A., Mateos, S., Calothy, G., and Lamballe, F. (2000). Spatiotemporal expression of noncatalytic TrkC NC2 isoform during early and late CNS neurogenesis: A comparative study with TrkC catalytic and p75NTR receptors. *Eur. J. Neurosci.* **12,** 3211–3223.

Miyamoto, Y., Yamauchi, J., Tanoue, A., Wu, C., and Mobley, W. C. (2006). TrkB binds and tyrosine-phosphorylates Tiam1, leading to activation of Rac1 and induction of changes in cellular morphology. *Proc. Natl. Acad. Sci. USA.* **103,** 10444–10449.

Nie, Z., Hirsch, D. S., and Randazzo, P. A. (2003). Arf and its many interactors. *Curr. Opin. Cell. Biol.* **15,** 396–404.

Nie, Z., and Randazzo, P. A. (2006). Arf GAPs and membrane traffic. *J. Cell Sci.* **119,** 1203–1211.

Radhakrishna, H., Klausner, R. D., and Donaldson, J. G. (1996). Aluminum fluoride stimulates surface protrusions in cells overexpressing the ARF6 GTPase. *J. Cell Biol.* **134,** 935–947.

Radhakrishna, H., Al-Awar, O., Khachikian, Z., and Donaldson, J. G. (1999). ARF6 requirement for Rac ruffling suggests a role for membrane trafficking in cortical actin rearrangements. *J. Cell Sci.* **112**(Pt 6), 855–866.

Ren, X. D., Kiosses, W. B., and Schwartz, M. A. (1999). Regulation of the small GTP-binding protein Rho by cell adhesion and the cytoskeleton. *EMBO J.* **18**, 578–585.

Tessarollo, L., Tsoulfas, P., Donovan, M. J., Palko, M. E., Blair-Flynn, J., Hempstead, B. L., and Parada, L. F. (1997). Targeted deletion of all isoforms of the trkC gene suggests the use of alternate receptors by its ligand neurotrophin-3 in neuronal development and implicates trkC in normal cardiogenesis. *Proc. Natl. Acad. Sci. USA.* **94**, 14776–14781.

Tessarollo, L. (1998). Pleiotropic functions of neurotrophins in development. *Cytokine Growth Factor Rev.* **9**, 125–137.

Vicario-Abejon, C. (1997). Long-term culture of hippocampal neurons. *Current Protocols in Neuroscience.* 3.2.1–3.2.12.

Yamauchi, J., Chan, J. R., and Shooter, E. M. (2003). Neurotrophin 3 activation of TrkC induces Schwann cell migration through the c-Jun N-terminal kinase pathway. *Proc. Natl. Acad. Sci. USA.* **100**, 14421–14426.

Yamauchi, J., Chan, J. R., and Shooter, E. M. (2004). Neurotrophins regulate Schwann cell migration by activating divergent signaling pathways dependent on Rho GTPases. *Proc. Natl. Acad. Sci. USA.* **101**, 8774–8779.

Yamauchi, J., Chan, J. R., Miyamoto, Y., Tsujimoto, G., and Shooter, E. M. (2005a). The neurotrophin-3 receptor TrkC directly phosphorylates and activates the nucleotide exchange factor Dbs to enhance Schwann cell migration. *Proc. Natl. Acad. Sci. USA.* **102**, 5198–5203.

Yamauchi, J., Miyamoto, Y., Tanoue, A., Shooter, E. M., and Chan, J. R. (2005b). Ras activation of a Rac1 exchange factor, Tiam1, mediates neurotrophin-3-induced Schwann cell migration. *Proc. Natl. Acad. Sci. USA.* **102**, 14889–14894.

Zhang, Q., Cox, D., Tseng, C. C., Donaldson, J. G., and Greenberg, S. (1998). A requirement for ARF6 in Fcgamma receptor-mediated phagocytosis in macrophages. *J. Biol. Chem.* **273**, 19977–19981.

CHAPTER THIRTEEN

METHODS FOR ANALYSIS OF RAB27A/ MUNC13-4 IN SECRETORY LYSOSOME RELEASE IN HEMATOPOIETIC CELLS

Peter van der Sluijs, Maaike Neeft, Thijs van Vlijmen, Edo Elstak, *and* Marnix Wieffer

Contents

Department of Cell Biology, University Medical Center Utrecht, Utrecht, The Netherlands

Methods in Enzymology, Volume 438
ISSN 0076-6879, DOI: 10.1016/S0076-6879(07)38013-0

Abstract

Secretory lysosomes constitute a heterogeneous organelle of hematopoietic cells that combines the properties of regular lysosomes with those of secretory granules. Although secretory lysosomes serve essential functions, such as in the immune system and blood clotting, the mechanisms underlying the release of contents are incompletely understood. It is clear, however, that rab27a and the C2 domain protein munc13-4 serve essential functions. Mutations in these genes lead to immune disorders where the lytic granule function of cytotoxic T cells is jeopardized in humans. We identified munc13-4 as a rab27a binding protein from spleen. Munc13-4 is highly expressed in several hematopoietic cells including cytotoxic T cells and mast cells. We describe the molecular features of the interaction and requirements for localization, and show that munc13-4 is a positive regulator of secretory lysosome exocytosis.

1. INTRODUCTION

Secretory lysosomes, also known as lysosome-related organelles, constitute a hybrid organelle that combines the functions of lysosomes and secretory granules (Stinchcombe *et al.*, 2004). In accordance with properties of lysosomes, they have an acidic lumenal pH, a complement of lysosomal enzymes, and lysosome–associated membrane glycoproteins. They often contain lumenal structures, and at the ultrastructural level appear to be related to multivesicular bodies (Peters *et al.*, 1991). Secretory lysosomes occur in cells of the hematopoietic lineage, and have been investigated particularly in cytotoxic T cells (CTLs), platelets, mast cells, dendritic cells, and B cells. Stimulation of immune receptors such as the T-cell receptor triggers signaling pathways that generate output via a variety of downstream effector systems. These include the degranulation of secretory lysosomes, which releases bioactive compounds like granzymes and perforin in CTLs, and of serotonin in mast cells (Gilfillan and Tkaczyk, 2006; Radoja *et al.*, 2006).

Melanosomes, the organelles that produce and store pigment in melanocytes, and retinal pigment epithelium are also considered secretory lysosomes (Marks *et al.*, 2003). In spite of the relationship between melanosomes and secretory lysosomes, they employ distinct machinery for docking and fusion with the cell surface. This notion is dramatically demonstrated by the case of human disease mutations in melanophilin (Griscelli syndrome type 3, GS3) and myosin–Va (Griscelli syndrome type 1, GS1) that preclude the formation of a ternary complex with rab27a, and thereby interfere with peripheral distribution of melanosomes and transfer of pigment to keratinocytes. Patients with GS1 (Pastural *et al.*, 1997) and GS3 (Menasche *et al.*, 2003) have normal degranulation of secretory lysosomes in CTLs, while the lack of functional rab27a in Griscelli syndrome 2 (GS2) also prevents docking and degranulation of lytic granules (Menasche *et al.*, 2000), suggesting that secretion from lysosomes in CTLs requires other factors that cooperate with rab27a.

To find proteins involved in the release of secretory lysosomes, we performed preparative pull-down assays using rab27a-GTP and pig spleen cytosol. The rationale for this approach is that rab27a is highly expressed in the spleen (Seabra *et al.*, 1995), that spleen is enriched in hematopoietic cells, and that fresh spleens can easily be obtained from a slaughterhouse. Using this approach, we found munc13-4 (Neeft *et al.*, 2005), a distant relative of the MHD and C2 domain–containing munc13 proteins, that with the exception of a ubiquitously expressed munc13-2 splice variant, are predominantly expressed in the nervous system (Augustin *et al.*, 1999; Betz *et al.*, 1998; Koch *et al.*, 2000). Munc13-1 serves as a priming factor of glutamatergic synaptic vesicles (Rosenmund *et al.*, 2002), and perhaps as well for fusion of insulin granules in the β cells of the pancreas (Kwan *et al.*, 2006). In analogy with the function of the neuronal munc13 proteins, munc13-4 might serve as an essential priming factor for secretory lysosomes in hematopoietic cells. Strong support for this idea derives from the findings that mutations in human munc13-4 cause familial hemophagocytic lymphohistiocytosis type 3 (FHL3), an autoimmune disease that is phenotypically related to GS2 (Feldmann *et al.*, 2003). Munc13-4 might have additional functions upstream of a role in the biogenesis of secretory lysosomes in CTLs (Menager *et al.*, 2007).

This chapter describes the method we used to search for proteins regulating release from secretory lysosomes using rab27a as bait. It is a modification of a procedure originally developed to screen for rab5 effectors (Christoforidis and Zerial, 2000). We also provide assays to analyze the properties and function of munc13-4 in a mast cell model.

2. AFFINITY PURIFICATION OF RAB27A EFFECTOR PROTEINS

2.1. Solutions

Lysis buffer: PBS containing 200 μM GDP, 5 mM MgCl$_2$,10 μg/ml DNase, 10 μg/ml RNase, 5 mM 2-mercaptoethanol, 10 μg/ml lysozyme, 5 μg/ml leupeptin, 10 μg/ml aprotinin, 1 μg/ml pepstatin, 100 μM PMSF

Nucleotide exchange buffer (NE): 20 mM Na–HEPES, pH 7.5, 100 mM NaCl, 10 mM EDTA, 5 mM MgCl$_2$, 1 mM dithiothroitol (DTT) containing 10 μM GMP-PNP or 10 μM GDP (both Sigma)

Nucleotide stabilization (NS) buffer: 20 mM Na-HEPES, pH 7.5, 100 mM
  NaCl, 5 mM MgCl$_2$, 1 mM DTT, 10 μM GMP-PNP or 10 μM GDP
Elution buffer (EB): 20 mM Na-HEPES, pH 7.5, 1.5 M NaCl, 20 mM
  EDTA, 1 mM DTT, 1 mM GMP-PNP (for rab GDP column) or
  5 mM GDP (for rab GTP-column)
Homogenization buffer (HB): 20 mM Na-HEPES, pH 7.5, 100 mM NaCl,
  5 mM MgCl$_2$, 1 mM DTT, 5 μg/ml leupeptin, 10 μg/ml aprotinin,
  1 μg/ml pepstatin, 100 μM PMSF

2.2. Expression of GST-rab27a in *E. coli*

Human rab27a cDNA was cloned in the BamH1 site of pGEX2T (Neeft
et al., 2005). The plasmid encoding the fusion protein and as control pGEX
plasmid is transformed into *Escherichia coli* strain BL21DE3 and grown on
LB/agar plates containing 0.1 mg/ml ampicillin. A colony is grown at 30°
in LB containing 0.1 mg/ml ampicillin, and after 8 h the culture is diluted
10 to 20 times and grown overnight at 30°. The next morning the culture is
diluted 10 to 20 times and grown at 30° until OD600 ∼0.6, which usually
takes 2 to 3 h. Isopropyl-β-D-thiogalactopyranoside (IPTG) is added to
1 mM, and the culture is continued at the same temperature. After 4 h,
250-ml aliquots are harvested by centrifugation in a SLA-3000 rotor for
10 min at 4000×g and 4°. Pellets are washed with LB medium, snap frozen
in liquid nitrogen, and stored at −80° until further use.

2.3. Isolation of GST-rab27a and guanine nucleotide loading

Frozen bacteria pellets are thawed and resuspended in 0.05 (culture) volume
lysis buffer. The bacteria suspension is sonicated twice for 45 s on ice using a
Branson Probe sonicator operating at 70% of maximum output. The
homogenate is spun for 1 h at 4° at 100,000×g in a Ti45 rotor.
The supernatant is next incubated with 0.5 ml prewashed (in PBS) GSH-
sepharose 4B beads (Amersham) for 2 h under rotation at 4°. Beads are
washed three times with PBS containing 100 μM GDP, 5 mM MgCl$_2$, and
5 mM 2-mercaptoethanol, and used immediately or stored overnight at 4°
in the same buffer. For guanine nucleotide loading, beads are washed with
NE buffer, and then incubated with this buffer containing 1 mM GMP-
PNP for 30 min under rotation at room temperature (RT). This cycle is
repeated twice, whereupon the beads are washed once with NS buffer and
finally incubated with NS buffer containing 1 mM GMP-PNP for 20 min
under rotation at RT. The above procedure is for loading GST-rab27a with
a nonhydrolyzable GTP analog to screen for rab27a effectors with a putative
function in secretion from lysosomes that bind to the active form. The same
method is followed for GDP loading (except that GMP-PNP is replaced
with GDP) in order to find proteins that bind to rab27a-GDP. This matrix

also serves as a negative control for proteins that are retrieved on the GTP-loaded rab27a. Before use in the assay, an aliquot of the beads is run on a 10% SDS-PAA gel together with a calibration curve of BSA to normalize the GST-rab27a.GDP, GST-rab27a.GMP-PNP, and GST input.

2.4. Preparation of pig spleen cytosol

All steps in the preparation of cytosol are done on ice in the cold room. A fresh pig spleen is diced and either immediately processed, or frozen in liquid nitrogen and stored at $-80°$. Tissue is immersed in two volumes of ice-cold HB, and homogenized using a Waring blender operating in 10-s time intervals and maximum speed until the tissue chunks are broken. The resulting suspension is subsequently homogenized in a Kinematica tissue homogenizer operating at 20 to 30% of maximum power. The sample is then centrifuged for 40 min at $4°$ and $10,000 \times g$ in a Sorvall SLA 3000 rotor. The supernatant is retrieved and centrifuged for 1 h at $4°$ and $100,000 \times g$ in a Ti45 rotor to generate a cytosol fraction. Cytosol is next dialyzed overnight against 1000 volumes of NS buffer using dialysis tubing with a 3500–molecular weight cut-off (Spectrumlabs). The retentate is centrifuged for 1 h at $4°$ and $100,000 \times g$. Protein concentration in the cytosol supernatant is determined with the BCG assay (Pierce) and routinely amounts to 25 to 30 mg/ml.

2.5. Affinity isolation of cytosolic proteins on GST-rab27a.GMP-PNP beads

Beads are incubated with 15 ml of cytosol in the presence of 100 μM GMP-PNP under rotation at $4°$. After 4 h, beads are washed twice with 2 ml NS buffer and 10 μM GMP-PNP, followed by two washes with 2 ml NS buffer, 250 mM NaCl, and 10 μM GMP-PNP, and two washes with 1 ml 20 mM Na-HEPES, pH 7.5, 250 mM NaCl, and 1 mM DTT. Bound proteins are eluted for 20 min at RT in a shaker with 20 mM Na-HEPES, pH 7.5, 1.5 M NaCl, 20 mM EDTA, 1 mM DTT, and 5 mM GDP. For elution of proteins bound to rab27aGDP, we replace GDP with 1 mM GMP-PNP. The elution is repeated once to increase the yield. Results of a typical affinity purification are shown in Fig.13.1A. Eluates are denatured for 5 min at $100°$ in reducing Laemmli buffer, and proteins are resolved on 6% SDS-PAA gels that are subsequently subjected to silver or Coomassie Brilliant Blue staining. Bands of interest are excised and in-gel digested using modified trypsin (Roche) in 50 mM ammonium bicarbonate. Digests are analyzed by nano-flow liquid chromatography–tandem mass spectrometry (LC-MS/MS), by using an electrospray ionization quadrupole time-of-flight mass spectrometer operating in positive ion mode. A nano-flow liquid chromatography system is coupled to the quadrupole time-of-flight,

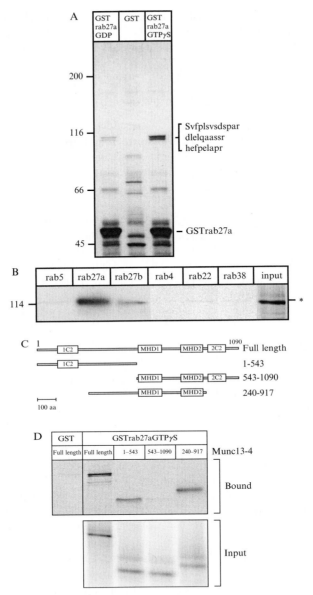

Figure 13.1 Characterization of munc13-4/rab27a complex on secretory lysosomes. (A) Affinity purification of munc13-4 from spleen on rab27a column. Spleen cytosol was incubated with GSH beads containing GST (negative control) and GST-rab27a charged with GDP or nonhydrolyzable GTP analog. Note the enrichment of 110-kDa protein on the GST-rab27aGMP-PNP beads. (B) Specific and direct binding of munc13-4 to rab27a. A panel of GMP-PNP–loaded GST-rab proteins was assayed for binding to ^{35}S-labeled munc13-4 produced in an *in vitro* transcription translation reaction. Note

essentially as described previously (Yatsuda *et al.*, 2003). Peptide mixtures are delivered to the system using a Famos autosampler (LC Packings) at 3 μl/min and trapped on an AquaTM C18RP column of 1 cm×100 μm (Phenomenex). After flow splitting down to 150 to 200 nl/min, peptides are transferred to an analytical column of 25 cm×50 μm (PepMap, LC Packings) in a gradient of acetonitrile (1%/min). Database searches are performed using Mascot software.

2.6. Comments

Although we used frozen pig spleen cytosol successfully, we get best results from freshly prepared material. Instead of cytosol, we also used 1% TX-100 lysates prepared from spleen. Since guanine nucleotide is not stably associated with most rab proteins in the presence of this concentration TX-100, we dilute the detergent in the binding reaction to 0.2%. For best results, we use the guanine nucleotide–loaded rab proteins and the dialyzed cytosol within 1 day after preparation. The preparative affinity isolations are done with an amount of GST-rab27a fusion protein that derived from 1 liter of *E. coli* culture immobilized on 0.5 ml GSH beads. We use two types of negative controls in the assay: (1) beads with the same amount of GST-rab27a fusion protein but now charged with GDP-nucleotide, and (2) beads with only GST.

3. DIRECT BINDING ASSAY WITH RECOMBINANT PROTEINS

To determine specificity of the candidate effector toward a collection of GST-rab proteins (Fouraux *et al.*, 2003; Neeft *et al.*, 2005), we employ a slightly modified binding assay in which the effector is produced as a

---

specificity in binding of munc13-4 to rab27. (C) Domain organization of munc13-4. Munc13-4 contains N-terminal and C-terminal C2 domains and two Munc13 homology domains (MHD) interspersed between them. This architecture is conserved in all munc13 proteins, and likely contains the minimal information required for function (Basu *et al.*, 2005; Stevens *et al.*, 2005). Indicated are munc13-4 truncations that were used for determining the rab27a binding domain and the requirements for cellular localization. (D) Binding domain of rab27a on munc13-4. Binding assay of ^{35}S-labeled His[6]-munc13-4 truncations produced in an *in vitro* transcription translation reaction, and GMP-PNP–loaded GST-rab27a. From these experiments we inferred that the rab27a binding domain is situated between the first C2 and first MHD.

^{35}S-labeled polypeptide in a coupled *in vitro* transcription–translation reaction. This material is then used in a binding assay with GST-rab proteins.

3.1. Solutions

Binding buffer (BB): 50 mM Na-HEPES, pH 8.0, 150 mM NaCl, 0.05%
    Triton X-100, 5 mM 2-mercaptoethanol, 10 μM GMP-PNP
^{35}S methionine PRO-MIX (Amersham)
TNT T7 Quick master mix (Promega)

3.2. Constructs

pGEX-rab27a, pGEX-rab27b, pGEX-rab4a, pGEX-rab5a, pGEX-rab22,
    pGEX-rab38, pcDNA3.1HisB-munc13-4, pcDNA3.1HisB-munc13-4
    (1–543), pcDNA3.1HisB-munc13-4(543–1090), pcDNA3.1HisB-
    munc13-4(240–917), pcDNA3.1HisB-munc13-4(Δ608–611)

3.3. Protocol

To determine specificity of the candidate effector toward rab proteins, we employ a slightly modified binding assay in which the effector is produced as a ^{35}S-labeled polypeptide in a coupled *in vitro* transcription–translation reaction. The ^{35}S-labeled effector protein is then incubated with a panel of GTP-loaded GST-rab proteins. In a typical reaction for five incubations, 1 μg of a T7-driven expression construct is mixed with 2 μl ^{35}S methionine PRO-MIX (Amersham) and 40 μl TNT T7 Quick master mix (Promega). Double-distilled H$_2$O is added to a final volume of 50 μl, and the mixture is incubated 1 h at 30°. In the meantime, beads containing immobilized, guanine nucleotide–charged GST-rab27a are washed once with BB.

Washed beads are then incubated for 2 h under rotation at 4° with 10 μl *in vitro* transcription–translation reaction product in 300 μl BB and 1 mM GMP-PNP. Beads are washed 4 times with BB, 0.2% Triton X-100, and 10 μM GMP-PNP. Bound protein is released by boiling for 5 min in reducing Laemmli buffer, resolved on a 10% SDS–PAA gel, and analyzed by phosphor imaging. Results of a specificity assay are shown in Fig. 13.1B.

3.4. Comment

The assay is used to establish or confirm guanine nucleotide–dependent binding to GST-rab27a. In addition, we employed the assay to map the rab27a binding domain using munc13-4 truncation constructs (Fig. 13.1C and D), and to analyze munc13-4 patient mutants for rab27a binding (Fig. 13.3C). For each binding reaction, we use 50-μl GSH beads and GST-rab27a isolated from a 50-ml bacteria culture.

4. Generation of an Antibody Against Munc13-4

A cDNA encoding munc13-4(904–1021) was generated by PCR and subcloned into the XhoI site of the bacterial expression plasmid pGEX4T3. The construct was transformed into *E. coli* BL21 (DE3) and grown on LB/agar plates containing 0.1 mg/ml ampicillin. A colony is grown at 30° in LB containing 0.1 mg/ml ampicillin, and after 8 h the culture is diluted 20 times and grown overnight at 30°. The next morning the culture is diluted 10 to 20 times and grown at 30° until OD600 ∼0.6. IPTG is added to 1 mM, and the culture is continued at the same temperature. After 4 h, bacteria are harvested by centrifugation in a SLA-3000 rotor for 10 min at 4000×g and 4°. Pellets are washed with LB medium, snap-frozen in liquid nitrogen, and stored at −80° until further use. Frozen bacteria pellets are thawed, resuspended in 60 ml lysis buffer (see previous discussion), and broken in two sonication steps of 45 s on ice. The bacteria are centrifuged for 1 h at 4° at 100,000×g, and then the supernatant is retrieved and incubated with 0.5 ml GSH sepharose 4B beads (washed PBS) for 2 h under rotation at 4°. Unbound material is removed by centrifugation, and the beads are washed four times with 1 ml PBS, and 5 mM 2-mercaptoethanol. GST-munc13-4(904–1021) was eluted at 4° in two steps of 45 min, with each 0.5 ml of 50 mM Na-HEPES, pH 8.0, 25 mM GSH, and 5 mM 2-mercaptoethanol. The eluates are then combined and split into two equal fractions. One of these was boiled for 5 min in the presence of 0.1% SDS and 10 mM DTT (final concentrations), followed by addition of N-ethylmaleimide to 20 mM alkylate free sulfhydryl groups. Finally, the fusion proteins were dialyzed overnight at 4° against 1000 volumes of PBS. This procedure yielded ∼1 mg of native and denatured GST-munc13-4(904–1021). Rabbits were then immunized with a mixture of the two forms to obtain antibodies against native and denatured protein. These antibodies were used to screen expression of munc13-4 in hematopoietic cell lysates, which revealed (among other things) high expression in the RBL-2H3 mast cell model and CD8-positive T cells (not shown).

4.1. Cell culture and transfection

4.1.1. Plasmids

pECFP-rab27a, pEYFP-munc13-4, pEGFP-munc13-4, pEGFP-munc13-4 (1–543), pEGFP-munc13-4(543–1090), pEGFP-munc13-4(240–917), pEGFP-munc13-4(Δ608–611)

4.1.2. Transfection of RBL cells and morphological assays

The antibodies against rab27a and munc13-4 are not good enough to detect the endogenous proteins by morphological methods. To determine localization of munc13-4 and requirements for membrane association, we expressed YFP of GFP-tagged munc13-4 constructs in RBL-2H3 cells for fluorescence microscopy. RBL-2H3 cells are cultured in DMEM, supplemented with 10% fetal calf serum, L-glutamine, and 50 μM 2-mercaptoethanol, and passed every 2 days. For transfection, 5×10^6 cells were harvested by trypsinization, resuspended in 250 μl DMEM, transferred to a 4-mm electroporation cuvette (BioRad), and mixed with 5 μg of plasmid DNA. A pulse of 1000 μF and 240 V was applied using a gene pulser (BioRad), and cells were immediately transferred to dishes containing prewarmed complete culture medium. With this method, ∼5% of the cells are transfected. Using this protocol we found that munc13-4 and rab27a extensively colocalized on serotonin containing secretory lysosomes (Fig.13.2A), and that membrane localization of munc13-4 critically relied on the region containing the MHDs (see Figs. 13.2C and 13.3A). Higher transfection efficiencies of 20 to 30% are achieved with nucleofaction (Amaxa GmbH). In a typical protocol, 1×10^6 RBL-2H3 cells are resuspended in 100 μl T-buffer (Amaxa). The cell suspension and 1 μg plasmid DNA are transferred to the Amaxa cuvette and transfected with program X-001 in the Amaxa nucleofactor. Cells are transferred to 12 ml prewarmed culture medium and plated out into a six-well plate.

For immunofluorescence and imaging experiments, cells are grown on 10-mm #1 coverslips and used 1 day after transfection. Cells are washed once with PBS, fixed with 3% paraformaldehyde for 30 min, washed once with PBS, and incubated for 5 min with PBS containing 50 mM NH$_4$Cl. Fixed cells are washed once with PBS and incubated for 1 h in PBS containing 0.5% BSA and 0.1% saponin (block buffer). Incubation with a monoclonal serotonin antibody (DakoCytomation) is done for 1 h, followed by three 5-min washes with block buffer. Staining with appropriately labeled secondary antibodies is done for 1 h, followed by three 5-min washes with block buffer and two washes with PBS. Coverslips are mounted with 3 μl Moviol (Hoechst). Cells stably expressing YFP-munc13-4 were fixed with 2% paraformaldehyde in 0.1 M sodium phosphate buffer for immuno–electron microscopy. After fixative removal, the cells were embedded in 10% gelatin, and prepared for ultrathin cryosectioning and immunogold labeling (for detailed procedures, see Raposo *et al.*, 1997). Ultrathin cryosections were double labeled with a rabbit antibody against GFP, a mouse monoclonal antibody against the lysosomal membrane glycoprotein p80 (Bonifacino *et al.*, 1986), and various combinations of protein A gold particles. A rabbit anti-mouse Ig antibody (DAKO) was used as a bridging step in case of monoclonal antibodies. At the ultrastructural level, munc13-4 localized predominantly to the limiting membrane of

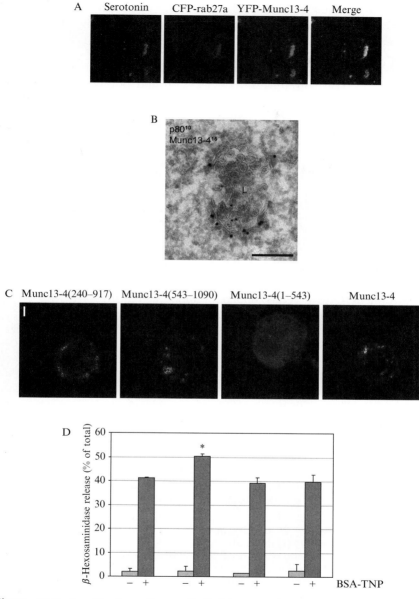

Figure 13.2 Localization of munc13-4/rab27a complex on secretory lysosomes. (A) Munc13-4/rab27a colocalize on granular structures in the cytoplasm. RBL-2H3 cells were transfected with pEYFP-munc13-4 (green) and pECFP-rab27a (blue), and labeled with an antibody against serotonin (red), a content marker of secretory lyso-somes. The merged image of this triple-label experiment shows extensive colocalization of munc13-4/rab27a with the marker on discrete cytoplasmic granules. Bar denotes 5 μm.

a multivesicular compartment in which the lumenal structures were decorated with an antibody against the lysosomal membrane protein p80 (Fig.13.2B).

4.2. Comment

Some RBL-2H3 clones grow in suspension and poorly adhere to glass coverslips. For microscopy experiments, we coat glass coverslips with poly-L-lysine (Sigma) before seeding the cells.

5. Degranulation Assay in RBL-2H3 Cells

5.1. Reagents

DMEM + pen/strep + 1% FCS (secretion medium)
IgE anti-TNP mouse hybridoma supernatant
BSA
BSA-TNP (Sigma)
Triton X-100
p-nitrophenyl-N-acetyl-α-D-glucosaminide (Sigma)
0.1 M Na-carbonate buffer, pH 10

5.2. Plasmids

pEYFP-munc13-4,   pEGFP,   pEYFP-munc13-4,   pEGFP-munc13-4
(543–1090), pEGFP-munc13-4(Δ608–611)

5.3. Method

We analyzed the function of munc13-4 in RBL-2H3 cells transfected with various munc13-4 constructs as described previously. Degranulation is induced by addition of IgE antibody and FcεR1 cross-linking with the antigen TNP-BSA. Lysosome release is then read out as β-hexosaminidase

---

(B) Munc13-4 localizes to the limiting membrane of multivesicular granules. Double labeling of YFP-munc13-4 (15 nm gold) and the lysosomal membrane glycoprotein p80 (10 nm gold) on cryosections prepared from RBL-2H3 transfectants. Bar denotes 200 nm. (C) Membrane localization of munc13-4 requires the MHDs. RBL-2H3 cells transfected with various GFP-tagged munc13-4 truncations were labeled with an antibody against serotonin. Note that only the truncations containing the region encompassing the MHDs localize to secretory lysosomes. Bar denotes 5 μm. (D) Munc13-4 is a positive regulator of secretory lysosome degranulation. RBL-2H3–expressing GFP-tagged munc13-4 constructs were assayed for their ability to release β-hexosaminidase in resting state and after stimulation. Full-length, munc13-4 enhanced antibody induced β-hexosaminidase secretion, while truncation mutants were without effect. (See color insert.)

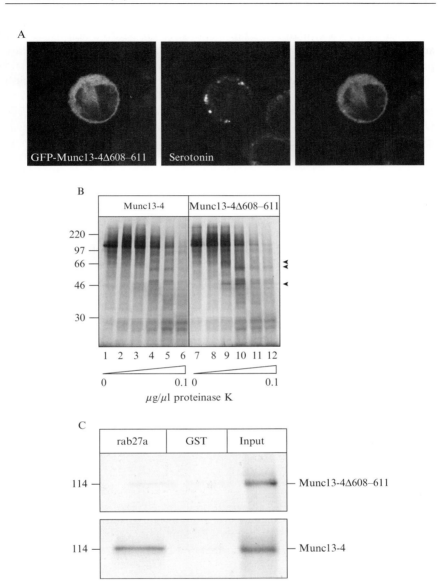

Figure 13.3 Analysis of FHL3 munc13-4Δ608–611 mutant. (A) Munc13-4Δ608–611 does not localize to secretory lysosomes. The FHL3 mutant (and control wildtype munc13-4) were transfected as GFP-tagged constructs in RBL-2H3 cells. Deletion of four amino acids in the first MHD creates a mutant that localizes to the cytoplasm, and is not recruited to serotonin containing secretory lysosomes. (B) Deletion of aa 608–611 causes misfolding of munc13-4. Conformational alterations of munc13-4Δ608–611 were investigated in a limited proteolysis assay. ^{35}S-labeled His$_6$munc13-4 were produced in a coupled transcription–translation reaction and submitted to limited proteolysis with

accumulation in the medium. Transfectants are grown to a density of 70% and then scraped in 0.5 ml secretion medium. Next, 5×10^6 cells are pelleted, resuspended in 0.5 ml IgE anti-TNP hybridoma supernatant, and subsequently incubated at 37°. After 20 min, cells are washed with secretion medium and resuspended in 0.5 ml of this medium. Aliquots of 75 μl cell suspension are pipetted in a 96-well plate and coincubated for 1 h at 37° with 75 μl BSA-TNP (4 ng/ml) or the same concentration of BSA (negative control). The 96-well plate is centrifuged at 1000 rpm for 5 min, and 10 μl supernatant is transferred to a second 96-well plate containing 50 μl 2 mM p-nitrophenyl-N-acetyl-α-D-glucosaminide and 40 μl PBS. The incubation is run for 1 h at 37°, and stopped by addition of 150 μl 0.1 M Na carbonate buffer, pH 10. The absorbance at 405 nm is then determined in a plate reader. An aliquot of cells is solubilized in 0.2% TX-100 to determine the total cellular content of β-hexosaminidase, and to calculate the extent of degranulation. Figure.13.2D shows that transfection of wild-type munc13-4 enhanced degranlation, while two mutants were without effect.

5.4. Comments

Since transfection efficiency is low in RBL-2H3 cells, we use stable cell lines for degranulation experiments. Stable transfectants were selected in the presence of 1 mg/ml G418. Cells can also be activated with a combination of 1 μM ionomycin (Calbiochem) and 80 nM phorbol 12-myristate 13-acetate (Sigma).

6. LIMITED PROTEOLYSIS OF 35S-LABELED MUNC13-4

To determine whether point mutants or short truncations that are found in FHL3 cause conformational alterations to munc13-4, we use a limited proteolysis assay. The approach capitalizes on the principle that accessibility of protease target sites is dictated by the structure of a protein.

protease K. The arrowheads denote differences between the digestion products of the wildtype and mutant proteins, showing that the mutation causes conformational alterations. (C) Deletion of aa 608–611 inhibits rab27a binding. Although the rab27a binding region is not in the MHD, it is necessary to determine whether rab27a binding is affected in the FHL3 mutant. GMP-PNP–charged GST-rab27a was used in a binding assay with 35S-labeled His6munc13-4Δ608–611. Note that deletion of aa 608–611 essentially precludes binding to rab27a, suggesting that the small deletion causes significant alterations to the overall conformation of munc13-4, which result in the failure to bind rab27a. (See color insert.)

Mutations that cause conformational changes should also give rise to distinct proteolytic fragments when the sample is analyzed on a SDS-PAA gel. Indeed, the digestion products of the patient mutant munc13-4Δ608–611 are distinct from those of wildtype munc13-4 both for protease K (Fig. 13.3B) and for endoproteinase Glu-C (not shown).

6.1. Plasmids

pcDNA3.1HisB–munc13-4, pcDNA3.1HisB–munc13-4Δ608–611

6.2. Reagents

^{35}S methionine, 10 μCi/μl (Amersham Biosciences)
TNT T7–driven coupled *in vitro* transcription–translation system (Promega)
Proteinase K (Roche)
Endoproteinase Glu-C (Roche)
PMSF (Calbiochem)
Cycloheximide (Sigma)

6.3. Procedure

For each protease digestion, 1 μl (0.5 μg/μl) cDNA is mixed with 7 μl *in vitro* transcription–translation mixture, 1 μl ^{35}S methionine, and 1 μl distilled water, and incubated at 30°. After 1 h, cycloheximide is added to 1 mM to inhibit ongoing protein synthesis. The reaction is next transferred to ice and incubated with proteinase K (0.25, 1, 5, 25, and 100 μg/ml) or endoproteinase Glu-C (5, 25, 100, and 500 μg/ml). After precisely 15 min, PMSF is added to 2.5 mM to stop the reaction. Samples subsequently receive one volume two times reducing Laemmli buffer, and are boiled for 5 min prior to SDS PAGE.

6.4. Comment

Although protease concentrations need to be optimized for each protein protease combination, the indicated concentrations have given good results for a range of proteins.

ACKNOWLEDGMENTS

This research was supported by grants from the Netherlands Organization for Chemical Research, Netherlands Proteomic Center and the Dutch Cancer Society (to P. vdS.). We thank René Scriwanek for help with preparation of the figures.

REFERENCES

Augustin, I., Rosenmund, C., Sudhof, T. C., and Brose, N. (1999). Munc13-1 is essential for fusion competence of glutamatergic synaptic vesicles. *Nature* **400**, 457–461.

Basu, J., Shen, N., Dulubova, I., Lu, J. C., Guan, R., Guryev, O., Grishin, N. V., Rosenmund, C., and Rizo, J. (2005). A minimal domain responsible for munc13 activity. *Nat. Struct. Mol. Biol.* **12**, 1017–1018.

Betz, A., Ashery, U., Rickmann, M., Augustin, I., Neher, E., Südhof, T. C., Rettig, J., and Brose, N. (1998). Munc13-1 is a presynaptic phorbol ester receptor that enhances neurotransmitter release. *Neuron* **21**, 123–136.

Bonifacino, J. S., Perez, P., Klausner, R. D., and Sandoval, I. V. (1986). Study of the transit of an integral membrane protein from secretory granules through the plasma membrane of secreting basophilic leukemia cells using a specific monoclonal antibody. *J. Cell Biol.* **102**, 516–522.

Christoforidis, S., and Zerial, M. (2000). Purification and identification of novel rab effectors using affinity chromatography. *Methods* **20**, 403–410.

Feldmann, J., Callebaut, I., Raposo, G., Certain, S., Bacq, D., Dumont, C., Lambert, N., Ouachee-Chardin, M., Chedeville, G., Tamary, H., Minard-Colin, V., Vilmer, E., *et al.* (2003). Munc13-4 is essential for cytolytic granules fusion and is mutated in a form of Familial Hemophagocytic Lymphohistiocytosis (FHL3). *Cell* **115**, 461–473.

Fouraux, M., Deneka, M., Ivan, V., van der Heijden, A., Raymackers, J., van Suylekom, D., van Venrooij, W. J., van der Sluijs, P., and Pruijn, G. J. M. (2003). rabip4' is an effector of rab5 and rab4 and regulates transport through early endosomes. *Mol. Biol. Cell.* **15**, 611–624.

Gilfillan, A. M., and Tkaczyk, C. (2006). Integrated signaling pathways for mast cell activation. *Nat. Rev. Immunol.* **6**, 218–229.

Koch, H., Hofmann, K., and Brose, N. (2000). Definition of munc13-homology domains and characterization of a novel ubiquitously expressed munc13 isoform. *Biochem. J.* **349**, 247–253.

Kwan, E., Xie, L., Sheu, L., Nolan, C. J., Prentki, M., Betz, A., Brose, N., and Gaisano, H. Y. (2006). Munc13-1 deficiency reduces insulin secretion and causes abnormal glucose tolerance. *Diabetes* **55**, 1421–1429.

Marks, M. S., Theos, A. C., and Raposo, G. (2003). Melanosomes and MHC class II antigen-processing compartments: A tinted view of intracellular trafficking and immunity. *Immunol. Res.* **27**, 409–426.

Menager, M. M., Menasche, G., Romao, M., Knapnougel, P., Ho, C. H., Garfa, M., Raposo, G., Feldmann, J., Fischer, A., and de Saint Basile, G. (2007). Secretory cytotoxic granule maturation and exocytosis require the effector protein hMunc13-4. *Nat. Immunol.* **8**, 257–267.

Menasche, H., Pastural, E., Feldmann, J., Certain, S., Ersoy, F., Dupuis, S., Wulffraat, N., Bianchi, D., Fischer, A., Le Deist, F., and de Saint Basile, G. (2000). Mutations in rab27a cause Griscelli's syndrome associated with haemophagocytic syndrome. *Nat. Genet.* **25**, 173–176.

Menasche, G., Ho, C. H., Sanal, O., Feldmann, J., Tezcan, I., Ersoy, F., Houdusse, A., Fischer, A., and de Saint Basile, G. (2003). Griscelli syndrome restricted to hypopigmentation results from a melanophilin defect (GS3) or a MYO5A F-exon deletion (GS1). *J. Clin. Invest.* **112**, 450–456.

Neeft, M., Wieffer, M., de Jong, A. S., Negroiu, G., Metz, C. H., van Loon, A., Griffith, J. M., Krijgsveld, J., Wulffraat, N., Koch, H., Heck, A. J., Brose, N., *et al.* (2005). Munc13-4 is an effector of rab27a and controls secretion of lysosomes in haematopoietic cells. *Mol. Biol. Cell.* **16**, 731–741.

Pastural, E., Barrat, F., Dufourcq-Lagelouse, R., Certain, S., Sanal, O., Jabado, N., Seger, R., Griscelli, C., Fischer, A., and de Saint Basile, G. (1997). Griscelli disease maps to chromosome 15q21 and is associated with mutations in the myosin-Va gene. *Nat. Genet.* **16,** 289–292.

Peters, P. J., Borst, J., Oorschot, V., Fukuda, M., Krahenbuhl, O., Tschopp, J., Slot, J. W., and Geuze, H. J. (1991). Cytotoxic T lymphocyte granules are secretory lysosomes, containing both perforin and granzymes. *J. Exp. Med.* **173,** 1099–1109.

Radoja, S., Frey, A. B., and Vukmanovic, S. (2006). T-cell receptor signaling events triggering granule exocytosis. *Crit. Rev. Immunol.* **26,** 265–290.

Raposo, G., Kleijmeer, M., Posthuma, G., Slot, J. W., and Geuze, H. J. (1997). "Immunogold labeling of ultrathin cryosections: Application in immunology." Malden: Blackwell Science, Malden.

Rosenmund, C., Sigler, A., Augustin, I., Reim, K., Brose, N., and Rhee, J. S. (2002). Differential control of vesicle priming and short-term plasticity by munc13 isoforms. *Neuron* **33,** 411–424.

Seabra, M. S., Ho, Y. K., and Anant, J. S. (1995). Deficient geranylgeranylation of ram/rab27 in choroideremia. *J. Biol. Chem.* **270,** 24420–24427.

Stevens, D. R., Wu, Z. X., Matti, U., Junge, H. J., Schirra, C., Becherer, U., Wojcik, S. M., Brose, N., and Rettig, J. (2005). Identification of the minimal protein domain required for priming activity of munc13-1. *Curr. Biol.* **15,** 2243–2248.

Stinchcombe, J., Bossi, G., and Griffiths, G. M. (2004). Linking albinism and immunity: The secrets of the secretory lysosome. *Science* **305,** 55–59.

Yatsuda, A. P., Krijgsveld, J., Cornelissen, A. W., Heck, A. J., and de Vries, E. (2003). Comprehensive analysis of the secreted proteins of the parasite Haemonchus contortus reveals extensive sequence variation and differential immune recognition. *J. Biol. Chem.* **278,** 16941–16951.

CHAPTER FOURTEEN

Analysis and Expression of Rab38 in Oculocutaneous Lung Disease

Kazuhiro Osanai* *and* Dennis R. Voelker[†]

Contents

Abstract

Rab38 is a low-molecular-weight G-protein highly expressed in melanocytes of the skin and alveolar type II cells in the lung. A point mutation in the postulated GTP/GDP-interacting domain of Rab38 has been identified as the genetic lesion responsible for oculocutaneous albinism (OCA) in *chocolate* (*cht*) mice. Another point mutation that prevents translation of Rab38 mRNA is the molecular basis of the *Ruby* gene mutation causing the phenotype of OCA and prolonged bleeding time in Fawn-Hooded and Tester-Moriyama rats. *Cht* mice show conspicuously enlarged lamellar bodies in alveolar type II cells and abnormal lung structure. Triton X-114 phase partitioning of *cht* mouse lung showed that

* Department of Respiratory Medicine, Kanazawa Medical University, Kahokugun, Ishikawa, Japan
† Program in Cell Biology, Department of Medicine, National Jewish Medical and Research Center, Denver, Colorado

Methods in Enzymology, Volume 438

ISSN 0076-6879, DOI: 10.1016/S0076-6879(07)38014-2

Rab38^{cht}-protein was recovered in the aqueous phase. We produced recombinant Rab38^{cht}-protein using a baculovirus/insect cell-protein expression system. The results demonstrate that Rab38^{cht}-protein is inactive due to reduced membrane binding and enhanced intracellular degradation. Rab38 is a new strong candidate gene for human Hermansky-Pudlak syndrome (HPS) that is characterized by OCA, bleeding diathesis, and lung disease.

1. INTRODUCTION

The original Rab38 cDNA was cloned from rat lung cDNA library and submitted to GenBank (M94043), and was later identified from a human melanoma cDNA library (Jager et al., 2000). Rab38 shows restricted tissue expression and is highly expressed in the rat lung, specifically in alveolar type II cells and Clara cells (Osanai et al., 2001). These lung-specific cells synthesize and secrete pulmonary surfactant, which lowers alveolar surface tension and keeps the alveolar lumen open for the purpose of gas exchange. Alveolar type II cells isolated from rats exhibit the highest mRNA abundance for Rab38, but rapidly lose this expression upon culture in vitro, which promotes dedifferentiation (Osanai et al., 2005).

Rab proteins are prenylated at their carboxyl termini, and some subsets are also modified by myristoylation and palmitoylation. The prenyl modifications are essential for biological activity, because these signals are required for localization to specific cellular organelles. Rab proteins can be found in both membrane-bound and cytosolic forms, and they mediate intracellular vesicle transport between restricted intracellular compartments. Each Rab family member is strictly localized to defined intracellular organelles.

A naturally occurring mutant mouse with oculocutaneous albinism, the *chocolate* (*cht*) mouse, was recently identified as having a point mutation in the *Rab38* gene (Loftus et al., 2002). Genomic DNA analysis indicated that a G146T mutation occurred in the region encoding the conserved GTP/GDP-interacting domain of Rab38. Mice homozygous for *cht* (*Rab38$^{cht/cht}$*) exhibit brown coats and eyes similar in color to *brown* mice that are a model for human oculocutaneous albinism (OCA) type 3. No other abnormality has been reported in *cht* mice. However, it seemed probable that the mutation in the *cht* locus should also cause lung disease, as well as oculocutaneous albinism, because the mRNA and its encoded protein are highly expressed in both the lung and the skin.

The *cht* mice have conspicuous changes in the morphology of lamellar bodies (LBs) (lysosome-related organelles for storage of lung surfactant), and develop emphysematous lung disease in addition to previously known OCA. The Rab38^{cht}-protein retains GTP-binding capacity, but does not undergo isoprenoid modification. These results suggest that functional loss of Rab38-protein causes multiple organ diseases including oculocutaneous lung disease (OCL). In this chapter, we describe methods to quantitatively examine the morphology of the lungs of *cht* mice and to characterize Rab38^{cht}-protein produced with the baculovirus/insect cell expression system.

2. MORPHOLOGICAL EXAMINATION OF MOUSE LUNGS

2.1. Mice

Male *cht/+* and female *cht/cht* mice were purchased from the Jackson Laboratory (Bar Harbor, ME). Specific-pathogen-free (SPF) animals were generated by rederivation with embryo transfer at Japan Charles River Inc. (Yokohama, Japan). Animals were shipped to and propagated in a specially designated clean room separated from other species of animals. There was no difference in animal propagation and growth between *wildtype* (*wt*) and *cht* up to 24 weeks of age. The genotypes of *cht* mice were determined by the DNA sequence of PCR products of amplified genomic DNA. The *cht* genotype was analyzed by PCR amplification of exon1 and subsequent DNA sequencing using ABI^{TM} Prism BigDye^{TM} Terminator Cycle Sequencing Ready Reaction Kit (Applied Biosystems). Rab38 primer pairs were designed to amplify the sequence flanked by the Ex1F (TAG-GAAGGAGGATTAAACCCG) and Ex1R (GAACTCCTCATGGCT-CACTCC), yielding a 428-bp product. The mutation was confirmed by using primers designed to amplify a 213-bp fragment surrounding the G146T sequence (*cht* Ex1F: GGCCTCCAGGATGCAGACACC and *cht* Ex1R: CCAGCAATGTCCCAGAGCTGC) (Loftus *et al.*, 2002).

2.2. Morphometry

Mice were anesthetized by intraperitoneal administration of 2 mg of sodium pentobarbital plus 66 units of sodium heparin, and exsanguinated by cutting the abdominal aorta. The lungs and heart were removed *en bloc* with care not to damage the organs. The lungs were inflated with 2.5% glutaraldehyde in 0.1 M cacodylate buffer (pH 7.4) at a constant hydrostatic pressure of 25 cmH$_2$O at the height of the carina in the upright position. At this stage, if the lungs leaked the fixative, they were not used for the analysis. The trachea was ligated, and the lungs and heart were kept for 48 h in the same fixative solution. Following fixation, the heart was cut off, and the lung weight was measured including the fixative. The right and left lungs were separated, embedded in paraffin, and cut into 5-μm–thick sections in the sagittal plane, so as to contain the largest lung area. Care was taken to cut the fixed lungs at the same thickness without any distortion. The right lungs were stained using hematoxylin and eosin and the left lungs were stained using toluidine blue.

An imaginary rectangle whose four sides touched the outline of a lung slice was defined on the microscopic stage, and was divided into more than 40 fields of the same size. Each field was photographed at $295 \times 223 \ \mu m^2$ and contained 35 equidistant test points. Two test lines connecting the opposite angles of the rectangle, both 370 μm in length, were drawn on the field (Fig. 14.1). After excluding inappropriate pictures with a dominant large airway and/or vessels, each animal had more than 34 fields with sufficient alveolar tissues available for morphometry. Every test point was classified into one of six categories based on its histological component: (1) alveolar air space, (2) air space in alveolar ducts and sacs, (3) alveolar wall, (4) bronchial and bronchiolar wall, (5) blood vessel wall, and (6) other. The numbers

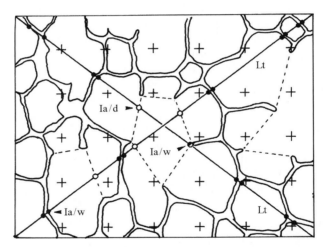

Figure 14.1 Schematic representation of lung morphometry. Lung tissues were photographed at $295 \times 223 \ \mu m^2$ that contained more than 30 equidistant test points; two test lines connecting the opposite angle, both 370 μm in lengths, were drawn on the field (Lt). Crosses indicate test points that fall into one of six categories based on its histological component alveolar air space, air space in alveolar ducts and sacs, alveolar wall, bronchial and bronchiolar wall, blood vessel wall, and other. Three kinds of intersections with respect to the drawn test lines were counted: visible intercepts between alveolar air and adjacent tissue, usually the alveolar wall (Ia/w, closed circle); visible intercepts between the alveolar ducts or sacs and the adjacent wall (Id/w, semiclosed circle); and invisible intercepts between the alveoli and the alveolar ducts (Ia/d, open circle). The three types of intercepts can be defined as: Iw = Ia/w + Id/w; Id = Ia/d + Id/w; and Ia = Ia/w + Ia/d. Using these intercepts, the mean thickness of the alveolar wall (τw), the mean linear intercept (Lm), the mean chord length of the alveoli (la), and the mean chord length of the ducts and sacs (ld) were calculated (see Table 14.1).

of test points from more than 34 fields were summed for each of the tissue components, and the volume proportion of the defined tissue component was expressed as a proportion of the total number of points according to the point-counting method (Weibel, 1979). Based on the modified point-counting method (Kawakami *et al.*, 1984), three different kinds of intersections with respect to the drawn test lines were counted: (1) visible intercepts between alveolar air and adjacent tissue, usually the alveolar wall (Ia/w) (see Fig. 14.1, closed circle); (2) visible intercepts between the alveolar ducts or sacs and the adjacent wall (Id/w) (see Fig. 14.1, semiclosed circle); and (3) invisible intercepts between the alveoli and the alveolar ducts (Ia/d) (see Fig. 14.1, open circle). The three types of intercepts can be defined as Iw = Ia/w + Id/w; Id = Ia/d + Id/w; Ia = Ia/w + Ia/d. Using these intercepts, the mean thickness of alveolar wall (τw), the mean linear intercept (Lm), the mean chord length of the alveoli (la), and the mean chord length of the ducts and sacs (ld) were calculated (Table 14.1).

2.3. Electron microscopy

The excised lungs were cut into small pieces and fixed for 2 h with fresh fixative containing 2.5% glutaraldehyde/0.1% picric acid/2% osmium tetroxide/4% sucrose/0.1 *M* cacodylate buffer (pH 7.4). Blocks were post-fixed with 1% aqueous uranyl acetate solution for 1 h, followed by

Table 14.1 Morphometric Parameters: Abbreviations and Calculations

Parameter	Abbreviation	Calculation
Volume proportion of		
Alveolar air	Vva	
Air in ducts and sacs	Vvd	
Alveolar wall	Vvw	
Bronchial and bronchiolar wall	Vvbw	
Blood vessel wall	Vvvw	
Mean thickness of alveolar walls, μm	τw	Vvs × Lt[a]/lw[b]
Mean linear intercept, μm	Lm	2 Lt/Iw
Mean chord length of alveoli, μm	la	2 Lt × Vva/la[b]
Mean chord length of ducts and sacs, μm	ld	2 Lt × Vvd/ld[b]
Surface area ratio of duct and sacs to alveolar wall	Rd/w	Id/lw

[a] Lt: Total test line length (μm)
[b] See text for explanation of these three intercepts.

dehydration in a graded series of ethanol and subsequent propylene oxide, and finally embedded in Epon. Thin sections (0.8 μm thick) were prepared and stained with toluidine blue while heating. Ultrathin sections (50 to 60 nm) were prepared and counterstained with 2% uranyl acetate, and then 2.66% lead nitrate/3.52% sodium citrate (pH 12). The sections were examined using a Hitachi H-7100 transmission electron microscope (Fig. 14.2).

2.4. Triton X-114 phase separation of mouse lung homogenates

Mice were anesthetized by intraperitoneal injection of 2 mg of sodium pentobarbital plus 66 units of sodium heparin, and exsanguinated by cutting the abdominal aorta. The left atrium was cut, and the lungs were perfused with ~5 ml of 10 mM Hepes-buffered saline (pH 7.4) from the right ventricle to left atrium to remove blood. The lungs and heart were excised *en bloc*. The lungs were weighed and cut into small pieces. Three volumes of 10 mM Hepes (pH 7.4) were added, and the tissue was homogenized with a Potter-Elvehjem–type homogenizer with the pestle driven by an electric drill for 15 strokes. MgCl$_2$ was added to 2 mM, and the homogenate centrifuged at 300×g for 10 min. The supernatant was recovered (~0.5 ml) and an equal volume of 2% Triton X-114/10 mM Hepes, pH 7.4/1.8% NaCl/2 mM MgCl$_2$) was added, incubated on ice for 30 min, and centrifuged at 100,000×g for 30 min. Triton X-114 containing samples

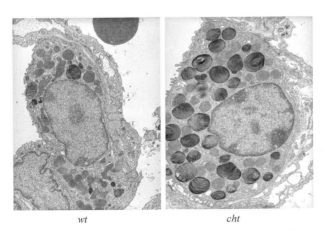

wt   cht

Figure 14.2   Electron microscopic appearance of alveolar type II cells. Lungs were cut into small pieces and fixed with 2% OsO$_4$/2.5% glutaraldehyde. The fixed lung pieces were embedded, sliced, and stained with uranyl acetate and lead nitrate. The alveolar type II cells of *cht* mice (right) are enlarged and have an increased size and number of LBs compared with those of wildtype mice (left). The structure of LBs in *cht* mouse looks normal. Original magnification ×5000.

were always kept at 4° to prevent phase separation unless otherwise noted. The supernatant was divided into two equal volumes. One was used as total lung lysate (~0.5 ml), and the other (~0.5 ml) was processed for Triton X-114 phase separation (Bordier, 1981). Approximately 0.5 ml of the sample was layered on the 0.3 ml of 6% sucrose/PBS/2 mM MgCl$_2$ / 0.06% Triton X-114 in a microcentrifuge tube and warmed in a 30° water bath for 10 min and centrifuged at 300×g for 10 min at room temperature in a fixed angle rotor. Micelles of Triton X-114 in solution aggregate above 20° and can be sedimented by low-speed centrifugation. The temperature of the centrifuge is important and should be equilibrated before use. The upper phase was collected and adjusted to 0.5% Triton X-114. The sample was gently mixed in the cold room for 30 min to dissolve the detergent completely. Next, the sample was warmed to 30° in a water bath for 10 min, and again layered on the previously described sucrose cushion (6% sucrose/PBS/2 mM MgCl$_2$/0.06% Triton X-114), and centrifuged at 300×g for 10 min at room temperature. The bottom detergent phase was recovered and diluted in 10 mM Hepes-buffered saline (pH 7.4)/ 2 mM MgCl$_2$ to adjust the concentration of Triton X-114 to 1%. The upper phase was also recovered and adjusted to 2% Triton X-114. Next, this upper phase was gently mixed in the cold room for 30 min, and warmed to 30° in a water bath for 10 min, and again centrifuged at 300×g for 10 min at room temperature. The final supernatant was recovered as the aqueous phase, and the detergent pellet from this step was discarded. Triton X-114 was added to the aqueous phase to make its final concentration 1%.

3. Characterization of Recombinant Rab38^{CHT}-Protein

3.1. Protein production and purification

The rat Rab38-cDNA originally cloned from a rat lung cDNA library in our laboratory was inserted into the pBlueBacHis-2A vector (Invitrogen) at the *Bam* HI and *Hind* III restriction enzyme sites. A forward primer that contains a *Bam* HI restriction site and initiation codon (5'-CCGGGATC-CATGCAGACA-3') and a reverse mutagenic PCR primer that carries the *cht*-mutation site and adjacent *Hae*II restriction site (5'-GTAGCGCTT-GATGATGCTGGTCTTGCCTACAACTA-3') was used to amplify the sequence, which contains the G146T (*cht*) mutation site. The PCR product was digested with *Bam*HI and *Hae*II, and inserted into the pBlueBacHis-2A-Rab38 previously digested with *Bam*HI and *Hae*II. Cationic liposomes (InsectinPlus, Invitrogen) were used to cotransfect the recombinant plasmid and modified baculovirus DNA (Bac-N-Blue DNA, Invitrogen) into Sf9 cells (Invitrogen) using the manufacturer's procedures. Sf9 cells were cultured in TNM-FH (Grace's insect media supplemented with lactalbumin hydrolysate and yeastolate) with 10% FBS and penicillin (50 U/ml) and streptomycin (50 μg/ml). The presence of recombinant virus that contained the Rab38 DNA was verified by both PCR and Western blot analysis. Virus titers were determined by plaque assays (O'Reilly *et al.*, 1992). Cells that were 80 to 90% confluent in 150-mm plastic dishes were infected with the recombinant virus at a multiplicity of infection (MOI) of 10, and cultured for 3 to 4 days. For one assay, we usually used 10 150-mm plates for Rab38^{wt}-protein, and 20 plates for Rab38^{cht}-protein, because of relatively low recovery of the protein. The cells were harvested, washed with PBS two times, and stored at $-80°$ until use. The frozen Sf9 cells were rapidly thawed and suspended in the lysis buffer (1% TritonX-114, 50 mM Hepes, pH 7.4, 150 mM NaCl, and 1.5 mM EGTA) containing protease inhibitors (1 mM PMSF, 10 μg/ml aprotinin, and 10 μg/ml leupeptin). In some cases, we used the Triton X-114 phase partitioning for purification of recombinant Rab38. However, in the case of Rab38^{cht}-protein, we did not use the detergent phase partitioning, because Rab38^{cht}-protein was only present in the aqueous phase. The cell suspension was centrifuged at $300 \times g$ for 10 min. Next, the supernatant was recovered and centrifuged at $100,000 \times g$ for 30 min. The supernatant was recovered and loaded on a 2-ml bed volume of Ni^{++}-charged affinity column (Probond, Invitrogen) under native conditions, according to the manufacturer's instructions. Imidazole step gradients were used to elute the column (2 ml of 50 mM, 2 ml of 200 mM, two successive 1 ml of 350 mM, and two successive 500 mM imidazoles in 20 mM phosphate buffer/0.5 M NaCl, pH 6.0). The effluents were collected in test tubes containing PBS/5% sodium cholate to make final 0.5% concentration of sodium cholate. The purified fractions were monitored with SDS-PAGE and subsequent Coomassie Blue staining and Western blot analysis. The second 350 mM imidazole elution and the first 500 mM imidazole elution contained highly purified Rab38-protein (70 to 80% purity by Coomassie blue staining), which were pooled, dialyzed against PBS/1.5 mM MgCl$_2$/0.5% sodium cholate, and stored at $-20°$ until use.

3.2. [^{32}P]GTP-binding assay

Between 2.5 and 12.5 μg of purified recombinant Rab38 was subjected to SDS-PAGE under reducing conditions. Subsequently, the protein was electrophoretically transferred to a nitrocellulose membrane. [α-^{32}P]-GTP binding to the protein immobilized on the membrane was determined (Lapetina and Reep, 1987). The membrane was incubated for 1.5 h at 25° in 100 ml of Tris-HCl buffer (50 mM Tris-HCl, pH 7.5, containing 5 mM MgCl$_2$, 0.1% Triton X-100, and 0.1% BSA). Next, the membrane was incubated with 60 nM [α-^{32}P]-GTP (specific activity 3000 Ci [111TBq]/mmol)

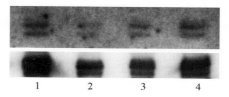

Figure 14.3   [α-^{32}P]-GTP binding to the purified recombinant Rab38 immobilized on the nitrocellulose membrane. (Upper panel) Purified proteins were loaded on SDS-PAGE under reducing conditions and electrophoretically transferred to a nitrocellulose membrane. The membrane was incubated with 60 nM [α-^{32}P]-GTP (specific activity 3000 Ci [111TBq]/mmol) in 10 ml of the Tris-HCl buffer for 1.5 h at 25 °. The membrane was then dried and autoradiographed at $-80°$ overnight. (Lower panel) The same membrane was washed with PBS and was processed for Western blotting for quantifying the protein amount. Lane 1: *wildtype*, 12.5 μg; lane 2: *cht*, 2.5 μg; lane 3: *cht*, 7.5 μg; lane 4: *cht*, 12.5 μg.

(Institute of Isotopes Co., Ltd., FP-208, Budapest) in 10 ml of Tris-HCl buffer for 1.5 h at 25° with gentle shaking on a platform shaker. The membrane was washed six times with 100 ml of Tris-HCl buffer for 15 min at 25° with shaking on a platform shaker. The membrane was dried and covered with a sheet of Saran-Wrap^{TM}, and autoradiographed on a Kodak BioMax MS film with BioMax MS screen at $-80°$ for 6 h to overnight. After development of the film, the membrane was washed with PBS, and then probed by Western blot for quantifying the amount of protein (Fig. 14.3).

3.3. Immunoprecipitation of [^{3}H]mevalonate-radiolabeled Rab38

Sf9 cells were seeded at 3×10^6 cells/60-mm dish. After 30 min, cells were infected with the recombinant baculovirus at a MOI of 10 and cultured for 1 day in TNM-FH media/10% FBS. Next, the cells were radiolabeled with 10 μCi/ml of [^{3}H]mevalonic acid (ARC ART-334, 1 mCi/ml) in Grace's insect media with 10% FBS (previously dialyzed against PBS) for 3 days (Lowe *et al.*, 1990). When used, 10{μg/ml of mevinolin (Calbio 438186) was added 3 h before adding radioisotope and replenished every day. The cells were harvested by vigorous pipetting, and concentrated by centrifugation at $500 \times g$ for 5 min. The cell pellets were washed twice by suspension in cold PBS and re-centrifugation. The final cell pellet was lysed with 0.5 ml of 1% Triton X-114/PBS/1.5 mM MgCl$_2$/antiproteases (PMSF, leupeptin, aprotinin). The cell lysis sample was centrifuged at $100,000 \times g$ for 30 min. The supernatant was precleared with 50 μl bed volume of normal rabbit serum–treated protein A sepharose (CL-4B) beads (Sigma P-3391). To the precleared cell lysate, a rabbit anti-rat Rab38 antibody was added at 10 μg/ml and the antigen–antibody complex was immunoprecipitated with 25 μl bed volume of normal protein A sepharose (CL-4B) beads. The beads were eluted with 40μl of two times the sample buffer: 0.125 M Tris-HCl, pH 7.4/ 20% (v/v) glycerol/4% (w/v) SDS/0.004% (v/v) bromophenol blue/0.4 M dithiothreitol. This step typically yielded the total ~60-μl eluate sample. A 40 μl-aliquot of the sample was used for autoradiography, and the remaining 20 μl was used for Western blot for evaluation of the amount of immunoprecipitated Rab38.

3.4. Immunoprecipitation of [³⁵S]methionine, [³⁵S]cysteine-radiolabeled Rab38

Sf9 cells were incubated and infected as described above. The cells were radiolabeled with 100 μCi/ml of [³⁵S]-methionine /[³⁵S]-cysteine (ARC, ARS 110 Met³⁵SLabel, methionine, specific activity 1175 Ci/mmol) in methionine-, cysteine-deficient insect cell media (Gibco21012) with 10% FBS (previously dialyzed against PBS) for 2 to 3 days, with or without 10 μg/ml of mevinolin. The medium was supplemented with 10% complete Grace's medium. When used, mevinolin was added 3 h before adding radioisotope. The cells were lysed with cold 1% Triton X-114/PBS/ 1.5 mM MgCl₂/antiproteases (PMSF, leupeptin, aprotinin) and divided into two samples. One sample was used as the total cell lysate. The other sample was phase separated by Triton X-114 partitioning. Both samples were immunoprecipitated with 10 μg/ml of a rabbit anti-rat Rab38 antibody (Osanai et al., 2005), and were subjected to SDS-PAGE under reducing conditions, followed by autoradiography (Fig. 14.4).

3.5. Western blot

Protein concentrations were determined with a BCA microprotein assay kit (Pierce). When used, deoxycholate (DOC)-trichloroacetic acid (TCA) precipitation was performed before BCA assay. Sample volume was adjusted to 1 ml with PBS in a 1.5-ml microtube, and 100 μl of 0.15% DOC added and mixed, and then left to set for 10 min at room temperature. Next, 50 μl of a saturated TCA solution was added and mixed, and subsequently centrifuged at 10,000×g for 5 min. The supernatant was removed and pellet dissolved in 100 μl of 0.1N NaOH. SDS-PAGE was performed under reducing conditions with 1-mm thick, 8 to 16% Tris-glycine gradient, precast minigels (Novex). After electrophoresis, the proteins separated in the gel were electrophoretically transferred to a nitrocellulose membrane. The membrane was blocked in 3% skim milk/1% Triton X-100/PBS for 30 min. The primary antibody used to detect recombinant Rab38^{wt} and Rab38^{cht} was an affinity-purified rabbit anti-rat Rab38 polyclonal antibody (1:5000 dilution) (Osanai et al., 2005), a mouse anti-Xpress monoclonal antibody (1:5000 dilution) (Invitrogen), or a mouse anti-6×HisG monoclonal

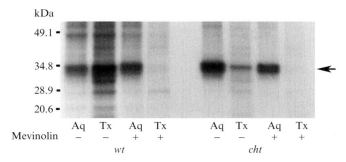

Figure 14.4 Characterization of the recombinant Rab38 expressed in Sf9 cells. Sf9 cells (3 × 10⁶/60-mm dish) infected with the recombinant baculovirus carrying Rab38^{wt} or Rab38^{cht} were radiolabeled with 100 μCi/ml of [³⁵S]-methionine and [³⁵S]-cysteine (with or without 10 μg/ml of mevinolin). Cells were lysed with 1% Triton X-114 and the postnuclear supernatants were phase separated. All samples were immunoprecipitated with a polyclonal rabbit anti-rat Rab38 antibody, and subjected to SDS-PAGE and autoradiography. Arrow indicates recombinant Rab38.

antibody (1:2500 dilution) (Invitrogen), and reacted overnight at 4°. The blots were washed four times for 15 min each. The washing and incubations with antibodies were all performed in 3% skim milk/1% Triton X-100/PBS. The secondary antibodies used were horseradish peroxidase–conjugated goat anti-rabbit IgG antibody (1:10,000 dilution) (BioRad) or horseradish peroxidase–conjugated goat anti-mouse IgG antibody (1:10,000 or 1:5,000 dilution) (BioRad) as appropriate for the different primary antibodies, and reacted for 1 h at room temperature. Subsequently, the blots were processed by three 15-min washings with 1% Triton X-100/PBS and two 15-min washings with PBS. Chemiluminescent detection assays were performed using a commercial kit (SuperSignal™ West Pico Chemiluminescent Substrate, Pierce). The same nitrocellulose membrane was used for blotting of a series of first antibodies after stripping off the previous antibody (Restore™ Western Blot Stripping Buffer, Pierce). The membrane was exposed to a Kodak BioMax Light film within 3 min of development.

4. DISCUSSION

Oculocutaneous albinism and lung disease are closely related to Hermansky-Pudlak syndrome (HPS). HPS is clinically characterized by OCA, a bleeding diathesis, and life-limiting pulmonary fibrosis, although the clinical picture is highly heterogeneous (Huizing et al., 2000). HPS is an autosomal recessive disease resulting from heterogeneous genetic abnormalities. Eight genetically distinct forms of HPS have so far been identified in humans—HPS1 through 8. Most of the implicated genes encode proteins that participate in vesicle trafficking associated with lysosome-related organelles. The mutations affect proteins involved in the biogenesis of lysosome-related organelle complexes (BLOC) -1, -2, and -3 (Wei, 2006). The only exception is HPS-2; this protein product is a component of adaptor protein -3 (AP-3), specifically AP-3β3A. The most critical problem for patients with HPS is lung fibrosis, which may lead to death between the fourth and fifth decades of life. The observed pulmonary changes in *cht* mice are similar to those reported in the mouse homologues of the human HPS, *pale ear* (*ep*), *pearl* (*pe*), and the double-mutant mice *ep/ep, pe/pe* (Lyerla *et al.*, 2003), although the extent of pulmonary changes is different. The double-mutant *ep/ep, pe/pe* mice show integrated pulmonary changes with striking abnormalities of alveolar type II cells. Type II cells and LBs of these mutant mice are remarkably enlarged, and the LBs are engorged with surfactant. *Cht* mice lack the bleeding diathesis phenotype of HPS, which is due to a platelet storage granule deficiency. *Cht* mice do not show prolonged bleeding time.

In rats, two HPS models are known, Fawn-Hooded (FH) rats and Tester-Moriyama (TM) rats. FH rats show several diseases including oculocutaneous albinism, bleeding diathesis, pulmonary hypertension, systemic hypertension, renal failure, depression, and alcoholism. OCA and the bleeding diathesis found in FH rats have been attributed to the pleiotropic effects of a single locus, *Ruby* (*R*) (also referred to as "*red-eyed dilution*" and "*red-eyed yellow*"), located on rat chromosome 1 in close proximity to other FH loci that contribute to renal disease, *Rf-1, Rf-2*, and to hypertension susceptibility, *Bpfh-1*. Recently, the *Ruby* locus has been identified as the *Rab38* gene, establishing that rat *Ruby* is homologous to mouse *cht* (Oiso *et al.*, 2004). The *Rab38* Met1Ile substitution found in FH rats and TM rats is a protein-null defect, which will abolish translation from the mutant allele resulting in complete loss of Rab38. Thus, it is likely that complete deficit of Rab38 causes a platelet pool storage deficiency and prolonged bleeding time.

5. CONCLUSION

Rab38 is a strong candidate gene for HPS, and should be used to screen patients with HPS who show no known genetic defect.

ACKNOWLEDGMENTS

The authors would like to thank Chiharu Tsuchihara, Rieko Hatta, and Tomoko Miwa for their technical assistance. This work was supported by a grant-in-aid for scientific research ((C)(2)15590833) from the Japan Society for the Promotion of Science, and a grant for collaborative research from Kanazawa Medical University (S2005-9), Japan.

REFERENCES

Bordier, C. (1981). Phase separation of integral membrane proteins in Triton X-114 solution. *J. Biol. Chem.* **256,** 1604–1607.

Huizing, M., Anikster, Y., and Gahl, W. A. (2000). Hermansky-Pudlak syndrome and related disorders of organelle formation. *Traffic* **1,** 823–835.

Jager, D., Stockert, E., Jager, E., Gure, A. O., Scanlan, M. J., Knuth, A., Old, L. J., and Chen, Y. T. (2000). Serological cloning of a melanocyte rab guanosine 5′-triphosphate-binding protein and a chromosome condensation protein from a melanoma complementary DNA library. *Cancer Res.* **60,** 3584–3591.

Kawakami, M., Paul, J. L., and Thurlbeck, W. M. (1984). The effect of age on lung structure in male BALB/cNNia inbred mice. *Am. J. Anat.* **170,** 1–21.

Lapetina, E. G., and Reep, B. R. (1987). Specific binding of [α-^{32}P]GTP to cytosolic and membrane-bound proteins of human platelets correlates with the activation of phospholipase C. *Proc. Natl. Acad. Sci. USA.* **84,** 2261–2265.

Loftus, S. K., Larson, D. M., Baxter, L. L., Antonellis, A., Chen, Y., Wu, X., Jiang, Y., Bittner, M., Hammer, J. A., 3rd, and Pavan, W. J. (2002). Mutation of melanosome protein RAB38 in chocolate mice. *Proc. Natl. Acad. Sci. U. S. A.* **99,** 4471–4476.

Lowe, P. N., Sydenham, M., and Page, M. J. (1990). The Ha-ras protein, p21, is modified by a derivative of mevalonate and methyl-esterified when expressed in the insect/baculovirus system. *Oncogene* **5,** 1045–1048.

Lyerla, T. A., Rusiniak, M. E., Borchers, M., Jahreis, G., Tan, J., Ohtake, P., Novak, E. K., and Swank, R. T. (2003). Aberrant lung structure, composition, and function in a murine model of Hermansky-Pudlak syndrome. *Am. J. Physiol. Lung Cell Mol. Physiol.* **285,** L643–L653.

Oiso, N., Riddle, S. R., Serikawa, T., Kuramoto, T., and Spritz, R. A. (2004). The rat *Ruby* (*R*) locus is *Rab38*: Identical mutations in Fawn-hooded and Tester-Moriyama rats derived from an ancestral Long Evans rat sub-strain. *Mamm. Genome.* **15,** 307–314.

O'Reilly, D. R., Miller, L. K., and Luckow, V. A. (1992). Virus methods. *In* "Baculovirus Expression Vectors: A Laboratory Manual." New York: W. H. Freeman and Company, New York.

Osanai, K, Iguchi, M., Takahashi, K., Nambu, Y., Sakuma, T., Toga, H., Ohya, N., Shimizu, H., Fisher, J. H., and Voelker, D. R. (2001). Expression and localization of a novel Rab small G protein (Rab38) in the rat lung. *Am. J. Pathol.* **158,** 1665–1675.

Osanai, K., Takahashi, K., Nakamura, K., Takahashi, M., Ishigaki, M., Sakuma, T., Toga, H., Suzuki, T., and Voelker, D. R. (2005). Expression and characterization of Rab38, a new member of the Rab small G protein family. *Biol. Chem.* **386,** 143–153.

Wei, M. L. (2006). Hermansky-Pudlak syndrome: a disease of protein trafficking and organelle function. *Pigment Cell Res.* **19,** 19–42.

Weibel, E. R. (1979). Practical methods for biological morphometry. *In* "Stereological methods," Vol. 1, pp. 30–161. Academic Press, London.

CHAPTER FIFTEEN

ANALYSIS OF RAB1 FUNCTION IN CARDIOMYOCYTE GROWTH

Catalin M. Filipeanu, Fuguo Zhou, *and* Guangyu Wu

Contents

Abstract

Protein transport between intracellular organelles is coordinated by Rab GTPases. As an initial approach to defining the function of Rab GTPases in cardiomyocytes, our laboratory focused on Rab1, which regulates protein transport specifically from the endoplasmic reticulum (ER) to the Golgi apparatus. Our studies have demonstrated that adenovirus-driven expression of Rab1 promotes cell growth of primary cultures of neonatal cardiomyocytes *in vitro* and that transgenic expression of Rab1 in the myocardium induces cardiac hypertrophy in mouse hearts *in vivo*. These data provide strong evidence implicating that ER-to-Golgi protein transport functions as a regulatory site for control of cardiomyocyte growth. Here we describe a sets of methods used in our laboratory to characterize the function of Rab1 GTPase in modulating cardiac myocyte growth.

Department of Pharmacology and Experimental Therapeutics, Louisiana State University Health Sciences Center, New Orleans, Louisiana

Methods in Enzymology, Volume 438
ISSN 0076-6879, DOI: 10.1016/S0076-6879(07)38015-4

1. INTRODUCTION

Cardiac hypertrophy is the compensatory increase in myocardial mass in response to mechanical injury or persistent increased hemodynamic loading. In cardiac hypertrophy, cardiac myocytes exhibit hypertrophic growth, which involves increases in protein synthesis and subsequent transport of newly synthesized proteins to their functional destinations. It has been well documented that intracellular protein trafficking is governed by Rab GTPases, the largest branch of the Ras-related small GTPases superfamily (Balch, 1990; Martinez and Goud, 1998). However, in contrast to the small GTPases Ras, Rac, and Rho, which function as components of many cellular signal transduction pathways and have been well investigated in cardiac hypertrophy and failure (Clerk and Sugden, 2000; Hunter *et al.*, 1995; Sussman *et al.*, 2000; Sah *et al.*, 1999), the function of Rab proteins in cardiac myocytes remain largely unknown.

As an initial approach to determine the possible role of Rab GTPases in cardiomyocytes, we investigated the function of Rab1 in myocyte growth both in primary cultures of neonatal rat ventricular myocytes *in vitro* and in the mouse heart *in vivo* (Wu *et al.*, 2001; Filipeanu *et al.*, 2004, 2006a; Duvernay *et al.*, 2005). Rab1 has two isoforms, Rab1a and Rab1b, with greater than 90% identity in their amino acid sequences. Rab1 is specifically localized in the ER and the Golgi apparatus and regulates export of newly synthesized proteins from the endoplasmic reticulum (ER) and subsequent anterograde transport from the ER to the Golgi (Allan *et al.*, 2000; Dugan *et al.*, 1995; Moyer *et al.*, 2001; Nuoffer *et al.*, 1994; Plutner *et al.*, 1991; Tisdale *et al.*, 1992; Wu *et al.*, 2003; Yoo *et al.*, 2002). Our studies have demonstrated that adenovirus-mediated expression of Rab1 in rat ventricular cardiomyocytes induces hypertrophic growth with an increase in the total protein synthesis, cell sizes, and sarcomeric organization at the cellular level (Filipeanu *et al.*, 2004). Consistent with the increased myocyte growth *in vitro*, transgenic mice overexpressing Rab1a specifically in the myocardium develops cardiac hypertrophy with progression to heart failure (Wu *et al.*, 2001). These data strongly suggest that protein transport between the ER and the Golgi may function as a regulatory site for control of cardiomyocyte growth. Here we describe the methods utilized in our laboratory to generate the adenoviruses expressing Rab1 and to investigate the role of Rab1 in protein synthesis, cell size, and sarcomeric organization in neonatal rat ventricular myocytes. Finally, we will briefly discuss cardiac phenotypes observed in transgenic mice overexpressing Rab1.

2. Generation of Adenoviruses Expressing Rab1 GTPase

To manipulate the function of endogenous Rab1 GTPase in primary cultures of cardiomyocytes, it is necessary to achieve efficient expression of Rab1, which will allow biochemical and morphological analysis of the entire cell population. It is difficult to achieve efficient expression of proteins in primary cultures of neonatal cardiomyocytes when using commonly used transfection reagents. Viral vectors including adenovirus, retrovirus, and lentivirus have been used to successfully deliver proteins of interest into primary cultures of cardiomyocytes. In our studies, we generated the recombinant adenovirus to drive expression of Rab1a in cardiomyocytes. Rab1a (AF226873) is cloned from a mouse cardiac cDNA library (Wu *et al.*, 2001). For easy detection of Rab1a protein expression and subcellular localization, Rab1a is tagged with the FLAG epitope at its N-terminus by polymerase chain reaction (PCR) using a primer GACTACAAGGACGAC-GATGACAAG coding a peptide DYKDDDDK (Wu *et al.*, 2003). The FLAG epitope has been widely used to label proteins, including small GTPases without altering their characteristics. FLAG-Rab1a is inserted into the pShuttle-CMV vector at the Sal I/Hind III restriction sites and transformed into the bacteria INVαF'. The recombinant plasmid is amplified and digested with Sal I and Hind III to confirm the presence of the insert. The purified plasmid is digested with Pme I and the linearized plasmid is co-transformed into *Escherichia coli* strain BJ5183 together with the viral DNA plasmid pAdEasy-1. The recombinant adenoviral constructs are screened by restriction endonuclease digestion, digested with Pac I and transfected into 911 cells to produce viral particles. Subsequent processing and purification of adenovirus are performed as described by He *et al.* (1999).

2.1. Adenoviral Rab1 infection of primary cultures of neonatal rat ventricular myocytes

In our laboratory we have used the neonatal cardiac myocytes isolation kit from Worthington (http://www.worthington-biochem.com/NCIS/cat.html) to isolate neonatal cardiac myocytes. Briefly, 1- to 2-day-old Sprague Dawley rat pups are anesthetized and killed by decapitation. The beating hearts are quickly removed from 8 to 12 pups and placed into cold Ca^{2+}/Mg^{2+}-free Hanks balanced salt solution (HBSS). The conjunctive tissue and atria are removed under a dissecting microscope and the remaining ventricular tissue is chopped into small pieces (1 to 2 mm^3) and washed repeatedly with fresh HBSS solution. The ventricular tissue is then digested overnight at 4° with 5 ml of HBSS containing trypsin at a concentration of

0.5 mg/ml. The digested ventricle is incubated with 5 ml of Leibovitz L-15 media containing 0.2 mg/ml collagenase at 37° for 30 to 45 min in a 360-degree rotator. The cardiac myocytes are gently released by repeated trituration (20 to 30 times) with a 10-ml pipette. The cell mixture is filtered through a cell strainer (diameter 70 μm) to remove the nondigested tissue. The cells are collected by centrifugation at 100 g for 5 min and resuspended in Media 199 supplemented with 15% fetal bovine serum (FBS) and anti-biotics (100 units/ml penicillin and 100 μg/ml streptomycin). The fibro-blasts are removed by preplating for 1 h at 37°. The remaining nonattached cells are ventricular myocytes and used for the subsequent experiments. In our hands, with this procedure, the purity of cardiac myocytes is greater than 95% as demonstrated by α-actinin antibodies staining.

The isolated myocytes are plated at different densities for different experiments. The myocytes are cultured in DMEM containing 10% FBS and antibiotics for 24 h and then infected with control and Rab1 adeno-viruses for 6 h. The myocytes are divided into three groups each in tripli-cate: no infection, infection with control parent adenovirus, and infection with Rab1 adenovirus. The infected myocytes are cultured in DMEM without FBS and antibiotics for 2 days before experiments. To optimize the infection condition, in the preliminary studies, isolated cardiac myocytes are infected with increasing concentrations of adenoviruses from 5 to 100 multiplicity of infection (MOI) for different periods of time (1 to 4 days).

3. ANALYSIS OF ADENOVIRUS-DRIVEN RAB1 EXPRESSION IN CARDIOMYOCYTES

To determine if Rab1 is successfully delivered into primary cultures of ventricular myocytes with adenoviruses, cardiomyocytes are cultured on 35-mm dishes at a density of 5×10^5 cells per dish and infected with the adenoviruses as above. The Rab1-infected myocytes are collected and solubilized with 300 μl of 1X SDS-gel loading buffer. Thirty μl of whole cardiomyocyte lysate is separated by 12% SDS-PAG and both anti-FLAG antibody M2 (Sigma) and Rab1a (Santa Cruz Biotechnology) antibodies are used to evaluate FLAG-Rab1a expression in myocytes by Western blot analysis. Rab1a antibodies detect both endogenous Rab1a and overex-pressed FLAG-Rab1a, whereas FLAG antibodies detect only overexpressed FLAG-Rab1a. Therefore, Western blots using Rab1a antibodies will deter-mine the relative amount of endogenous and infected Rab1a. Using this protocol, Rab1 and its mutant Rab1N124I were successfully delivered into neonatal rat ventricular myocytes (Filipeanu *et al.*, 2004), as well as cardiac myocytes isolated from adult rats (Zhou, F., and Wu, G., unpublished data), and the maximal FLAG-Rab1a expression was achieved after infection at 20 MOI for 2 days (Fig. 15.1A).

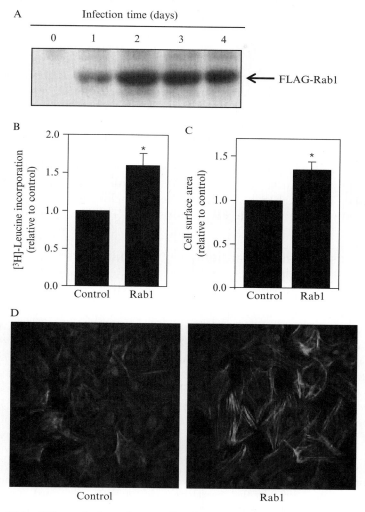

Figure 15.1 Effect of adenovirus-mediated expression of Rab1a on hypertrophic growth in neonatal cardiomyocytes. (A) Adenovirus-mediated expression of FLAG-Rab1a in neonatal rat ventricular myocytes. Neonatal myocytes were infected with empty adenoviral vector (control) or recombinant FLAG-Rab1a adenovirus for 1, 2, 3, and 4 days at 20 MOI. FLAG-Rab1a expression was detected by immunoblotting with anti-FLAG antibody M2 following SDS-PAGE. (B) The effect of adenovirus-driven FLAG-Rab1 expression on total protein synthesis. Cardiomyocytes were infected with control or Rab1 adenoviruses (20 MOI), and incubated with 1 μCi [3H]-leucine for 12 h.

3.1. Assessment of infection efficiency by Rab1 adenoviruses in cardiomyocytes

To determine the efficiency of adenoviral infection, cardiomyocytes are grown on coverslips in six-well dishes and infected as described above. Infected myocytes are fixed with a mixture of 4% paraformaldehyde and 4% sucrose in PBS for 15 min, permeabilized with PBS containing 0.2% Triton X-100 for 5 min, and blocked with 5% normal donkey serum for 1 h. The cells are then incubated with primary anti-FLAG antibody at 1:50 dilution for 1 h. After washing with PBS (3×5 min), the cells are incubated with Alexa Fluor 594–labeled secondary antibody at 1:2000 dilution for 1 h at room temperature. The coverslips are mounted, and fluorescence detected with a fluorescence microscope. In general, greater than 95% of the cardiomyocytes are infected (Filipeanu *et al.*, 2004). Importantly, consistent with the localization of endogenous Rab1 and its function in regulating protein transport between the ER and the Golgi, exogenously expressed FLAG-Rab1a is localized to the perinuclear region of the infected myocytes, presumably in the ER and the Golgi (Filipeanu *et al.*, 2004).

3.2. Measurement of cardiomyocyte viability after Rab1 infection

We have used the calcein AM retention method to measure myocyte viability after Rab1 infection (Yakovlev *et al.*, 2000). Calcein is a polyanionic fluorescein derivative whose acetoxy-methyl-ester form is plasma membrane permeable. Once inside the cell, the ester is hydrolyzed by intracellular esterases to release the fluorescent calcein molecules. The free calcein is trapped intracellularly only in the cells with intact plasma membrane and thus will indicate exclusively the viable cells. Briefly, the isolated ventricular myocytes are cultured on 24-well plates at a density of 1×10^5 cells per well in DMEM containing 10% FBS and antibiotics for 24 h. After 6 h of adenoviral infection and 40 h of culture, the myocytes are washed

---

The data are shown as the *x*-fold increase over the control and represent the means (plus or minus standard error) of five separate experiments, each performed in duplicate. (C) The effect of Rab1a on the cell surface area. Cardiomyocytes were stained with phalloidin for F-actin after a 2-day infection. The cell surface area was measured by using the NIH Image program. At least 100 myocytes from 20 randomly selected fields in three separate experiments were measured. $*p < 0.05$ versus cardiomyocytes infected with control adenovirus. (D) The effect of Rab1 on sarcomeric organization. The myocytes were stained with phalloidin and fluorescence visualized by a confocal microscope. (Figures adapted from Filipeanu, C. M., Zhou, F., Claycomb, W. C., and Wu, G. (2004). Regulation of the cell surface expression and function of angiotensin II type 1 receptor by Rab1-mediated endoplasmic reticulum-to-Golgi transport in cardiac myocytes. *J. Biol. Chem.* **279,** 41077–41084.) (See color insert.)

twice with PBS containing 145 mM NaCl, 5 mM KCl, 0.5 mM MgSO$_4$, 0.5 mM CaCl$_2$, 10 mM glucose, and 10 mM HEPES (pH 7.4), and incubated with 5 μM per well of calcein AM (Molecular Probes) at 37° for 1 h. The cells are washed twice to remove the extracellular calcein AM and allowed for dye hydrolysis for 30 min in the dark. The fluorescence is then monitored with a CytoFluor 4000 fluorimeter using 480 nm as excitation wavelength and 520 nm as emission wavelength.

4. EFFECT OF RAB1 INFECTION ON THE TOTAL PROTEIN SYNTHESIS IN NEONATAL CARDIOMYOCYTES

We used the [^{3}H]-leucine incorporation into the newly synthesized proteins to measure the effect of Rab1 on the total protein synthesis in cardiac myocytes (Filipeanu *et al.*, 2004). The isolated cardiac myocytes are cultured on 12-well dishes at a density of 2×10^5 cells per well, and infected with control and Rab1 adenoviruses at 20 MOI as above. After infection for 6 h and culture in DMEM without FBS for 2 days, the myocytes are incubated with 0.5 ml of DMEM containing 1 μCi [^{3}H]-leucine for 12 h. The myocytes are washed twice with 1 ml of ice-cold 5% trichloraceticacid (TCA) solution and then incubated with 1 ml of TCA solution in ice for 1 h. This step is critical to remove the cytosolic [^{3}H]-leucine, which are not incorporated into the newly synthesized proteins. After removing TCA solution, the myocytes are digested with 1 ml of 1 M NaOH for 4 to 6 h at 37°. The solution is collected into scintillation tubes and neutralized with 1 ml of 1 M HCl. The radioactivity is counted with 5 ml of Ecoscint A scintillation solution in a β-scintillation counter. Our data have demonstrated that adenoviral expression of Rab1 markedly augments total protein synthesis by 59% as measured by [^{3}H]-leucine incorporation in neonatal rat ventricular myocytes (Filipeanu *et al.*, 2004) (see Fig. 15.1B).

5. EFFECT OF RAB1 INFECTION ON THE CELL SIZE AND SARCOMERIC ORGANIZATION OF CARDIOMYOCYTES

The effect of adenoviral expression of Rab1 on the sizes of cardiac myocytes is determined by measuring cell surface area (Filipeanu *et al.*, 2004). Isolated myocytes are grown on glass coverslips in six-well dishes at low density (1×10^4 cells/well) and made quiescent in DMEM without FBS for 48 h and infected with control and Rab1 adenoviruses. The myocytes are fixed in 4% paraformaldehyde and 4% sucrose, and stained with Alexa Fluor-594 conjugated phalloidin to visualize F-actin. Cell images from 20 randomly chosen fields of 100 cardiac myocytes are visualized (40× objective) on each coverslip and their size is quantified using National Institutes of Health (NIH) image software. Only myocytes that are completely in the field are considered. Our data have demonstrated that adenoviral expression of Rab1 significantly enlarge the cell surface of neonatal cardiomyocytes (see Fig. 15.1C). Furthermore, the number of actin filaments as revealed by phalloidin staining is also largely increased by Rab1 infection in neonatal cardiomyocytes (Filipeanu *et al.*, 2004) (see Fig. 15.1D). Together with increased protein synthesis obtained in Rab1–infected myocytes, these results indicate that adenovirus–mediated Rab1 expression induces cellular hypertrophy in neonatal rat ventricular myocytes. These data suggest that augmentation of Rab1-mediated ER-to-Golgi transport plays an important role in regulating cardiomyocyte growth.

6. CARDIAC PHENOTYPES OF TRANSGENIC MICE OVEREXPRESSING RAB1A IN THE MYOCARDIUM

To determine the role of Rab1 GTPase in cardiac myocyte growth and function *in vivo*, we generated transgenic mice with cardiac specific expression of Rab1a (Wu *et al.*, 2001). A 1431-bp DNA fragment encompassing 28 bp of 5′-untranslated region, the mouse Rab1a coding region,

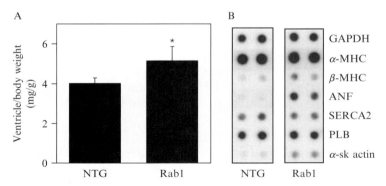

Figure 15.2 Overexpression of Rab1a specifically in the myocardium induces cardiac hypertrophy with increased ventricular weight (A) and cardiac gene expression (B). The data are obtained from six 6-week-old nontransgenic (NTG) and transgenic mice expressing Rab1a at a medium level. ANF, atrial natriuretic peptide; MHC, myosin heavy chain; SERCA, sarcoplasmic reticulum ATPase; PLB, phospholamban. $\star p < 0.05$ versus NTG sibling control. (Figures adapted from Wu, G., Yussman, M. G., Barrett, T. J., Hahn, H. S., Osinska, H., Hilliard, G. M., Wang, X., Toyokawa, T., Yatani, A., Lynch, R. A., Robbins, J., and Dorn, G. W. 2nd. (2001). Increased myocardial Rab GTPase expression: A consequence and cause of cardiomyopathy. *Circ. Res.* **89**, 1130–1137.)

and 785 bp of 3'-untranslated region is inserted into Sal I and Hind III restriction sites of the mouse α-myosin heavy-chain (MHC) promoter, which will drive Rab1 expression specifically in the myocardium (Subramaniam et al., 1991; Wu et al., 2000). Consistent with Rab1 function in promoting cardiomyocyte growth at the cellular level, Rab1 overexpression in the myocardium results in cardiac hypertrophy with an increase in whole heart and ventricular mass, an enlargement of myocyte size and a upregulation of fetal genes including atrial natriuretic peptide, β-myosin heavy chain (MHC), and α-skeletal actin (Fig. 15.2). Rab1 transgenic mice develop heart failure as cardiac fractional shortening measured by echocardiography and myocyte contractility are markedly impaired in Rab1 transgenic mice (Wu et al., 2001).

The most striking cardiac phenotype observed in transgenic mice overexpressing Rab1a, as compared with other transgenic mouse models of cardiac hypertrophy and failure (Filipeanu et al., 2006b; Koch et al., 2000), is the marked enlargement of the ER and the Golgi stacks and increased secretory atrial natriuretic peptide granules in ventricular myocytes as revealed by ultrastructural analysis (Wu et al., 2001). These data are consistent with the functional role of Rab1 GTPase in modulating overall newly synthesized protein transport between the ER and the Golgi. These results indicate that altered Rab1 function in cardiomyocytes may contribute to cardiac pathology.

ACKNOWLEDGMENT

This work was supported by the National Institutes of Health grants R01-GM076167 and P20-R018766.

REFERENCES

Allan, B. B., Moyer, B. D., and Balch, W. E. (2000). Rab1 recruitment of p115 into a cis-SNARE complex: Programming budding COPII vesicles for fusion. *Science* **289,** 444–448.

Balch, W. E. (1990). Small GTP-binding proteins in vesicular transport. *Trends Biochem. Sci.* **15,** 473–477.

Clerk, A., and Sugden, P. H. (2000). Small guanine nucleotide-binding proteins and myocardial hypertrophy. *Circ. Res.* **86,** 1019–1023.

Dugan, J. M., deWit, C., McConlogue, L., and Maltese, W. A. (1995). The Ras-related GTP-binding protein, Rab1B, regulates early steps in exocytic transport and processing of β-amyloid precursor protein. *J. Biol. Chem.* **270,** 10982–10989.

Duvernay, M. T., Filipeanu, C. M., and Wu, G. (2005). The regulatory mechanisms of export trafficking of G protein-coupled receptors. *Cell. Signal* **17,** 1457–1465.

Filipeanu, C. M., Zhou, F., Claycomb, W. C., and Wu, G. (2004). Regulation of the cell surface expression and function of angiotensin II type 1 receptor by Rab1-mediated endoplasmic reticulum-to-Golgi transport in cardiac myocytes. *J. Biol. Chem.* **279,** 41077–41084.

Filipeanu, C. M., Zhou, F., Fugetta, E. K., and Wu, G. (2006a). Differential regulation of the cell-surface targeting and function of beta- and alpha1-adrenergic receptors by Rab1 GTPase in cardiac myocytes. *Mol. Pharmacol.* **69,** 1571–1578.

Filipeanu, C. M., Zhou, F., Lam, M. L., Kerut, K. E., Claycomb, W. C., and Wu, G. (2006b). Enhancement of recycling and signaling of β-adrenergic receptors by Rab4 in cardiac myocytes. *J. Biol. Chem.* **281,** 11097–11103.

He, T. C, Zhou, S., da Costa, L. T., Yu, J., Kinzler, K. W., and Vogelstein, B. (1998). "A simplified system for generating of recombinant adenoviruses." *Proc. Natl. Acad. Sci. USA* **95,** 2509–2514.

Hunter, J. J., Tanaka, N., Rockman, H. A., Ross, J. Jr, and Chien, K.R. (1995). Ventricular expression of a MLC-2v-ras fusion gene induces cardiac hypertrophy and selective diastolic dysfunction in transgenic mice. *J. Biol. Chem.* **270,** 23173–23178.

Koch, W. J., Lefkowitz, R. J., and Rockman, H. A. (2000). Functional consequences of altering myocardial adrenergic receptor signaling. *Annu. Rev. Physiol.* **62,** 237–260.

Martinez, O., and Goud, B. (1998). Rab proteins. *Biochim. Biophys. Acta* **1404,** 101–112.

Moyer, B. D., Allan, B. B., and Balch, W. E. (2001). Rab1 interaction with a GM130 effector complex regulates COPII vesicle cis——Golgi tethering. *Traffic* **2,** 268–276.

Nuoffer, C., Davidson, H. W., Matteson, J., Meinkoth, J., and Balch, W. E. (1994). A GDP-bound of rab1 inhibits protein export from the endoplasmic reticulum and transport between Golgi compartments. *J. Cell Biol.* **125,** 225–237.

Plutner, H., Cox, A. D., Pind, S., Khosravi-Far, R., Bourne, J. R., Schwaninger, R., Der, C. J., and Balch, W. E. (1991). Rab1b regulates vesicular transport between the endoplasmic reticulum and successive Golgi compartments. *J. Cell Biol.* **115,** 31–43.

Sah, V. P., Minamisawa, S., Tam, S. P., Wu, T. H., Dorn, G. W., 2nd, Ross, J. Jr, Chien, K. R., and Brown, J. H. (1999). Cardiac-specific overexpression of RhoA results in sinus and atrioventricular nodal dysfunction and contractile failure. *J.Clin Invest.* **103,** 1627–1634.

Subramaniam, A., Jones, W. K., Gulick, J., Wert, S., Neumann, J., and Robbins, J. (1991). Tissue-specific regulation of the α-myosin heavy chain gene promoter in transgenic mice. *J. Biol. Chem.* **266,** 24613–24620.

Sussman, M. A., Welch, S., Walker, A., Klevitsky, R., Hewett, T. E., Price, R. L., Schaefer, E., and Yager, K. (2000). Altered focal adhesion regulation correlates with cardiomyopathy in mice expressing constitutively active rac1. *J. Clin. Invest.* **105,** 875–886.

Tisdale, E. J., Bourne, J. R., Khosravi-Far, R., Der, C. J., and Balch, W. E. (1992). GTP-binding mutants of rab1 and rab2 are potent inhibitors of vesicular transport from the endoplasmic reticulum to the Golgi complex. *J. Cell Biol.* **119,** 749–761.

Wu, G., Toyokawa, T., Hahn, H., and Dorn, G. W., II. (2000). ε Protein kinase C in pathological myocardial hypertrophy: Analysis by combined transgenic expression of translocation modifiers and Gαq. *J. Biol. Chem.* **275,** 29927–29930.

Wu, G., Yussman, M. G., Barrett, T. J., Hahn, H. S., Osinska, H., Hilliard, G. M., Wang, X., Toyokawa, T., Yatani, A., Lynch, R. A., Robbins, J., and Dorn, G. W., 2nd (2001). Increased myocardial Rab GTPase expression: A consequence and cause of cardiomyopathy. *Circ. Res.* **89,** 1130–1137.

Wu, G., Zhao, G., and He, Y. (2003). Distinct pathways for the trafficking of angiotensin II and adrenergic receptors from the endoplasmic reticulum to the cell surface: Rab1-independent transport of a G protein-coupled receptor. *J. Biol. Chem.* **278,** 47062–47069.

Yakovlev, A. G., Wang, G., Stoica, B. A., Boulares, H. A., Spoonde, A. Y., Yoshihara, K., and Smulson, M. E. (2000). A role of the Ca2+/Mg2+-dependent endonuclease in apoptosis and its inhibition by poly(ADP-ribose) polymerase. *J. Biol. Chem.* **275,** 21302–21308.

Yoo, J. S., Moyer, B. D., Bannykh, S., Yoo, H. M., Riordan, J. R., and Balch, W. E. (2002). Non-conventional trafficking of the cystic fibrosis transmembrane conductance regulator through the early secretory pathway. *J. Biol. Chem.* **277,** 11401–11409.

CHAPTER SIXTEEN

Regulation of the Trafficking and Function of G Protein-Coupled Receptors by Rab1 GTPase in Cardiomyocytes

Guangyu Wu

Contents

Abstract

G protein–coupled receptors (GPCRs) play a crucial role in regulating cardiac growth and function under normal and diseased conditions. It has been well documented that the precise function of GPCRs is controlled by intracellular trafficking of the receptors. Compared with the extensive studies on the events of the endocytic pathway, molecular mechanism underlying the transport process of GPCRs from the endoplasmic reticulum (ER) through the Golgi to the cell surface and regulation of receptor signaling by these processes in cardiac myocytes remain poorly defined. This chapter describes the methods to

Department of Pharmacology and Experimental Therapeutics, Louisiana State University Health Sciences Center, New Orleans, Louisiana

Methods in Enzymology, Volume 438
ISSN 0076-6879, DOI: 10.1016/S0076-6879(07)38016-6

characterize the function of Rab1 GTPase, which modulates protein transport from the ER to the Golgi apparatus, in the trafficking and signaling of angiotensin II type 1 receptor (AT1R), α_1-adrenergic receptor (AR), and β-AR, and in hypertrophic growth in response to agonist stimulation in neonatal cardiac myocytes.

1. INTRODUCTION

G protein–coupled receptors (GPCRs) constitute a superfamily of cell surface receptors which, through coupling to distinct heterotrimeric G proteins, regulate downstream effectors such as adenylyl cyclases, phospholipases, protein kinases, and ion channels (Lefkowitz, 1996; Wess, 1998; Armbruster and Roth, 2005; Wu et al., 1997, 1998, 2000a; Dong et al., 2007). It has been well documented that GPCRs play a central role in regulating cardiac growth and function under both physiological and pathological conditions and a number of signaling molecules in the GPCR system are targets for treating cardiac disease (Post et al., 1999; Lefkowitz et al., 2000; Wu et al., 2000b; Zhong and Neubig, 2001). The precise function of GPCRs and the magnitude of the agonist-elicited response are elaborately regulated by intracellular trafficking of the receptors (Duvernay et al., 2005). However, compared with the extensive studies on the events involved in the endocytic pathway, the transport process of GPCRs from the endoplasmic reticulum (ER), where they are synthesized and properly assembled, through the Golgi to the cell surface and regulation of receptor signaling by export trafficking remain poorly defined. As an initial approach to understanding the molecular mechanisms underlying the export trafficking of GPCRs, we have investigated the function of Rab1 GTPase in regulating the cell-surface expression and signaling of α_{2B}-adrenergic receptor (AR), α_1-AR, β_2-AR and angiotensin II (Ang II) type 1A receptor (AT1R) (Wu et al., 2003; Filipeanu et al., 2004, 2006a), which couple to different heterotrimeric G-proteins and initiate distinct signal transduction pathways.

Rab1 is specifically localized in the ER and the Golgi apparatus and coordinates the export of newly synthesized proteins from the ER and subsequent anterograde transport from the ER to and through the Golgi stacks (Allan et al., 2000; Dugan et al., 1995; Moyer et al., 2001; Nuoffer et al., 1994; Plutner et al., 1991; Tisdale et al., 1992; Yoo et al., 2002). There are two isoforms, Rab1a and Rab1b, with greater than 90% identity in their amino acid sequences. We first used the dominant-negative Rab1 mutants and small interfering RNA as tools to explore Rab1 function in GPCR transport from the ER to the cell surface in HEK293T cells (Wu et al., 2003). Our studies were then expanded to determine if the transport and signaling of endogenous GPCRs could be modified through manipulating

Rab1 function in cardiomyocytes (including HL-1 atrial cardiomyocytes, primary cultures of neonatal cardiomyocytes, and the mouse heart genetically overexpressing Rab1a (Wu *et al.*, 2001; Filipeanu *et al.*, 2004, 2006a). In this chapter, we describe the methods to study the role of Rab1 GTPase in regulating the trafficking and signaling of GPCRs, particularly AT1R, α_1-AR, and β-AR, and the hypertrophic response to their agonists in primary cultures of neonatal rat cardiomyocytes. At least two subtypes of β-ARs (β_1- and β_1-AR), two subtypes of α_1-ARs (α_{1A}- and α_{1B}-AR), and two subtypes of Ang II receptors (the Ang II type 1 receptor [AT1R] and the Ang II type 2 receptor (AT2R) have been identified in the mammalian heart. These receptors belong to the GPCR superfamily and play a crucial role in regulating cardiac growth and function under normal and diseased conditions (Post *et al.*, 1999; Filipeanu *et al.*, 2006b).

2. MATERIALS

Human β_2-AR, human α_{1B}-AR and rat AT1R were kindly provided by John D. Hildebrandt (Medical University of South Carolina), Kenneth P. Minneman, and Kenneth E. Bernstein (Emory University, Atlanta, GA), respectively. The receptors are tagged with green fluorescent protein (GFP) at their C-termini as described in Wu *et al.* (2003) and Hague *et al.* (2004). GFP as a tag is particularly useful for directly visualizing the subcellular localization and trafficking of proteins in the cell and has no substantial effect on the function and localization of many GPCRs (Kallal and Benovic, 2000). Rab1a is cloned from a mouse cardiac cDNA library (Wu *et al.*, 2001). The dominant-negative guanine nucleotide binding–deficient mutant Rab1aN124I is generated using QuikChange site-directed mutagenesis (Stratagene). Rab1a is tagged with the FLAG epitope at its N-terminus by polymerase chain reaction (PCR) using a primer GACTA-CAAGGACGACGATGACAAG coding a peptide DYKDDDDK (Wu *et al.*, 2003). [7-methoxy-^{3}H]-prazosin (specific activity = 70 Ci/mmol), [^{3}H]-Ang II (50.5 Ci/mmol), [^{3}H]-CGP12177 (51 Ci/mmol), and [^{3}H]-leucine (173 Ci/mmol) are purchased from PerkinElmer Life Sciences (Boston, MA). Human Ang II is obtained from Calbiochem. Brefeldin A (BFA), PD123319, phenylephrine (PE), isoproterenol (ISO), atenolol, ICI 118,551, niguldipine, chloroethylclonidine (CEC), and anti-FLAG M2 monoclonal antibodies are from Sigma. Alexa Fluor 594-labeled phalloidin and 4,6-diamidino-2-phenylindole (DAPI) are obtained from Molecular Probes (Eugene, OR). Antibodies against Rab1 and phospho-ERK1/2 are purchased from Santa Cruz Biotechnology. Antibodies against ERK1/2 are from Cell Signaling Technology.

2.1. Rab1 regulation of GPCR expression at the plasma membrane in neonatal rat ventricular myocytes

Isolation and culture of neonatal rat ventricular myocytes, generation of adenoviruses expressing FLAG-Rab1a, and infection of myocytes by Rab1 adenoviruses are described in Filipeanu et al. (2007). Briefly, neonatal ventricular myocytes are isolated from the hearts of 1- to 2-day-old Sprague Dawley rats and cultured on 12-well plates at a density of 2×10^5 cells/well in DMEM medium supplemented with 10% fetal bovine serum (FBS) and antibiotics for 24 h. The myocytes are infected with control parent adenovirus or adenovirus expressing Rab1 or its dominant-negative mutant Rab1N124I at a multiplicity of infection (MOI) of 20 for 6 h, and then cultured in serum-free DMEM for 2 days. The number of receptors at the plasma membrane is measured by radioligand binding in intact live cardiomyocytes.

To measure the number of AT1R at the plasma membrane, the infected cardiomyocytes are incubated with PBS containing radiolabeled Ang II overnight. There are two types of radiolabeled Ang II, [^{3}H]-Ang II and [^{125}I]-Ang II. [^{125}I]-Ang II has a higher affinity for the Ang II receptors than [^{3}H]-Ang II (Bouscarel et al., 1988; Matsubara et al., 1994). Here we describe the method to use [^{3}H]-Ang II as a ligand to measure AT1R at the plasma membrane. There are several important points for measuring AT1R expression at plasma membrane by intact cell binding. First, the concentration of [^{3}H]-Ang II used for incubation should be optimized in initial experiments. This is true for all the ligand binding experiment. In our experiments, [^{3}H-Ang II dose–dependent binding curve (2.5 to 20 nM) in cardiomyocytes infected with control parent adenovirus reveals that [^{3}H]-Ang II binding increases linearly at concentrations of 2.5 to 15 nM, and reaches maximal binding at the concentration of 15 nM. Therefore, cell-surface expression of AT1R in neonatal cardiomyocytes is measured at a saturating concentration of 20 nM. Second, as AT1R undergoes internalization in the presence of its agonist Ang II, the ligand binding assay is performed at 4°. Third, to exclude the contribution of AT2R to ligand binding, all solutions are supplemented with the specific AT2R antagonist PD123319 at a concentration of 1 μM. Finally, the nonspecific binding is determined in the presence of 10 μM of nonradioactive Ang II.

To measure the number of β-AR at the plasma membrane, the cells are incubated with PBS containing 10 nM of [^{3}H]-CGP12177, a hydrophobic β-AR ligand, for 90 min at room temperature. The nonspecific binding is determined in presence of alprenolol (20 μM). To measure expression of individual β-AR subtypes, myocytes are preincubated with the β_2-AR-selective antagonists ICI 118,551 or β_1-AR-selective antagonist atenolol (10 μM) for 30 min.

To measure the number of α_1-AR at the plasma membrane, the infected myocytes are incubated with PBS containing 10 nM of [^{7}-Methoxy-^{3}H]-prazosin for 2 h at room temperature. To measure expression of individual α_1-AR subtypes, myocytes are preincubated with the α_{1A}-AR-selective antagonist niguldipine or α_{1B}-AR–selective antagonist CEC at a concentration of 10 μM for 30 min. Nonspecific binding is determined in the presence of phentolamine (20 μM).

At the end of incubation, the myocytes are washed twice with 1 ml ice-cold PBS and the cell-surface–bound radioligand is extracted by mild acid treatment (2 × 5 min with 0.5 ml buffer containing 50 mM glycine, pH 3, and 125 mM NaCl) (Hunyady et al., 2002). Alternatively, the cells are digested with 1 ml of 1 M NaOH. The radioactivity is counted by liquid scintillation spectrometry in 6 ml of Ecoscint A scintillation solution (National Diagnostics, Inc., Atlanta, GA).

It has been well demonstrated that treatment with BFA, a fungal metabolite that disrupts the structures of the Golgi (Klausner et al., 1992), blocks protein transport from the ER to the Golgi. To determine if BFA treatment can attenuate Rab1-mediated receptor transport in cardiac myocytes, the Rab1-infected cardiomyocytes are incubated with BFA at a concentration of 5 μg/ml for 8 h. Our data indicate BFA treatment significantly inhibits the Rab1-induced increase in AT1R expression at the cell surface.

In contrast to the ligand binding of membrane preparations which has been extensively used to quantify GPCR expression, intact cell ligand binding has at least two advantages. One is that intact cell ligand binding has the ability to accurately measure the number of receptors at the plasma membrane, compared with possible contamination with intracellular receptors of membrane preparations. The other is that intact cell ligand binding uses a relatively smaller quantity of cells, which is particularly important for primary cultures of cardiomyocytes. However, one issue associated with the radioligand binding of intact live cardiomyocytes is that radiolabeled ligands are often able to induce their receptor internalization (such as [^{3}H]-Ang II for AT1R). One strategy to limit receptor internalization upon stimulation with the radiolabeled agonists is to carry out the experiment at low temperature as described for [^{3}H]-Ang II binding. Low-temperature incubation (4°) limits AT1R internalization induced by the ligand Ang II during the binding as demonstrated by the fact that the amounts of radioligand obtained after NaOH digestion (to obtain total bound ligand) and acidic washing (to obtain cell-surface radioligand) are not significantly different. However, low-temperature incubation may reduce ligand binding to the receptors. For example, prazosin is a nonspecific α_1-antagonist that undergoes internalization together with the receptor. However, our initial determinations performed at 4° failed to detect specific [^{3}H]-prazosin binding (<500 cpm) in neonatal cardiomyocytes. In a typical experiment performed at room temperature, the specific binding of [³H]-prazosin obtained by digesting the cells with NaOH is about 2800 cpm, and nonspecific binding about 300 cpm, measured in the presence of phentolamine in neonatal cardiomyocytes. The specific binding of [³H]-prazosin recovered by successive acidic washings (representing the cell surface ligand) is about 2600 cpm. These data suggest that α_1-AR internalization induced by the incubation with [³H]-prazosin (<10% of the total receptor) does not dramatically influence the cell-surface number of the receptor.

2.2. Rab1 regulation of the subcellular distribution of GPCRs in neonatal cardiomyocytes

Cardiomyocytes are grown on coverslips in six-well dishes and infected with control or Rab1 adenoviruses as above. After 6 h infection, the medium is removed and the myocytes are transiently transfected using LipofectAMINE 2000 reagent (Invitrogen) as described previously (Filipeanu *et al.*, 2004; Wu *et al.*, 2003). One microgram of GFP-tagged receptor is diluted into 125 μl of serum-free Opti-MEM in a tube. In another tube, 5 μl of LipofectAMINE is diluted into 125 μl of serum-free Opti-MEM. Five min later both solutions are mixed and incubated for another 20 min. The transfection mixture is then added to culture dishes containing 0.8 ml of fresh Dulbecco's modified Eagle's medium. After transfection 36 to 48 h, the myocytes are fixed with a mixture of 4% paraformaldehyde and 4% sucrose in PBS for 15 min. The coverslips are mounted, and fluorescence is detected with a fluorescence microscope.

Based on the GFP signal, approximately 5 to 10% myocytes are transfected by this plasmid transfection protocol. As almost all the cardiomyocytes are infected by Rab1 adenovirus, the number of myocytes expressing GFP-tagged receptor achieved by transient transfection is sufficient to measure the effect of Rab1 on the subcellular localization of receptors in individual myocytes.

 **3. Effects**

3.1. Effect of Rab1 on the signaling of GPCRs in cardiac myocytes

It is interesting to determine whether Rab1 is capable of regulating receptor function by modifying receptor trafficking. GPCRs activate multiple signaling pathways in cardiac myocytes, and any of these signaling systems can be chosen for this purpose. In our studies, we have determined the effect of Rab1 on the activation of the MAPK pathway, which is potently stimulated by AT1R, α_1-AR, and β-AR in cardiac myocytes. Neonatal cardiomyocytes are cultured in six-well plates at a density of 5×10^5 cells/well and infected with Rab1 constructs as described previously, and cultured in DMEM without serum for 48 h. The Rab1-infected myocytes are stimulated with 100 nM Ang II for 2 min (for AT1-mediated signaling), 10 μM PE for 8 min (for α_1-AR) or 10 μM ISO for 8 min (for β-AR).

To block AT2R response to the Ang II stimulation, the myocytes are preincubated with 10 μM PD123319 for 5 min. Similarly, to measure ERK1/2 activation by individual adrenergic receptors, the myocytes are preincubated with atenolol, ICI 118,551, niguldipine, and CEC at a concentration of 10 μM for 30 min to block the function of β_1-AR, β_2-AR, α_{1A}-AR, and α_{1B}-AR, respectively.

The reaction is terminated by the addition of 600 μl of 1 $\times$ SDS gel loading buffer. After solubilizing the cells, 30 μl of total cell lysate is separated by 10% SDS–PAGE and ERK1/2 activation measured by immunoblotting to determine phosphorylation with phospho-specific antibodies. The membranes are stripped and reprobed with anti-ERK1/2 antibodies to determine the total amount of kinases and to confirm equal loading of proteins.

3.2. Effect of Rab1 on the agonist-mediated protein synthesis in cardiomyocytes

The [^{3}H]-leucine incorporation is used to measure total protein synthesis in cardiac myocytes (Filipeanu et al., 2007). Briefly, neonatal cardiomyocytes are cultured on 12-well plates at a density of 2×10^5/well in DMEM supplemented with 10% FBS. After infection with the control and Rab1 adenoviruses, the myocytes are made quiescent by incubation for 48 h in DMEM without FBS. The cardiomyocytes are then incubated with [^{3}H]-leucine (1 μCi) for 12 h in the presence or absence of individual receptor agonists. The final concentrations of the receptor agonists are 100 nM Ang II, 10 μM ISO and 10 μM PE. Similar to the measurement of ERK1/2 activation, the AT12R antagonist PD123319 is added together with Ang II at a final concentration of 10 μM to block AT2R activation by Ang II stimulation. To block the function of individual adrenergic receptor β_1-AR, β_2-AR, α_{1A}-AR, and α_{1B}-AR, the myocytes are preincubated with atenolol, ICI 118,551, niguldipine, and CEC at a concentration of 10 μM for 30 min, respectively. The reaction is terminated and protein synthesis determined by liquid scintillation spectrometry. Since Rab1 influences protein synthesis (Filipeanu et al., 2004, 2007), the effect of Rab1 on receptor agonist–stimulated protein synthesis can be calculated using the following formula:

$$\frac{[\text{Ang II and Rab1}] - [\text{Rab1}]}{[\text{Ang II and control adenovirus}] - [\text{control adenovirus}]}$$

3.3. Effect of Rab1 on the agonist-mediated increase in size and sarcomeric organization in cardiomyocytes

The effect of adenoviral expression of Rab1 on the sizes of cardiac myocytes is determined by measuring cell surface area (Filipeanu *et al.*, 2004, 2007). The isolated myocytes are cultured on coverslips in six-well dishes at a density of 1×10^4 cells/well, and made quiescent in DMEM without FBS for 48 h and infected with control and Rab1 adenoviruses. The myocytes are stimulated with the receptor agonists: 100 nM Ang II, 10 μM ISO, and 10 μM PE for 24 h. After the myocytes are fixed and stained with Alexa Fluor-594 conjugated phalloidin, the cell surface is measured (Filipeanu *et al.*, 2007). The effect of Rab1 on the receptor agonist–mediated increase in cell size is calculated using the same formula as for the protein synthesis described previously.

4. Experimental Results and Concluding Remarks

Using the methods described in this chapter, we have demonstrated that adenoviral expression of the dominant negative mutant Rab1N124I attenuates the cell-surface expression of all GPCRs examined including AT1R, α_{1A}-AR, α_{1B}-AR, β_1-AR, and β_2-AR in neonatal cardiac myocytes. Interestingly, augmentation of Rab1 function by adenoviral expression of Rab1 facilitates the cell-surface expression of AT1R, α_{1A}-AR, and α_{1B}-AR, without altering β_1-AR and β_2-AR expression (Fig. 16.1A) (Filipeanu *et al.*, 2004, 2006a). These data indicate that the transport to the cell surface of endogenous GPCRs is differentially regulated by Rab1 in cardiac myocytes. Subcellular localization analysis revealed that the receptors were unable to transport to the cell surface and were trapped inside the cells infected with Rab1N124I adenoviruses (see Fig. 16.1B) (Filipeanu *et al.*, 2004, 2006a). Consistent with Rab1 effect on the receptor expression at the cell surface, ERK1/2 activation in response to stimulation with receptor agonists is inhibited in neonatal cardiomyocytes infected with adenovirus expressing Rab1N124I, and adenoviral expression of Rab1 selectively promotes ERK1/2 activation by Ang II and PE, but not ISO (Filipeanu *et al.*, 2004, 2006a). Similarly, Rab1N124I attenuates cardiomyocyte hypertrophic response to all the agonists, whereas Rab1 selectively promotes cardiomyocyte hypertrophic response to Ang II (Fig. 16.2) and PE, but not ISO (Filipeanu *et al.*, 2004, 2006a). These data provide strong evidence implicating that GPCR function can be modulated through manipulating their traffic from the ER to the Golgi, which in turn alters the responsiveness of cardiomyocytes to the receptor stimulation.

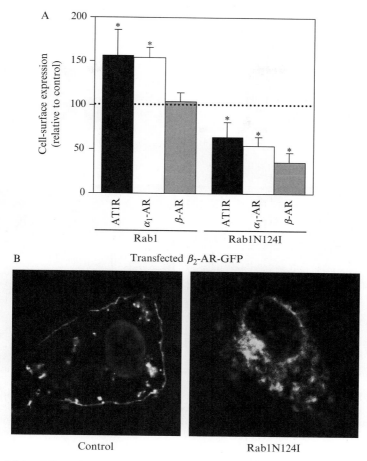

Figure 16.1   Effect of adenovirus-mediated expression of Rab1 and its dominant negative mutant Rab1N124I on the cell surface expression (A) and subcellular localization
(B) of GPCRs in neonatal cardiomyocytes. (A) Cardiomyocytes were cultured
and infected with control, Rab1, or Rab1N124I adenovirus for 2 days at a multiplicity
of infection of 20. The cell-surface expression of AT1R, α_1-AR, and β-AR was
determined by the radioligand [³H]-Ang II, [³H]-prazosin, and [³H]-CGP12177,
respectively. The data shown are the percentage of the mean value obtained from the
cardiomyocytes infected with control adenovirus (the dotted line). $\star p < 0.05$ versus
control. (B) Cardiomyocytes were grown on coverslips, infected with control
or Rab1N124I adenoviruses, and transiently transfected with GFP-tagged β_2-AR.
Two days after infection, the myocytes were fixed and stained with 4,6-diamidino-2-
phenylindole (nuclear). Blue, nuclear stained by 4,6-diamidino-2-phenylindole; green,
GFP-β_2-AR. (From Filipeanu, C. M., Zhou, F., Claycomb, W. C., and Wu, G. (2004).
Regulation of the cell surface expression and function of angiotensin II type 1 receptor
by Rab1-mediated endoplasmic reticulum-to-Golgi transport in cardiac myocytes.
J. Biol. Chem. **279,** 41077–41084; and Filipeanu, C. M., Zhou, F., Fugetta, E. K., and
Wu, G. (2006a). Differential regulation of the cell-surface targeting and function of
β- and $\alpha1$-adrenergic receptors by Rab1 GTPase in cardiac myocytes. *Mol. Pharmacol.*
69, 1571–1578.) (See color insert.)

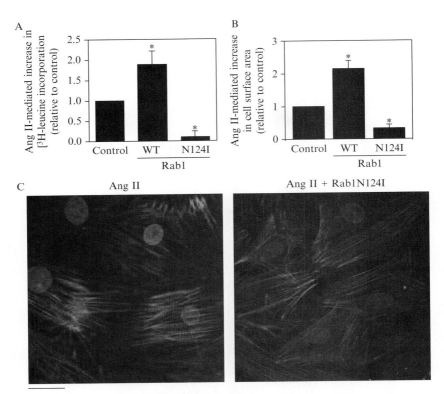

Figure 16.2 Effect of adenovirus–mediated expression of Rab1 and its dominant negative mutant Rab1N124I on Ang II–stimulated hypertrophic response in neonatal cardiomyocyte. To reflect the effect of Rab1 on Ang II–mediated stimulation, the contribution of Rab1 itself to protein synthesis and cell surface area was subtracted. Cardiomyocytes were cultured and infected with control, Rab1, and Rab1N124I adenoviruses for 2 days, and then stimulated with Ang II (100 nM) for 24 h. (A) Effect of Rab1 on Ang II–stimulated total protein synthesis measured by [^{3}H]-leucine incorporation. (B) Effect of Rab1 on Ang II–mediated increase in cell surface area. (C) Effect of Rab1N124I on Ang II–stimulated sarcomeric organization revealed by staining with phalloidin for F-actin. $\star p<0.05$ versus control. (Adapted from From Filipeanu, C. M., Zhou, F., Claycomb, W. C., and Wu, G. (2004). Regulation of the cell surface expression and function of angiotensin II type 1 receptor by Rab1-mediated endoplasmic reticulum-to-Golgi transport in cardiac myocytes. *J. Biol. Chem.* **279**, 41077–41084.) (See color insert.)

ACKNOWLEDGMENTS

This work was supported by the National Institutes of Health grants R01-GM076167 and P20-R018766. The author thanks Stephen M. Lanier, John D. Hildebrandt (Medical University of South Carolina), Kenneth P. Minneman, and Kenneth E. Bernstein (Emory University) for sharing reagents.

REFERENCES

Allan, B. B., Moyer, B. D., and Balch, W. E. (2000). Rab1 recruitment of p115 into a cis-SNARE complex: Programming budding COPII vesicles for fusion. *Science* **289**, 444–448.

Armbruster, B. N., and Roth, B. L. (2005). Mining the receptorome. *J. Biol. Chem.* **280**, 5129–5132.

Bouscarel, B., Blackmore, P. F., and Exton, J. H. (1988). Characterization of the angiotensin II receptor in primary cultures of rat hepatocytes. Evidence that a single population is coupled to two different responses. *J. Biol. Chem.* **263**, 14913–14919.

Dong, C., Filipeanu, C. M., Duvernay, M. T., and Wu, G. (2007). Regulation of G protein–coupled receptor export trafficking. *Biochim. Biophys. Acta* **1768**, 853–870.

Dugan, J. M., deWit, C., McConlogue, L., and Maltese, W. A. (1995). The Ras-related GTP-binding protein, Rab1B, regulates early steps in exocytic transport and processing of β-amyloid precursor protein. *J. Biol. Chem.* **270**, 10982–10989.

Duvernay, M. T., Filipeanu, C. M., and Wu, G. (2005). The regulatory mechanisms of export trafficking of G protein–coupled receptors. *Cell. Signal.* **17**, 1457–1465.

Filipeanu, C. M., Zhou, F., Claycomb, W. C., and Wu, G. (2004). Regulation of the cell surface expression and function of angiotensin II type 1 receptor by Rab1-mediated endoplasmic reticulum-to-Golgi transport in cardiac myocytes. *J. Biol. Chem.* **279**, 41077–41084.

Filipeanu, C. M., Zhou, F., Fugetta, E. K., and Wu, G. (2006a). Differential regulation of the cell-surface targeting and function of β- and α1-adrenergic receptors by Rab1 GTPase in cardiac myocytes. *Mol. Pharmacol.* **69**, 1571–1578.

Filipeanu, C. M., Zhou, F., Lam, M. L., Kerut, K. E., Claycomb, W. C., and Wu, G. (2006b). Enhancement of recycling and signaling of β-adrenergic receptors by Rab4 in cardiac myocytes. *J. Biol. Chem.* **281**, 11097–11103.

Filipeanu, C. M., Zhou, F., and Wu, G. (2008). Analysis of Rab1 function in cardiomyocyte growth. *Methods Emzymol.* **438**, 217–226.

Hague, C., Uberti, M. A., Chen, Z., Hall, R. A., and Minneman, K. P. (2004). Cell surface expression of α1D-adrenergic receptors is controlled by heterodimerization with α1B-adrenergic receptors. *J. Biol. Chem.* **279**, 15541–15549.

Hunyady, L., Baukal, A. J., Gaborik, Z., Olivares-Reyes, J. A., Bor, M., Szaszak, M., Lodge, R., Catt, K. J., and Balla, T. (2002). Differential PI 3-kinase dependence of early and late phases of recycling of the internalized AT1 angiotensin receptor. *J. Cell Biol.* **157**, 1211–1222.

Kallal, L., and Benovic, J. L. (2000). Using green fluorescent proteins to study G-protein–coupled receptor localization and trafficking. *Trends Pharmacol. Sci.* **21**, 175–180.

Klausner, R. D., Donaldson, J. G., and Lippincott-Schwartz, J. (1992). Brefeldin A: Insights into the control of membrane traffic and organelle structure. *J. Cell Biol.* **116**, 1071–1080.

Lefkowitz, R. J. (1996). G protein-coupled receptors and receptor kinases: From molecular biology to potential therapeutic applications. *Nat. Biotechnol.* **14**, 283–286.

Lefkowitz, R. J., Rockman, H. A., and Koch, W. J. (2000). Catecholamines, cardiac β-adrenergic receptors, and heart failure. *Circulation* **101**, 1634–1637.

Matsubara, H., Kanasaki, M., Murasawa, S., Tsukaguchi, Y., Nio, Y., and Inada, M. (1994). Differential gene expression and regulation of angiotensin II receptor subtypes in rat cardiac fibroblasts and cardiomyocytes in culture. *J. Clin. Invest.* **93**, 1592–1601.

Moyer, B. D., Allan, B. B., and Balch, W. E. (2001). Rab1 interaction with a GM130 effector complex regulates COPII vesicle cis–Golgi tethering. *Traffic* **2**, 268–276.

Nuoffer, C., Davidson, H. W., Matteson, J., Meinkoth, J., and Balch, W. E. (1994). A GDP-bound of rab1 inhibits protein export from the endoplasmic reticulum and transport between Golgi compartments. *J. Cell Biol.* **125**, 225–237.

Plutner, H., Cox, A. D., Pind, S., Khosravi-Far, R., Bourne, J. R., Schwaninger, R., Der, C. J., and Balch, W. E. (1991). Rab1b regulates vesicular transport between the endoplasmic reticulum and successive Golgi compartments. *J. Cell Biol.* **115,** 31–43.

Post, S. R., Hammond, H. K., and Insel, P. A. (1999). β-Adrenergic receptors and receptor signaling in heart failure. *Annu. Rev. Pharmacol. Toxicol.* **39,** 343–360.

Tisdale, E. J., Bourne, J. R., Khosravi-Far, R., Der, C. J., and Balch, W. E. (1992). GTP-binding mutants of rab1 and rab2 are potent inhibitors of vesicular transport from the endoplasmic reticulum to the Golgi complex. *J. Cell Biol.* **119,** 749–761.

Wess, J. (1998). Molecular basis of receptor/G-protein-coupling selectivity. *Pharmacol. Ther.* **80,** 231–264.

Wu, G., Krupnick, J. G., Benovic, J. L., and Lanier, S. M. (1997). Interaction of arrestins with intracellular domains of muscarinic and α2-adrenergic receptor. *J. Biol. Chem.* **272,** 17836–17842.

Wu, G., Benovic, J. L., Hildebrandt, J. D., and Lanier, S. M. (1998). Receptor docking sites for G-protein βγ subunits: Implications for signal regulation. *J. Biol. Chem.* **273,** 7197–7200.

Wu, G., Bogatkevich, G. S., Mukhin, Y. V., Benovic, J. L., Hildebrandt, J. D., and Lanier, S. M. (2000a). Identification of Gβγ-binding sites in the third intracellular loop of the muscarinic receptors and their role in receptor regulation. *J. Biol. Chem.* **275,** 9026–9034.

Wu, G., Toyokawa, T., Hahn, H., and Dorn, G. W., II. (2000b). ε Protein kinase C in pathological myocardial hypertrophy: Analysis by combined transgenic expression of translocation modifiers and Gαq. *J. Biol. Chem.* **275,** 29927–29930.

Wu, G., Yussman, M. G., Barrett, T. J., Hahn, H. S., Osinska, H., Hilliard, G. M., Wang, X., Toyokawa, T., Yatani, A., Lynch, R. A., Robbins, J., and Dorn, G. W. (2001). 2nd.Increased myocardial Rab GTPase expression: A consequence and cause of cardiomyopathy. *Circ. Res.* **89,** 1130–1137.

Wu, G., Zhao, G., and He, Y. (2003). Distinct pathways for the trafficking of angiotensin II and adrenergic receptors from the endoplasmic reticulum to the cell surface: Rab1-independent transport of a G protein-coupled receptor. *J. Biol. Chem.* **278,** 47062–47069.

Yoo, J. S., Moyer, B. D., Bannykh, S., Yoo, H. M., Riordan, J. R., and Balch, W. E. (2002). Non-conventional trafficking of the cystic fibrosis transmembrane conductance regulator through the early secretory pathway. *J. Biol. Chem.* **277,** 11401–11409.

Zhong, H., and Neubig, R. R. (2001). Regulator of G protein signaling proteins: Novel multifunctional drug targets. *J. Pharmacol. Exp. Ther.* **297,** 837–845.

CHAPTER SEVENTEEN

Regulation of Endosome Dynamics by Rab5 and Huntingtin-HAP40 Effector Complex in Physiological versus Pathological Conditions

Arun Pal,* Fedor Severin,[†] Sebastian Höpfner,[‡] *and* Marino Zerial*

Contents

Abstract

Vesicular transport of signaling molecules, specifically neurotrophins, in neurons is essential for their differentiation, survival, and plasticity. Neurotrophins such as neuron growth factor (NGF) and brain-derived neurotrophic factor (BDNF) are internalized by receptor-mediated endocytosis at synaptic terminals and loaded into endosomes for microtubule-based transport along axons to the cell body where they exert their signaling function in the nucleus. The molecular

* Max Planck Institute of Molecular Cell Biology and Genetics, Dresden, Germany
[†] A. N. Belozersky Institute of Physico-Chemical Biology, Moscow State University, Russia
[‡] Kanzlei Dr. Volker Vossius, Munich, Germany

Methods in Enzymology, Volume 438   © 2008 Elsevier Inc.
ISSN 0076-6879, DOI: 10.1016/S0076-6879(07)38017-8   All rights reserved.

mechanisms underlying this intracellular transport are not only relevant from a basic knowledge viewpoint, but have also important implications for neurodegenerative diseases. Defects in trafficking are increasingly implicated in the pathology of Huntington's disease (HD) and other neurodegenerative disorders. The small GTPases Rab5 and Rab7 play important roles in the endocytic trafficking of neurotrophins. We have recently identified Huntingtin (Htt) and Huntingtin associated protein of 40 kDa (HAP40) as a novel Rab5 effector complex that regulates endosome motility. In HD, we detected higher HAP40 protein levels compared with normal cells. Such increase causes an augmented recruitment of Htt onto Rab5-positive early endosomes that drastically reduces their motility by "switching" these organelles from microtubules to F-actin. These findings suggest a mechanism by which impaired Rab5-mediated trafficking of neurotrophic factors may be a key event of the pathogenetic process leading to neurodegeneration in HD. To dissect the mechanisms by which Htt, HAP40, and Rab5 function in early endosome interactions with the cytoskeleton, we developed assays to investigate endosome–cytoskeleton interactions that can be applied to normal and pathological conditions. We provide here detailed protocols for, first, an assay that measures binding of early endosomes to microtubules and F-actin. Second, we describe an improved protocol for a cell-free assay that recapitulates the motility of early endosomes along microtubules *in vitro*. These assays provide mechanistic insights into the dysfunction of endosome motility occurring in HD as well as other neurodegenerative disorders.

1. INTRODUCTION

In endocytosis, nutrients and growth factors are internalized in budding vesicles from the cell surface and targeted to a series of intracellular compartments where they are sorted and routed to the recycling and degradative pathways (Conner and Schmid, 2003; Gruenberg and Maxfield, 1995; Mellman, 1996; Pelkmans and Helenius, 2003). Endocytic transport allows signaling cargo to transduce signals to the nucleus (Lanzetti *et al.*, 2000; Miaczynska *et al.*, 2004). To accomplish these trafficking tasks, endocytic organelles such as early endosomes undergo heterotypic fusion with incoming clathrin coated vesicles and also homotypic fusion with each other, transform their geometry through membrane fission and tubulation, and move over long distances to transfer cargo to other organelles. In these dynamics, membrane interactions with the tubulin and actin cytoskeleton play pivotal roles at all the levels of trafficking events (Apodaca, 2001; Gasman *et al.*, 2003; Hoepfner *et al.*, 2005; McPherson, 2002; Nielsen *et al.*, 1999), particularly in neurons where vesicular transport of essential neurotrophins from axonal terminals over long distances to the nucleus (Deinhardt *et al.*, 2006; Hibbert *et al.*, 2006; Morfini *et al.*, 2005) calls for a coordinated interplay of vesicles with different types of motors, static anchors, and cytoskeletal fibers.

Alterations of this complex trafficking scenario are implicated in neuro-degeneration in several inherited disorders such as Huntington's disease (HD), Kennedy's disease (spinal and bulbar muscular atrophy [SBMA]), spinocerebellar ataxia type 1c (SCA-1), Machado-Joseph disease (MJD) or SCA-3, and dentatorubral-pallidoluysian atrophy (DRPA) (Morfini *et al.*, 2005), to name a few. HD is a good example of a trafficking dysfunction. Its pathogenesis is triggered by an expanded, amino-terminal polyglutamine sequence (polyQ) (>35 for mutant Htt) (Gusella and MacDonald, 2000) in Htt, a large 348-kDa protein interacting with a wide panel of Htt interact-ing (HIPs) and associated partners (HAPs) via domains adjacent to the polyQ stretch. The expansion of the polyQ sequence in HD is thought to cause a conformational change that alters the binding affinity of mutant Htt for HIPs and HAPs (Harjes and Wanker, 2003). The wide number of HIPs and HAPs that implicate Htt in diverse cellular functions and the complex-ity of aberrant protein–protein interactions (Goehler *et al.*, 2004) hamper the identification of the most critical players in HD pathology. Recent work has pointed to defects in endocytosis (Trushina *et al.*, 2006) and intraneur-onal transport of neurotrophins (Block-Galarza *et al.*, 1997; Engelender *et al.*, 1997; Gauthier *et al.*, 2004; Li and Li, 2005; Pal *et al.*, 2006; Rong *et al.*, 2006) as primary trigger for HD pathology. Particularly compromised microtubule-dependent vesicular transport of BDNF across cortical-striatal afferents is considered to be a key event in HD pathology (Altar *et al.*, 1997; Conner *et al.*, 1997; Gauthier *et al.*, 2004; Zuccato and Cattaneo, 2007). Clearly, a better understanding of HD pathology also requires dissecting the underlying regulation of endocytosis.

1.1. Rab Proteins are key players in the coordination of endocytic events

Several small GTPases of the Rab family are key regulators of endocytosis. They serve as membrane organizers, regulating sequential transport along endocytic transport routes (Zerial and McBride, 2001). Rab5 is among the best-characterized key players of endocytosis. Functional studies of numer-ous Rab effectors have broadened our view of the role of Rab5 in early endocytic transport (Bucci *et al.*, 1995; Singer-Krüger *et al.*, 1994, 1995), but also of Rab GTPases in general. By recruiting distinct subsets of downstream effectors onto the membrane, Rab5 protein is capable of assembling multiple domains for different functions (Zerial and McBride, 2001).

One important aspect in the spatiotemporal regulation of endocytic events is the capability of Rab5 to coordinate the homotypic fusion of early endo-somes with their active transport along microtubules through the recruitment of several effectors such as EEA1, Rabenosyn-5, and Rabankyrin-5 for fusion

(Schnatwinkel *et al.*, 2004; Zerial and McBride, 2001), and the early endosome-specific kinesin motor KIF16B for motility (Hoepfner *et al.*, 2005; Nielsen *et al.*, 1999). However, it remains unclear which parameters determine the particular effector subset that is recruited and activated at a given time and location. We gained initial clues when we functionally characterized Htt and HAP40, which we isolated by affinity chromatography on immobilized active Rab5 (Christoforidis and Zerial, 2000; Pal *et al.*, 2006). Htt/HAP40 complex is recruited by active Rab5 as novel downstream effector onto early endosomes. In HD, HAP40 protein levels are ~10-fold increased by a yet unknown mechanism, leading to augmented Htt/HAP40 amounts on Rab5-positive endosomal membranes and, in turn, to a "switch" of these organelles from microtubules to F-actin (Pal *et al.*, 2006). The consequence is a drastic reduction of early endosome motility along microtubules due to the preferential association with F-actin. Because we can mimic these perturbations by ectopically overexpressing HAP40, we conclude that the protein level of HAP40 is a critical key parameter in determining whether Rab5 recruits those effector subsets required for early endosome tethering and fusion (via EEA1, Rabenosyn-5, Rabankyrin-5) and motility along microtubules (via KIF16B) or for the association with F-actin (via Htt/HAP40). In line with this view, we found KIF16B displaced from early endosomes through elevated levels of recruited Htt/HAP40 (our unpublished observation). Remarkably, ablation of HAP40 by RNAi can rescue the defects in Rab5 dynamics in HD cells, suggesting that excess of HAP40 is both necessary and sufficient to block endosome motility, independently of the polyQ motif in Htt (Pal *et al.*, 2006).

Collectively, the upregulation of HAP40 naturally occurring in HD provides an interesting model system to explore the role of Rab5, Htt, and HAP40 in the regulation of early endosome motility. A better understanding of the underlying membrane–cytoskeletal interactions will also help assessing the contribution of compromised endosome transport in HD pathology and, more in general, neurodegeneration.

1.2. Outlook: Feasible candidates for functional characterizations

How does excess of Htt/HAP40 on early endosomes effectuate the switch from microtubules to F-actin? For melanosomes, Gross and coworkers (2002) documented a coordinated activation/deactivation of microtubule- (dynein, kinesins) and F-actin–based motors (myosins) to regulate the motility of these vesicles along microtubules and F-actin as well as their switching between types of filaments. Because we observe a displacement of the endosomal motor KIF16B (Hoepfner *et al.*, 2005) from early endosomes through elevated levels of HAP40 (our unpublished observation; see section 1.1.), we suggest a Rab5 effector machinery with Htt and HAP40 functioning in a similar interplay of different motors and putative anchors

(Walenta *et al.*, 2001). Some of our previous findings already hint to possible candidates in such a "switch": Gasman and coworkers (2003) reported that the small GTPase RhoD recruits hDia2C, a splice variant of human Diaphanous 2, onto Rab5-positive early endosomes, and activates c-Src kinase, causing an alignment of early endosomes along F-actin and a reduction of their motility. The striking resemblance of this phenotype with the perturbations caused through HAP40 upregulation in HD (Pal *et al.*, 2006) suggests a functional interaction between Rab5/Htt/HAP40 on the one hand, and RhoD/hDia2C/c-Src on the other. We thus propose to use the assays here to test the role of regulatory molecules in endosome–cytoskeleton interactions.

2. Assays for Functional Characterization of Rab5, Htt, and HAP40

Because trafficking of organelles along microtubules and F-actin and their switching between filament types represent very complex regulatory and mechanical processes, we sought to begin our dissection of the phenotypic alterations observed in HD by recapitulating basic interactions between purified early endosomes and cytoskeletal components *in vitro*. Specifically, the switch of early endosomes from microtubules to F-actin corresponds to a decreased binding of early endosomes to microtubules and increase to F-actin. This may also include an inactivation of microtubule-dependent motor proteins. Thus, we provide here detailed protocols for (1) a biochemical spin-down assay to measure binding of early endosomes to microtubules and F-actin; and (2) an *in vitro* assay that recapitulates the movement of early endosomes along microtubules. Both assays were used to explore the role of Htt and HAP40 in Rab5-dependent interactions of early endosomes with the cytoskeleton (Pal *et al.*, 2006). Candidate regulators can be added as recombinant (e.g., HAP40) or native proteins to test for their ability to alter endosome binding and motility. Alternatively, cytosol prepared from cells can be added to the reaction mixture with the candidate of interest immunoblocked (e.g., Htt) or knocked down by RNAi. The latter approach has the disadvantage of working with a less defined system due to the complexity of cytosol. These assays can help unraveling the mechanisms underlying endosome dynamics under physiological versus pathological conditions. Defects in axonal transport also occur in other polyQ-expansion diseases (Morfini *et al.*, 2005), such as in Kennedy's disease (SBMA), SCA-1, SCA-3, and DRPA, pointing to a common trigger. We thus propose to explore the protocols here for diagnostic tests, scoring for perturbed interactions of cytoskeletal filaments with vesicles prepared from such disease models.

3. SPIN-DOWN BINDING ASSAYS

3.1. Basic principles

The protocol was adapted from Nielsen and coworkers (1999) to allow for improved accuracy in the quantification by fluorimetry. The assay exploits the differential sedimentation characteristics of cytoskeletal polymers and membrane organelles. Specifically, microtubules and F-actin pellet in 35% sucrose whereas early endosomes float under these conditions. To measure binding, purified early endosomes and freshly polymerized microtubules are mixed with candidate proteins to be tested, placed on top of a 35% sucrose layer, and centrifuged. Unbound early endosomes remain in the top layer, while microtubule-bound material is forced to co-pellet. The amount of early endosomes in the pellet is quantified and provides a measure for binding affinity. Analysis of pelleted proteins by Western blotting as described (Nielsen *et al.*, 1999) provides a rather tedious and semiquantitative means of binding measurements. Our comparative analysis described in Pal *et al.* (2006) clearly required a better accuracy to reveal more subtle effects of candidate regulators on membrane–cytoskeleton binding. To this end, we pulsed HeLa cells with rhodamine-conjugated transferrin to label early endosomes prior to their isolation (Nielsen *et al.*, 2001). Following centrifugation, microtubule pellets were washed and lysed, and the rhodamine content measured in a fluorimeter to reveal the amount of bound early endosomes. Only very modest (maximum ±5% standard deviation of mean) variations were detected between triplicate samples and replica experiments (Pal *et al.*, 2006).

3.2. Materials and reagents

3.2.1. Spin-down binding assays

All buffers and stock solution should be prepared in demineralized water and sterile-filtered or autoclaved.

1. Early endosomes and cytosol prepared from HeLa cells, pulsed with rhodamine transferrin: refer to Nielsen *et al.* (2001) for detailed protocol. HeLa cell cytosol is prepared in KHMG (see following) that is identical with KEHM buffer in the original protocol except for glycerin replacing the ATP (ATP is added as part of the energy mix to the binding reaction, as discussed below). The protocol can be easily adapted to prepare cytosol with a candidate protein depleted by RNAi. Store early endosomes and cytosol aliquoted at −80°.

2. Candidate protein stocks in KHMG buffer. We prepared HAP40 as carboxyterminal-tagged GST-fusion protein from *Escherichia coli* lysates, and ditto for GDI and RN-tre (see section 3.3.3.2), but with the GST-tag

cleaved off. Preparations to obtain functional proteins must be optimized for each novel candidate. Store after aliquoting at −80°.

3. Affinity-purified candidate effector mix isolated from immobilized GTPγS-bound Rab5 (Rab5 eluate): refer to Christoforidis and Zerial (2000) for detailed protocol. Store aliquoted material at −80°.

4. Antibodies for functional blocks (e.g., against Htt from Chemicon, MAB 2166).

5. KCl stock, 1 M in water; autoclave and store at room temperature.

6. N-2-hydroxyethylpiperazine-N'-2-ethanesulfonic acid (HEPES) stock, 1 M; autoclave and store at room temperature.

7. MgCl$_2$ stock, 1 M in water; autoclave and store at room temperature.

8. Glycerol stock, 80% in water; autoclave and store at room temperature.

9. Creatine phosphate stock, 1 M in water; filter-sterilize and store aliquoted at −20°.

10. ATP stock, 1 M in water; filter-sterilize and store aliquoted at −20°.

11. GTP stock, 1 M in water; filter-sterilize and store aliquoted at −20°.

12. Imidazole, 1 M in water; filter-sterilize and store aliquoted at −20°.

13. KHMG buffer (must be sodium-free): 110 mM KCl from 1 M stock; 50 mM HEPES from 1 M stock, pH adjusted to 7.4 with KOH, 2 mM MgCl$_2$ from 1 M stock, 10% glycerol from 80% stock; store sterile at room temperature.

14. Energy mixture: 75 mM creatine phosphate from 1 M stock, 10 mM ATP from 1 M stock, 10 mM GTP from 1 M stock, and 20 mM MgCl$_2$ from 1 M stock, in KHMG buffer; store sterile and aliquoted at −20°.

15. Non-specific rabbit serum, heat-inactivated; filter-sterilize and store aliquoted at −20°.

16. Lysis buffer: KHMG buffer supplemented with 1% sodium desoxycholate; store sterile at room temperature.

17. 35% SIM buffer: 35% sucrose (v/v) solution, 40.3 g sucrose, 3 mM imidazole from 1 M stock, 1 mM MgCl$_2$ from 1 M stock in 100 ml water (prepare the night before).

18. Beckman Coulter Optima^{TM} Max tabletop ultracentrifuge with TLA 100 rotor or similar suitable for 100,000 rcf.

19. Beckman 7 × 20-mm polycarbonate tubes #343775 for maximum 200 μl.

20. Perkin Elmer LS50B fluorimeter or similar with 100-μl sample cuvette (plastic or quartz glass).

3.2.2. Preparing taxol-stabilized tubulin

1. MgCl$_2$ stock, 1 M; see section 3.2.1.

2. GTP stock, 1 M; see section 3.2.1.

3. EGTA stock, 1 M in water, use acid or dipotassium salt if available, and adjust pH with KOH; filter-sterilize and store at room temperature.

4. BRB80 buffer (must be sodium-free): 80 mM piperazine-1,4-bis [2-ethanesulfonic acid] (PIPES), pH adjusted to 6.8 with KOH, 1 mM MgCl$_2$ from 1 M stock, 1 mM EGTA from 1 M stock; store sterile at room temperature.
5. Taxol stock, 10 mM in dimethylsulfoxide (DMSO); store 1-μl aliquots at −80°.
6. Purified, cycled, unlabeled tubulin (e.g., from Cytoskeleton Inc.) stock, 35 mg/ml in BRB80; store 10-μl aliquots at −80°.
7. For final ultracentrifugation, see section 3.2.1.

3.2.3. Preparing F-actin

1. Lyophilized G-actin (e.g., from Cytoskeleton Inc.).
2. Potassium iodide (KI), dry salt.
3. ATP stock, 1 M; see section 4.2.1.
4. DTT stock, 1 M in water, filter-sterilize, and store aliquoted at −20°.
5. MgCl$_2$ stock, 1 M; see section 4.2.1.
6. CaCl$_2$ stock, 1 M in water; autoclave and store at room temperature.
7. Tris-HCl stock, 1 M in water; filter-sterilize and store at room temperature.
8. Polymerization buffer: 2 mM Tris-HCl from 1 M stock, adjust pH to 8.0 with KOH, 0.5 mM ATP from 1 M stock, 0.5 mM CaCl$_2$ from 1 M stock, 0.2 mM DTT from 1 M stock; store sterile at −20°.
9. Stabilizing buffer: 50 mM Tris-HCl from 1 M stock, adjust pH to 8.0 with 1 M KOH, 1.2 M KI, freshly weighted dry salt, 3 mM CaCl$_2$ from 1 M stock; make fresh.
10. For final ultracentrifugation, see section 4.2.1.

3.3. Protocols for spin-down binding assays

3.3.1. Preparing microtubules

1. Prewarm TLA 100 rotor on top of a 37° water bath.
2. Prepare 1 ml of 1 mM GTP in BRB80 from 1 M stock, keep at room temperature. Immediately freeze GTP stock again.
3. Prepare 10 μM taxol by diluting 1-μl aliquot of stock (10 mM) with 1 ml BRB80; keep at room temperature.
4. Gently thaw 10-μl stock aliquot of tubulin (35 mg/ml) at 4°; put on ice once thawed.
5. For polymerization, mix 20 μl of 1 mM GTP with tubulin aliquot, incubate for 20 min in 37° water bath.
6. For stabilization, add 170 μl of 10 μM taxol, pipette up and down, and centrifuge for 5 min at 100,000 rcf in TLA100 rotor at 22°.

7. Carefully remove and discard the supernatant, resuspend the pellet in 50 μl of 10 μM taxol to obtain the final 7 μg/ml microtubule stock. The pellet should be 3 to 4 mm in diameter and of jelly-like appearance. Can be stored at room temperature for up to 1 week. Microtubules will depolymerize below 16°!

3.3.2. Preparing F-actin

1. Dissolve G-actin in polymerization buffer at 5 mg/ml; avoid bubbles. Proceed with 180-μl aliquot.
2. Trigger polymerization by adding KCl to final 100 mM from 1 M stock and $MgCl_2$ to final 2 mM from 1M stock. Mix and incubate for 1 h at room temperature.
3. Centrifuge in TLA 100 rotor at 100,000 rcf for 1 h at 4°.
4. Carefully remove and discard supernatant.
5. Soak pellet (F-actin) for 1 h in 100 μl stabilization buffer at 4°, and then gently resuspend by pipetting up and down.
6. Remove KI in stabilization buffer by centrifugation as described previously, and wash pellet three times in 200 μl polymerization buffer. Alternatively, dialyze against polymerization buffer overnight and centrifuge.
7. After final centrifugation, resuspend in 180 μl polymerization buffer to obtain the final 5 mg/ml F-actin stock. Can be stored at 4° for up to 1 month. Do not freeze!

3.3.3. Binding of early endosomes to microtubules or F-actin (spin-down assays)

1. Allow TLA 100 rotor to equilibrate to room temperature.
2. For each binding reaction, mix the following:
   - 15 μg prepared early endosomes (according to protein determination)
   - 5 μl energy mix
   - 1 μl nonspecific rabbit serum
   - Optional: candidate factor(s), such as HAP40-GST, 1 μM final concentration each
   - Optional: Rab5 eluate or similar at final 1 mg/ml
   - Optional: antibody against candidate, such as 1 μl undiluted Chemicon MAB 2166 against Htt
   - HeLa cell cytosol to final 1 mg/ml
3. Add KHMG buffer to final 40 μl. Incubate at room temperature for 20 min.
4. Add 16 μg microtubules or 10 μg F-actin from prepared stock, and further KHMG buffer to final 50 μl. Incubate at room temperature for 10 min.

5. Carefully pipette reaction mixture on top of a 100-μl layer of 35% SIM buffer in centrifugation tubes, and centrifuge at 22° for 20 min at 100,000 rcf.
6. Carefully remove supernatant and keep for analysis of unsedimented material (optional). Wash pellet twice in cytosol-free KHMG by centrifugation as described above.
7. Lyse pellet thoroughly in 150 μl lysis buffer.
8. Transfer to cuvette and place in fluorimeter. For rhodamine, measure at 550 nm excitation, 582 nm emission, and 10 nm slit widths.

3.3.3.1. *Cytosol in binding reactions*   Rab5-dependent interactions of prepared early endosomes with microtubules (binding and motility) can be recapitulated *in vitro* without cytosol (Hoepfner *et al.*, 2005). However, when we tested early endosome binding to F-actin without cytosol, we observed only poor interactions with no effect of any added candidate protein. Only upon addition of 1 mg/ml cytosol, did we achieve appreciable basal binding levels under control conditions that could be modulated through the addition of candidates. Hence, the basal binding of early endosomes to F-actin required cytosolic factors. In addition, removal of active Rab5 from the prepared early endosomes with GDI or RN-tre (see section 3.3.3.2) did not reduce binding below control levels. We concluded that the basal binding was independent from Rab5, while the increase in binding through HAP40 and Htt clearly required active Rab5 on endosomal membranes (Pal *et al.*, 2006).

Collectively, to ensure consistent experimental conditions we had to assay binding of early endosomes to both types of filaments in the presence of cytosol.

3.3.3.2. *Controls in binding reactions*

* Autofluorescence: The assay should be performed with a reaction mixture as specified in section 3.3.3, but without early endosomes and filaments. The final fluorescence readout provides the basal autofluorescence.
* Sedimentation control: Omission of filaments in the reaction mixture must result in a final fluorescence readout indistinguishable from the autofluorescence; otherwise, the sucrose concentration is incorrect and/or the early endosome preparation used is heavily contaminated with other labeled membranous material that sediments without binding to filaments.
* Omission of early endosomes: Sedimentation in the absence of labeled early endosomes will reveal the extra contribution of sedimented filaments to the basal autofluorescence. In our hands, neither microtubules

nor F-actin led to a significant increase in basal autofluorescence (Pal *et al.*, 2006).

- ATP-dependency: The molecular nature of early endosome binding to cytoskeletal filaments is certainly complex and not completely understood. However, previous findings (Nielsen *et al.*, 1999) support our working model of functional Rab5 microdomains that recruit kinases (e.g., hVps34; see section 1.1) (Zerial and McBride, 2001) to produce the early endosome-specific lipid phosphoinositol-3-phosphate. Further downstream effectors mediating membrane–cytoskeleton interactions bind specifically to this lipid as shown for the early endosome kinesin motor KIF16B (Hoepfner *et al.*, 2005). KIF16B alone in the cytosol-free reaction was sufficient for binding of early endosomes to microtubules. Therefore, omission of ATP or its substitution by the nonhydrolysable adenylyl-imidodiphosphate (AMP-PNP) analog will block kinase activity and, thus, compromise the recruitment of downstream effectors; hence, the observed reduction in binding (Pal *et al.*, 2006). By testing whether the effect of a novel candidate on binding is ATP dependent, the investigator can gain first mechanistic insights.
- Rab5-dependency: In line with our working model portrayed above, the level of active Rab5 on early endosomal membranes should affect their binding to cytoskeletal fibers. Indeed, extracting active Rab5 from prepared early endosomes through addition of recombinant Rab-GDP-dissociation inhibitor (GDI) (Ullrich *et al.*, 1994), or RN-tre, a GTPase activating protein (GAP) for Rab5 (Lanzetti *et al.*, 2000), resulted in a decrease of binding to microtubules (Pal *et al.*, 2006). These controls have shown that the effect of Htt/HAP40 on the binding of early endosomes to F-actin is Rab5 dependent and, thus, confirmed Htt/HAP40 as downstream effector of this small GTPase. Moreover, evidencing Rab5-dependency of changes in binding raises confidence levels in ruling out nonspecific co-pelleting effects.
- Specificity of candidate proteins: To confirm that the effect of a novel candidate on binding is specific, we recommend adding the protein—for instance, as GST-tagged variant to the binding reaction, and setting up a control with an equimolar amount of GST protein that should have no effect on binding (Pal *et al.*, 2006).

3.3.3.3. *Data processing*
Since the readout of the fluorimeter is in arbitrary units, we first set up a dilution series for calibration with prepared early endosomes to express binding in mass units for total endosomal proteins bound to filaments. In our hands, fluorescence emission correlated linearly with the concentration of prepared early endosomes in lysis buffer in the range from 0 to at least 35 μg/ml. The basal autofluorescence (see section 3.3.3.2.) should be used for background correction. A typical equation obtained by linear regression was y = 4548.4x − 2653.1, with $R^2 = 0.9879$, where y

is the emission in arbitrary units, and x the concentration of EE in protein mass units (micrograms per milliliter).

3.3.3.4. Complementary controls by western blotting

It is important to confirm that changes in fluorescence correspond specifically to changes in the amount of early endosomes bound to cytoskeletal filaments, and not of other membrane contaminants. Organelle specificity is ensured by restricting the internalization of fluorescent transferrin by HeLa cells to 5 min. Other subsequent compartments (i.e., late and recycling endosomes) are normally not entered by the label under this condition (Mellman, 1996). A more elegant way to ensure organelle specificity is to prepare early endosomes from a source expressing—for instance, GFP-tagged Rab5. We recommend confirming organelle specificity by Western blot analysis of obtained pellets. Any changes in fluorescence must correspond to changes of band intensities on blots probed for specific early endosome markers (Pal *et al.*, 2006). Conversely, probing the blots for nonearly endosome markers such as LAMP1 (for late endosomes/lysosomes) (Eskelinen *et al.*, 2003) and the Golgi marker GM130 (for Golgi) (Nakamura *et al.*, 1995) will probably reveal contaminants of early endosome preparations. By confirming that band intensities of nonearly endosome markers do not change throughout all experimental conditions, the investigator can elegantly document (1) organelle specificity and (2) rule out nonspecific co-pelleting of any membranous material in the dense cytoskeletal filament mesh (Pal *et al.*, 2006). The latter point is important to address because novel candidate proteins could lead to a bundling of filaments that will increase such nonspecific co-pelleting. In such a scenario, the investigator would misinterpret the increase of the fluorescence readout as increased binding affinity. Conversely, novel candidate factors might depolymerize filaments in the binding reaction, thus leading to less early endosomes co-pelleting. In such a scenario, the investigator would misinterpret the decrease of the fluorescence readout as reduced binding affinity. To rule out this possibility, we recommend probing blots for tubulin/actin to confirm equal amount of sedimented filaments throughout all experimental conditions.

We recommend the following antibodies for complementary Western blot analysis:

- Anti-EEA1: Cytostore, mouse monoclonal, catalog no. MAB-004; diluted 1:1000
- Anti-transferrin receptor: Invitrogen, mouse monoclonal, catalog no. 13–6800; diluted 1:1000
- Anti-LAMP1: BD Biosciences, mouse monoclonal, catalog no. 611042; diluted 1:1000

- Anti–GM130: BD Biosciences, mouse monoclonal, catalog no. 610823; diluted 1:1000
- Anti–β-tubulin: BD Biosciences, mouse monoclonal, catalog no. 556321, diluted 1:1000
- Anti-actin: Sigma Diagnostics, rabbit polyclonal, catalog no. A20660; diluted 1:200
- Anti–HAP40: Chemicon, rabbit polyclonal, catalog no. AB5872; diluted 1:200
- Anti–Htt: Chemicon, mouse monoclonal, catalog no. MAB2166; diluted 1:500

However, other validated antibodies should be equally suitable.

3.3.3.5. Critical parameters and assay optimization The ratio of early endosomes to filaments is critical in the assay: a high concentration of filaments in the binding reactions will increase nonspecific binding and co-pelleting that can obscure all specific changes in binding. Therefore, the investigator might only obtain nearly constant high basal fluorescence emission throughout all experimental conditions. Conversely, using very low concentrations of filaments will lead to a saturation of binding sites for early endosome that, again, will not allow revealing subtle changes in binding. Particularly the binding of early endosomes to F–actin requires preliminary titration series to establish the optimal ratio within a working window because these filaments tend to form fine, sticky meshes and bundles. Important parameters in this regard are the type of buffer, pH, salt concentration, and concentration of added cytosol. In our hands, $10\mu g$ F-actin and 15 μg early endosomes were optimal in the reaction mixture specified in section 3.3.3., but variations among preparations might require optimizing again.

4. IN VITRO MOTILITY ASSAYS

4.1. Introduction

Our original protocol (Nielsen *et al.*, 2001) allowed gaining first clues about the important role of Rab5 in early endosome motility along microtubules (Nielsen *et al.*, 1999). In brief, taxol-stabilized microtubules are perfused into a microscopy chamber and allowed to bind to the glass surface. Next, prepared early endosomes labeled with fluorescent rhodamine-conjugated transferrin are added with ATP, candidate proteins to be tested and antifade reagents. Binding to and motility events of early endosomes along microtubules are then imaged by fluorescence video microscopy.

Optimal assay conditions are on microtubule tracks evenly and loosely scattered across the field. The addition of cytosol, however, can lead to a bundling of microtubules and, hence, creation of suboptimal conditions that reduces the number of motility events and velocity, presumably by steric hindrance (our unpublished observations). Thus, we have sought to improve the assay to allow for (1) better motility, (2) a simplified setup of the microscopy perfusion chamber, and (3) omission of cytosol for a defined composition and elimination of microtubule bundling. We found that replacing the original BRB80 by KHMG buffer (see sections 3.2.1 and 3.2.2) in the assay allowed recording *in vitro* motility without cytosol, but only for plus-end motility while activity toward the minus-end of microtubules was lost. Apparently, a minimal motility machinery enriched in plus-end microtubule motors remains preserved on early endosome in the preparation. Under these experimental conditions, cytosol decreases *in vitro* motility in general, but is required to restore minus-end activity. After having confirmed that the inhibitory effect of Htt and HAP40 on early endosome motility along microtubules was similar with and without cytosol (our unpublished observations), we studied these candidates in the absence of cytosol to benefit from the improvements described above. However, we recommend testing for the requirement of cytosol for each novel candidate. Thus, cytosol is listed as an optional ingredient in the protocol provided here.

4.2. Material and reagents for *in vitro* motility

4.2.1. Preparing fluorescently labeled early endosomes
Refer to section 3.2.1.

4.2.2. Preparing microtubules
See section 3.2.2.

4.2.3. Video microscopes and data acquisition
See Nielsen *et al.* (2001).

4.2.4. Reconstituting early endosome motility along microtubules

1. Energy mix: see section 3.2.1.
2. Glucose stock, 1 M in water; autoclave and store sterile at room temperature.
3. MgCl$_2$ stock, 1 M stock in water; autoclave and store sterile at room temperature.
4. Glucose oxidase stock, 10 mg/ml in KHMG buffer; filter-sterilize and store aliquoted at $-20°$.
5. Catalase stock, 10 mg/ml in KHMG (see section 3.2.1) buffer; filter-sterilize and store aliquoted at $-20°$.

6. Antifade buffer: 10 μM taxol from 10 mM stock (one 1-μl aliquot) (see section 3.2.2), 10 mM glucose from 1 M stock, 4 mM MgCl$_2$ from 1 M stock, 50 μg/ml glucose oxidase from 10 mg/ml stock, 50 μg/ml catalase from 10 mg/ml stock, 0.1% 2-mercaptoethanol in total 1 ml KHMG buffer (see section 3.2.1); make fresh.

7. Two times antifade buffer; ditto but with all ingredients at doubled concentration except of KHMG buffer (remains 1×).

8. Bovine hemoglobin, 30 mg/ml in KHMG; filter-sterilize and make fresh.

9. 10% nonspecific, heat-inactivated rabbit serum in antifade buffer; make fresh.

10. 10% nonspecific, heat-inactivated rabbit serum in 2× antifade buffer; make fresh.

4.2.5. Preparing the microscope perfusion chamber

1. Microscope slides (76 mm × 26 mm × 0.8/1 mm thick, Select Micro Slides, washed; Chance Propper Ltd., Warley, UK, catalog no. KTH 360).

2. Glass coverslips (18 mm × 18 mm, no.1, Clay Adams, Gold Seal, catalog no. 3305)

3. Standard office tape, adhesive on both sides (any manufacturer).

4.3. Protocols for *in vitro* motility assay

4.3.1. Preparing the microscope perfusion chamber

1. Place two strips of office tape in parallel about 1.5 cm apart on microscope slide. Place coverslip on top so that tape serves as spacer between glass layers. Perfusion can then be performed by capillary force. To fill the chamber, 5 μl should be sufficient; otherwise, vary space between tape strips accordingly.

2. Perfuse with 5 μl of previously prepared microtubules in BRB80 buffer (see section 3.2.2) by adding the solution to one side of the chamber using an air-displacement micropipette. Allow nonspecific adherence of microtubules at room temperature over 5 min, and then proceed to section 4.3.2.

4.3.2. Reconstituting early endosome motility along microtubules

1. To set up reaction (10 μl in total), mix the following and incubate for 5 min at room temperature:
   - 1 μl of fluorescently labeled, prepared early endosomes (~5 mg/ml)
   - 1 μl energy mix
   - 1 μl candidate protein(s), optional (e.g., to final 10 μM for HAP40-GST or to final 1 mg/ml for Rab5 eluate)

- 1 μl antibody against candidate, optional (e.g., 1 μl undiluted Chemicon MAB 2166 against Htt)
- 1 μl HeLa cell cytosol to final 2 mg/ml, optional
- 1 μl bovine hemoglobin
- 4 μl of 10% nonspecific serum in 2× antifade buffer

Omission of optional ingredients should be compensated by equal volumes of water.

2. Meanwhile, perfuse chamber prepared with microtubules (see section 4.3.1) with 10 μl of 10% nonspecific serum in antifade buffer to block nonspecific binding sites and to remove BRB80 buffer. Perfuse by placing the new solution as a drop at one side of the chamber and aspirating 10 μl with a micropipette from the other side.
3. Finally, perfuse chamber with prepared reaction mix.
4. Place a drop of immersion oil on the glass coverslip, and visualize fluorescent early endosomes with appropriate filters on a time-lapse video microscope. Examples of early endosome motility on *in vitro* synthesized microtubules are shown in Pal *et al.* (2006) and Nielsen *et al.* (2001).

4.3.2.1. *Controls*

1. Candidate specificity: As for section 3.3.3.2, the tag part of fusion proteins should be added alone in a control reaction to test for candidate specificity.
2. Antibody specificity: Because the reaction contains an excess of nonspecific immunoglobulins (10% of nonspecific rabbit serum), it is unlikely that addition of insignificantly more antibodies against candidates can cause appreciable changes in motility through nonspecific interactions. However, we still recommend validating the use of any antibody against candidate proteins by testing other specific antibodies recognizing nonrelevant epitopes.
3. Validation of affinity-purified Rab effector mix (e.g., Rab5 eluate): Candidate proteins difficult to purify (e.g., Htt due to its large size and numerous interacting partners) can be added in mixture with other co-purified proteins. To address a particular candidate selectively, specific antibodies can be added to test for a rescue of any alteration in motility caused by, for example, the Rab5 eluate (as done for Htt with Chemicon MAB 2166) (Pal *et al.*, 2006). However, the complex nature of these purified effector mixtures calls for additional controls, that is, with eluate from Rab affinity columns preloaded with GDP instead of GTPγS (Pal *et al.*, 2006).

ACKNOWLEDGMENTS

The authors would like to thank Marisa McShane and Thierry Galvez for helpful discussions and advice. This work was supported by the Max Planck Society and grants from the Human Frontier Science Program (RG-0260/1999-M), the European Union (HPRN-CT-2000–00081), and the High Q Foundation.

REFERENCES

Altar, C. A., Cai, N., Bliven, T., Juhasz, M., Conner, J. M., Acheson, A. L., Lindsay, R. M., and Wiegand, S. J. (1997). Anterograde transport of brain-derived neurotrophic factor and its role in the brain. *Nature* **389,** 856–860.

Apodaca, G. (2001). Endocytic traffic in polarized epithelial cells: Role of the actin and microtubule cytoskeleton. *Traffic* **2,** 149–159.

Block-Galarza, J., Chase, K. O., Sapp, E., Vaughn, K. T., Vallee, R. B., DiFiglia, M., and Aronin, N. (1997). Fast transport and retrograde movement of huntingtin and HAP 1 in axons. *Neuroreport* **8,** 2247–2251.

Bucci, C., Lütcke, A., Steele-Mortimer, O., Olkkonen, V. M., Dupree, P., Chiariello, M., Bruni, C. B., Simons, K., and Zerial, M. (1995). Co-operative regulation of endocytosis by three Rab5 isoforms. *FEBS Lett.* **366,** 65–71.

Christoforidis, S., and Zerial, M. (2000). Purification and identification of novel Rab effectors using affinity chromatography. *Methods* **20,** 403–410.

Conner, J. M., Lauterborn, J. C., Yan, Q., Gall, C. M., and Varon, S. (1997). Distribution of brain-derived neurotrophic factor (BDNF) protein and mRNA in the normal adult rat CNS: evidence for anterograde axonal transport. *J. Neurosci.* **17,** 2295–2313.

Conner, S. D., and Schmid, S. L. (2003). Regulated portals into the cell. *Nature* **422,** 37–44.

Deinhardt, K., Salinas, S., Verastegui, C., Watson, R., Worth, D., Hanrahan, S., Bucci, C., and Schiavo, G. (2006). Rab5 and Rab7 control endocytic sorting along the axonal retrograde transport pathway. *Neuron* **52,** 293–305.

Engelender, S., Sharp, A. H., Colomer, V., Tokito, M. K., Lanahan, A., Worley, P., Holzbaur, E. L., and Ross, C. A. (1997). Huntingtin-associated protein 1 (HAP1) interacts with the p150Glued subunit of dynactin. *Hum. Mol. Genet.* **6,** 2205–2212.

Eskelinen, E. L., Tanaka, Y., and Saftig, P. (2003). At the acidic edge: Emerging functions for lysosomal membrane proteins. *Trends Cell Biol.* **13,** 137–145.

Gasman, S., Kalaidzidis, Y., and Zerial, M. (2003). RhoD regulates endosome dynamics through Diaphanous-related formin and Src tyrosine kinase. *Nat. Cell Biol.* **5,** 195–204.

Gauthier, L. R., Charrin, B. C., Borrell-Pagès, M., Dompierre, J. P., Rangone, H., Cordelieres, F. P., De Mey, J., MacDonald, M. E., Lessmann, V., Humbert, S., and Saudou, F. (2004). Huntingtin controls neurotrophic support and survival of neurons by enhancing BDNF vesicular transport along microtubules. *Cell* **118,** 127–138.

Goehler, H., Lalowski, M., Stelzl, U., Waelter, S., Stroedicke, M., Worm, U., Droege, A., Lindenberg, K. S., Knoblich, M., Haenig, C., Herbst, M., Suopanki, J., *et al.* (2004). A protein interaction network links GIT1, an enhancer of huntingtin aggregation, to Huntington's disease. *Mol. Cell* **15,** 853–865.

Gross, S. P., Tuma, M. C., Deacon, S. W., Serpinskaya, A. S., Reilein, A. R., and Gelfand, V. I. (2002). Interactions and regulation of molecular motors in *Xenopus* melanophores. *J. Cell Biol.* **156,** 855–865.

Gruenberg, J., and Maxfield, F. R. (1995). Membrane transport in the endocytic pathway. *Curr. Opin. Cell Biol.* **7,** 552–563.

Gusella, J. F., and MacDonald, M. E. (2000). Molecular genetics: unmasking polyglutamine triggers in neurodegenerative disease. *Nat. Rev. Neurosci.* **1**, 109–115.

Harjes, P., and Wanker, E. E. (2003). The hunt for huntingtin function: Interaction partners tell many different stories. *Trends Biochem. Sci.* **28**, 425–433.

Hibbert, A. P., Kramer, B. M., Miller, F. D., and Kaplan, D. R. (2006). The localization, trafficking and retrograde transport of BDNF bound to p75NTR in sympathetic neurons. *Mol. Cell. Neurosci.* **32**, 387–402.

Hoepfner, S., Severin, F., Cabezas, A., Habermann, B., Runge, A., Gillooly, D., Stenmark, H., and Zerial, M. (2005). Modulation of receptor recycling and degradation by the endosomal kinesin KIF16B. *Cell* **121**, 437–450.

Lanzetti, L., Rybin, V., Malabarba, M. G., Christoforidis, S., Scita, G., Zerial, M., and DiFiore, P. P. (2000). The Eps8 protein coordinates EGF receptor signalling through Rac and trafficking through Rab5. *Nature* **408**, 374–377.

Li, X. J., and Li, S. H. (2005). HAP1 and intracellular trafficking. *Trends Pharmacol. Sci.* **26**, 1–3.

Lindsay, R. M., Wiegand, S. J., Altar, C. A., and DiStefano, P. S. (1994). Neurotrophic factors: From molecule to man. *Trends Neurosci.* **17**, 182–190.

McPherson, P. S. (2002). The endocytic machinery at an interface with the actin cytoskeleton: a dynamic, hip intersection. *Trends Cell Biol.* **12**, 312–315.

Mellman, I. (1996). Endocytosis and molecular sorting. *Annu. Rev. Cell Dev. Biol.* **12**, 575–625.

Miaczynska, M., Christoforidis, S., Giner, A., Shevchenko, A., Uttenweiler-Joseph, S., Habermann, B., Wilm, M., Parton, R. G., and Zerial, M. (2004). APPL proteins link Rab5 to nuclear signal transduction via an endosomal compartment. *Cell* **116**, 445–456.

Morfini, G., Pigino, G., and Brady, S. T. (2005). Polyglutamine expansion diseases: Failing to deliver. *Trends Mol. Med.* **11**, 64–70.

Nakamura, N., Rabouille, C., Watson, R., Nilsson, T., Hui, N., Slusarewicz, P., Kreis, T. E., and Warren, G. (1995). Characterization of a cis-Golgi matrix protein, GM130. *J. Cell Biol.* **131**, 1715–1726.

Nielsen, E., Severin, F., Backert, J. M., Hyman, A. A., and Zerial., M. (1999). Rab5 regulates motility of early endosomes on microtubules. *Nat. Cell Biol.* **1**, 376–382.

Nielsen, E., Severin, F., Hyman, A. A., and Zerial, M. (2001). *In vitro* reconstitution of endosome motility along microtubules. *Methods Mol. Biol.* **164**, 133–146.

Pal, A., Severin, F., Lommer, B., Shevchenko, A., and Zerial, M. (2006). Huntingtin-HAP40 complex is a novel Rab5 effector that regulates early endosome motility and is up-regulated in Huntington's disease. *J. Cell Biol.* **172**, 605–618.

Pelkmans, L., and Helenius, A. (2003). Insider information: What viruses tell us about endocytosis. *Curr. Opin. Cell Biol.* **15**, 414–422.

Schnatwinkel, C., Christoforidis, S., Lindsay, M. R., Uttenweiler-Joseph, S., Wilm, M., Parton, R. G., and Zerial, M. (2004). The Rab5 effector Rabankyrin-5 regulates and coordinates different endocytic mechanisms. *PLoS Biol.* **2**, 1363–1380E261.

Singer-Krüger, B., Stenmark, H., Düsterhöft, A., Philippsen, P., Yoo, J.-S., Gallwitz, D., and Zerial, M. (1994). Role of three Rab5-like GTPases, Ypt51p, Ypt52p and Ypt53p, in the endocytic and vacuolar protein sorting pathway of yeast. *J. Cell Biol.* **125**, 283–298.

Singer-Krüger, B., Stenmark, H., and Zerial, M. (1995). Yeast Ypt51p and mammalian Rab5: counterparts with similar function in the early endocytic pathway. *J. Cell Sci.* **108**, 3509–3521.

Trushina, E., Singh, R. D., Dyer, R. B., Cao, S., Shah, V. H., Parton, R. G., Pagano, R. E., and McMurray, C. T. (2006). Mutant huntingtin inhibits clathrin-independent endocytosis and causes accumulation of cholesterol in vitro and in vivo. *Hum. Mol. Genet.* **15**, 3578–3591.

Ullrich, O., Horiuchi, H., Bucci, C., and Zerial, M. (1994). Membrane association of Rab5 mediated by GDP-dissociation inhibitor and accompanied by GDP/GTP exchange. *Nature* **368,** 157–160.

Walenta, J. H., Didier, A. J., Liu, X., and Kramer, H. (2001). The Golgi-associated hook3 protein is a member of a novel family of microtubule-binding proteins. *J. Cell Biol.* **152,** 923–934.

Zerial, M., and McBride, H. M. (2001). Rab proteins as membrane organizers. *Nat. Rev. Mol. Cell Biol.* **2,** 107–117.

Zuccato, C., and Cattaneo, E. (2007). Role of brain-derived neurotrophic factor in Huntington's disease. *Prog. Neurobiol* Epub 2007 Feb 9 ahead of print.

CHAPTER EIGHTEEN

BIOCHEMICAL AND BIOPHYSICAL ANALYSES OF RAS MODIFICATION BY UBIQUITIN

Sonia Terrillon *and* Dafna Bar-Sagi

Contents

Abstract

Ras proteins are small GTPases that play key roles in the regulation of several cellular processes such as growth, differentiation, and transformation. Although Ras signaling was thought to occur uniformly on the inner leaflet of the plasma membrane, a growing body of evidence indicates that Ras activation happens dynamically within defined plasma membrane microdomains and at other specific intracellular compartments, thus ensuring the generation of distinct signal outputs. Yet the mechanisms that control the spatiotemporal segregation of Ras proteins remain poorly characterized. We have recently shown that the differential modification of Ras proteins by ubiquitination is a crucial factor that controls Ras intracellular trafficking and signaling potential. To better understand the process of Ras ubiquitination, it is important to establish assays that not only provide information about the nature of the

Department of Biochemistry, New York University School of Medicine, New York, New York

Methods in Enzymology, Volume 438

ISSN 0076-6879, DOI: 10.1016/S0076-6879(07)38018-X

ubiquitin modification involved, but also enable the monitoring of the dynamics of this process. In this chapter, we will describe biochemical and biophysical methodologies, namely immunoprecipitation, nickel-chelate affinity chromatography, and bioluminescence resonance energy transfer (BRET), for monitoring the ubiquitination of Ras proteins. Although the description focuses on Ras, the assays described can in principle be applied to the study of a range of proteins of interest that may be subject to ubiquitination, and the use of the different methods in parallel should provide new insights into the nature and dynamics of protein ubiquitination.

1. INTRODUCTION

Ras proteins (HRas, KRas, and NRas) are small GTPases that mediate many biological processes through their ability to cycle between an inactive GDP-bound and an active GTP-bound form. They play a crucial role in the regulation of cell growth, proliferation, and differentiation, and their abnormal activation contributes to the development of several types of human cancers (Downward, 2003). Despite their identical effector-binding domains, the Ras isoforms generate distinct signal outputs, attributed in part to differences in localization within specific plasma membrane microdomains as well as in other internal cell membranes (e.g., endoplasmic reticulum, Golgi, and endosomes) (Hancock, 2003; Silvius, 2002). However, how Ras proteins are segregated in different cellular compartments remains unclear. It is now becoming widely accepted that the mode of membrane anchoring of the Ras isoforms is largely responsible for their localization in different plasma membrane microdomains as well as different endomembrane compartments. This spatial regulation is dictated by the C-terminal hypervariable region (HVR) that encompasses the last 25 amino acids, and is unique for each isoform. Thus, whereas the HVR region includes farnesylation and palmitoylation sites for HRas and NRas, it contains a farnesylation site and a polybasic region for KRas. Acylation of HRas and NRas with two and one palmitoyl group(s), respectively, is responsible for their traffic through the Golgi apparatus during the maturation process from the endoplasmic reticulum (ER) to the plasma membrane (Apolloni et al., 2000; Choy et al., 1999), and their specific localization in lipid rafts (Prior et al., 2001). HRas and NRas are also subject to a constitutive depalmitoylation/repalmitoylation cycle, which leads to their continuous shuttling between the plasma membrane and the Golgi complex (Rocks et al., 2005). The polybasic stretch of lysine residues in the KRas HVR mediates electrostatic interactions with the plasma membrane. Phosphorylation within this polybasic region is responsible for KRas relocalization from the plasma membrane to other endomembrane compartments, including ER, Golgi, and mitochondria (Bivona et al., 2006). The differential modification of Ras proteins by ubiquitination constitutes an additional key mechanism involved in the control of their location-specific signaling activities. Indeed, contrary to KRas, HRas is subject to a constitutive mono- and di–ubiquitination process, which leads to its redistribution from the plasma membrane to endosomes and a modulation of its ability to activate the Raf/MAPK signaling pathway (Jura *et al.*, 2006).

Ubiquitination is the covalent attachment of a 76–amino acid ubiquitin peptide to lysine residues of target proteins (Passmore and Barford, 2004; Pickart, 2001). A similar isopeptide linkage can be formed between lysine residues within ubiquitin and the C-terminus of another ubiquitin molecule to form polyubiquitin chains. This rapid and reversible post-translational protein modification has been implicated in the regulation of a broad range of eukaryotic cell functions (e.g., proteasomal degradation, protein subcellular localization, protein-protein interaction and intracellular signal transduction). The type of ubiquitin modification involved (i.e., mono- vs. polyubiquitination) subsequently determines the regulatory effect that ubiquitin exerts on the target protein. For instance, whereas polyubiquitination through lysine-48 targets the substrates for proteasomal degradation (Nandi *et al.*, 2006; Pickart and Cohen, 2004), polyubiquitination through lysine-63 or monoubiquitination does not signal to the 26S proteasome, but instead regulates other functions (Haglund and Dikic, 2005), such as internalization and endosomal sorting of many membrane proteins (Hicke and Dunn, 2003). The functional consequences of ubiquitination on the regulation of Ras trafficking and function have been discussed elsewhere (Jura and Bar-Sagi, 2006; Rodriguez-Viciana and McCormick, 2006). Here, we describe methodologies to probe the molecular determinants and dynamics of Ras ubiquitination. The first two methods, immunoprecipitation and nickel-chelate affinity chromatography (Ni-NTA), are cell-based biochemical assays to determine whether a protein is ubiquitinated and the nature of the modification involved (i.e., mono- vs. polyubiquitination and branching of the ubiquitin moieties within the polymeric chains). The third method, bioluminescence resonance energy transfer (BRET), is a biophysical approach that enables the monitoring of protein ubiquitination in living cells with a high temporal resolution, thus allowing the study of the dynamic nature of the ubiquitination/deubiquitination cycles.

2. RAS UBIQUITINATION MONITORED BY IMMUNOPRECIPITATION

2.1. Method overview

A common strategy used to study protein ubiquitination involves the isolation of the putative substrate by cellular lysis and immunoprecipitation, followed by immunoblotting of the immune precipitate with antibodies that specifically recognize ubiquitin. The existence of antibodies capable of precipitating the substrate and detecting ubiquitin by Western blot is therefore a prerequisite to ensure the success of this method. Alternatively, when these critical reagents are not available, one can introduce an epitope tag in frame with the coding sequence for the protein of interest and/or ubiquitin using a PCR-based strategy. These epitope-tagged proteins can then be monitored with commercially available antibodies whose immunoprecipation and immunodetection capabilities are well established.

The immunoprecipitation approach can be also used to characterize the molecular determinants involved in the ubiquitination process by coupling it with site directed mutagenesis approaches through which residues comprising functional determinants within the protein of interest or ubiquitin are altered. For example, by mutating specific lysine residues or domains within the protein of interest and analyzing their potential effect on ubiquitination by immunoprecipitation, one can obtain information about the site of ubiquitination and the functional domains that regulate this process. In addition, the use of mutated versions of ubiquitin where a specific arginine residue substitutes for different lysines helps to ascertain whether a protein of interest is modified by mono-, poly- or even multi-ubiquitination (i.e., when several Lys residues of the target protein are tagged with single ubiquitin molecules), and the type of linkage that mediates the formation of ubiquitin chains. Indeed, whereas these ubiquitin mutants retain their ability to be conjugated to target proteins through their C-terminus, they cannot form polyubiquitin chains through the specific mutated lysine residue(s). Immunoprecipitation studies carried out with a ubiquitin mutant where all of the seven lysine residues are substituted with arginines (UbiK0), constitute a useful means to address the question of polyubiquitination versus poly-mono-ubiquitination. Similarly, ubiquitin mutants where all of the lysine residues but one are substituted with arginines represent key tools to determine the nature of the chains linking the molecules of ubiquitin in the case of polyubiquitination.

Overexpressing the protein of interest and/or ubiquitin, whether mutated or not, considerably facilitates the detection and analysis of protein ubiquitination. However, an important caveat of tagged protein overexpression is that it can lead to artifactual ubiquitination, most likely caused by ubiquitination of a misfolded fraction of the overexpressed protein. It is therefore important to assess the physiological relevance of protein ubiquitination, by establishing that this modification applies to the endogenous protein as well. In principle, the demonstration of endogenous protein ubiquitination can be achieved by using specific antibodies to immunoprecipitate the endogenous protein and by following an optimized protocol, as illustrated by the monitoring of endogenous Ras ubiquitination (Jura et al., 2006). However, the endogenous levels of many substrates are often very low and only a subpopulation of the protein may be modified by ubiquitin. Thus, studying the process of ubiquitination of endogenous proteins can represent a significant challenge and requires very high quality antibodies to both ubiquitin and the protein of interest.

2.2. Experimental details

2.2.1. Expression constructs

For the analysis of Ras ubiquitination by immunoprecipitation, the following constructs are used:

- T7-HRas, T7-KRas, and T7-NRas: full-length Ras fused on their N-terminus with a T7 epitope tag (MASMTGGQQMG)
- Ubiquitination-deficient T7-HRas8RK mutant: lysines 5, 42, 88, 101, 147, 167, 170, and 185 substituted with arginines in HRas wildtype (WT)
- Farnesylation-deficient T7-HRasC186S mutant: CAAX cysteine substituted with a serine in HRas WT
- Palmitoylation deficient T7-HRas(C181S,C184S) mutant: cysteines 181 and 184 substituted with serines in HRas WT
- T7-HRas-Ktail chimera: HVR domain of HRas (from residue 166 to residue 189) replaced with HVR domain of KRas (from residue 166 to residue 188) in HRas WT
- T7-KRas-Htail chimera: HVR domain of KRas replaced with the HVR domain of HRas in KRas WT
- HA-ubiquitin: full-length ubiquitin fused on its N-terminus with an HA epitope tag (YPYDVPDYA)
- Ubiquitin HA-UbiK0 mutant: all the lysine residues substituted with arginines in ubiquitin WT
- Ubiquitin HA-UbiK0R63K mutant: arginine 63 restored to lysine in UbiK0
- Ubiquitin HA-UbiK48R mutant: lysine 48 substituted with an arginine in ubiquitin WT

2.2.2. Cell culture and ectopic expression

For immunoprecipitation studies based on the ectopic coexpression of the protein of interest and ubiquitin differentially tagged by an epitope, the following considerations need to be taken into account: transfection efficiency, level of expression and cell type. Thus, like most biochemical assays, it is critical to achieve a high efficiency of transfection, which can be ensured by optimizing the conditions of transfection (choice of cell type and transfection reagent). The amount of each vector used for transfection also needs to be empirically determined to identify the lowest DNA concentration at which a quantifiable signal is obtained by Western blot. Indeed, although it is possible to increase the signal by overexpressing the ubiquitin construct and/or the protein of interest, a risk to this approach is that it can lead to artifactual ubiquitination as discussed above. Finally, it is important to assess the potential ubiquitination of a protein of interest in different cell lines because this modification may be impacted by the nature and levels of proteins endogenously expressed (in particular the enzymes required for substrate ubiquitination).

For immunoprecipitation experiments based on overexpressed Ras and ubiquitin constructs, cells are seeded in 60-mm Petri dishes and transient transfections performed 24 h later with the FuGene6 transfection reagent (Roche, Indianapolis, IN) as described in the manufacturer's instructions. The differential susceptibility of Ras isoforms to modification by ubiquitin is determined by generating cell populations coexpressing both HA-tagged ubiquitin and T7-tagged Ras protein (HRas, KRas, or NRas), as well as negative control populations expressing HA-ubiquitin only. Additionally, the molecular determinants of HRas ubiquitination are assessed by generating populations of cells coexpressing: (1) HA–ubiquitin with the ubiquitination-deficient mutant (T7-HRas8RK), farnesylation-deficient mutant (T7-HRasC186S), palmitoylation deficient mutant (T7-HRasC181S,C184S), or HRas/KRas chimeric constructs (T7-HRas-Ktail or T7-KRas-Htail); (2) T7-HRas with the ubiquitin mutant that can support only monoubiquitination (UbiK0), the ubiquitin mutant that cannot bear polyubiquitination through lysine 48 (UbiK48R), or the ubiquitin mutant that can support polyubiquitination only through lysine 63 (UbiK0R63K). For experiments addressing the dependence of ubiquitination on growth factor stimulation, cells are serum starved for ~16 h before the immunoprecipitation assay. For the study of endogenous Ras protein ubiquitination, cells are seeded in 100-mm Petri dishes 24 h before the experiment and serum starved under the same conditions.

2.2.3. Preparation of cell lysates

Twenty-four h post-transfection, cells are washed twice with ice-cold phosphate-buffered saline (PBS) solution and lysed in 500 μl ice-cold lysis buffer containing 50 mM HEPES (pH 7.4), 1% Triton X-100, 100 mM NaCl, 5 mM MgCl$_2$, 1 mg/ml bovine serum albumin (BSA), 1 mM NaF, 1 mM Na$_3$VO$_4$, 1 mM phenyl-methanesulfonyl fluoride (PMSF), 1% aprotinin, 10 mM benzamidine, 10 μg/ml pepstatin, 10 μg/ml leupeptin, and 10 μg/ml trypsin inhibitor. For studies examining the dependence of Ras ubiquitination on growth factor stimulation, cells are first treated with or without 50 ng/ml EGF at 37° for different periods of time before being washed with cold PBS and lysed in the same conditions described above. It is important to note that several key factors could make the modification of a protein by ubiquitination difficult to detect. For instance, the deubiquitination activity is often high in cell lysates, which may lead to the removal of the ubiquitin moieties from the protein of interest, thus preventing the detection of ubiquitinated species. If so, such a problem can be overcome by preparing the lysis buffer with an inhibitor of the deubiquitinating enzymes (Dubs) such as N-ethylmaleimide (NEM), which blocks the critical cysteine residue present in the active site of most Dubs. Another factor to consider is the possible rapid proteasomal degradation of the protein of interest after its ubiquitination, thus making this transient event difficult to detect by most biochemical approaches unless precautions are taken to preserve the ubiquitinated species. To do so, one can pretreat the cells with an inhibitor of the proteasome before performing the experiment (e.g., MG132 at 5 to 10 μM for 2 to 4 h).

The lysed cells are scraped off the dishes and transferred to eppendorf tubes. The lysate is homogenized with five strokes through a 26⅜ gauge needle, before being clarified by centrifugation at 14,000×g for 15 min at 4°. The clarified supernatant is transferred to a fresh tube and used for subsequent immunoprecipitation experiments. In order to control for the expression level of each protein, 40 μl of the lysate is kept for immunoblotting analysis. For the specific study of endogenous Ras ubiquitination for which only a small fraction of Ras appears to be modified by di-ubiquitination, the detection of a quantifiable signal is favored by pooling at least two 100-mm Petri dishes per condition and supplementing the lysis buffer with 3 mM NEM to inhibit the activity of Dubs, as discussed above.

2.2.4. Immunoprecipitation of ubiquitinated Ras

Immunoprecipitation of Ras proteins is carried out using specific anti-Ras antibodies. Typically, we use the Y13–259 rat monoclonal antibody, which is prepared by culturing rat hybridoma Y13–259 cells. The antibody is precipitated with 45% ammonium sulfate, and the pellet is resuspended in PBS and then dialyzed against PBS to yield a 2 mg/ml stock solution of antibody. Clarified cell lysates are incubated with anti-Ras antibody Y13–259 (20 μg/ml final) for 1 h at 4° with constant rotation. In parallel, protein A sepharose (Sigma, St. Louis, MO) is swollen for 1 h at 4° in 50 mM Tris-HCl (pH 7.5), washed three times with 50 mM Tris-HCl (pH 7.5), and resuspended in 50 mM Tris-HCl (pH 7.5) (50% v/v slurry) before being incubated with a rabbit anti-rat immunoglobulin-G antiserum (Rockland, Gilbertsville, PA) (20 mg/ml final of proteins) for 1 h at 4° with constant rotation. The protein A sepharose beads precoated with rabbit anti-rat IgG are then thoroughly washed and resuspended in 50 mM Tris-HCl (pH 7.5) (50% v/v slurry) before being added to the immune complexes (50 μl per sample). It is important to note that, as exemplified here with the case of the anti-Ras antibody, some antibodies are not bound effectively by the protein A commonly used for immunoprecipitation experiments. To overcome this problem, one can precouple the protein A beads with a secondary antibody that recognizes the primary antibody immunoprecipitating the substrate. One alternate yet more expensive option is to use protein G beads that bind to a wider range of antibodies than protein A beads. Another rather expensive option allowing rapid and convenient precipitation of the target protein is to use antibodies already covalently coupled to the beads, when commercially available. For precipitation of endogenous Ras for which the stoichiometry of ubiquitination is low, it is necessary to use an excess of antibody to favor the detection of this modification. In addition, the anti–Ras antibody Y13–259 is cross-linked to protein A sepharose precoupled to rabbit anti-rat IgG using dimethyl pimelimidate•2 HCl (DMP) (Pierce Biotechnology, Rockford, IL) according to the manufacturer's protocol. This additional cross-linking step greatly enhances the detection of endogenous HRas di-ubiquitinated form, which otherwise migrates just below the heavy chain of the immunoprecipitating antibodies. Following incubation for 1 h at 4°, the immunoprecipitation complexes are collected by centrifugation and washed five times with an ice-cold buffer containing 50 mM HEPES (pH 7.4), 0.1% Triton X-100, 100 mM NaCl, and 5 mM MgCl$_2$. The final pellet is resuspended in sample buffer (80 mM Tris-HCl [pH 6.8], 2% SDS, 15% glycerol, 100 mM dithiothreitol and 0.01% bromophenol blue) and heated at 95° for 5 min.

2.2.5. SDS polyacrylamide gel electrophoresis (SDS-PAGE) analysis

Proteins are resolved by SDS-PAGE before being transferred to nitrocellulose membrane (Schleicher and Schuell, Dassel, Germany). When the protein of interest is modified by polyubiquitination, the immunoblot signals often appear as smears of high-molecular-weight species. To ensure an efficient transfer of the species, lower percentage polyacrylamide gels can be run. For experiments based on heterologous coexpression of T7–Ras and HA–ubiquitin constructs, the membranes are incubated with either the mouse monoclonal anti–HA 12CA5 antibody (1:5000) (purified from ascite fluids) to specifically detect the Ras ubiquitinated forms, or the mouse monoclonal anti–T7 antibody (1:10,000) (Novagen, Madison, WI) to quantify the total expression level of Ras proteins. For studying the ubiquitination status of endogenous Ras proteins, the following primary antibodies are used: mouse monoclonal anti–ubiquitin, P4D1 1:1000 (Santa Cruz Biotechnology, Santa Cruz, CA) and mouse monoclonal anti–Ras, clone RAS10 1:1,000 (Upstate, Charlottesville, VA). Several polyclonal and monoclonal antibodies directed to either free ubiquitin or ubiquitin protein conjugates are now available. However, one should be careful when interpreting the data because certain antibodies, particularly those raised against ubiquitin-protein conjugates, may cross–react and detect ubiquitin cross-reactive protein (UCRP), a 15-kDa protein that has sequence homology with ubiquitin (Ahrens et al., 1990). The immunoreactivity is then revealed using a secondary goat anti–mouse antibody coupled to horseradish peroxidase (1:10,000) (Cappel Laboratories, Cochranville, PA). Enhanced chemiluminescence reagents (Perkin Elmer Life Sciences, Boston, MA) are used to visualize immunoreactive bands.

3. RAS Ubiquitination Monitored by Ni-NTA Affinity Chromatograhy

3.1. Method overview

An alternative biochemical method to assay the ubiquitination of a protein of interest is based on the overexpression of ubiquitin with a polyhistidine (His) tag that allows the purification of ubiquitinated proteins using nickel Ni^{2+}-chelate affinity chromatography. Purified ubiquitinated proteins are then resolved by SDS-PAGE and probed with an antibody specific for the protein of interest. The potential ubiquitination of the protein is revealed by the presence of specific immunoreactive bands with molecular weights that should match the expected sizes for the various ubiquitinated forms of the protein. When specific antibodies to the protein of interest are available, this approach can be applied to the study of endogenous proteins by using cells over-expressing only the His–ubiquitin construct. Otherwise, the protein of interest needs to be tagged with an epitope different from His and experiments are carried out with cells coexpressing the two differentially epitope tagged proteins. As described for the immunoprecipitation method, this approach can also be very helpful to characterize the molecular determinants involved in the ubiquitination process by expressing mutants of His–ubiquitin and/or mutants of the protein of interest. As discussed above, one must exercise caution when interpreting results using overexpressed tagged proteins.

This strategy has several advantages over other biochemical approaches, such as immunoprecipitation, because His–ubiquitin can be purified in the presence of strong protein denaturants (buffers containing chaiotropic salts such as urea or guanidine-HCl) without affecting the binding capacity to the Ni–NTA affinity resins. The highly denaturing conditions under which the lysate preparation and purification are performed present a first main advantage to limit any Dub activity, thus preserving the ubiquitination status during the entire process. In addition, strong protein denaturants can remove cellular proteins noncovalently associated with either the protein of interest or the construct His–ubiquitin, or even the Ni–NTA resin, thus ensuring that the ubiquitinated species detected by this approach truly correspond to forms of the protein covalently modified by ubiquitin. The solubilization of proteins can be more efficient under denaturing conditions, which may help to improve the yield for detection of protein ubiquitination.

3.2. Experimental details

Cells are cultured and transfected using similar procedures to the ones described for immunoprecipitation studies. Hence, the same general considerations have to be taken into account (see previous section).

Experiments are performed with cells coexpressing His–ubiquitin (Treier *et al.*, 1994) and T7-HRas, as well as with cells expressing each of the constructs alone as negative controls. Cells expressing untagged ubiquitin also constitute an important internal control to verify the specificity of the assay. Twenty-four h post-transfection, cells are washed twice with ice-cold PBS and harvested in PBS before being resuspended in 1 ml of ice-cold lysis buffer containing 6 M guanidine-HCl, 0.1 M sodium-phosphate buffer (pH 8), 10 mM imidazole, 1 mM Na$_3$VO$_4$, 1 mM phenyl-methanesulfonyl fluoride (PMSF), 1% aprotinin, 10 mM benzamidine, 10 μg/ml pepstatin, 10 μg/ml leupeptin, and 10 μg/ml trypsin inhibitor. As in the immunoprecipitation experiments, a preincubating step with proteasome inhibitors (such as MG132) prior to cell lysis may favor the detection of the ubiquitinated species by increasing their steady-state levels if the protein of interest is subject to degradation by the proteasome. Also, the alkylating agent NEM can be included in the lysis buffer in order to block Dub activity. The lysate is then sonicated for 30 s to reduce viscosity, before being mixed with 50 μl of Ni-NTA beads (Invitrogen, Carlsbad, CA) prewashed with lysis buffer, and incubated for 3 h at room temperature (RT) with constant rotation. As already noted, carrying out the successive steps under denaturing conditions limits the activity of proteases and Dubs, which would usually be very high at RT. The resin is then washed twice with 1 ml of lysis buffer, twice with 1 ml of lysis buffer diluted 1:3 in 25 mM Tris-HCl (pH 6.8)/20 mM imidazole, and once with 1 ml of 25 mM Tris-HCl (pH 6.8)/20 mM imidazole. Purified proteins are eluted by boiling the beads in SDS sample buffer supplemented with 200 mM imidazole, and then analyzed by SDS-PAGE followed by immunoblotting with an antibody to the endogenous protein or to a tag present on an expressed substrate, as described above for the immunoprecipitation method.

4. RAS UBIQUITINATION MONITORED BY BRET IN LIVING CELLS

4.1. Method overview

BRET, which relies on nonradiative energy transfer between luciferase-coupled donors and GFP-coupled acceptors, is emerging as a useful tool for analyzing dynamic protein–protein interactions in living cells. In this assay, the proteins of interest are fused to either a donor molecule, luciferase, or to an acceptor molecule, a variant of GFP (Pfleger and Eidne, 2006; Pfleger *et al.*, 2006). After their heterologous coexpression, the luciferase is activated by its natural substrate (coelenterazine). If the coexpressed partners are far apart from each other (>100 Å), only one signal, emitted by the luciferase, can be detected. If the two proteins interact or are in close proximity (<100 Å), an energy transfer occurs between the donor and the acceptor, and the emission of luminescence by the luciferase is accompanied by a signal of fluorescence by the GFP variant protein. This resultant acceptor energy emission, reflective of the interaction or close proximity of the two proteins, can then be detected relative to the donor emission by using a suitable detection instrument. Using βarrestin as a model of study, Perroy *et al.* (2004) recently demonstrated that the changes in ubiquitination state of a protein can be monitored in real time by BRET. In order to study HRas ubiquitination in living cells and address the question of dynamics of this phenomenon, Rluc and YFP are fused to the N-terminus of HRas and ubiquitin. The occurrence of ubiquitination is then assessed by determining the energy transfer between Rluc-HRas and YFP-ubiquitin or between Rluc-ubiquitin and YFP-HRas, upon addition of the Rluc substrate, coelenterazine.

4.2. Experimental details

4.2.1. BRET constructs
For the analysis of Ras ubiquitination by BRET, the following constructs are used:

- Rluc-HRas: full-length HRas fused on its N-terminus with Renilla-luciferase (Rluc)
- YFP-HRas: full-length HRas fused on its N-terminus with enhanced yellow fluorescent protein (YFP)
- Ubiquitination-deficient Rluc-HRas8RK mutant: lysines 5, 42, 88, 101, 147, 167, 170, and 185 substituted with arginines in HRas WT
- Rluc-ubiquitin: full-length ubiquitin fused on its N-terminus with Rluc
- YFP-ubiquitin: full-length ubiquitin fused on its N-terminus with YFP
- YFP-ubiquitinAA mutant: the two C-terminal glycines 75 and 76 substituted with alanines in ubiquitin WT

4.2.2. General BRET assay protocol
The generic aspects of ectopic expression described previously for the method of immunoprecipitation should also be taken into account for BRET experiments. Twenty-four h before transfection with FuGene6, cells are seeded in six-well cell culture plates. Rluc- and YFP-fused HRas and ubiquitin are transfected to generate cell populations coexpressing both donor- and acceptor-tagged proteins (Rluc-HRas + YFP-ubiquitin or Rluc-ubiquitin + YFP-HRas), as well as the donor-tagged proteins only (Rluc-HRas or Rluc-ubiquitin, respectively) at similar expression levels to those in the samples coexpressing both Rluc- and YFP-tagged proteins. In parallel, negative control populations coexpressing Rluc-HRas + YFP-ubiquitinAA or Rluc-HRas8RK + YFP-ubiquitin need to be

generated. After transfection, cells are serum starved for at least 16 h before being prepared for BRET assay.

Routine BRET experiments are carried out in cell suspension in PBS solution at RT ($\sim$22°). Twenty-four h post-transfection, cells are washed twice with PBS before being detached with PBS containing 5 mM EDTA and resuspended in PBS containing 1 g/liter glucose. To control for the amount of cells in the assays, total protein concentration of the samples are determined using a Bradford assay kit (BioRad, Hercules, CA) and BSA dilutions as a standard. The equivalent of 20 μg of cell suspension is then distributed in duplicate in 96-well white microplates with opaque bottoms (90 μl of cell suspension per well) to carry out the BRET measurements. Total fluorescence and luminescence signals are also determined for the same samples to assess the level of expression of the YFP and Rluc fusion constructs. The YFP fluorescence is measured using a fluorescence/luminescence spectrometer (such as the LS-55 instrument from PerkinElmer, Wellesley, MA) with an excitation wavelength of 485 nm and an emission wavelength of 530 nm, using the following parameters: excitation and emission slit widths of 2.5 nm, emission filter cut-off of 515 nm, photomultiplier PMT of type standard with a voltage fixed at 800 V, and read time of 0.50 s. Fluorescence is quantified as fold over the background, which is measured using cells expressing the donor Rluc only. For assessing the expression of Rluc-fused constructs, the same samples are incubated for 10 min with 5 μM of the substrate coelenterazine h (Nanolight Technology, Prolume Ltd., Pinetop, AZ), and the luminescence is measured in the absence of emission filter using a multidetector plate reader (such as the MITHRAS LB940 from Berthold Technologies, Bad Wildbad, Germany). The substrate coelenterazine h can be stored over several months as a stock solution in ethanol (1 mM final) at -20° protected from light. The substrate stock is then freshly diluted to an intermediate solution of 50 μM in PBS/glucose (10$\times$ final concentration) before adding to samples ready to be used for luminescence or BRET measurement.

4.2.3. BRET detection and analysis

The BRET signal is detected 1 min following the addition of the substrate coelenterazine h (5 μM final) to the cells distributed in a 96-well plate. A highly sensitive instrument capable of sequentially or simultaneously detecting the energy emitted in two distinct wavelength windows is required to measure a specific BRET signal. For example, one can use the Mithras plate reader from Berthold Technologies which is not only equipped with a very low-noise photomultiplier, but also includes temperature control, injectors, and a kinetics software well adapted for real-time BRET measurements in living cells, as discussed below. The use of appropriate filter combination allows the sequential integration of the signals detected in the 485 ($\pm$20) nm and 530 ($\pm$20) nm windows for Rluc and YFP light emissions, respectively (1 s per filter). Given the stability of the coelenterazine h (Hamdan *et al.*, 2005), the BRET readings are repeated twice for each sample performed in duplicate. The BRET signal is determined by calculating the ratio of the light intensity emitted by the acceptor YFP-fused protein over the light intensity emitted by the donor Rluc-fused protein. The values are corrected by subtracting the background BRET signal detected when the donor fusion protein is expressed alone. The BRET ratio is calculated using the following equation:

$$\left(\frac{\text{Em } (530/20 \text{ nm})}{\text{Em } (485/20 \text{ nm})}\right)^{\text{Rluc}- \text{ and YFP}-\text{fused proteins co}-\text{expressed}}$$
$$- \left(\frac{\text{Em } (530/20 \text{ nm})}{\text{Em } (485/20 \text{ nm})}\right)^{\text{Rluc}-\text{fused protein expressed alone}}$$

4.2.4. Specificity of the BRET signal detected

Studies of constitutive interactions using BRET must provide evidence for interaction specificity since random or nonspecific interactions likely to occur with high ectopic expression levels could potentially lead to a "bystander" transfer of energy as a result of random collisions. To limit this effect, the experiments are performed at low expression levels of Rluc- and YFP-fused HRas and ubiquitin, as assessed by measuring the luminescence and fluorescence, respectively. However, additional experiments are required to exclude the possibility that the BRET signal observed could result from overexpression in a heterologous system. First, BRET experiments are carried out with cell populations expressing increasing total protein at a constant donor/acceptor ratio. In the case of random collisions between the BRET partners, the bystander BRET increases as a function of the total protein concentration (Mercier *et al.*, 2002). In contrast, if the BRET ratio is found to be independent of the total expression level over the range used, it indicates that the transfer of energy is specific and does not result from spurious interactions due to overexpression of the BRET partners. In addition, suitable BRET negative controls are required to support the specificity of the energy transfer, such as replacing one of the proteins of interest with a mutated form that does not interact with the second protein of interest. As far as the study of HRas ubiquitination by BRET is concerned, two different negative controls can be designed: (1) a YFP-tagged ubiquitin mutant that is unable to take part in the ubiquitination process as a result of the replacement of its last two glycine residues by alanines, YFP-ubiquitinAA (Perroy *et al.*, 2004); and (2) a Rluc-fused ubiquitination-deficient HRas mutant lacking the Lys residues that could serve as potential ubiquitination sites, Rluc-HRas8RK (Jura *et al.*, 2006). Control BRET experiments are then performed with cells coexpressing Rluc-HRas + YFP-ubiquitinAA or Rluc-HRas8RK + YFP-ubiquitin under similar experimental conditions (same fluorescence and luminescence) with cells coexpressing Rluc-HRas + YFP-ubiquitin. Importantly, since the BRET ratio is influenced by the relative expression of the donor Rluc– and acceptor YFP–fused proteins, one should systematically measure the fluorescence and luminescence signals of the cells coexpressing the different BRET partners to insure the performance of the comparison study not only at low total expression level (Rluc- + YFP-fused proteins), but also at the same fluorescence/luminescence ratio. However, the BRET is also dependent on several other transfer efficiency factors; thus, the amplitude of BRET signal may not reflect the affinity between two proteins. Consequently, "single-point" assays in which BRET is monitored in cells coexpressing the same donor/acceptor ratio (luminescence/fluorescence ratio) are a good indication for the specificity of interaction, but such data should only be considered qualitative.

To further confirm the specificity of the BRET signal between Rluc-HRas and YFP-ubiquitin, quantitative BRET titration assays (Mercier *et al.*, 2002) are carried out with cells coexpressing Rluc-HRas + YFP-ubiquitin, Rluc-HRas + YFP-ubiquitinAA, and Rluc-HRas8RK + YFP-ubiquitin. For this purpose, the Rluc construct is maintained constant, whereas the concentration of the YFP partner is gradually increased and the BRET ratios are expressed as a function of the acceptor/donor ratio. The lowest expression level of Rluc donor-tagged protein that provides a reliable luminescence signal is used. A BRET signal that increases hyperbolically before reaching a saturation level with high concentrations of acceptor YFP-fused protein, indicates a specific interaction between the BRET partners (Mercier *et al.*, 2002). In contrast, a BRET signal that increases in a quasi-linear fashion with increasing amounts of acceptor-tagged protein, eventually saturating only at very high acceptor/donor ratios, indicates a nonspecific interaction resulting from random collisions (bystander BRET) (Mercier *et al.*, 2002). One should also verify that the BRET detection is not influenced by the relative fusion of the donor and acceptor on the proteins by conducting experiments with the reciprocal constructs, such as Rluc-ubiquitin and YFP-HRas.

4.2.5. Measurement of BRET in real time

A major advantage of BRET is that it can be used to monitor the dynamics of protein–protein interactions in live cells in real time. Indeed, modulation of these interactions upon addition of external agents (such as agonists) could result in quantitative changes in the BRET ratio. However, it is important to note that, as with all approaches based on resonance energy transfer, the efficacy of transfer in BRET is highly dependent on both the relative distance and orientation between the donor and acceptor fluorophores (Xu et al., 1999). Consequently, changes in BRET signal may not only represent an increase or decrease in the number and/or rate of interactions between the proteins of interest, but may also result from conformational changes that influence the relative positioning of the donor and acceptor molecules. Thus, it is important to take this into consideration when interpreting the data. In any case, modulation of the energy transfer in response to a specific drug provides additional evidence for the specificity of the interactions. Alternatively, the lack of effect of external agents on the BRET detected between the proteins of interest does not necessary reflect the constitutive nature of their interaction in total absence of dynamic regulation. Indeed, it should be emphasized that the BRET level observed is a function of the efficacy of energy transfer that is given by the following equation: Efficacy $= R_0^6/(R_0^6 + R^6)$, where R_0 is the Förster distance at which the efficacy is equal to 0.5, and R is the distance between the donor and acceptor (Hovius et al., 2000). It follows that the changes in distance have a dramatic effect around the R_0, but can be undetectable for distances significantly smaller than R_0 where the transfer efficacy is reaching its maximum ($\sim$1). In fact, changes in distance can be assessed accurately only when the distance between donor and acceptor lies between 0.5 R_0 and 1.5 R_0. The potential lack of effect of external agents on a constitutive signal of BRET could therefore indicate that the distance between the two interacting proteins fused to Rluc and YFP is already smaller than 0.5 R_0 and cannot be accurately detected.

In order to monitor the dynamic regulation of HRas ubiquitination in living cells in real time, BRET measurements are performed after treatments with saturating concentrations of EGF in cells coexpressing Rluc-HRas + YFP–ubiquitin. These experiments can be performed both in cell suspension and in adherent cells maintained at 37°. For assays based on cells in suspension, cells are transfected, serum-deprived, and prepared as previously described for routine BRET measurements. The equivalent of 20 μg of cell suspension in PBS/glucose is then distributed in 96-well white microplates with opaque bottoms (80 μl of cell suspension per well) to carry out the BRET measurements. For BRET experiments performed on attached cells, a slightly different protocol is used. Cells are seeded in 60-mm Petri dishes 24 h before being transfected with FuGene6 to generate cell populations coexpressing both Rluc- and YFP-fused BRET partners (Rluc-HRas + YFP-ubiquitin, Rluc-HRas + YFP-ubiquitinAA or Rluc-HRas8RK + YFP-ubiquitin), as well as the donor-tagged proteins only (Rluc-HRas or Rluc-HRas8RK respectively), at similar expression levels to those in the samples coexpressing both Rluc- and YFP-tagged proteins. Twenty-four h post-transfection, cells are transferred into 96-well white microplates with clear bottoms that allow direct microscopic viewing. Cells are allowed to set for 8 h before being serum starved for at least 16 h prior to the BRET experiment. Forty-eight h post-transfection, cell culture medium is replaced with PBS containing 1 g/l glucose (80 μl per well), and the bottom of the plate covered with a white adhesive seal for 96-well and 384-well microplates. For long EGF incubation times (from 5 to 60 min), "standard" BRET assays are performed and both the EGF and the substrate coelenterazine h are added by pipetting. The various cell populations maintained at 37° are thus first stimulated or not with 10 μl of EGF (50 ng/ml final) in PBS/glucose for different periods of time. Ten microliters of coelenterazine h (5 μM final) is then added in the continuous presence of EGF and the BRET measurements are taken 40 s after the addition of the luciferase substrate. For accurate measurement of very early time points ($<$ 5 min), one can take advantage of the injection capability of the MITHRAS microplate reader and the kinetic Micro-Win2000 software (Berthold) (Gales *et al.*, 2005), which allows rapid measurement after EGF addition. The substrate coelenterazine h is added before the injection of either EGF or PBS/glucose and the readings collected at 2.5-s intervals. Injection of EGF or PBS/glucose is included within the kinetic program to allow baseline recording followed by real-time recording of the potential BRET changes. Real-time EGF-promoted modulations in BRET can then be expressed as the difference in the absolute BRET values obtained in the presence and in the absence of EGF.

5. CONCLUDING REMARKS

Most strategies used for the study of protein ubiquitination have been based on immunoprecipitation or Ni-NTA affinity chromatography techniques followed by Western blot analysis. Although these conventional biochemical approaches can be used to determine whether a protein is ubiquitinated, they are poorly quantitative with limited temporal resolution and no possibility of monitoring transient changes in ubiquitination. To overcome these obstacles and enable the identification of the temporal components of protein ubiquitination in living cells, one can take advantage of BRET technology, which allows monitoring of the changes in the ubiquitination state of a protein in real time (Perroy *et al.*, 2004). However, contrary to the immunoprecipitation approach, the BRET technique, as well as the Ni-NTA affinity chromatography approach, are not suitable for investigating the endogenous process of protein ubiquitination because of the need for fusion protein ectopic expression. In addition, whereas the BRET method does not permit one to draw a conclusion about the nature of the ubiquitination process involved (i.e., determine whether the Ras proteins are subject to either mono-, di- or polyubiquitination), one can easily discern these molecular determinants with the methods of immunoprecipitation or Ni–NTA affinity chromatography. Consequently, given the potential of the biochemical and biophysical techniques described here, the combined use of these approaches should give a new prospective to the study of protein ubiquitination. However, it is important to note that, unless cells are fractionated, neither of these methods can provide information about subcellular localization of the ubiquitinated protein forms. To circumvent this limitation, one can use alternative methods such as fluorescence resonance energy transfer (FRET) or ubiquitin-mediated fluorescence complementation (UbFC) (Fang and Kerppola, 2004), which are beyond the scope of this chapter.

ACKNOWLEDGMENTS

We are grateful to William Parrish, Natalia Jura, and Laura Taylor for critical reading of the manuscript. S. T. is supported by a fellowship from Human Frontier Science Program. D. B.-S. acknowledges the support of National Institutes of Health grant CA055360.

REFERENCES

Apolloni, A., Prior, I. A., Lindsay, M., Parton, R. G., and Hancock, J. F. (2000). H-Ras but not K-Ras traffics to the plasma membrane through the exocytic pathway. *Mol. Cell Biol.* **20,** 2475–2487.

Ahrens, P. B., Besancon, F., Memet, S., and Ankel, H. (1990). Tumour necrosis factor enhances induction by beta-interferon of a ubiquitin cross-reactive protein. *J. Gen. Virol.* **71**(Pt 8), 1675–1682.

Bivona, T. G., Quatela, S. E., Bodemann, B. O., Ahearn, I. M., Soskis, M. J., Mor, A., Miura, J., Wiener, H. H., Wright, L., Saba, S. G., Yim, D., Fein, A., *et al.* (2006). PKC regulates a farnesyl-electrostatic switch on K-Ras that promotes its association with Bcl-XL on mitochondria and induces apoptosis. *Mol. Cell* **21,** 481–493.

Choy, E., Chiu, V. K., Silletti, J., Feoktistov, M. M., Morimoto, T., Michaelson, D., Ivanov, I. E., and Philips, M. R. (1999). Endomembrane trafficking of Ras: The CAAX motif targets proteins to the ER and Golgi. *Cell* **98,** 69–80.

Downward, J. (2003). Targeting RAS. Signalling Pathways in Cancer Therapy. *Nat. Rev. Cancer* **3,** 11–22.

Fang, D., and Kerppola, T. K. (2004). Ubiquitin-mediated fluorescence complementation reveals that Jun ubiquitinated by itch/AIP4 is localized to lysosomes. *Proc. Natl. Acad. Sci. USA* **101,** 14782–14787.

Gales, C., Rebois, R. V., Hogue, M. M., Trieu, P., Breit, A., Hebert, T. E., and Bouvier, M. (2005). Real-time monitoring of receptor and G-protein interactions in living cells. *Nat. Methods* **2,** 177–184.

Haglund, K., and Dikic, I. (2005). Ubiquitylation and cell signaling. *EMBO J.* **24,** 3353–3359.

Hancock, J. F. (2003). Ras proteins: Different signals from different locations. *Nat. Rev. Mol. Cell Biol.* **4,** 373–384.

Hamdan, F., Audet, M., Garneau, P., Pelletier, J., and Bouvier, M. (2005). High-throughput screening of G protein-coupled receptor antagonists using a bioluminescence resonance energy transfer 1–based beta-arrestin2 recruitment assay. *J. Biomol. Screen.* **10,** 463–475.

Hicke, L., and Dunn, R. (2003). Regulation of membrane protein transport by ubiquitin and ubiquitin-binding proteins. *Annu. Rev. Cell Dev. Biol.* **19,** 141–172.

Hovius, R., Vallotton, P., Wohland, T., and Vogel, H. (2000). Fluorescence techniques: Shedding light on ligand-receptor interactions. *Trends Pharmacol. Sci.* **21,** 266–273.

Jura, N., and Bar-Sagi, D. (2006). Mapping cellular routes of Ras: A ubiquitin trail. *Cell Cycle* **5,** 2744–2747.

Jura, N., Scotto-Lavino, E., Sobczyk, A., and Bar-Sagi, D. (2006). Differential modification of Ras proteins by ubiquitination. *Mol. Cell* **21,** 679–687.

Mercier, JF., Salahpour, A., Angers, S., Breit, A., and Bouvier, M. (2002). Quantitative assessment of beta 1- and beta 2-adrenergic receptor homo- and heterodimerization by bioluminescence resonance energy transfer. *J. Biol. Chem.* **277,** 44925–44931.

Nandi, D., Tahiliani, P., Kumar, A., and Chandu, D. (2006). The ubiquitin-proteasome system. *J. Biosci.* **31,** 137–155.

Passmore, L. A., and Barford, D. (2004). Getting into position: The catalytic mechanisms of protein ubiquitylation. *Biochem. J.* **379,** 513–525.

Perroy, J., Pontier, S., Charest, P. G., Aubry, M., and Bouvier, M. (2004). Real-time monitoring of ubiquitination in living cells by BRET. *Nat. Methods* **1,** 203–208.

Pfleger, K. D., and Eidne, K. A. (2006). Illuminating insights into protein–protein interactions using bioluminescence resonance energy transfer (BRET). *Nat. Methods* **3,** 165–174.

Pfleger, K. D., Seeber, R. M., and Eidne, K. A. (2006). Bioluminescence resonance energy transfer (BRET) for the real-time detection of protein-protein interactions. *Nat. Protocols* **1,** 337–345.

Pickart, C. M. (2001). Mechanisms underlying ubiquitination. *Annu. Rev. Biochem.* **70,** 503–533.

Pickart, C. M., and Cohen, R. E. (2004). Proteasomes and their kin: Proteases in the machine age. *Nat. Rev. Mol. Cell Biol.* **5,** 177–187.

Prior, I. A., Harding, A., Yan, J., Sluimer, J., Parton, R. G., and Hancock, J. F. (2001). GTP-dependent segregation of H-Ras from lipid rafts is required for biological activity. *Nat. Cell Biol.* **3,** 368–375.

Rocks, O., Peyker, A., Kahms, M., Verveer, P. J., Koerner, C., Lumbierres, M., Kuhlmann, J., Waldmann, H., Wittinghofer, A., and Bastiaens, P. I.. (2005). An acylation cycle regulates localization and activity of palmitoylated Ras isoforms. *Science* **307,** 1746–1752.

Rodriguez-Viciana, P., and McCormick, F. (2006). Ras ubiquitination: Coupling spatial sorting and signal transmission. *Cancer Cell* **9,** 243–244.

Silvius, J. R. (2002). Mechanisms of Ras protein targeting in mammalian cells. *J. Membr. Biol.* **190,** 83–92.

Treier, M., Staszewski, L. M., and Bohmann, D. (1994). Ubiquitin-dependent C-Jun degradation *in vivo* is mediated by the delta domain. *Cell* **78,** 787–798.

Xu, Y., Piston, D. W., and Johnson, C. H. (1999). A bioluminescence resonance energy transfer (BRET) system: Application to interacting circadian clock proteins. *Proc. Natl. Acad. Sci. USA* **96,** 151–156.

CHAPTER NINETEEN

Biochemical Characterization of Novel Germline BRAF and MEK Mutations in Cardio-Facio-Cutaneous Syndrome

Pablo Rodriguez-Viciana* *and* Katherine A. Rauen*,[†]

Contents

Abstract

Cardio-facio-cutaneous syndrome (CFC) is a sporadic, complex developmental disorder involving characteristic craniofacial features, cardiac defects, ectodermal abnormalities, growth deficiency, hypotonia, and developmental delay. CFC is caused by alteration of activity through the mitogen-activated protein kinase (MAPK) pathway due to heterogeneous de novo germline mutations in B-Raf mutant proteins, MEK1 and MEK2. Approximately 75% of individuals with CFC have mutations in BRAF. *In vitro* functional studies demonstrate that many of these mutations confer increase activity upon the mutant protein as compared to the wildtype protein. However, as is seen cancer, some of the B-Raf mutant proteins are kinase impaired. Western blot analyses corroborate kinase assays as determined by mutant proteins phosphorylating downstream effectors MEK

* UCSF Helen Diller Family, Comprehensive Cancer Center and Cancer Research Institute, University of California, San Francisco, California
† Department of Pediatrics, Division of Medical Genetics, University of California, San Francisco, California

Methods in Enzymology, Volume 438
ISSN 0076-6879, DOI: 10.1016/S0076-6879(07)38019-1

and ERK. Approximately 25% of individuals with CFC have mutations in either MEK1 or MEK2 that lead to increased MEK kinase activity as judged by increased phosphorylation of its downstream effector ERK. Unlike BRAF, no somatic mutations have ever been identified in MEK genes. The identification of novel germline BRAF and MEK mutations in CFC will help understand the pathophysiology of this syndrome. Furthermore, it will also provide insight to the normal function of B-Raf and MEK, and contribute to the knowledge of the role of the MAPK pathway in cancer. Since the MAPK pathway has been studied intensively in the context of cancer, numerous therapeutics that specifically target this pathway may merit investigation in this population of patients.

1. INTRODUCTION

The Raf/MEK/ERK MAPK cascade is one of the downstream pathways from Ras, and is critically involved in cell proliferation, differentiation, motility, apoptosis, and senescence. Extracellular stimuli, through the activation of Ras, lead to the activation of Raf (A-Raf, B-Raf, and/or C-Raf-1) the most upstream kinase of the cascade. Raf then phosphorylates and activates MEK1 and/or MEK2 (MAPK kinase), which in turn phosphorylates and activates ERK1 and/or ERK2 (MAPK). ERK, once activated, has numerous cytosolic and nuclear substrates (Yoon and Seger, 2006). The MAPK cascade has been intensely studied in the context of cancer. Upregulation of this pathway plays a key role in the pathogenesis and progression of many cancers. Hyperactivated ERK is found in approximately 30% of human cancers with pancreas, colon, lung, ovary, and kidney demonstrating the highest levels of ERK activation (Hoshino et al., 1999). Altered signaling through the MAPK pathway in cancer results from mutations in upstream components of ERK, including K-Ras, N-Ras, H-Ras, C-Raf-1, and B-Raf.

A fascinating class of medical genetic syndromes has emerged which are due to mutations in genes associated with the Ras pathway. Some of these syndromes include Noonan syndrome (NS), LEOPARD syndrome, neurofibromatosis 1, Costello syndrome (CS), and cardio-facio-cutaneous syndrome (CFC) (Kratz et al., 2007). CFC has many phenotypic features that overlap with these syndromes, especially NS and CS. CFC is a sporadic, complex developmental disorder that has characteristic craniofacial features, cardiac defects with the most prevalent being pulmonic stenosis, atrial septal defects and hypertrophic cardiomyopathy, and ectodermal abnormalities of the hair and skin. Neurologic complications are many and are present in all individuals studied to date. Universal findings include hypotonia and developmental delay, which varies from mild to severe (Yoon G. et al., 2007). Individuals with CFC may also have musculoskeletal, ophthalmologic, audiologic, lymphatic, gastrointestinal, and/or renal abnormalities.

To date, there is one report of a CFC individual who developed acute lymphoblastic leukemia in early childhood. The nature of the association as coincidental or causal is unknown (Niihori *et al.*, 2006).

After determining that CFC and CS were genetically distinct entities, we hypothesized that the CFC gene(s) may encode a protein participating in signal transduction of the Ras/MAPK pathway (Estep *et al.*, 2006). In a cohort of 23 CFC individuals, we sequenced the coding and flanking intronic regions of Ras family members (KRAS, NRAS and MRAS), as well as downstream effectors of Ras in the MAPK pathway (BRAF, CRAF, MEK1, and MEK2). We identified germline mutations within BRAF in the majority of our CFC cohort (Rodriguez-Viciana *et al.*, 2006). Another group has also independently identified mutations in BRAF in CFC patients (Niihori *et al.*, 2006). In addition, we determined that approximately 20% of CFC individuals have germline MEK1 or MEK2 mutations (Rodriguez-Viciana *et al.*, 2006).

Somatic mutations in BRAF have been reported at high frequency in numerous cancers including melanoma, thyroid, colorectal, and ovarian cancer. Approximately 70 missense mutations affecting 34 codons have been reported with the majority of the mutations occurring in exon 11 and exon 15 of the protein kinase domain (www.sanger.ac.uk/genetics/CGP/cosmic). One mutation, ^{V600E}B-Raf (exon 15), which has increased kinase activity, accounts for over 90% of BRAF mutations identified in human cancer. Interestingly, somatic ^{V600E}B-Raf mutations are found in the majority of benign nevi, as well as primary and metastatic melanoma, suggesting that MAPK activation is important in melanocytic neoplasia, but that in isolation, is insufficient for tumorigenesis (Pollock *et al.*, 2003). Some BRAF mutations identified in CFC are similar to somatic mutations that have been identified in cancer. However, in contrast to the mutation spectrum seen in cancer, the majority of BRAF mutations we identified were novel and found in exons 6, 11, 12, 13, 14, 15, and 16 (Rauen, 2006; Rodriguez-Viciana *et al.*, 2006; Rauen, unpublished data). Unlike BRAF, prior to the identification of germline MEK mutations, no somatic mutations have ever been reported in MEK genes. Here we describe the biochemical characterization of these BRAF and MEK mutations.

2. METHODS

2.1. DNA isolation

Genomic DNA is isolated from peripheral blood lymphocytes using a QIAamp DNA Blood Midi kit (Qiagen, Valencia, CA). In addition, genomic DNA is extracted from buccal cells using a DNeasy Tissue kit (Qiagen, Valencia, CA).

2.2. Sequence analysis

The entire KRAS, NRAS, MRAS, BRAF, CRAF, MEK1, and MEK2 coding regions are sequenced for mutations using direct bidirectional sequencing. Exons and intronic flanking regions are amplified by PCR. DNA sequencing are performed using a Big Dye v3.1 Cycle Sequencing Kit (Applied Biosystems, Foster City, CA) according to the manufacturer's recommendations and run on an ABI3730xl or ABI3700 capillary sequencing instrument. Data are extracted and analyzed with Sequencer Analysis Software version 3.7 (Applied Biosystems, Foster City, CA). Accession numbers are as follow: KRAS (NM 004985), NRAS (NM 002524), MRAS (NM 012219), HRAS (NM 0176795), PTPN11 (NM 002834), BRAF (NM 004333), CRAF (NM 002880), MEK1 (NM 002755), and MEK2 (NM 030662).

2.3. Plasmids for studies

Human BRAF cDNA with a Myc and His tags at the 3' end was a kind gift from Dr. Martin McMahon. Human MEK1 cDNA was cloned by RT-PCR from fetal brain mRNA and purchased from Clontech (Mountain View, CA). Human MEK2 cDNA was purchased from Origene (Rockville, MD). All three cDNAs are cloned into the pENTR vector (Invitrogen, Carlsbad, CA). Point mutations are generated using the Quick-Change Site-Directed Mutagenesis Kit (Stratagene, La Jolla, CA), and are verified by direct sequencing at the UCSF Comprehensive Cancer Center Genome Core Facility. BRAF cDNAs (wildtype and mutants) are transferred into a Gateway-compatible pcDNA3-FLAG vector (pcDNA3 with a Flag-tag at the N-terminus) by Gateway-mediated recombination according to manufacturer's protocols (Invitrogen, Carlsbad, CA). Similarly, MEK1 and MEK2 cDNAs (wildtype and mutants) are transferred by Gateway-mediated recombination to a Gateway-compatible pcDNA3-MYC vector (Myc-tag at the N-terminus).

2.4. Transient transfections and Western blot analysis of MAPK pathway activation by B-Raf and MEK mutants

293 or 293T cells (a variant expressing SV40 large T antigen) are a convenient and commonly used cell type to analyze activation of the MAPK pathway by exogenously expressed genes because of their high transfection efficiency (up to 90%) and the low basal level of activation of the endogenous MAPK pathway. Activation of endogenous MEK and/or ERK is measured in total cell lysates after transient transfection of B-Raf and MEK plasmids. Phospho-specific antibodies are used to measure MEK

and/or ERK phosphorylation by Western blotting. The kinase activity of the B-Raf mutants can be measured by immunoprecipitation of the transfected B-Raf from the same cell lysates (see below).

293T cells are maintained in high-glucose Dulbecco's modified Eagle's medium (DMEM) supplemented with 10% fetal bovine serum (FBS) without antibiotics. For transfection, seed 1×10^6 293T cells per well in six-well dishes. Incubate cells overnight and subsequently transfect with Lipofectamine 2000 (Invitrogen, Carlsbad, CA) according to manufacturer's instructions. Briefly, add 2 μg total amount of plasmid DNA to 100 μl OPTIMEM (Invitrogen, Carlsbad, CA). In a separate tube, add 5 μl of Lipofectamine 2000 to 100 μl OPTIMEM, vortex, and allow to stand 5 min. Combine, vortex and add to cells after 20 min. The next day, serum-starve cells to decrease endogenous signaling through the MAPK pathway by replacing the media with DMEM with 0.5% FBS.

Two days after transfection, the cells are lysed in 350 μl/well of TNE (20 mM Tris, pH 7.5, 150 mM NaCl, 1 mM EDTA), 1% Triton X-100, 1 mM DTT, with protease and phosphatase inhibitor cocktails (Sigma, St. Louis, MO). Clear the lysate by microfuge centrifugation at 4° for 10 min. Transfer 30 μl of the supernatant to tube with 10 μl of 4× sample buffer to check for expression levels and MEK and ERK phosphorylation status by Western blot. The remaining lysate can be used for immunoprecipitations on the same day or it can be snap-frozen in dry ice/ethanol bath and stored at −80° for future use after standardization of expression levels.

Expression levels of transfected proteins are analyzed by Western blotting with antibodies against the Tag (Myc or Flag). The phosphorylation status of endogenous ERK and MEK proteins is analyzed by Western blotting with phospho-specific antibodies. Flag-M2 antibody is purchased from Sigma, Myc (A-14) from Santa Cruz Biotechnology (Santa Cruz, CA), Myc (9B11), Phospho-ERK, Total ERK, and Phospho-MEK antibodies are from Cell Signaling Technology (Danvers, MA).

2.5. Immunoprecipitation and Raf kinase assays

The kinase activity of the transfected Flag-tagged B-Raf variants is measured in Flag-immunoprecipitates by performing a coupled MEK/ERK2 kinase assay using myelin basic protein (MBP) as the final substrate as previously described (Alessi et al., 1995). Because of variable levels of expression of some of the B-Raf mutant proteins, we standardize for B-Raf protein levels by initially analyzing levels of the transfected B-Raf mutants by Western blot analysis. After tentatively quantifying relative levels of expression, corresponding amounts of cell lysates are used for subsequent inmunoprecipitation. As an alternative to freezing cell lysates, washed immunoprecipitates can also be frozen. After the last wash, add sample buffer to an aliquot of the Flag-beads and freeze the remainder in dry-ice/ ethanol bath and store at −80°.

Because of the high levels of expression of our B-Raf constructs in 293T cells, care has to be taken not use excessive amounts of B-Raf protein, as this may saturate the subsequent kinase assay (see following discussion). The amount of immunoprecipitated B-Raf can be estimated by running an aliquot of the immunoprecipitates in SDS-PAGE gels and staining with Coomassie Blue. To immunoprecipitate B-Raf, transfer the cleared supernatant to a tube with 10 μl packed Flag-M2 agarose beads (Sigma, St. Louis, MO). If different amounts of B-Raf containing lysates are used for various mutants, the amount of total cell lysate is made the same by addition of lysate from cells transfected with empty vector. Incubate 1 to 3 h by tumbling at 4°. Wash beads twice with TNE and 1% Triton X100, and once with TNE without detergent. Drain and add 15 μl of cold kinase buffer (25 mM Tris, pH 7.5, 50 mM NaCl, 1 mM EDTA).

To each tube, add 10 μl of 50 mM MgCl$_2$, 0.5 mM ATP containing 0.25 μg of purified recombinant MEK1 (14–205, Upstate Biotechnology, Lake Placid, NY) and 2 μg of purified recombinant GST-ERK2 (The pGEX vector expressing GST-ERK2 was a kind gift of Chris Marshall. Alternatively, ERK2 can also be purchased from Upstate Biotechnology, Lake Placid, NY). Place the tube at 30° for 20 min. Transfer 2 μl of the kinase reaction to a tube with 15 μl of cold kinase buffer (done in duplicate). Start second kinase reaction by adding 10 μl of 50 mM MgCl$_2$, 0.5 mM ATP containing 20 μg of MBP (Sigma, St. Louis, MO), and 2 μCi of γ-32P-ATP (Amersham, Piscataway, NJ). Incubate for 10 min at 30°. Stop the reactions by adding 75 μl of cold 50 mM EDTA. Add 70 μl to pre-cut squares of p81 Whatman paper and place into a beaker with 75 mM H$_3$PO$_4$. Wash three times for 5 min with 75 mM H$_3$PO$_4$ and once with acetone. Air dry the paper squares and measure the radioactivity by Cerenkov on a liquid scintillation counter. The counts per minute of a blank kinase reaction performed in the absence of Flag-beads are subtracted from all remaining reactions. Kinase activity of B-Raf immunoprecipitates are compared to immunoprecipitates of the lysates transfected with an empty vector control.

3. RESULTS AND DISCUSSION

To date there have been numerous germline BRAF and MEK1 and MEK2 mutations reported in the CFC syndrome (Table 19.1). The biochemical activity of B-Raf mutations found in CFC is similar to the

Table 19.1 Mutations in BRAF, MEK1, and MEK2 reported in CFC syndrome with comparison to reported mutations in cancer

Gene	Exon	Amino acid substitution	Patients reported	Somatic mutations[a]	References[b]
BRAF	6	T244P	1	No	1
	6	A246P	2	No	2
	6	Q257R	14	No	1, 2, 3, 4
	6	Q257K	1	No	4
	11	Δ462–469	1	No	1
	11	G464V	2	No	1
	11	G466R	1	G599E/R/V	1
	11	S467A	1	No	3
	11	F468S	3	F468C	1,3
	11	G469E	6	G469E/A/R/S/V	1,2,3
	11	Δ T470	1	No	1
	12	L485S	1	No	1
	12	L485F	2	No	2,3
	12	V487G	1	No	4
	12	K499E	2	No	2,3
	12	K499N	1	No	1
	12	E501K	2	No	2,3
	12	E501G	4	No	1, 2, 3, 4
	12	E501V	1	No	1
	13	G534R	1	No	5
	14	N580D	1	No	4
	14	N581D	4	N581S	2,3
	15	F595L	1	F595L/S	3
	15	G596V	4	G596R	1,3
	15	T599R	1	T599I	1
	16	D638E	1	No	5
MEK1	2	F53S	1	No	3
	2	ΔK59	1	No	1
	3	P124L	1	No	4
	3	G128V	1	No	1
	3	Y130N	1	No	1
	3	Y130C	6	No	1,3,4
MEK2	2	F57C	1	No	3
	2	F57I	1	No	6
	2	K61E	1	No	4

(continued)

Table 19.1 (*continued*)

Gene	Exon	Amino acid substitution	Patients reported	Somatic mutations[a]	References[b]
	3	P128R	1	No	4
	3	G132V	1	No	4
	3	Y134C	1	No	1
	7	K273R	1	No	4

[a] Amino acid alterations of somatic mutations found in B–Raf as listed in the Sanger Institute Catalogue of Somatic Mutations in Cancer (www.sanger.ac.uk/genetics/CGP/cosmic).
[b] References: (1) Rauen, Unpublished data (2) Niihori *et al.*, 2006; (3) Rodriguez-Viciana *et al.*, 2006; (4) Narumi *et al.*, 2007; (5) Rauen, 2006; (6) Kratz *et al.*, 2007.

types of mutations found in cancer, those with high kinase and kinase-impaired activities (Wan *et al.*, 2004). The kinase activity of selected CFC B–Raf mutants compared to the wildtype protein (^{WT}B–Raf) and several cancer-derived mutants vary (Fig. 19.1A). Five CFC B–Raf mutants—^{Q257R}B–Raf, ^{S467A}B–Raf, ^{L485F}B–Raf, ^{K499E}B–Raf, and ^{G534R}B–Raf—have increased kinase activity compared with ^{WT}B–Raf, and activity was as high as that of the ^{V600E}B–Raf mutant found in cancer. Three CFC B–Raf mutants—^{E510G}B–Raf, ^{G469V}B–Raf, and ^{D638E}B–Raf—have lower activity in our assay than ^{WT}B–Raf, and appear to be kinase impaired. In addition, CFC B–Raf mutants activate downstream effectors, as determined by measuring phosphorylated species of MEK and ERK by Western blotting (see Fig. 19.1B). Both cancer and CFC-associated B–Raf mutants with elevated kinase activity induce higher levels of MEK and ERK phosphorylation compared with ^{WT}B–Raf, whereas kinase-impaired B–Raf mutants are impaired in their ability to induce phosphorylation of MEK and ERK.

B–Raf mutants found in CFC, like those found in cancer, can have either higher kinase activity or be kinase-impaired when compared to wildtype B–Raf. Wildtype B–Raf is known to possess high basal activity (at least when overexpressed). In our assays, we have been unable to demonstrate more than a two- to three-fold activation over wildtype by activating mutations, including ^{V600E}B–Raf. Another group has reported a 500-fold higher activity for ^{V600E}B–Raf (Wan *et al.*, 2004). This may be due to higher concentrations of cold ATP in their kinase assays, but we have not seen significant differences when using the same higher ATP concentrations. It is noteworthy, however, that Western blot analysis measuring activation of the downstream MEK and ERK proteins, as determined by their phosphorylation on the same cell lysates, correlates well with B–Raf kinase activity, and we see a two- to three-fold increase in ERK phosphorylation by activating B–Raf mutations, including ^{V600E}B–Raf. The Marais lab also reports that kinase-impaired B–Raf mutants can activate MEK

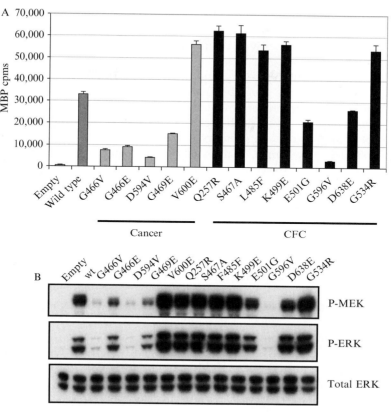

Figure 19.1 Characterization of B-Raf and MEK mutant proteins identified in CFC. (A) Kinase activities of selected B-Raf missense CFC mutations are compared to known B-Raf mutations found in cancer. Empty vector, wildtype B-Raf (^{WT}B-Raf) or the indicated B-Raf point mutants were transfected in 293T cells, and B-Raf activity was measured on Flag-immunoprecipitates using a coupled MEK-ERK-MBP phosphorylation assay. G466V, G466E, D594V, G469E, and V600E are B-Raf mutant proteins that have been previously reported in cancer (light gray bars). Q257R, S467A, L485F, K499E, E501G, G596V, D638E, and G534R are novel missense mutations identified in CFC individuals (black bars). Five of the CFC mutants evaluated had increased kinase activity over that of WTB-Raf, and as high as that of the ^{V600E}B-Raf, the most common activating BRAF mutation found in cancer. Three CFC mutant - proteins—G466V, G466E, and D594V which have also been reported in tumors—have lower kinase activity as compared to ^{WT}B-Raf, and thus, fall in the category of kinase-impaired mutations. (B) MEK and ERK activation by B-Raf mutants. Lysates from 293T cells transfected with empty vector, ^{WT}B-Raf or B-Raf mutants were separated by SDS-PAGE. MEK and ERK (p44 ERK1 and p42 ERK2) phosphorylation were assayed by Western blotting using phospho-specific antibodies. Good correlation is observed between kinase activity and phosphorylation for both high and low activity mutants. Total ERK is shown as a loading control.

and ERK signaling through interaction with and activation of endogenous c-Raf-1. In our hands, however, the reduced activity of kinase-impaired B-Raf mutants correlates with their impaired ability to stimulate MEK and ERK phosphorylation.

The reasons why the inactivating mutations found in cancer and CFC, which presumably are a gain-of-function mutant, would have lower activity than the wildtype protein are not clear, but may be related to the nature of the assay and to the fact that the activity of overexpressed wildtype B-Raf may not accurately reflect its activity when expressed at physiological levels. It is possible, for example, that an unknown regulatory mechanism may negatively regulate B-Raf activity *in vivo*, and that wildtype B-Raf overcomes this negative regulation upon overexpression. Clearly, more work is needed to understand the mechanism by which B-Raf mutations with impaired kinase activity result in gain of function.

Missense mutations in MEK1 and MEK2, which encode downstream effectors of B-Raf, also cause CFC syndrome (Rodriguez-Viciana et al., 2006). MEK1 and MEK2 are threonine/tyrosine kinases with both isoforms having the ability to activate ERK1 and ERK2. MEK1 and MEK2 do not serve redundant purposes. Genetic evidence utilizing mouse models has demonstrated that MEK1 is vital for embryonic development (Giroux et al., 1999), whereas MEK2 is dispensable (Belanger et al., 2003). Missense MEK mutations are identified in about 25% of CFC individuals who are BRAF-mutation negative. Unlike B-Raf, MEK1 and MEK2 mutations have not been reported in cancer (Bansal et al., 1997; www.sanger.ac.uk/genetics/CGP/cosmic). However, activation of MEK is necessary for transformation through the MAPK cascade (Cowley et al., 1994), and constitutively active MEK mutants are capable of promoting transformation of mammalian cells *in vitro* and *in vivo* (Mansour et al., 1994). The functional properties of these novel CFC MEK mutations, as determined by Western blotting demonstrate that CFC mutants—[F53S]MEK1, [Y130C]MEK1, [F57C]MEK2, and [Y134C]MEK2—are more active than wildtype MEK in stimulating ERK phosphorylation (Fig. 19.2).

The relationship of how these novel CFC germline mutations exert their pathophysiologic effect remains to be elucidated. In addition to its critical role in tumorigenesis, the MAPK pathway has also important roles on development, differentiation, and cellular homeostasis. ERK1 and ERK2, the final kinases of the cascade, exert their function on a large number of downstream molecules, both nuclear and cytosolic, which may vary depending on cell type and mechanism of pathway activation. Activated ERK1/2 phosphorylates numerous substrates including nuclear components, transcription factors, membrane proteins, cytoskeletal proteins, and protein kinases (Yoon and Seger, 2006). Research will resolve the extent to which germline mutations in the MAPK pathway perturb signaling. Understanding the pathogenetics may allow the use of possible

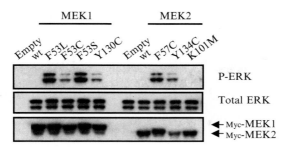

Figure 19.2 Western blot analysis of novel CFC MEK1 and MEK2 mutant proteins. Empty vector, wildtype human MEK1 and MEK2 or the indicated MEK1 and MEK2 mutant proteins were transfected into 293T cells, and ERK phosphorylation was assayed on cell lysates by Western blot with a phospho-specific antibody. The levels of total ERK and transfected myc-tagged MEK1 and MEK2 proteins were also measured. ^{F53S}MEK1 and ^{Y130C}MEK1 CFC mutants and ^{F57C}MEK2 and ^{Y134C}MEK2 mutants led to increased MEK activity compared to the wildtype proteins as measured by the ability to stimulate ERK phosphorylation. In addition, ^{F53L}MEK1 and ^{F53C}MEK1 (the comparable amino acid substitution to the ^{F57C}MEK2 mutant) were produced and their activity compared to known CFC MEK1 mutants. Both amino acid substitutions inferred increased activity over wildtype. ^{K101M}MEK2 is a kinase inactive mutant due to alteration of the ATP site and is used as a negative control.

therapeutics to modulate signaling. Because the MAPK pathway has been studied intensively in the context of cancer, numerous therapeutics that specifically target this pathway are in development (Sebolt-Leopold and Herrera, 2004). Inhibitors of Raf and MEK are being evaluated in clinical trials and appear to be well tolerated (Sebolt-Leopold and Herrera, 2004). In addition, one report indicates that cells with activated B-Raf have enhanced, selective sensitivity to MEK inhibitors (Solit *et al.*, 2006). Since CFC has an evolving phenotype, systemic therapies that reduce MAPK activity may merit investigation in this population of patients.

ACKNOWLEDGMENTS

The authors thank the families of CFC International (www.cfcsyndrome.org) for their ongoing support of research in genetic medicine, and Dr. W. Tidyman for critical comments. The authors apologize for not citing all relevant references due to space limitations. This work was supported in part by National Institutes of Health grant HD048502 (K. A. R.).

REFERENCES

Alessi, D. R., Cohen, P., Ashworth, A., Cowley, S., Leevers, S. J., and Marshall, C. J. (1995). Assay and expression of mitogen-activated protein kinase, MAP kinase kinase, and Raf. *Methods Enzymol.* **255,** 279–290.

Bansal, A., Ramirez, R. D., and Minna, J. D. (1997). Mutation analysis of the coding sequences of MEK-1 and MEK-2 genes in human lung cancer cell lines. *Oncogene* **14,** 1231–1234.

Belanger, L. F., Roy, S., Tremblay, M., Brott, B., Steff, A. M., Mourad, W., Hugo, P., Erikson, R., and Charron, J. (2003). Mek2 is dispensable for mouse growth and development. *Mol. Cell Biol.* **23,** 4778–4787.

Cowley, S., Paterson, H., Kemp, P., and Marshall, C. J. (1994). Activation of MAP kinase kinase is necessary and sufficient for PC12 differentiation and for transformation of NIH 3T3 cells. *Cell* **77,** 841–852.

Estep, A. L., Tidyman, W. E., Teitell, M. A., Cotter, P. D., and Rauen, K. A. (2006). HRAS mutations in Costello syndrome: Detection of constitutional activating mutations in codon 12 and 13 and loss of wild-type allele in malignancy. *Am. J. Med. Genet. A* **140,** 8–16.

Giroux, S., Tremblay, M., Bernard, D., Cardin-Girard, J. F., Aubry, S., Larouche, L., Rousseau, S., Huot, J., Landry, J., Jeannotte, L., and Charron, J. (1999). Embryonic death of Mek1-deficient mice reveals a role for this kinase in angiogenesis in the labyrinthine region of the placenta. *Curr. Biol.* **9,** 369–372.

Hoshino, R., Chatani, Y., Yamori, T., Tsuruo, T., Oka, H., Yoshida, O., Shimada, Y., Arii, S., Wada, H., Fujimoto, J., and Kohno, M. (1999). Constitutive activation of the 41-/43-kDa mitogen-activated protein kinase signaling pathway in human tumors. *Oncogene* **18,** 813–822.

Kratz, C. P., Niemeyer, C. M., and Zenker, M. (2007). An unexpected new role of mutant Ras: perturbation of human embryonic development. *J. Mol. Med.* **85,** 223–231.

Mansour, S. J., Matten, W. T., Hermann, A. S., Candia, J. M., Rong, S., Fukasawa, K., Vande Woude, G. F., and Ahn, N. G. (1994). Transformation of mammalian cells by constitutively active MAP kinase kinase. *Science* **265,** 966–970.

Narumi, Y., Aoki, Y., Niihori, T., Neri, G., Cave, H., Verloes, A., Nava, C., Kavamura, M. I., Okamoto, N., Kurosawa, K., Hennekam, R. C., and Wilson, L. C., *et al.* (2007). Molecular and clinical characterization of cardio-facio-cutaneous (CFC) syndrome: Overlapping clinical manifestations with Costello syndrome. *Am. J. Med. Genet. A* **143,** 799–807.

Niihori, T., Aoki, Y., Narumi, Y., Neri, G., Cave, H., Verloes, A., Okamoto, N., Hennekam, R. C., Gillessen-Kaesbach, G., Wieczorek, D., Kavamura, M. I., Kurosawa, K., *et al.* (2006). Germline KRAS and BRAF mutations in cardio-facio-cutaneous syndrome. *Nat. Genet.* **38,** 294–296.

Pollock, P. M., Harper, U. L., Hansen, K. S., Yudt, L. M., Stark, M., Robbins, C. M., Moses, T. Y., Hostetter, G., Wagner, U., Kakareka, J., Salem, G., Pohida, T., *et al.* (2003). High frequency of BRAF mutations in nevi. *Nat. Genet.* **33,** 19–20.

Rauen, K. A. (2006). Distinguishing Costello versus cardio-facio-cutaneous syndrome: BRAF mutations in patients with a Costello phenotype. *Am. J. Med. Genet. A* **140,** 1681–1683.

Rodriguez-Viciana, P., Tetsu, O., Tidyman, W. E., Estep, A. L., Conger, B. A., Cruz, M. S., McCormick, F., and Rauen, K. A. (2006). Germline mutations in genes within the MAPK pathway cause cardio-facio-cutaneous syndrome. *Science* **311,** 1287–1290.

Sebolt-Leopold, J. S., and Herrera, R. (2004). Targeting the mitogen-activated protein kinase cascade to treat cancer. *Nat. Rev. Cancer* **4,** 937–947.

Solit, D. B., Garraway, L. A., Pratilas, C. A., Sawai, A., Getz, G., Basso, A., Ye, Q., Lobo, J. M., She, Y., Osman, I., Golub, T. R., Sebolt-Leopold, J., *et al.* (2006). BRAF mutation predicts sensitivity to MEK inhibition *Nature* **439,** 358–362.

Wan, P. T., Garnett, M. J., Roe, S. M., Lee, S., Niculescu-Duvaz, D., Good, V. M., Jones, C. M., Marshall, C. J., Springer, C. J., Barford, D., and Marais, R. (2004). Mechanism of activation of the RAF-ERK signaling pathway by oncogenic mutations of B-RAF. *Cell* **116,** 855–867.

Yoon, G., Rosenberg, J., Blaser, S., and Rauen, K. A. (2007). Neurological complications of the cardio–facio–cutaneous syndrome (CFC). *Developmental Medicine and Child Neurology* **49,** 894–899.

Yoon, S., and Seger, R. (2006). The extracellular signal-regulated kinase: Multiple substrates regulate diverse cellular functions. *Growth Factors* **24,** 21–44.

CHAPTER TWENTY

Biochemical and Biological Characterization of Tumor-Associated Mutations of p110α

Adam Denley, Marco Gymnopoulos, Jonathan R. Hart, Hao Jiang, Li Zhao, *and* Peter K. Vogt

Contents

Abstract

Signaling by class I phosphatidylinositol 3-kinase (PI3K) controls cell growth, replication, motility, and metabolism. The PI3K pathway commonly shows gain of function in cancer. Two small GTPases, Rheb (Ras homolog enriched in brain) and Ras (rat sarcoma viral oncogene), play important roles in PI3K signaling. Rheb activates the TOR (target of rapamycin) kinase in a GTP-dependent manner; it links TOR to upstream signaling components, including the tuberous sclerosis complex (TSC) and Akt (homolog of the Akt8 murine lymphoma viral oncoprotein). Constitutively active, GTP-bound Rheb is oncogenic in cell culture, and activity that requires farnesylation. Ras activates PI3K by recruitment to the plasma membrane and possibly by inducing a conformational change in the catalytic subunit p110 of PI3K. In return, Ras signaling through the MAP kinase (MAPK) pathway is activated by PIP$_3$, the product of PI3K. Loss of Ras function can interfere with PI3K signaling. Various lines of evidence suggest complementary roles for PI3K and MAPK signaling in oncogenesis.

Department of Molecular and Experimental Medicine, The Scripps Research Institute, La Jolla, California

Methods in Enzymology, Volume 438
ISSN 0076-6879, DOI: 10.1016/S0076-6879(07)38020-8

Abbreviations: 4E-BP, eukaryotic initiation factor 4E binding protein; AKT, cellular homolog of murine thymoma virus *akt8* oncogene; CEF, chicken embryo fibroblasts; EGFR, epidermal growth factor receptor; elF4E, eukaryotic initiation factor 4E; FOXO, forkhead box O transcription factor; GEF, guanine exchange factor; GSK3β, glycogen synthase kinase-3 beta; GTP, guanosine triphosphate; hTERT, human telomerase reverse transcriptase; IRS, insulin receptor substrate; MAPK, mitogen-activated protein kinase; Myc, cellular homolog of avian myelocytoma retroviral oncogene; PDK1, phosphatidylinositol-dependent kinase; PI3K, phosphatidylinositol 3-kinase; PIP$_2$, phosphatidylinositol (4,5)-bisphosphate; PIP$_3$, phosphatidylinositol (3,4,5)-trisphosphate; PI, phosphatidylinositol; PIPP, proline-rich inositol phosphatase; PTEN, phosphatase and tensin homolog deleted on chromosome 10; RAL, cellular homolog of simian leukemia virus oncogene *ral*; RAPTOR, regulatory-associated protein of mTOR; Ras, rat sarcoma viral oncogene; RBD, Ras binding domain; RCAS, replication-competent, avian leucosis long-terminal repeat with splice acceptor; RHEB, Ras homolog enriched in brain; RICTOR, rapamycin-insensitive companion of mTOR; S6, S6 ribosomal protein; S6K, p70S6 kinase; SHIP, Src homology-2 domain-containing inositol polyphosphate 5'-phosphatase; SOS, son of sevenless; TOR, target or rapamycin; TSC, tuberous sclerosis complex

1. CANONICAL PATHWAY

Phosphatidylinositol 3-kinases (PI3Ks) catalyze the transfer of a phosphate group from ATP to the 3 position of phosphatidylinositol (PI) (Vanhaesebroeck *et al.*, 1997). PI3Ks are found throughout eukaryotes, and are partitioned into classes, I, II, and III, which differ in structure and function (Vanhaesebroeck and Waterfield, 1999). This review is confined to class I enzymes because of their prominent and well-studied role in cellular signaling. Class I enzymes are heterodimeric proteins comprised of a catalytic subunit, p110, and a regulatory subunit, most commonly p85. The principal function of class I enzymes is to catalyze the conversion of phosphatidylinositol 4,5-bisphosphate (PIP$_2$) to phosphatidylinositol 3,4,5-trisphosphate (PIP$_3$) (Vanhaesebroeck *et al.*, 1997). The reverse of this reaction, the conversion of PIP$_3$ to PIP$_2$, is catalyzed by the phosphatase PTEN (Maehama and Doxin, 1998).

PI3K-mediated signals regulate cell growth (Foukas *et al.*, 2006), motility (Vanhaesebroeck *et al.*, 1999), and metabolism (Hara *et al.*, 1994). Class I PI3K contains four isoforms, α, β, γ, and δ, which carry out nonredundant signaling functions (Hawkins *et al.*, 2006). The α and β isoforms are ubiquitously expressed; α is linked upstream mainly to receptor tyrosine kinase, whereas β can accept signals from both G-protein–coupled receptors and from receptor tyrosine kinases (Hooshmand-Rad *et al.*, 2000). The γ and δ isoforms are expressed primarily in lymphocytes and play important roles in the regulation of immune responses (Okkenhaug *et al.*, 2002; Sasaki *et al.*, 2000).

Feedback loops contribute to the regulation of the PI3K pathway. The downstream target of PI3K, S6 kinase (S6K), can phosphorylate the insulin receptor substrate-1 (IRS-1), thereby attenuating the signal from insulin to PI3K (Li *et al.*, 1999). An example of positive feedback in the pathway is the phosphorylation of Akt by the TOR-Rictor complex (Guertin *et al.*, 2006; Sarbassov *et al.*, 2006). PI3K signaling activates TOR, including the TOR-Rictor complex, which mediates the activating phosphorylation of Akt on serine 473 (Fig. 20.1).

2. CANCER LINK

A gain of function in PI3K signaling is common in many types of human cancer and includes amplification of *PIK3CA*, the gene encoding p110α, mutations in p110α, amplification or enhanced activation of Akt, and loss of PTEN function (Brader and Eccles, 2004; Downward, 2004; Parsons, 2004; Parsons and Simpson, 2003; Samuels and Ericson, 2006). Of particular interest are the cancer-specific mutations in p110α (Samuels and Velculescu, 2004; Samuels *et al.*, 2004), which result in elevated catalytic activity and confer *in vivo* and *in vitro* oncogenic activity onto p110α (Bader *et al.*, 2006; Ikenoue *et al.*, 2005; Isakoff *et al.*, 2005; Kang *et al.*, 2005a; Zhao *et al.*, 2005). This oncogenicity requires functional Akt and depends on active TOR kinase. Genetic changes firmly link the p110α isoform to the causation of cancer. But the nonalpha isoforms of p110 may also have oncogenic potential. The p110β isoform shows consistent elevated expression in colon and bladder carcinoma (Benistant *et al.*, 2000). The expression of the p110δ isoform is elevated in acute myeloid leukemia (Sujobert *et al.*, 2005) and in glioblastoma (Knobbe *et al.*, 2005; Mizoguchi *et al.*, 2004). Several reports have suggested a role for p110γ in the initiation and progression of chronic myeloid leukemia (Hickey and Cotter, 2006; Skorski *et al.*, 1997). These findings are of significance, because the wild-type proteins of the nonalpha isoforms of p110 induce oncogenic transformation in cell culture by mere overexpression (Kang *et al.*, 2006).

3. CANCER-SPECIFIC HOT-SPOT AND RARE MUTATIONS IN P110α

The cancer-specific mutations in p110α are single-nucleotide substitutions that lead to single–amino acid substitutions (Samuels *et al.*, 2004). Mutations in p110α occur with different frequencies, depending on cancer type.

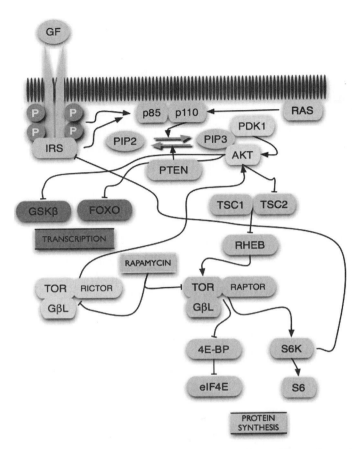

Figure 20.1 Example of canonical PI3K signaling pathway. The pathway is divided into four functionally distinct, color-coded modules. The basic enzymatic activity of PI3K is shown in green. Binding of an extracellular growth factor to a receptor tyrosine kinase leads to recruitment and activation of PI3K. Activated PI3K phosphorylates PIP_2 to PIP_3; the phosphatase PTEN catalyzes the reverse reaction. PIP_3 acts as a second messenger and initiates the signal by recruiting the serine threonine kinase Akt and its activating kinase, PDK1, shown in yellow. Akt is additionally activated by the TOR/rictor complex. Rapamycin is a specific inhibitor of TOR with a stronger effect on the TOR/raptor complex than on the TOR/rictor complex. Activated Akt controls gene transcription and translation, shown in magenta and orange, respectively. Akt phosphorylation negatively regulates the growth-attenuating targets FOXO and GSKβ. Activated Akt affects protein synthesis through TOR signaling. Akt disrupts the TSC1/TSC2 complex, resulting in activation of Rheb and of the TOR/raptor complex. This complex controls the initiation of protein synthesis by activating S6 kinase and inhibiting 4E-BP. Several of the PI3K signaling components have oncogenic potential: the p85 and p110 subunits of PI3K, Akt, Rheb, and eIF4E. The interactions of Ras and PI3K are described in the text. (See color insert.)

They are most common in cancers of the breast (Bachman *et al.*, 2004), colon (Frattini *et al.*, 2005) and endometrium (Oda *et al.*, 2005) (about 30% of the primary tumors in these tissues carry mutated p110α). In cancers of the liver (see catalog of somatic mutations, www.sanger.ac.uk/genetics/cgp/cosmic/), ovary (Wang *et al.*, 2005), and pancreas (Schonleben *et al.*, 2006), the frequency of mutated p110α is between 5 and 10%, and in cancers of the lung (Samuels *et al.*, 2004) and brain, (Broderick *et al.*, 2004) mutated p110α is rare. About 80% of the mutations occur in three hot spots, suggesting that they confer a selective advantage to the cell carrying the mutation. The hot-spot mutations in p110α are E542K, E545K, and H1047R (Samuels *et al.*, 2004) (Fig. 20.2).

The three-dimensional structure of p110α has not yet been determined, but a partial model of the protein can be built based on the known structure of p110γ (Walker *et al.*, 1999). The model defines several structural and functional domains of p110α. The N-terminal region of the protein forms a domain required for binding to the regulatory subunit p85, which is followed by the Ras binding domain and the C2 domain. The latter participates in the recruitment of p110α to the plasma membrane. Adjacent to the C2 domain are the helical and the C-terminal kinase domains. In this model, E542K and E545K map it to the helical domain, and H1047R is located in the kinase domain (Gymnopoulos *et al.*, 2007; Ikenoue *et al.*, 2005; Kang *et al.*, 2005a; Stephens *et al.*, 2005).

The mutations confer *in vitro* and *in vivo* oncogenic potential on p110α (Bader *et al.*, 2006; Ikenoue *et al.*, 2005; Isakoff *et al.*, 2005; Kang *et al.*, 2005a; Zhao *et al.*, 2005). The mutated proteins also increase enzymatic activity and constitutively activate downstream signaling components, including Akt, TOR, S6K, and 4E-BP. The requirement for TOR activation

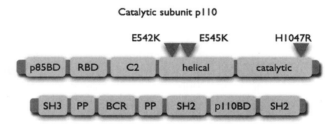

Figure 20.2   Domain structure of catalytic subunit p110 and regulatory subunit p85 of PI3K. The catalytic subunit contains the p85-binding domain (p85BD), Ras-binding domain (RBD), C2 domain (C2), helical domain, and catalytic domain. The three hot-spot mutations identified in human cancer are marked. The regulatory subunit contains three src homology domains (two SH2 and one SH3), polyproline regions (PP), BCR homology domain (BCR), and p110-binding domain (p110BD). Dimensions of the domains are not to scale. (See color insert.)

is reflected in the sensitivity of mutant-induced oncogenic transformation to rapamycin. Several cell culture systems have been used to detect the oncogenic activity of p110α mutants. Human mammary epithelial cells are susceptible to transformation by gain-of-function p110α. These are spontaneously immortalized cells, such as the MCF-10A cell line, or immortalized with hTERT (human telomerase reverse transcriptase); in addition, they have impaired function of p53 and enhanced expression of Myc (Isakoff *et al.*, 2005; Zhao *et al.*, 2005). These cells can be stably transfected with retroviral expression vectors of p110α; expression of the hot-spot mutants induces morphological changes, growth-factor independence, anchorage-independent growth, and resistance to anoikis. NIH-3T3 mouse fibroblasts similarly become transformed after stable transfection with p110α mutant-expressing vectors (Ikenoue *et al.*, 2005). They show changes in growth pattern and acquire anchorage independence. Primary chicken embryo fibroblasts (CEF) are highly susceptible to the oncogenicity of mutant p110α (Gymnopoulos *et al.*, 2007; Kang *et al.*, 2005a). In this cell culture system, stably transfected cells are generated with the replicating retroviral expression vector RCAS; they do not have to be selected with a drug-resistance marker. As a consequence of this direct transformation of primary cells by a single oncogenic protein, the RCAS-mediated transfection and infection assays in CEF are quantitative, and can be used to compare relative oncogenic potencies of various mutants.

In addition to the hot-spot mutations numerous rare cancer-specific mutations are widely distributed over the coding sequence of p110α (Bader *et al.*, 2005; Samuels *et al.*, 2004). Fifteen of these have been studied, and all except one show a gain of function (Gymnopoulos *et al.*, 2007). However, the gain induced by the rare mutations is smaller than that seen with the hot-spot mutations. The difference is evident in quantitative measurements of oncogenic transformation. Hot-spot mutations generate significantly more transformed cell foci per unit of transfected DNA than rare mutations. In other signaling properties, all cancer-specific mutants are similar: they induce anchorage-independent growth, cause constitutive phosphorylation of Akt and S6K, and their oncogenicity depends on TOR function (Bader *et al.*, 2006; Gymnopoulos *et al.*, 2007; Ikenoue *et al.*, 2005; Isakoff *et al.*, 2005; Kang *et al.*, 2005a; Zhao *et al.*, 2005). Mapping of hot-spot and rare cancer-specific mutations on the modeled structure of p110α suggests at least three different molecular mechanisms for the gain of function (Gymnopoulos *et al.*, 2007).

First, mutations in the C2 domain are located on a surface patch that interacts with the plasma membrane. These mutations substitute acidic residues with basic residues. This change of p110α surface properties is likely to enhance the affinity of the protein for lipid membranes. By increasing recruitment to the plasma membrane, the C2 domain mutations would bring p110α into the immediate proximity of its substrates and thus enhance PI3K activity. Second, mutations in the helical domain are also located on the protein surface and delineate a congruent area that could mediate interaction with other proteins or other domains of p110α. The amino acid substitutions in this domain again tend from acidic or neutral to basic.

Third, the mutations in the kinase domain pack closely against the hinge region of the activation loop. They could affect the movement or position of the activation loop, perhaps by locking the activation loop in the "on" position. The gain-of-function mutations derived from cancer tissues stand in contrast to mutations that have been arbitrarily introduced into p110α. Four such mutants show no phenotype (Gymnopoulos et al., 2007). Their map positions are distant from the critical domains defined by the gain-of-function mutations or they are located in the interior of the protein. However, the fact that many of the gain-of-function mutations map to the protein surface and increase the positive charge can be used to construct targeted mutations that satisfy these putative criteria for increased function (Gymnopoulos et al., 2007). Two out of three mutations constructed along these lines activate the oncogenic potential of p110α in cell culture but this gain of function is not associated with constitutive activation of Akt.

The cancer-specific mutants of p110α are attractive therapeutic targets (Kang et al., 2005b; Samuels and Ericson, 2006; Samuels and Velculescu, 2004). Mutant-specific small molecule inhibitors of p110α could have anti-cancer activity without causing the side effects that are expected from interference with normal physiological functions of PI3K signaling. The structural changes resulting from a single amino acid substitution could be significant, affecting conformation or, as in the case of the kinase domain mutations, changing the position of the activation loop. These changes could be exploited to generate mutant-specific inhibitors. Such efforts would be greatly aided by structural information on wildtype and mutant p110α.

4. ONCOGENICITY OF NONALPHA ISOFORMS

Cancer-specific mutations have been found only in the α isoform of class I PI3K. Signaling through p110α is coupled to input from receptor tyrosine kinases, and p110α is the isoform most active in the control of cell growth and replication (Fan et al., 2006; Foukas et al., 2006). Yet the wildtype nonalpha isoforms of p110 have an oncogenic potential that is revealed by overexpression. Overexpressed wildtype p110β, γ and δ induce oncogenic transformation in cell culture; in contrast, wild-type p110α lacks this ability (Kang et al., 2006). The transforming activity of p110β and p110γ requires binding to Ras, and is not associated with constitutive phosphorylation of Akt. In contrast, oncogenic transformation induced by p110δ and the myristylated, constitutively active form of p110α is independent of Ras binding and results in constitutive phosphorylation of Akt. Despite these differences in Akt signaling, the oncogenicity of all p110 isoforms is highly sensitive to rapamycin and hence dependent on TOR function. The observed overexpression of wild-type p110β and p110δ in colon and acute myeloid leukemia, respectively, suggests that the nonalpha class I isoforms may therefore be of importance in determining the oncogenic cellular phenotype in these cancer (Benistant *et al.*, 2000; Sujobert *et al.*, 2005).

5. FUNCTIONAL IMPORTANCE OF PIP₃

Class I PI3Ks can utilize the nonphosphorylated phosphatidylinositol (PI), as well as the monophosphate (PI(4)P) and the biphosphate (PI(4,5)P$_2$), also referred to as PIP$_2$, as substrates in phosphorylation reactions giving rise to PI(3)P, PI(3,4)P$_2$, and PI(3,4,5)P$_3$, also referred to as PIP$_3$. Cellular levels of PIP$_3$ increase sharply after stimulation with growth factors, and this increase results in the activation of Akt and its downstream targets (Auger *et al.*, 1989). PIP$_3$ recruits Akt and its primary activating kinase, PDK1, to the cell membrane, and this proximity of enzyme and substrate results in the activating phosphorylation of Akt at threonine 308 (Andjelkovic *et al.*, 1997; Bellacosa *et al.*, 1998; Watton and Downward, 1999).

A burning question in PI3K cancer biology is which phospholipid product mediates the oncogenic signal of mutant p110α and of wildtype nonalpha p110 proteins. To answer this question, we took two complementary approaches. First, we created mutants of p110γ that are restricted in their substrate usage and hence generate only a limited set of phospholipid products as has been described previously (Bondeva *et al.*, 1998). Only a mutant that could produce PIP$_3$ showed oncogenic activity (Denley *et al.*, unpublished data, 2007). Second, PIP$_3$ can be converted into PI(3,4)P$_2$ by an inositol 5-phosphatase such as SHIP (SH2-containing inositol phosphatase) Damen *et al.*, 1996; Lioubin *et al.*, 1996) or PIPP (proline-rich inositol phosphatase) (Ooms *et al.*, 2006). Expression of PIPP induces a specific cellular resistance to transformation by all four isoforms of class I p110, suggesting that PIP$_3$ is required for oncogenic signaling (Denley *et al.*, unpublished, 2007).

The cellular levels of PIP$_3$ are also reduced by the inositol 3-phosphatase PTEN (Maehama and Dixon, 1998). Loss-of-function mutations in PTEN are observed in various malignancies, resulting in elevated levels of PIP$_3$ (Li *et al.*, 1997; Liaw *et al.*, 1997; Steck *et al.*, 1997). The generation of PIP$_3$ is likely to be an important requirement in oncogenic PI3K signaling. Only class I PI3Ks have the ability to produce this critical second messenger molecule.

6. Involvement of Ras in PI3K Signaling

Ras is a monomeric GTPase that regulates a variety of pathways including Raf-MAPK, RalGEF-Ral, and the PI3K-Akt pathways (Repasky et al., 2004). Ras is located at the plasma membrane by virtue of a C-terminal farnesyl moiety (Casey et al., 1989). All class I PI3K isoforms can interact with Ras-GTP, and all isoforms of Ras can interact with and activate PI3K (Rodriguez-Viciana et al., 2004; Suire et al., 2002). Ras serves two functions in PI3K signaling, illustrated by data on p110γ. Ras can induce a GTP-dependent translocation of p110γ to the plasma membrane; recruitment to the plasma membrane brings PI3K in the proximity of its substrates and thereby increases PI3K activity. Ras can also directly activate the kinase function of p110γ by binding to the enzyme and inducing a conformational change (Chan et al., 2002; Pacold et al., 2000). Each class I p110 isoform has a Ras binding domain (RBD), and disruption of certain residues critical for interaction with Ras completely abrogates Ras binding (Bondeva et al., 1998; Kang et al., 2006; Rodriguez-Viciana et al., 1996). Studies with mutants in the RBD of the four class I P110 isoforms show that oncogenic transformation by p110β and p110γ requires interaction with Ras, whereas p110δ and the cancer-specific mutant H1047R of p110α are oncogenic in the absence of Ras binding (Kang et al., 2006). In contrast to the Ras independence of the H1047R mutant, which maps to the kinase domain of p110α, the cancer-derived E542K and E545K mutants, located in the helical domain, are significantly compromised in their oncogenicity in the absence of Ras binding (Zhao et al., 2008). Mutations in the helical domain and in the kinase domain probably induce a gain of function by different molecular mechanisms that are reflected in different levels of dependence on Ras. Although gain-of-function mutations in p110α are widely distributed over the coding sequence of this protein, no mutations have been found within the Ras-binding domain, suggesting that mutations in this domain may lead to a loss of function (Forbes et al., 2006). Additional indications for a role of Ras in p110-induced oncogenic transformation come from experiments with dominant negative H-Ras (N17). Dominant negative H-Ras (N17) interferes with the oncogenic activities of p110β, p110γ, and the cancer-derived E542K, E545K, and H1047R mutants of p110α. However, wildtype p110δ and the constitutively active myristylated p110α can at least partially overcome this cellular resistance (Zhao et al., unpublished, 2007). Sequencing of the cancer genome has revealed that 30% of the cancer-specific mutations in p110α occur together with mutations in KRAS, BRAF, and EGFR (Thomas et al., 2007). The PI3K and MAPK pathway may therefore have complementary, nonredundant functions, and hyperactivation of both pathways may be important in tumor formation. Previous studies have shown that activation of Raf alone is not sufficient in mediating the oncogenic properties of Ras, but specific roles for the MAPK and the PI3K pathways have not been defined (Repasky *et al.*, 2004). Expression of the RBD mutants of p110γ and p110β has no effect on MAPK signaling, suggesting that the interaction of Ras and p110 affects the activation of targets downstream of PI3K, not downstream of Ras (Denley *et al.*, 2007). In the human breast epithelial cell line MCF-10A, expression of the cancer-specific mutants H1047R and E545K of p110α induces constitutively elevated phosphorylation of Erk (Isakoff *et al.*, 2005). This effect could result from an activation of Ras by PIP$_3$; PIP$_3$ can stimulate the GEF activity of the Ras SOS (Nimnual *et al.*, 1998).

These observations on possible crosstalk between PI3K and Ras signaling are still fragmentary, but are indicative of important interactions that require further study.

7. ONCOGENICITY OF RHEB

Rheb is a small GTPase that connects Akt and the tuberous sclerosis complex proteins TSC1/TSC2 to TOR (Garami *et al.*, 2003; Inoki *et al.*, 2003; Tee *et al.*, 2003; Zhang *et al.*, 2003). Rheb binds to TOR and activates TOR in a GTP-dependent manner (Long *et al.*, 2005). Expression of wildtype Rheb increases the size of *Drosophila* cells, but has no significant effect on the morphology or growth of vertebrate cells (Clark *et al.*, 1997; Saucedo *et al.*, 2003; Stocker *et al.*, 2003). A mutant of Rheb, Q64L, is similar to the Ras mutant Q61L, keeping Rheb charged with GTP and constitutively active. As measured by the kinase activity of its downstream target TOR, Rheb Q64L shows a gain of function compared to the wildtype protein (Inoki *et al.*, 2003; Li *et al.*, 2004; Long *et al.*, 2005). Expression of the mutant Rheb Q64L in primary avian fibroblasts induces robust oncogenic transformation evident in the appearance of focal tumors in the cell monolayer and induction of anchorage-independent growth (Jiang and Vogt, unpublished, 2007). The transformed cells show a 25% increase in volume and a 30% increase in protein content as compared to normal fibroblasts. The oncogenic transformation induced by Rheb Q64L is correlated with constitutive phosphorylation of S6K and 4E-BP and is highly sensitive to rapamycin, in accord with the central role played by TOR in PI3K signaling. The oncogenic activity of constitutively active Rheb is readily demonstrable in cultures of CEF using the replication-competent retroviral expression vector RCAS. In contrast, the NIH3T3 murine cell line, the standard system for detecting oncogencitiy, is not susceptible to the oncogenic effects of activated Rheb (Clark *et al.*, 1997). The molecular basis for this difference is not known. High levels of expression, mediated by the RCAS vector may be a contributing factor.

However, avian cells are generally more sensitive indicators of oncogenic activity than mammalian cells. Unlike primary mammalian cells, primary avain cells can be transformed by single oncoproteins. The oncogenic potential of Rheb is significant in view of recent observations that show consistent overexpression of Rheb in brain cancer (Momota and Holland, personal communication, 2007).

Like other members of the Ras family, Rheb contains a C-terminal farnesylation motif. Farnesylation of Rheb is essential for cell cycle progression in *Schizosaccharomyces pombe* (Yang *et al.*, 2001), and is also required for arginine uptake in *Saccharomyces cerevisiae* (Urano *et al.*, 2000). In mammalian cells, loss of farnesylation markedly reduces the ability of Rheb to induce TOR-dependent phosphorylation of S6K1 (Castro *et al.*, 2003; Tee *et al.*, 2003). Mutations in Rheb Q64L that inactivate the farnesylation signal also abolish the oncogenic activity and constitutive phosphorylation of S6K and 4E-BP (Jiang and Vogt, unpublished, 2007). These activities can be restored in the mutants with an N-terminal myristylation signal, demonstrating that the important function of farnesylation is in mediating the localization of Rheb to the cell membrane and that this localization is essential for Rheb activity.

8. CONCLUSION

PI3K has moved from being an obscure lipid kinase to one of the most important signaling molecules in cancer. The catalytic subunit of PI3K is itself a promising drug target, and the cancer-specific mutations in the protein offer the possibility of devising a molecularly targeted therapy. Additional attractive features of p110α as a drug target include the facts that it is an enzyme, thus belonging to a category of proteins that is readily affected by small molecule inhibitors, and that it undergoes a gain of function. It is far easier to attenuate excess activity than to restore lost activity of protein. The singaling pathways emerging from PI3K are multiple and complex, involving numerous phosphorylation events and protein–protein interactions that are only incompletely understood. It is likely that this signaling network contains additional targets for effective therapeutic intervention in cancer.

While this review was in press, the structure of p110a becomes available (Hung *et al.*, 2007). It differs significantly from the model but the main conclusions concerning the mutants are also in accord with the newly determined structure.

ACKNOWLEDGMENTS

This work is supported by grants from the National Cancer Institute and the Stein Fund. This is manuscript number 18855 of the Scripps Research Institute. We thank Sohye Kang for a critical reading of the manuscript and for numerous valuable suggestions.

REFERENCES

Andjelkovic, M., *et al.* (1997). Role of translocation in the activation and function of protein kinase B. *J. Biol. Chem.* **272,** 31515–31524.

Auger, K. R., *et al.* (1989). PDGF-dependent tyrosine phosphorylation stimulates production of novel polyphosphoinositides in intact cells. *Cell* **57,** 167–175.

Bachman, K. E., *et al.* (2004). The PIK3CA gene is mutated with high frequency in human breast cancers. *Cancer Biol. Ther.* **3,** 772–775.

Bader, A. G., *et al.* (2005). Oncogenic PI3K deregulates transcription and translation. *Nat. Rev.* **5,** 921–929.

Bader, A. G., *et al.* (2006). Cancer-specific mutations in PIK3CA are oncogenic *in vivo. Proc. Natl. Acad. Sci. USA* **103,** 1475–1479.

Bellacosa, A., *et al.* (1998). Akt activation by growth factors is a multiple-step process: The role of the PH domain. *Oncogene* **17,** 313–325.

Benistant, C., *et al.* (2000). A specific function for phosphatidylinositol 3-kinase alpha (p85alpha-p110alpha) in cell survival and for phosphatidylinositol 3-kinase beta (p85alpha-p110beta) in *de novo* DNA synthesis of human colon carcinoma cells. *Oncogene* **19,** 5083–5090.

Bondeva, T., *et al.* (1998). Bifurcation of lipid and protein kinase signals of PI3Kgamma to the protein kinases PKB and MAPK. *Science* **282,** 293–296.

Brader, S., and Eccles, S. A. (2004). Phosphoinositide 3-kinase signalling pathways in tumor progression, invasion and angiogenesis. *Tumori* **90,** 2–8.

Broderick, D. K., *et al.* (2004). Mutations of PIK3CA in anaplastic oligodendrogliomas, high-grade astrocytomas, and medulloblastomas. *Cancer Res.* **64,** 5048–5050.

Casey, P. J., *et al.* (1989). p21ras is modified by a farnesyl isoprenoid. *Proc. Natl. Acad. Sci. USA* **86,** 8323–8327.

Castro, A. F., *et al.* (2003). Rheb binds tuberous sclerosis complex 2 (TSC2) and promotes S6 kinase activation in a rapamycin- and farnesylation-dependent manner. *J. Biol. Chem.* **278,** 32493–32496.

Chan, T. O., *et al.* (2002). Small GTPases and tyrosine kinases coregulate a molecular switch in the phosphoinositide 3-kinase regulatory subunit. *Cancer Cell* **1,** 181–191.

Clark, G. J., *et al.* (1997). The Ras-related protein Rheb is farnesylated and antagonizes Ras signaling and transformation. *J. Biol. Chem.* **272,** 10608–10615.

Damen, J. E., *et al.* (1996). The 145-kDa protein induced to associate with Shc by multiple cytokines is an inositol tetraphosphate and phosphatidylinositol 3,4,5-triphosphate 5-phosphatase. *Proc. Natl. Acad. Sci. USA* **93,** 1689–1693.

Denley, A., *et al.* (2007). Oncogenic signaling of class I PI3K isoforms. *Oncogene.*

Downward, J. (2004). PI 3-kinase, Akt and cell survival. *Semin. Cell Dev. Biol.* **15,** 177–182.

Fan, Q. W., *et al.* (2006). A dual PI3 kinase/mTOR inhibitor reveals emergent efficacy in glioma. *Cancer Cell* **9,** 341–349.

Forbes, S., *et al.* (2006). COSMIC 2005. *Br. J. Cancer* **94,** 318–322.

Foukas, L. C., *et al.* (2006). Critical role for the p110alpha phosphoinositide-3-OH kinase in growth and metabolic regulation. *Nature* **441,** 366–370.

Frattini, M., *et al.* (2005). Phosphatase protein homologue to tensin expression and phosphatidylinositol-3 phosphate kinase mutations in colorectal cancer. *Cancer Res.* **65,** 112–127.

Garami, A., *et al.* (2003). Insulin activation of Rheb, a mediator of mTOR/S6K/4E-BP signaling, is inhibited by TSC 1 and 2. *Mol. Cell* **11,** 1457–1466.

Guertin, D.A, *et al.* (2006). Ablation in mice of the mTORC components raptor, rictor, or mLST8 reveals that mTORC2 is required for signaling to Akt-FOXO and PKCalpha, but not S6K1. *Dev. Cell* **11,** 859–871.

Gymnopoulos, M., et al. (2007). Rare cancer-specific mutations in PIK3CA show gain of function. Proc. Natl. Acad. Sci. USA **104**, 5569–4474.

Hara, K., et al. (1994). 1-Phosphatidylinositol 3-kinase activity is required for insulin-stimulated glucose transport but not for RAS activation in CHO cells. Proc. Natl. Acad. Sci. USA **91**, 7415–7419.

Hawkins, P. T., et al. (2006). Signalling through class I PI3Ks in mammalian cells. Biochem. Soc. Trans. **34**, 647–662.

Hickey, F. B., and Cotter, T. G. (2006). BCR-ABL regulates phosphatidylinositol 3-kinase-p110gamma transcription and activation and is required for proliferation and drug resistance. J. Biol. Chem. **281**, 2441–2450. (Epub 2005 Nov 16.)

Huang, C. H., et al. (2007). The structure of a human p110alpha/p85alpha complex elucidates the effects of oncogenic PI3Kalpha mutations. Science **318**, 1744-1748.

Hooshmand-Rad, R., et al. (2000). The PI 3-kinase isoforms p110(alpha) and p110(beta) have differential roles in PDGF- and insulin-mediated signaling. J. Cell Sci. **113**(Pt 2), 207–214.

Ikenoue, T., et al. (2005). Functional analysis of PIK3CA gene mutations in human colorectal cancer. Cancer Res. **65**, 4562–4567.

Inoki, K., et al. (2003). Rheb GTPase is a direct target of TSC2 GAP activity and regulates mTOR signaling. Genes Dev. **17**, 1829–1834.

Isakoff, S. J., et al. (2005). Breast cancer-associated PIK3CA mutations are oncogenic in mammary epithelial cells. Cancer Res. **65**, 10992–11000.

Kang, S., et al. (2005a). Phosphatidylinositol 3-kinase mutations identified in human cancer are oncogenic. Proc. Natl. Acad. Sci. USA **102**, 802–807(Epub 2005 Jan 12.)

Kang, S., et al. (2005b). Mutated PI 3-kinases: Cancer targets on a silver platter. Cell Cycle **4**, 578–581.

Kang, S., et al. (2006). Oncogenic transformation induced by the p11 Obeta, -gamma, and -delta isoforms of class I phosphoinositide 3-kinase. Proc. Natl. Acad. Sci. USA **103**, 1289–1294.

Knobbe, C. B., et al. (2005). Genetic alteration and expression of the phosphoinositol-3-kinase/Akt pathway genes PIK3CA and PIKE in human glioblastomas. Neuropathol. Appl. Neurobiol. **31**, 486–490.

Li, J., et al. (1997). PTEN, a putative protein tyrosine phosphatase gene mutated in human brain, breast, and prostate cancer. Science **275**, 1943–1947.

Li, J., et al. (1999). Modulation of insulin receptor substrate-1 tyrosine phosphorylation by an AkUphosphatidylinositol 3-kinase pathway. J. Biol. Chem. **274**, 9351–9356.

Liaw, D., et al. (1997). Germline mutations of the PTEN gene in Cowden disease, an inherited breast and thyroid cancer syndrome. Nat. Genet. **16**, 64–67.

Lioubin, M. N., et al. (1996). p150Ship, a signal transduction molecule with inositol polyphosphate-5-phosphatase activity. Genes Dev. **10**, 1084–1095.

Long, X., et al. (2005). Rheb binds and regulates the mTOR kinase. Curr. Biol. **15**, 702–713.

Maehama, T., and Dixon, J. E. (1998). The tumor suppressor, PTEN/MMAC1, dephosphorylates the lipid second messenger, phosphatidylinositol 3,4,5-trisphosphate. J. Biol. Chem. **273**, 13375–13378.

Mizoguchi, M., et al. (2004). Genetic alterations of phosphoinositide 3-kinase subunit genes in human glioblastomas. Brain Pathol. (Zurich) **14**, 372–377.

Nimnual, A. S., et al. (1998). Coupling of Ras and Rac guanosine triphosphatases through the Ras exchanger Sos. Science **279**, 560–563.

Oda, K., et al. (2005). High frequency of coexistent mutations of PIK3CA and PTEN genes in endometrial carcinoma. Cancer Res. **65**, 10669–10673.

Okkenhaug, K., et al. (2002). Impaired B and T cell antigen receptor signaling in p110delta PI 3-kinase mutant mice. Science **297**, 1031–1034.

Ooms, L. M., *et al.* (2006). The inositol polyphosphate 5-phosphatase, PIPP, is a novel regulator of phosphoinositide 3-kinase-dependent neurite elongation. *Mol. Biol. Cell* **17**, 607–622. (Epub 2005 Nov 9.)

Pacold, M. E., *et al.* (2000). Crystal structure and functional analysis of Ras binding to its effector phosphoinositide 3-kinase gamma. *Cell* **103**, 931–943.

Parsons, R. (2004). Human cancer, PTEN and the PI-3 kinase pathway. *Semin. Cell Dev. Biol.* **15**, 171–176.

Parsons, R., and Simpson, L. (2003). PTEN and cancer. *Methods Mol. Biol.* **222**, 147–166.

Repasky, G. A., *et al.* (2004). Renewing the conspiracy theory debate: Does Raf function alone to mediate Ras oncogenesis? *Trends Cell Biol.* **14**, 639–647.

Rodriguez-Viciana, P., *et al.* (1996). Activation of phosphoinositide 3-kinase by interaction with Ras and by point mutation. *EMBO J.* **15**, 2442–2451.

Rodriguez-Viciana, P., *et al.* (2004). Signaling specificity by Ras family GTPases is determined by the full spectrum of effectors they regulate. *Mol. Cell Biol.* **24**, 4943–4954.

Samuels, Y., and Ericson, K. (2006). Oncogenic PI3K and its role in cancer. *Curr. Opin. Oncol.* **18**, 77–82.

Samuels, Y., and Velculescu, V. E. (2004). Oncogenic mutations of PIK3CA in human cancers. *Cell Cycle* **3**, 1221–1224.

Samuels, Y., *et al.* (2004). High frequency of mutations of the PIK3CA gene in human cancers. *Science* **304**, 554.

Sarbassov, D. D., *et al.* (2006). Prolonged rapamycin treatment inhibits mTORC2 assembly and Akt/PKB. *Mol. Cell* **22**, 159–168.

Sasaki, T., *et al.* (2000). Function of PI3Kgamma in thymocyte development, T cell activation, and neutrophil migration. *Science* **287**, 1040–1046.

Saucedo, L. J., *et al.* (2003). Rheb promotes cell growth as a component of the insulin/TOR signalling network. *Nat. Cell Biol.* **5**, 566–571.

Schonleben, F., *et al.* (2006). PIK3CA mutations in intraductal papillary mucinous neoplasm/carcinoma of the pancreas. *Clin. Cancer Res.* **12**, 3851–3855.

Skorski, T., *et al.* (1997). Transformation of hematopoietic cells by BCR/ABL requires activation of a PI-3k/Akt-dependent pathway. *EMBO J.* **16**, 6151–6161.

Steck, P. A., *et al.* (1997). Identification of a candidate tumour suppressor gene, MMAC 1, at chromosome 10q23.3 that is mutated in multiple advanced cancers. *Nat. Genet.* **15**, 356–362.

Stephens, L., *et al.* (2005). Phosphoinositide 3-kinases as drug targets in cancer. *Curr. Opin. Pharmacol.* **5**, 357–365.

Stocker, H., *et al.* (2003). Rheb is an essential regulator of S6K in controlling cell growth in *Drosophila*. *Nat. Cell Biol.* **5**, 559–565.

Suire, S., *et al.* (2002). Activation of phosphoinositide 3-kinase gamma by Ras. *Curr. Biol.* **12**, 1068–1075.

Sujobert, P., *et al.* (2005). Essential role for the p110delta isoform in phosphoinositide 3-kinase activation and cell proliferation in acute myeloid leukemia. *Blood* **106**, 1063–1066. (Epub 2005 Apr 19.)

Tee, A. R., *et al.* (2003). Tuberous sclerosis complex gene products, Tuberin and Hamartin, control mTOR signaling by acting as a GTPase-activating protein complex toward Rheb. *Curr. Biol.* **13**, 1259–1268.

Thomas, R. K., *et al.* (2007). High-throughput oncogene mutation profiling in human cancer. *Nat. Genet* **39**, 347–351. (Epub 2007 Feb 11.)

Urano, J., *et al.* (2000). The *Saccharomyces cerevisiae* Rheb G-protein is involved in regulating canavanine resistance and arginine uptake. *J. Biol. Chem.* **275**, 11198–11206.

Vanhaesebroeck, B., and Waterfield, M. D. (1999). Signaling by distinct classes of phosphoinositide 3-kinases. *Exp. Cell Res.* **253**, 239–254.

Vanhaesebroeck, B., *et al.* (1997). Phosphoinositide 3-kinases: A conserved family of signal transducers. *Trends Biochem. Sci.* **22,** 267–272.

Vanhaesebroeck, B., *et al.* (1999). Distinct PI(3)Ks mediate mitogenic signalling and cell migration in macro phages. *Nat. Cell Biol.* **1,** 69–71.

Walker, E. H., *et al.* (1999). Structural insights into phosphoinositide 3-kinase catalysis and signalling. *Nature* **402,** 313–320.

Wang, Y., *et al.* (2005). PIK3CA mutations in advanced ovarian carcinomas. *Hum. Mutat.* **25,** 322.

Watton, S. J., and Downward, J. (1999). AktlPKB localisation and 3' phosphoinositide generation at sites of epithelial cell–matrix and cell–cell interaction. *Curr. Biol.* **9,** 433–436.

Yang, W., *et al.* (2001). Failure to farnesylate Rheb protein contributes to the enrichment of GO/G1 phase cells in the *Schizosaccharomyces pombe* farnesyltransferase mutant. *Mol. Microbiol.* **41,** 1339–1347.

Zhang, Y., *et al.* (2003). Rheb is a direct target of the tuberous sclerosis tumour suppressor proteins. *Nat. Cell Biol.* **5,** 578–581.

Zhao, J. J., *et al.* (2005). The oncogenic properties of mutant p110alpha and p110beta phosphatidylinositol 3-kinases in human mammary epithelial cells. *Proc. Natl. Acad. Sci. USA* **102,** 18443–18448.

Zhao, L., and Vogt, P. K. (2008). Helical domain and kinase domain mutations in p110a of phosphatidylinositol 3-kinase induce gain of function by different mechanisms. *Proc. Natl. Acad. Sci. USA* (in press).

CHAPTER TWENTY-ONE

CHARACTERIZATION OF THE RHEB-mTOR SIGNALING PATHWAY IN MAMMALIAN CELLS: CONSTITUTIVE ACTIVE MUTANTS OF RHEB AND mTOR

Tatsuhiro Sato, Akiko Umetsu, *and* Fuyuhiko Tamanoi

Contents

Abstract

Rheb (Ras homolog enriched in brain) is a GTPase conserved from yeast to human and belongs to a unique family within the Ras superfamily of GTPases. Rheb plays critical roles in the activation of mTOR, a serine/threonine kinase that is involved in the activation of protein synthesis and growth. mTOR forms two distinct complexes, mTORC1 and mTORC2. While mTORC1 is implicated in the regulation of cell growth, proliferation, and cell size in response to amino acids and growth factors, mTORC2 is involved in actin organization. However, the mechanism of activation is not fully understood. Therefore, studies to elucidate the Rheb-mTOR signaling pathway are of great importance. Here we describe methods to characterize this pathway and to evaluate constitutive active mutants of Rheb and mTOR that we recently identified. Constitutive activity of the mutants can be demonstrated by the phosphorylation of ribosomal protein S6 kinase 1 (S6K1) and eIF4E-binding protein 1 (4E-BP1) both

Department of Microbiology, Immunology and Molecular Genetics, Jonsson Comprehensive Cancer Center, Molecular Biology Institute, University of California, Los Angeles, California

Methods in Enzymology, Volume 438
ISSN 0076-6879, DOI: 10.1016/S0076-6879(07)38021-X

in vivo and *in vitro* after starving cells for amino acids and growth factors. In addition, formation and activity of mTORC1 and mTORC2 can be measured by immunoprecipitating these complexes and carrying out *in vitro* kinase assays. We also describe a protocol for rapamycin treatment, which directly inhibits mTOR and can be used to investigate the mTOR signaling pathway in cell growth, cell size, etc.

1. INTRODUCTION

Small GTPases bind guanine nucleotides and serve as a molecular switch to regulate a number of physiological processes such as cell growth and morphology (Bourne *et al.*, 1990). Rheb, a small GTPase that belongs to a unique family within the Ras superfamily of GTPases, controls cell growth and proliferation as well as cell size (Aspuria and Tamanoi, 2004; Patel *et al.*, 2003; Yamagata *et al.*, 1994; Yu *et al.*, 2005). Unlike most small GTPases that are predominantly in an inactive GDP bound state, Rheb exists in a high activated state (Im *et al.*, 2002), presumably due to a low intrinsic GTPase activity as well as to a limiting amount of Tsc1/Tsc2 GAP protein inside the cell.

Regulation of Rheb is catalyzed by tuberous sclerosis 2 (Tsc2), which acts as a GTPase activating protein (GAP) that enhances the hydrolysis of GTP to GDP in Rheb (Castro *et al.*, 2003; Garami *et al.*, 2003; Inoki *et al.*, 2003; Tee *et al.*, 2003; Zhang *et al.*, 2003). Tsc2 forms a complex with tuberous sclerosis 1 (Tsc1) and directly inhibits the Rheb activation. It has been found that the negative regulation of Rheb by the Tsc1/2 complex is controlled by insulin. Insulin binding to its receptor triggers the activation of the class I PI3-kinase/Akt pathway. The activated Akt then increases Tsc2 phosphorylation at serine 939 and 981 (Cai *et al.*, 2006), leading to the dissociation of the Tsc1/2 complex.

Recent studies suggest that Rheb is involved in the activation of mTOR, a serine/threonine kinase that belongs to the family of PI3-kinase–related kinases. This family of kinases shares common features that include the presence of the HEAT domain, FAT domain, kinase domain, and FATC domain (Abraham, 2004). In addition, TOR kinases contain the FRB domain where FKBP/rapamycin complex binds. mTOR acts as a central protein that controls cell growth and proliferation through transcriptional and translational mechanisms in response to amino acids and growth factors such as insulin. However, amino acids and insulin use two distinct pathways to activate mTOR. Vps34, a class III PI3-kinase, but not class I PI3-kinase, is activated by amino acid stimulation (Byfield *et al.*, 2005;

Nobukuni *et al.*, 2005). On the other hand, growth factors activate the class I PI3-kinase/Akt signaling pathway, which then inactivates Rheb GAP, Tsc1/2, as described previously (Gao and Pan, 2001; Inoki *et al.*, 2002). It has been reported that mTOR forms two distinct complexes, which respond to amino acids or insulin (Jacinto *et al.*, 2004; Sarbassov *et al.*, 2004). mTOR complex 1 (mTORC1) is rapamycin sensitive and contains Raptor, GβL/mlST8, and PRAS40 (Hara *et al.*, 2002; Sabatini, 2006). This complex phosphorylates S6K1 and 4E-BP1, and plays an essential role in the regulation of cell growth and proliferation (Kim *et al.*, 2002, 2003). Within the complex, Raptor acts as a scaffold protein that connects mTOR to its substrates. PRAS40 is a negative regulator of mTOR that is affected by insulin (Haar *et al.*, 2007; Sancak *et al.*, 2007). GβL is also involved in mTORC1 activity, but its mechanism remains to be elucidated (Kim *et al.*, 2003). On the other hand, mTOR complex 2 (mTORC2) is relatively rapamycin insensitive and contains Rictor, Sin1, and GβL (Frias *et al.*, 2006; Jacinto *et al.*, 2004; Sarbassov *et al.*, 2004; Yang *et al.*, 2006). mTORC2 is involved in actin organization and cell survival, and mediates insulin signal to Akt by the phosphorylation at serine 473 (Hresko and Mueckler, 2005; Jacinto *et al.*, 2006; Sarbassov *et al.*, 2004; Sarbassov *et al.*, 2005; Yang *et al.*, 2006). Interestingly, inhibition of mTORC2 decreases the phosphorylation level of Akt substrates, forkhead transcription factor, FOXO1/3 proteins, but not other Akt substrates such as Tsc2 and GSK3β, suggesting that mTORC2 preferentially affects downstream events mediated by Akt (Guertin *et al.*, 2006; Jacinto *et al.*, 2006).

Our genetic analysis of the Tsc/Rheb/TOR signaling pathway in fission yeast led to the identification of novel Rheb and TOR mutants. In the case of Rheb, we first developed screening assays to identify active Rheb mutants in yeast. Screening of a random mutant library of Rheb identified a number of yeast Rheb mutants that showed phenotypes similar to those exhibited by the cells lacking the Tsc1/2 complex, which negatively regulates Rheb (Urano *et al.*, 2005). Comparison of Rheb sequences from different organisms led to the identification of other active mutants of human Rheb (Yan *et al.*, 2006). These mutants will be valuable in elucidating Rheb function and the activation mechanism for the Rheb-mTOR signaling. Constitutive active mutants of Tor2p have also been identified from the analysis of fission yeast signaling (Urano *et al.*, 2007). Altogether, 22 single amino acid changes have been identified in Tor2p. Introduction of some of these mutations to mTOR conferred nutrient-independent activation of mTOR.

In this chapter, we present methods to characterize the constitutive active mutants of Rheb and mTOR. We also describe methods to detect activation of mTOR and to characterize mTOR complexes. Finally, rapamycin sensitivity will be examined.

2. Methods

2.1. Detection of mTOR activation by examining phospho-S6K1 or phospho-4E-BP1

Activation of mTOR is detected by examining phosphorylation of down-stream proteins. We usually examine phosphorylation of S6K1 at Ser 389 and/or phosphorylation of 4E-BP1 at Thr 37/46 or Thr 70 for this experiment. To enhance the sensitivity of detection, genes encoding these proteins are transfected. Figure 21.1 shows an example of detecting mTOR activation after amino acid addition. Briefly, cells are transfected with FLAG-tagged S6K1 and then starved for both serum and amino acids. Then amino acid mixture containing glucose is added and incubated for 30 min. Cells are collected and the level of phospho-S6K1 is examined by using an antibody specific for phosphorylated S6K1. The total level of S6K1 is examined by using anti–FLAG antibody. Phosphorylation of S6K1 is shut down after serum and amino acid starvation (PBS lane). However, the addition of amino acids and glucose leads to the appearance of phospho-S6K1 band (+glucose +AA lane). We cannot detect the increased phosphorylation of S6K1 in HEK293 cells when treated with glucose (+glucose lane) or amino acid mixture only (not shown).

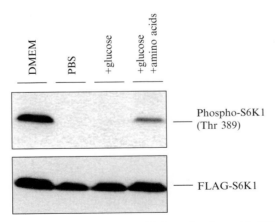

Figure 21.1   Activation of mTOR by the addition of amino acids. HEK293 cells were transfected with FLAG-S6K1 and starved for serum for 24 h (DMEM). These cells were then cultured in D-PBS for 1 h (PBS) and further incubated in D-PBS containing 4.5 g/l of glucose (+glucose) or 4.5 g/l of glucose and 1 × amino acid mixture (+glucose, +amino acids) for 30 min. The amount of total or phosphorylated S6K1 was analyzed by Western blot.

2.1.1. Cell culture and transfection

HEK293 and HeLa cells are maintained in Dulbecco's Modified Eagle's Medium (DMEM, Invitrogen) supplemented with 10% (v/v) fetal bovine serum and cultured at 37° in a 5% CO_2 incubator. Transfection in these cells can be performed either by calcium phosphate method or lipofection method. Here we show the former method. The latter method requires reagents that are commercially available from several sources and transfection is performed according to the manufacturer's protocol.

1. Prepare following reagents for the calcium phosphate transfection.
   0.25 M $CaCl_2$: Dissolve $CaCl_2$ in water to the concentration of 0.25 M, and filter through a 0.45-μm membrane filter.
   2× BBS: Make a 2× BBS solution (50 mM BES (N, N-bis [2-hydro-xyethyl]-2-aminoethane sulfonic acid), 280 mM NaCl, 1.5 mM Na_2HPO_4), adjust the pH to 6.95 with NaOH, and filter through a 0.45-μm membrane filter.
   Cell: Plate 1 × 10^6 HEK293 cells in a 3.5-cm dish the day before transfection. Growth medium is replaced before adding transfection mixture.
2. Add 20 μg of plasmid DNA in 500 μl of 0.25 M $CaCl_2$ and vortex well.
3. Add 500 μl of 2× BBS, vortex well, and incubate for 20 min at room temperature.
4. Pour this transfection mixture directly to each dish, mix gently, and incubate cells at 37°.
5. Replace transfection medium with fresh medium at 18 to 24 h after transfection.
6. Incubate at 37° for a total of 48 to 72 h until target proteins are expressed.

2.1.2. Serum and amino acid starvation

1. Incubate cells in DMEM containing 0.1% bovine serum albumin at 37° in a 5% CO_2 incubator for 24 h.
2. Remove medium and wash cells two times with Dulbecco's Phosphate-Buffered Saline (D-PBS, Invitrogen) containing 100 mg/l each $CaCl_2$ and $MgCl_2$. Incubate them in D-PBS for 1 h.

2.1.3. Preparation and addition of amino acid mixture

To make amino acid mixture, add several amino acids in D-PBS at the following concentrations: L-Arg, 84 mg/l; L-Cys, 48 mg/l; L-Glu, 584 mg/l; L-His, 42 mg/l; L-Ile, 105 mg/l; L-Leu, 105 mg/l; L-Lys, 146 mg/l; L-Met, 30 mg/l; L-Phe, 66 mg/l; L-Thr, 95 mg/l; L-Trp, 16 mg/l; L-Tyr, 72 mg/l; L-Val, 94 mg/l. Stir until dissolved and then filter through 0.22-μm membrane filter. To activate mTOR, 4.5 g/l of glucose is mixed in this mixture.

For mTOR stimulation, cells are incubated in this amino acid mixture containing 4.5 g/l of glucose for 30 min at 37° in 5% CO_2 after serum and amino acid starvation.

2.1.4. Detection of phospho-S6K1 and phospho-4E-BP1

Lyse the cells with lysis buffer (1% Triton X-100, 20 mM Tris–HCl (pH 7.4), 150 mM NaCl, 1 mM EDTA, 50 mM β-glycerophosphate, 1× protease inhibitor cocktail from Roche. After protein quantification, add 1 volume of 2× SDS sample buffer (6% SDS, 10% glycerol, 124.7 mM Tris–HCl) (pH 6.7), 2% 2-mercaptoethanol, 0.02% bromophenol blue), and incubate them at 95° for 5 min. These samples are resolved by 10% polyacrylamide gel for S6K1 detection or 14% polyacrylamide gel for 4E–BP1 detection, and analyzed by Western blot. The anti-phospho-S6K at Ser 389, anti-phospho-4E-BP1 at Thr 37/46, and anti-phospho-4E-BP1 at Thr 70 antibodies are available from several sources. In addition, since hyperphosphorylated 4E-BP1 is separated from less- or non-phosphorylated 4E-BP1 in SDS-polyacrylamide gel electrophoresis, total 4E-BP1 can be resolved into three bands; α, β, and γ from the top. Transiently expressed S6K1 and 4E-BP1 are used because the expression and phosphorylation levels of endogenous S6K1 and 4E-BP1 are low in HEK293 cells.

2.2. Overexpression of wildtype Rheb or constitutive active mTOR mutants confer mTOR activation in the absence of nutrients

Rheb activates mTOR and its downstream proteins in the presence of amino acids. Therefore, as described above, no activation is observed after nutrient starvation. However, overexpression of the wildtype or active mutant Rheb, Rheb-N153T, can induce the activation of mTOR that is identified by the high phosphorylation level of mTOR substrates, S6K1 or 4E-BP1, even in the absence of amino acids (Fig. 21.2).

It is important to note that a similar level of mTOR activation is observed with the wildtype and with the mutant Rheb. A possible reason why one can observe mTOR activation after the overexpression of the wildtype Rheb is that the overexpressed wildtype Rheb contains a high GTP level. Presumably, Rheb GAP activity is limiting in some types of cells, including HEK293. In fact, it is reported that the level of GTP bound to Rheb increases as more Rheb DNA is transfected (Im *et al.*, 2002). Therefore, the overexpressed wildtype Rheb contains a high GTP level to begin with, and the differences between the wildtype and the mutant Rheb are not observed in this setting. However, if Tsc1/2 is overexpressed, clear differences are observed (Yan *et al.*, 2006). In this case, mTOR activation is observed only with the mutant Rheb.

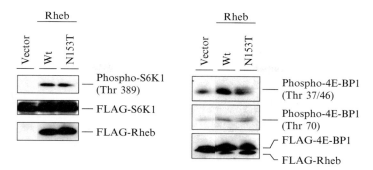

Figure 21.2 Rheb overexpression confers phosphorylation of mTOR substrates in the absence of nutrients. HEK293 cells were transfected with pcDNA3 as a control (vector), Rheb wildtype (wt), or N153T active mutant. To detect the phosphorylation of S6K1 or 4E-BP1, FLAG-S6K1 or FLAG-4E-BP1 was cotransfected. After serum and amino acid starvation, the cell lysates were analyzed by Western blot. (Adapted from Urano, J., Sato, T., Matsuo, T., Otsubo, Y., Yamamoto, M., and Tamanoi, F. (2007). Point mutations in TOR confer Rheb-independent growth in fission yeast and nutrient-independent mammalian TOR signaling in mammalian cells. *Proc. Natl. Acad. Sci. USA* **104**, 3514–3519.)

The situation is different with mTOR. Activation of mTOR signaling in the absence of nutrients can only be observed when constitutive active mTOR mutants are overexpressed (Fig. 21.3). In contrast, little activation is observed with the wildtype mTOR. This result is consistent with the fact that mTOR has strict requirement for amino acids to be active.

Protocols for the experiments are similar to those described above except that the Rheb or mTOR construct is cotransfected with S6K1 or 4E-BP1 construct.

2.2.1. Rheb mutants

A variety of novel mutations in Rheb that confer constitutive activity have been identified (Urano *et al.*, 2005, 2007; Yan *et al.*, 2006). We commonly use Rheb–N153T as an active mutant. This mutant shows a low GTP-binding activity and higher GTP-bound level than those of the wildtype protein in mammalian cells. Rheb-K120R is also reported to show constitutive activity, but it is unstable in mammalian cells and its expression level is low. Lamb and colleagues found the new active mutants, Rheb-S16N and –S16H, that show high GTP-bound levels and mTOR activation even in the cells overexpressing Tsc1/2 (Yan *et al.*, 2006). These mutants are expected to facilitate analysis of Rheb structure and elucidation of the activation mechanism of mTOR by Rheb. While the Rheb-Q64L mutant, analogous to the H–Ras Q61L, is reported to display a high basal GTP level (Inoki *et al.*, 2003; Li *et al.*, 2004), it is unclear whether Rheb-Q64L is an active mutant, as Rheb-Q64L is sensitive to Tsc2GAP, and its GTP-bound level is decreased by the overexpression of Tsc1/2 (Li *et al.*, 2004).

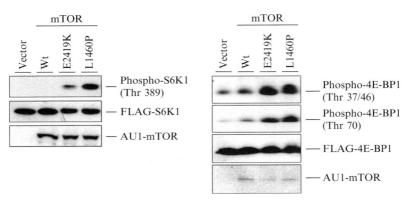

Figure 21.3 Overexpression of mTOR mutants but not mTOR wildtype confers phosphorylation of mTOR substrates in the absence of nutrients. HEK293 cells were transfected with pcDNA3 as a control (vector), mTOR wildtype, E2419K active mutant, or L1460P active mutant. To detect the phosphorylation of S6K1 or 4E-BP1, FLAG-S6K1 or FLAG-4E-BP1 was cotransfected. After serum and amino acid starvation, the cell lysates were analyzed by Western blot. (Adapted from Urano, J., Sato, T., Matsuo, T., Otsubo, Y., Yamamoto, M., and Tamanoi, F. (2007). Point mutations in TOR confer Rheb-independent growth in fission yeast and nutrient-independent mammalian TOR signaling in mammalian cells. *Proc. Natl. Acad. Sci. USA* **104**, 3514–3519.)

2.2.2. mTOR mutants

We recently found a number of active mutants of yeast Tor2p (Urano *et al.*, 2007). Introducing these mutations to mTOR enabled identification of two mammalian TOR mutants, E2419K and L1460P. They exhibit mTORC1 activity in HEK293 cells starved for serum and amino acids. On the other hand, these mutants do not confer constitutive activity of mTORC2 toward Akt in HEK293 cells, since comparable amounts of phospho-Akt at Ser 473 are observed with the wildtype and mutant proteins. Interestingly, most of the active mutations identified in yeast Tor2p occur at residues conserved between yeast and human proteins. Further analysis may provide insight into the activation mechanism of mTOR. In addition, the finding that mTOR-activating mutants can be identified gives rise to the possibility that mutations in mTOR are involved in the uncontrolled growth of cancer cells.

2.3. Analysis of mTOR complexes and their *in vitro* kinase assay

The components of mTOR complexes can be examined by immunoprecipitating mTOR and performing Western blot analysis. Figure 21.4A shows detection of mTOR binding proteins; Raptor, a mTORC1 component, and Rictor, a mTORC2 component. Since mTOR complexes are unstable and disrupted in the presence of some detergents such as 1% NP-40 or Triton X-100, mTOR binding proteins cannot be observed in the mTOR

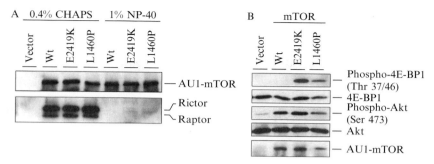

Figure 21.4 Immunoprecipitation and *in vitro* kinase activity of mTOR complexes. (A) HEK293 cells were transfected with pcDNA3 as a control (vector), mTOR wildtype (wt), E2419K active mutant, or L1460P active mutant. After serum and amino acid starvation, mTOR complexes were immunoprecipitated using anti-AU1 antibody from each cell lysates containing 0.4% CHAPS or 1% NP-40 as indicated and the immuno-precipitates were analyzed by Western blot. (B) mTOR immunoprecipitates were divided in two, and used for *in vitro* kinase assay with 4E-BP1 as a mTORC1 substrate, and Akt as a mTORC2 substrate. Phosphorylation of substrates was analyzed by Western blot. (Adapted from Urano, J., Sato, T., Matsuo, T., Otsubo, Y., Yamamoto, M., and Tamanoi, F. (2007). Point mutations in TOR confer Rheb-independent growth in fission yeast and nutrient-independent mammalian TOR signaling in mammalian cells. *Proc. Natl. Acad. Sci. USA* **104**, 3514–3519.)

immunoprecipitates containing these detergents (see Fig. 21.4A). Therefore, we commonly use lysis buffer containing CHAPS detergent to lyse cells and detect mTOR complexes. Endogenous mTOR can also be immunopreci-pitated with anti-mTOR antibody.

mTOR complexes exist as a dimer or multimer in mammalian cells, and dimeric mTOR is reported to be the major form that responds to insulin (Takahara *et al.*, 2006; Wang *et al.*, 2006). The mTOR dimer is observed by the expression of two distinct epitope-tagged mTOR proteins. One epitope is used to immunoprecipitate mTOR, and then the other epitope is detected by the Western blot analysis of immunoprecipitates.

To directly assess the activity of mTOR complexes, *in vitro* kinase assay is performed. The *in vitro* kinase activity of wildtype mTOR, E2419K, and L1460P mutants are shown in Fig. 21.4B. This *in vitro* kinase assay is examined using mTOR immunoprecipitates and recombinant proteins, 4E-BP1, and Akt, which are used as substrates for mTORC1 and mTORC2, respectively. Active mutants of mTOR show higher activity with 4E-BP1 as the substrate compared with the wildtype mTOR in HEK293 cells under nutrient-starved condition. On the other hand, mTORC2 activities of mTOR mutant proteins are similar to that of wildtype mTOR protein. Alternatively, we can measure *in vitro* kinase activity after immunoprecipitation of mTORC1 or mTORC2, separately. Immunoprecipitation with anti-Raptor and anti-Rictor antibodies can isolate mTORC1 and mTORC2, respectively.

2.3.1. Immunoprecipitation of mTOR complex

1. Lyse 1×10^7 HEK293 cells with lysis buffer for immunoprecipitation (0.4% CHAPS, 50 mM Tris-HCl [pH 7.4], 150 mM NaCl, 10 mM MgCl$_2$, 50 mM β-glycerophosphate, $1\times$ protease inhibitor). These lysates are incubated for 30 min at $4°$ and then centrifuged at 15,000 rpm for 15 min at $4°$.

2. Put the supernatants into new tubes and add 5 μg of anti-mTOR antibody (Santa Cruz, N-19), or other antibodies that recognize epitope-tagged mTOR as well as 30 μl of protein-G sepharose 4 fast-flow beads (GE Healthcare).

3. Rotate the samples at $4°$ for 2 h.

4. Centrifuge at 3000 rpm for 2 min at $4°$, and then wash the beads three times with 1 ml of washing buffer (0.4% CHAPS, 50 mM Tris-HCl [pH 7.4], 150 mM NaCl, 10 mM MgCl$_2$, 50 mM β-glycerophosphate). After the wash, add 30 μl of $2\times$ SDS sample buffer (6% SDS, 10% glycerol, 124.7 mM Tris-HCl (pH 6.7), 2% 2-mercaptoethanol), and incubate them at $95°$ for 5 min. These samples are resolved by 8% polyacrylamide gel to detect mTOR, Raptor, and Rictor, and 13% gel to detect GβL and PRAS40. The antibodies for Western blot analysis can be obtained from the following vendors: anti-mTOR and anti-Raptor antibody, Cell Signaling; anti-GβL antibody, BETHYL Lab; and anti-PRAS40 antibody, Biosource.

2.3.2. Dimer formation

To examine the formation of dimeric mTOR, two distinct epitope-tagged mTORs are coexpressed in HEK 293 cells. After one epitope-tagged mTOR is immunoprecipitated using antibody against its epitope, the presence of the other epitope-tagged mTOR is analyzed by Western blot. Coimmunoprecipitation of the two epitope-tagged mTORs indicate the dimer formation. The protocols for transfection and immunoprecipitation are same as above.

2.3.3. *In vitro* mTORC1 and mTORC2 kinase assay

1. Immunoprecipitate mTOR complex using anti-Raptor antibody (BETHYL Lab) for mTORC1 or anti-Rictor antibody (BETHYL Lab) for mTORC2.

2. After the wash, immunoprecipitates are washed one time with $1\times$ kinase buffer (20 mM Tris-HCl [pH 7.4], 10 mM MgCl$_2$, 0.2 mM ATP)

3. Add 8 μl of $5\times$ kinase buffer and 1 μg of recombinant 4E-BP1 for mTORC1 kinase assay or 1 μg of recombinant unactive Akt (Millipore) for mTORC2 kinase assay, and bring the total volume to 40 μl with sterile water.

4. Incubate immunoprecipitates at 37° for 30 min. To stop the reaction, immediately add 40 μl of 2× SDS sample buffer and incubate at 95° for 5 min.

5. These samples are resolved by 10% polyacrylamide gel for Akt detection or 14% polyacrylamide gel for 4E-BP1 detection, and analyzed by Western blot. The anti-phospho-Akt at Ser 473, anti-Akt, anti-phospho-4E-BP1 at Thr 37/46, and anti-4E-BP1 antibodies are available from several sources. Phosphorylation of 4E-BP1 at Thr 70 is not detected in this assay.

2.4. Rapamycin treatment

Rapamycin suppresses mTOR kinase activity by the formation of a ternary complex with FKBP12 and mTOR (Chen et al., 1995; Choi et al., 1996). It has been suggested that rapamycin inhibits mTORC1 activity, but does not affect mTORC2 activity. However, recent studies show that a prolonged treatment of rapamycin can suppress mTORC2 as well as mTORC1 in some cells such as PC3 and Jurkat cells (Sarbassov et al., 2006). The mechanism by which rapamycin inhibits mTOR activity is still unknown, but the rapamycin binding to mTOR may perturb the mTOR complexes and lead to the disruption of mTOR interaction with its substrates.

Before treating with rapamycin, cells are cultured in DMEM containing 0.1% bovine serum albumin at 37° in a 5% CO_2 incubator for 24 h. The cells are then treated with 20 nM rapamicin in DMEM containing 0.1% bovine serum albumin. Although mTORC1 kinase activity is suppressed by 1 h of treatment, 24 h of treatment are required for the inhibition of mTORC2 kinase activity. Decreased phosphorylation levels of S6K1 and 4E-BP1 are observed in these cells.

3. CONCLUSION

The Rheb-mTOR signaling pathway has been extensively studied from yeast to human. This pathway plays a pivotal role in the regulation of cell growth, proliferation, cell size, etc. We presented here the protocols commonly used to examine the Rheb-mTOR signaling pathway. In addition, we showed protocols to evaluate Rheb and mTOR mutants that we recently identified as constitutively active mutants (Urano et al., 2005, 2007). The findings of these activating mutants are significant as it raises the possibility that one mutation in these genes may disrupt homeostasis such as contact inhibition in mammalian cells and cause tumor progression. In fact, activation of the mTOR pathway is implicated in a number of human diseases associated with benign tumors, including tuberous sclerosis. Inhibitors of mTOR, including rapamycin and its derivatives, are under clinical evaluation as anticancer drugs. Further studies on the Rheb-mTOR signaling pathway may provide important insights into the activation mechanism of the Rheb-mTOR signaling pathway.

REFERENCES

Abraham, R. T. (2004). PI 3-kinase related kinases: "Big" players in stress-induced signaling pathways. *DNA Repair (Amst.)* **3,** 883–887.

Aspuria, P. J., and Tamanoi, F. (2004). The Rheb family of GTP-binding proteins. *Cell Signal.* **16,** 1105–1112.

Bourne, H. R., Sanders, D. A., and McCormick, F. (1990). The GTPase superfamily: A conserved switch for diverse cell functions. *Nature* **348,** 125–132.

Byfield, M. P., Murray, J. T., and Backer, J. M. (2005). hVps34 is a nutrient-regulated lipid kinase required for activation of p70 S6 kinase. *J. Biol. Chem.* **280,** 33076–33082.

Cai, S. L., Tee, A. R., Short, J. D., Bergeron, J. M., Kim, J., Shen, J., Guo, R., Johnson, C. L., Kiguchi, K., and Walker, C. L. (2006). Activity of TSC2 is inhibited by AKT-mediated phosphorylation and membrane partitioning. *J. Cell Biol.* **173,** 279–289.

Castro, A. F., Rebhun, J. F., Clark, G. J., and Quilliam, L. A. (2003). Rheb binds tuberous sclerosis complex 2 (TSC2) and promotes S6 kinase activation in a rapamycin- and farnesylation-dependent manner. *J. Biol. Chem.* **278,** 32493–32496.

Chen, J., Zheng, X. F., Brown, E. J., and Schreiber, S. L. (1995). Identification of an 11-kDa FKBP12-rapamycin–binding domain within the 289-kDa FKBP12-rapamycin–associated protein and characterization of a critical serine residue. *Proc. Natl. Acad. Sci. USA* **92,** 4947–4951.

Choi, J., Chen, J., Schreiber, S. L., and Clardy, J. (1996). Structure of the FKBP12-rapamycin complex interacting with the binding domain of human FRAP. *Science* **273,** 239–242.

Frias, M. A., Thoreen, C. C., Jaffe, J. D., Schroder, W., Sculley, T., Carr, S. A., and Sabatini, D. M. (2006). mSin1 is necessary for Akt/PKB phosphorylation, and its isoforms define three distinct mTORC2s. *Curr. Biol.* **16,** 1865–1870.

Gao, X., and Pan, D. (2001). TSC1 and TSC2 tumor suppressors antagonize insulin signaling in cell growth. *Genes Dev.* **15,** 1383–1392.

Garami, A., Zwartkruis, F. J., Nobukuni, T., Joaquin, M., Roccio, M., Stocker, H., Kozma, S. C., Hafen, E., Bos, J. L., and Thomas, G. (2003). Insulin activation of Rheb, a mediator of mTOR/S6K/4E-BP signaling, is inhibited by TSC1 and 2. *Mol. Cell* **11,** 1457–1466.

Guertin, D. A., Stevens, D. M., Thoreen, C. C., Burds, A. A., Kalaany, N. Y., Moffat, J., Brown, M., Fitzgerald, K. J., and Sabatini, D. M. (2006). Ablation in mice of the mTORC components raptor, rictor, or mlST8 reveals that mTORC2 is required for signaling to Akt-FOXO and PKCalpha, but not S6K1. *Dev. Cell* **11,** 859–871.

Haar, E. V., Lee, S. I., Bandhakavi, S., Griffin, T. J., and Kim, D. H. (2007). Insulin signalling to mTOR mediated by the Akt/PKB substrate PRAS40. *Nat. Cell Biol.* **9,** 316–323.

Hara, K., Maruki, Y., Long, X., Yoshino, K., Oshiro, N., Hidayat, S., Tokunaga, C., Avruch, J., and Yonezawa, K. (2002). Raptor, a binding partner of target of rapamycin (TOR), mediates TOR action. *Cell* **110,** 177–189.

Hresko, R. C., and Mueckler, M. (2005). mTOR.RICTOR is the Ser473 kinase for Akt/protein kinase B in 3T3-L1 adipocytes. *J. Biol. Chem.* **280,** 40406–40416.

Im, E., von Lintig, F. C., Chen, J., Zhuang, S., Qui, W., Chowdhury, S., Worley, P. F., Boss, G. R., and Pilz, R. B. (2002). Rheb is in a high activation state and inhibits B-Raf kinase in mammalian cells. *Oncogene* **21**, 6356–6365.

Inoki, K., Li, Y., Zhu, T., Wu, J., and Guan, K. L. (2002). TSC2 is phosphorylated and inhibited by Akt and suppresses mTOR signalling. *Nat. Cell Biol.* **4**, 648–657.

Inoki, K., Li, Y., Xu, T., and Guan, K. L. (2003). Rheb GTPase is a direct target of TSC2 GAP activity and regulates mTOR signaling. *Genes Dev.* **17**, 1829–1834.

Jacinto, E., Loewith, R., Schmidt, A., Lin, S., Ruegg, M. A., Hall, A., and Hall, M. N. (2004). Mammalian TOR complex 2 controls the actin cytoskeleton and is rapamycin insensitive. *Nat. Cell Biol.* **6**, 1122–1128.

Jacinto, E., Facchinetti, V., Liu, D., Soto, N., Wei, S., Jung, S. Y., Huang, Q., Qin, J., and Su, B. (2006). SIN1/MIP1 maintains rictor-mTOR complex integrity and regulates Akt phosphorylation and substrate specificity. *Cell* **127**, 125–137.

Kim, D. H., Sarbassov, D. D., Ali, S. M., King, J. E., Latek, R. R., Erdjument-Bromage, H., Tempst, P., and Sabatini, D. M. (2002). mTOR interacts with raptor to form a nutrient-sensitive complex that signals to the cell growth machinery. *Cell* **110**, 163–175.

Kim, D. H., Sarbassov, D. D., Ali, S. M., Latek, R. R., Guntur, K. V., Erdjument-Bromage, H., Tempst, P., and Sabatini, D. M. (2003). GbetaL, a positive regulator of the rapamycin-sensitive pathway required for the nutrient-sensitive interaction between raptor and mTOR. *Mol. Cell* **11**, 895–904.

Li, Y., Inoki, K., and Guan, K. L. (2004). Biochemical and functional characterizations of small GTPase Rheb and TSC2 GAP activity. *Mol. Cell. Biol.* **24**, 7965–7975.

Nobukuni, T., Joaquin, M., Roccio, M., Dann, S. G., Kim, S. Y., Gulati, P., Byfield, M. P., Backer, J. M., Natt, F., Bos, J. L., Zwartkruis, F. J., and Thomas, G. (2005). Amino acids mediate mTOR/raptor signaling through activation of class 3 phosphatidylinositol 3OH-kinase. *Proc. Natl. Acad. Sci. USA* **102**, 14238–14243.

Patel, P. H., Thapar, N., Guo, L., Martinez, M., Maris, J., Gau, C. L., Lengyel, J. A., and Tamanoi, F. (2003). *Drosophila* Rheb GTPase is required for cell cycle progression and cell growth. *J. Cell Sci.* **116**, 3601–3610.

Sabatini, D. M. (2006). mTOR and cancer: Insights into a complex relationship. *Nat. Rev. Cancer* **6**, 729–734.

Sancak, Y., Thoreen, C. C., Peterson, T. R., Lindquist, R. A., Kang, S. A., Spooner, E., Carr, S. A., and Sabatini, D. M. (2007). PRAS40 is an insulin-regulated inhibitor of the mTORC1 protein kinase. *Mol. Cell* **25**, 903–915.

Sarbassov, D. D., Ali, S. M., Kim, D. H., Guertin, D. A., Latek, R. R., Erdjument-Bromage, H., Tempst, P., and Sabatini, D. M. (2004). Rictor, a novel binding partner of mTOR, defines a rapamycin-insensitive and raptor-independent pathway that regulates the cytoskeleton. *Curr. Biol.* **14**, 1296–1302.

Sarbassov, D. D., Guertin, D. A., Ali, S. M., and Sabatini, D. M. (2005). Phosphorylation and regulation of Akt/PKB by the rictor-mTOR complex. *Science* **307**, 1098–1101.

Sarbassov, D. D., Ali, S. M., Sengupta, S., Sheen, J. H., Hsu, P. P., Bagley, A. F., Markhard, A. L., and Sabatini, D. M. (2006). Prolonged rapamycin treatment inhibits mTORC2 assembly and Akt/PKB. *Mol. Cell* **22**, 159–168.

Takahara, T., Hara, K., Yonezawa, K., Sorimachi, H., and Maeda, T. (2006). Nutrient-dependent multimerization of the mammalian target of rapamycin through the N-terminal HEAT repeat region. *J. Biol. Chem.* **281**, 28605–28614.

Tee, A. R., Manning, B. D., Roux, P. P., Cantley, L. C., and Blenis, J. (2003). Tuberous sclerosis complex gene products, Tuberin and Hamartin, control mTOR signaling by acting as a GTPase-activating protein complex toward Rheb. *Curr. Biol.* **13**, 1259–1268.

Urano, J., Comiso, M. J., Guo, L., Aspuria, P. J., Deniskin, R., Tabancay, A. P., Jr., Kato-Stankiewicz, J., and Tamanoi, F. (2005). Identification of novel single amino acid changes that result in hyperactivation of the unique GTPase, Rheb, in fission yeast. *Mol. Microbiol.* **58,** 1074–1086.

Urano, J., Sato, T., Matsuo, T., Otsubo, Y., Yamamoto, M., and Tamanoi, F. (2007). Point mutations in TOR confer Rheb-independent growth in fission yeast and nutrient-independent mammalian TOR signaling in mammalian cells. *Proc. Natl. Acad. Sci. USA* **104,** 3514–3519.

Wang, L., Rhodes, C. J., and Lawrence, J. C., Jr. (2006). Activation of mammalian target of rapamycin (mTOR) by insulin is associated with stimulation of 4EBP1 binding to dimeric mTOR complex 1. *J. Biol. Chem.* **281,** 24293–24303.

Yamagata, K., Sanders, L. K., Kaufmann, W. E., Yee, W., Barnes, C. A., Nathans, D., and Worley, P. F. (1994). rheb, a growth factor- and synaptic activity-regulated gene, encodes a novel Ras-related protein. *J. Biol. Chem.* **269,** 16333–16339.

Yan, L., Findlay, G. M., Jones, R., Procter, J., Cao, Y., and Lamb, R. F. (2006). Hyper-activation of mammalian target of rapamycin (mTOR) signaling by a gain-of-function mutant of the Rheb GTPase. *J. Biol. Chem.* **281,** 19793–19797.

Yang, Q., Inoki, K., Ikenoue, T., and Guan, K. L. (2006). Identification of Sin1 as an essential TORC2 component required for complex formation and kinase activity. *Genes Dev.* **20,** 2820–2832.

Yu, Y., Li, S., Xu, X., Li, Y., Guan, K., Arnold, E., and Ding, J. (2005). Structural basis for the unique biological function of small GTPase RHEB. *J. Biol. Chem.* **280,** 17093–17100.

Zhang, Y., Gao, X., Saucedo, L. J., Ru, B., Edgar, B. A., and Pan, D. (2003). Rheb is a direct target of the tuberous sclerosis tumour suppressor proteins. *Nat. Cell Biol.* **5,** 578–581.

CHAPTER TWENTY-TWO

CHARACTERIZATION OF RalB-Sec5-TBK1 FUNCTION IN HUMAN ONCOGENESIS

Yuchen Chien *and* Michael A. White

Contents

Abstract

The Ras-like GTPases, RalA and RalB, are key components of the oncogenic Ras signaling network. Recent evidence suggests that RalA and RalB collaborate to support tumorigenic transformation through distinct cell regulatory events. While RalA is apparently required to bypass normal restraints on cell proliferation, RalB is required to bypass normal restraints on cell survival. A direct Ral effector protein, Sec5, is a subunit of the exocyst complex, and is required to mediate RalB-dependent survival signals in transformed cells. Further analysis identified TBK1, a key mediator of the host defense response to viral challenge, as a novel Sec5 interacting protein essential for the capacity of RalB and Sec5 to deflect cell death in transformed cells. RalB activation promotes a direct interaction between Sec5 and TBK1 that results in TBK1 kinase activation via an unknown mechanism. Accordingly, both RalB and Sec5 are required for initiating host defense pathway activation upon virus infection. These observations revealed a novel relationship between molecular components of cell-autonomous innate immune signaling pathways and oncogenic transformation, and identified TBK1 as a potential target for therapeutic intervention in cancer. Here we describe details of methods, including protein complex analysis,

Department of Cell Biology, University of Texas Southwestern Medical Center, Dallas, Texas

Methods in Enzymology, Volume 438
ISSN 0076-6879, DOI: 10.1016/S0076-6879(07)38022-1

protein kinase assays, host defense–response pathway activation, and cell transformation analysis, that can be used to investigate the contribution of the RalB-Sec5-TBK1 signaling cascade to both innate immune signaling and cell transformation.

1. INTRODUCTION

Ral (Ras–like) GTPases, RalA and RalB, are members of the Ras GTPase superfamily (Chardin and Tavitian, 1986). RalA and RalB are 80% identical and like other small GTPases, they cycle between a GDP-bound inactive and a GTP-bound active conformation in cells. The GTP-bound conformation is required for Ral proteins to bind and presumably activate downstream effector proteins (Feig, 2003).

Loss of function analysis has indicated discrete cellular functions of RalA and RalB in support of oncogenic transformation, that is, RalA is required for matrix-independent cell proliferation while RalB is essential for survival of transformed cells (Chien and White, 2003; Chien et al., 2006). Several Ral effectors have been identified (Feig, 2003), and among them, depletion of Sec5 resulted in the same cellular phenotype as depletion of RalB, suggesting that Sec5 is the effector mediating survival signals downstream of RalB in transformed cells (Chien et al., 2006).

Since not much is known about Sec5 except its function as a structural component of exocyst complex (Kee et al., 1997), we approached this question by identifying novel exocyst binding proteins (Moskalenko et al., 2002; Chien et al., 2006). Using combined immunoprecipitation and mass spectrometry analysis, TBK1 was isolated as an exocyst-interacting protein. Further functional assay and biochemical analysis revealed TBK1 associates with Sec5 in intact cells and in vitro, and TBK1 is required for survival of transformed cells. Importantly, Ral activation promotes TBK1 and Sec5 complex assembly that leads to increased TBK1 kinase activity (Chien et al., 2006).

TBK1, as a major kinase phosphorylating transcription factor IRF3 in response to virus infection (Fitzgerald et al., 2003; McWhirter et al., 2004; Sharma et al., 2003), led us to investigate the contribution of RalB and Sec5 to initiating innate immune pathway. Notably, both RalB and Sec5 are required for mounting host defense responses, and their function is dependent on TBK1 (Kee et al., 1997).

The observation that TBK1-null mouse embryonic fibroblasts (MEF) displayed increased cell death, and decreased cell transformation upon oncogenic Ras expression, further underscores the role of TBK1 in tumorigenesis (Chien et al., 2006). In the following we describe methods used to study physical and functional relationships between RalB, Sec5 and TBK1.

2. Methods

2.1. Identification of TBK1-exocyst complexes

We have defined stimulus-evoked TBK1-exocyst association as a component of both the host defense response and oncogenic transformation (Chien *et al.*, 2006). Here, we describe methods to detect endogenous TBK1-exocyst complexes in cells and to examine direct interactions *in vitro*.

2.1.1. Reagents

Mouse monoclonal anti-TBK1 antibodies are obtained from Imgenex (use at 1:1000 for immunoblotting). Mouse monoclonal anti-Sec8 and rabbit polyclonal anti-Sec5 antibodies are as described (Grindstaff *et al.*, 1998; Moskalenko *et al.*, 2003) (use at 1:1000 for immunoblotting). Goat anti-mouse HRP and goat anti-rabbit HRP secondary antibodies (use at 1:10,000) are from Jackson Laboratories.

2.1.2. Procedures

Detection of endogenous TBK1-exocyst complexes Human bronchial epithelial cells (HBEC) (kindly provided by Jerry Shay, UT Southwestern Medical Center) are seeded on 10-cm dishes to be 80% confluent the next day. Each precipitation will require two plates. The next day, each plate of intact cells is scraped down into a prechilled eppendorf with 500-μl lysis buffer containing 1% NP-40, 10 mM Tris (pH7.6), 250 μM sodium deoxycholate, 1 mM MgCl$_2$, 1 mM EGTA, 10% glycerol, 150 mM NaCl, 50 mM NaF, 1 mM sodium vanadate, and 80 mM β-glycerol phosphate plus protease inhibitors. The lysates are homogenized by rotating for 30 min at 4° and cleared by centrifugation for 30 min at 17,000$\times g$ at 4°. The endogenous exocyst complex is immunoprecipitated from the supernatant with 20 μg of anti-Sec8 antibodies with gentle agitation and incubation overnight at 4°. Normal mouse IgG (Santa Cruz) is used as a negative control. Next, 30 μl of a 50% slurry of Protein A/G PLUS-Agarose (Santa Cruz) is added, and lysates are rotated for an additional 3 h. The agarose is pelleted and washed four times with 1 ml lysis buffer by gentle resuspension and centrifugation for 5 min at 5000$\times g$ at 4°. The final pellet is suspended in 50 μl of sample, and half of the volume is loaded on polyacrylamide gels for immunodetection by anti-Sec8 and anti-TBK1 antibodies.

To detect enhanced association between TBK1 and endogenous Sec5 upon Ral activation, HEK293 cells are transfected by Lipofectamine (Invitrogen) with Flag-TBK1, with or without the fast-exchange RalF39L mutant. At 48 h post-transfection, Flag-TBK1 is immunoprecipitated using the same lyses procedure as above, but with anti-Flag M2 agarose

(Sigma) by rotating 4 h at $4°$. The agarose is washed four times, and sample buffer is added to agarose followed by standard SDS-PAGE analysis and Western blotting with anti–Sec5 and anti–Flag M5 (Sigma) antibodies.

To detect induction of TBK1/Sec5 complex formation upon Sendai virus infection, HBEC are seeded on 10-cm dishes to be 80% confluent the next day. Each precipitation will require two plates. The next day, each plate of cells is transfected using 30 μl of Lipofectamine 2000 (Invitrogen) with 10 μg of Flag-TBK1 and 10 μg of full–length HA-Sec5. At 48 h post-transfection, HBEC are infected with 100 (HA) units/ml Sendai virus (Cantell strain, Charles River Laboratory) for 5 h at $37°$. Flag-TBK1 is immunoprecipitated with anti–Flag M2 agarose (Sigma) for 4 h at $4°$, and the associated Sec5 is detected by Western blotting with anti–HA antibodies (Santa Cruz).

In vitro binding assay pGEX-RLIP Ral-binding domain (pGEX-RLIPrbd) and pGEX-Sec5 Ral-binding domain (pGEX-Sec5rbd) are as previously described (Moskalenko *et al.*, 2003). GST-tagged RLIPrbd and Sec5rbd proteins are purified by standard procedures from BL21 *Escherichia coli*, and immobilized on glutathione-sepharose beads (Amersham). Protein concentrations are determined with DC Protein Assay (Bio-Rad), and confirmed by comparing to serial concentrations of bovine serum albumin on a coomassie blue-stained gel. To each binding reaction, equal amounts (10 μg) of each GST fusion proteins are incubated with 0.5 μg of recombinant TBK1 protein (Upstate) in 50 mM Tris (pH7.6) and 300 mM NaCl for 1 h at $4°$. Beads are then washed four times with 50 mM Tris (pH7.6), 300 mM NaCl, and 0.1% Triton X-100 prior to addition of sample buffer and SDS-PAGE analysis. The presence of TBK1 is detected with anti–TBK1 antibodies.

2.2. Activation of innate immune responses by a RalB-Sec5-TBK1 signaling cascade

TBK1 associates with Sec5 *in vitro* and in intact cells. In addition, both Ral activation and Sendai virus infection promote TBK1 and Sec5 interaction and concomitant TBK1 activation (Chien *et al.*, 2006). TBK1 plays a central role in host defense–pathway activation. In the following, we describe methods to detect activation of host defense pathways that include induction of TBK1 kinase activity, IRF3 nuclear translocation, and IFN-β and ISG56 gene expression.

2.2.1. Reagents
Mouse monoclonal anti–TBK1 antibodies are obtained from Imgenex (use 30 μg/ml for immunoprecipitation). Rabbit polyclonal anti–IRF3 antibodies are from Santa Cruz (use at 1:50 for immunofluorescence),

and rabbit polyclonal anti–NF-κB p65 subunit (relA) antibodies are from Biomol (use at 1:50 for immunofluorescence). Goat anti-rabbit rhodamine secondary antibodies (use at 1:100) are from Jackson Laboratories.

Synthetic siRNA targeting RalA, RalB, and Sec5 are designed by standard methods using the following sense sequences:

5′-CAGAGCUGAGCAGUGGAAUdTdT-3′ (RalA)
5′-GGUGAUCAUGGUUGGCAGCdTdT-3′ (RalB)
5′-GACUAUGAACCUACCAAAGdTdT-3′ (RalB)
5′-GGUCGGAAAGACAAGGCAGdTdT-3′ (Sec5)
5′-GGGUGAUUAUGAUGUGGUUdTdT-3′ (Sec5)

Gene-specific primer sequences for ISG56 include:

5′-ACA CCT GAA AGG CCA GAA TGA GGA-3′ (forward)
5′-TGT CTG GAT TTA AGC GGA CAG CCT-3′ (reverse)

Gene-specific primer sequences for IFN-β follow:

5′-TGG GAG GCT TGA ATA CTG CCT CAA-3′ (forward)
5′-TCC TTG GCC TTC AGG TAA TGC AGA-3′ (reverse)

Gene-specific primer sequences for GAPDH include:

5′-GAC CAC AGT CCA TGC CAT CAC-3′ (forward)
5′-CAT ACC AGG AAA TGA GCT TGA-3′ (reverse)

2.2.2. Procedures

TBK1 kinase assay Endogenous TBK1 proteins are immunoprecipitated from cells with anti–TBK1 mouse monoclonal antibodies (Imgenex) using the protocol described in section 2.1.2.1. TBK1 immunoprecipitates from two 10-cm plates at 80% confluence are used for each kinase activity assay in a buffer containing 25 mM Hepes (pH7.6), 20 mM β-glycerophosphate, 0.1 mM sodium vanadate, 10 mM MgCl$_2$, 50 mM NaCl, 50 μM ATP, and 5 μCi of [γ-32P]ATP. For each reaction, 1 μg of purified recombinant GST-tagged IRF3C' (kindly provided by Zhijian Chen, UT Southwestern Medical Center) is used as a substrate for TBK1 and reactions are incubated at 30° for 30 min. Reactions are terminated by the addition of sample buffer. [γ-32P]ATP incorporation in recombinant GST-tagged IRF3C' is visualized by SDS-PAGE followed by exposure to a phosphorimager plate. Equivalent protein concentrations are confirmed by immunoblotting with anti-TBK1 and anti-GST (Santa Cruz) antibodies.

To detect elevated TBK1 kinase activity upon Ral activation, HEK293 cells are transfected using Lipofectamine (Invitrogen) with Flag-TBK1, with or without fast-exchange RalF39L mutant. At 48 h post-transfection,

Flag-TBK1 is immunoprecipitated with anti–Flag M2 agarose (Sigma), and TBK1 kinase activity is measured as described above.

Immunofluorescence imaging of IRF3 and NF-κB p65 subunit (relA)   To detect endogenous IRF3 nuclear translocation upon Ral activation, HBEC are plated on glass coverslips the night before and then transfected with expression vectors for the fast-exchange RalF39L mutant using Lipofectamine 2000 (Invitrogen). At 48 h post-transfection, HBEC are washed twice with PBS and fixed with 3.7% formaldehyde at room temperature for 30 min. The cells are then washed twice with PBS, permeabilized with 0.25% Triton-X 100 in PBS for 1 h, and blocked with 1% BSA in PBS for 1 h. Antibody dilutions and subsequent washes are performed with the blocking solutions. Primary antibodies are used at 1:50 dilutions and secondary antibodies are used at 1:100. Coverslips are dipped several times in distilled water and mounted on uncoated glass slides with Immuno-Mount. All images are acquired at 40× using a Ziess microscope.

To assess the requirement of RalB and Sec5 for endogenous IRF3 and NF-κB p65 subunit (relA) nuclear translocation after Poly (I:C) stimulation, HBEC are transfected with synthetic siRNA targeting human Sec5, RalA, and RalB using Oligofectamine (Invitrogen). HBEC are seeded on 35-mm dishes to be 40 to 50% confluent for transfection the next day. For each plate, 6 μl of Oligofectamine is used to transfect 20 μM of annealed double-stranded siRNA and 1 μg of CMV-eGFP. The GFP-expressing plasmid is cotransfected to mark siRNA-transfected cells. Cell are first washed twice with PBS, and then treated with 500 μl of trypsin-EDTA (Invitrogen) for approximately 1 min. Trypsin is neutralized by addition of 500 μl of 1× Trypsin Inhibitor (Sigma), followed by addition of 500 μl of optimem (GIBCO) and preprepared siRNA and DNA/oligofectamine complexes. Fresh media is added at 16 h post-transfection. At 72 h post-transfection, cells are either treated with 100 μg/ml poly (I:C) (Amersham) for 90 min at 37° or left untreated. Localization of endogenous IRF3 and NF-κB p65 subunit (relA) are visualized by immunostaining using the protocol described above. For high efficiency siRNA transfection, HBEC are seeded on 35 mm dishes to be 40 to 50% confluent the next day. Cells are transfected as the same procedure above but without GFP plasmid. At 48 h post-transfection, a second transfection is performed to enhance transfection efficiency, and fresh media is added after an additional sixteen h. At 48 h after second transfection, cells are either treated with 100 μg/ml poly (I:C) (Amersham) for 90 min at 37°, or left untreated and subjected to immunostaining procedures as described previously.

2.3. Using RT-PCR to assess ISG56 and IFN-β gene expression

To measure ISG56 and IFN-β gene expression upon poly (I:C) stimulation or Sendai virus infection in Ral-depleted or Sec5-depleted cells, HBEC are transfected with synthetic siRNA targeting RalA, RalB, or Sec5 using the high-efficiency transfection protocol described above. At 48 h after the second transfection, cells are either treated with 100 μg/ml poly (I:C) for 90 min, infected with Sendai virus 100 (HA) units/ml for 17 h, or left untreated. Next, total RNA is isolated from HBEC cells using the High Pure RNA Isolation Kit (Roche), and subjected to RT-PCR using the SuperScript III One-Step RT-PCR system (Invitrogen) according to manufacturer's instructions to measure expression of ISG56, IFN-β, and GAPDH (as a control).

2.4. Contribution of TBK1 to oncogenic transformation

TBK1-null mouse embryonic fibroblasts (MEF) revealed the requirement of TBK1-dependent survival signals during oncogenic Ras-induced transformation (Chien et al., 2006). Next, we describe methods to perform retroviral transduction followed by assays for cell transformation, including soft-agar growth, BrdU incorporation, and TUNEL labeling.

2.4.1. Reagents

pBabe-Ras12V has been described (Chien and White, 2003). Mouse monoclonal anti-Ras antibodies (use at 1:1000 for immunoblotting) and mouse monoclonal anti-BrdU antibodies (use at 1:5 for immunofluorescence) are obtained from Becton Dickinson (BD). Goat anti-mouse rhodamine secondary antibodies (use at 1:200) are from Jackson Laboratories.

2.4.2. Procedures

Retroviral transduction Standard retroviral mediated gene transfer is employed using an ecotropic packaging system in HEK 293T cells. 293T cells are plated at 3×10^6 cells per 10-cm dish 16 h prior to transfection. Next, each plate of cells is transfected using Fugene 6 (Roche) with 4 μg of the pCLeco packaging vector and 4 μg of pBabe-Ras12V or pMDH-GFP as control. At 24 h post-transfection, an additional 7 ml of fresh medium is added. After 24 h, the media is collected and filtered through a 0.45-μm syringe filter. The viral supernatant is used to infect actively proliferating MEFs that are plated at 10^5 cells per 10-cm dish the night before. The highest viral titers are obtained by using supernatant collected directly from the packaging lines to infect the target cells. To enhance virus infection, 4 μg/ml of polybrene are included. At 48 h post-infection, MEFs are examined under the microscope for GFP-positive cells to evaluate infection efficiency. The protocol described here should generate virus that is capable of infecting exponentially growing MEFs at an efficiency of 75 to 100%.

Cell proliferation assays   For BrdU incorporation assays, MEFs, 48 h post-infection, are split and replated either onto glass coverslips or into 1%-agarose coated dishes. BrdU is then added at a final concentration of 30 μM. After incubation for 17 h, suspension cell cultures are washed twice with PBS and spun onto poly-L-lysine (Sigma) coated glass coverslips. Cells are fixed with 3.7% formaldehyde for 30 min at room temperature, permeabilized with acetone at $-20°$ for 5 min, washed with 2 M HCl for 10 min, and blocked with 1% BSA in PBS for 1 h. BrdU incorporation is visualized by staining with mouse monoclonal anti-BrdU antibodies and rhodamine-conjugated antimouse secondary antibodies.

To assess growth in soft agar, MEFs, at 48 h postinfection, are resuspended in DMEM containing 0.33% low-melting temperature agarose (Sigma) and 10% FBS, and 5000 cells per 30-mm plate are layered onto a hardened coating of 0.5% agarose in DMEM containing 10% FBS. Fresh DMEM containing 10% FBS is added weekly.

Apoptosis assays   To detect cell death, equal numbers of wildtype and TBK1-null MEFs are split onto glass coverslips before virus infection. At 48 h postinfection, cells are exposed to terminal deoxynucleotidyltransferase-mediated dUTP-biotin nick-end labeling (TUNEL), according to manufacturer's instructions (Becton Dickinson).

For detection of cell death using fluorescence-activated cell sorting (FACS), MEFs at 48 h postinfection are harvested by trypsinization, fixed with 50% ethanol at 4° for 1 h, and stained with propidium iodide (Becton Dickinson) at 37° for 30 min. Approximately 10,000 events are collected for each assay, and analyzed using Cell Quest software (Becton Dickinson).

3. Conclusions

We have presented here a series of protocols that can be used to study RalB-Sec5-TBK1 signaling complex assembly and function. Future studies may use these methods to help further elaborate mechanistic insights on initiation of this signaling cascade and on the downstream TBK1 targets mediating cell survival in transformed cells.

ACKNOWLEDGMENTS

This work was supported by grants from the Welch Foundation, the Susan G. Komen Breast Cancer Foundation, the CDMRP Breast Cancer Inititative, and the National Cancer Institute.

REFERENCES

Chardin, P., and Tavitian, A. (1986). The ral gene: A new ras related gene isolated by the use of a synthetic probe. *EMBO J.* **5,** 2203–2208.

Chien, Y., and White, M. A. (2003). RAL GTPases are linchpin modulators of human tumour-cell proliferation and survival. *EMBO Rep.* **4,** 800–806.

Chien, Y., Kim, S., Bumeister, R., Loo, Y. M., Kwon, S. W., Johnson, C. L., Balakireva, M. G., Romeo, Y., Kopelovich, L., Gale, M., Jr., Yeaman, C., Camonis, J. H., *et al.* (2006). RalB GTPase-mediated activation of the IkappaB family kinase TBK1 couples innate immune signaling to tumor cell survival. *Cell* **127,** 157–170.

Feig, L. A. (2003). The Ral GTPases: Approaching their 15 minutes of fame. *Trends Cell Biol.* **13,** 419–425.

Fitzgerald, K. A., McWhirter, S. M., Faia, K. L., Rowe, D. C., Latz, E., Golenbock, D. T., Coyle, A. J., Liao, S. M., and Maniatis, T. (2003). *Nat. Immunol.* **4,** 491–496.

Grindstaff, K. K., Yeaman, C., Anandasabapathy, N., Hsu, S. C., Rodriguez-Boulan, E., Scheller, R. H., and Nelson, W. J. (1998). Sec6/8 complex is recruited to cell–cell contacts and specifies transport vesicle delivery to the basal-lateral membrane in epithelial cells. *Cell* **93,** 731–740.

Kee, Y., Yoo, J. S., Hazuka, C. D., Peterson, K. E., Hsu, S. C., and Scheller, R. H. (1997). Subunit structure of the mammalian exocyst complex. *Proc. Natl. Acad. Sci. USA* **94,** 14438–14443.

McWhirter, S. M., Fitzgerald, K. A., Rosains, J., Rowe, D. C., Golenbock, D. T., and Maniatis, T. (2004). IFN-regulatory factor 3-dependent gene expression is defective in Tbk1-deficient mouse embryonic fibroblasts. *Proc. Natl. Acad. Sci. USA* **101,** 233–238.

Moskalenko, S., Henry, D. O., Rosse, C., Mirey, G., Camonis, J. H., and White, M. A. (2002). The exocyst is a Ral effector complex. *Nat. Cell Biol.* **4,** 66–72.

Moskalenko, S., Tong, C., Rosse, C., Mirey, G., Formstecher, E., Daviet, L., Camonis, J., and White, M. A. (2003). Ral GTPases regulate exocyst assembly through dual subunit interactions. *J. Biol. Chem.* **278,** 51743–51748.

Sharma, S., tenOever, B. R., Grandvaux, N., Zhou, G. P., Lin, R., and Hiscott, J. (2003). Triggering the interferon antiviral response through an IKK-related pathway. *Science* **300,** 1148–1151.

CHAPTER TWENTY-THREE

High-Throughput Lung Cancer Cell Line Screening for Genotype-Correlated Sensitivity to an EGFR Kinase Inhibitor

Ultan McDermott, Sreenath V. Sharma, *and* Jeffrey Settleman

Contents

Abstract

Human cancer cell lines that can be propagated and manipulated in culture have proven to be excellent models for studying many aspects of gene function in cancer. In addition, they can provide a powerful system for assessing the molecular determinants of sensitivity to anticancer drugs. They have also been used in recent studies to identify genomic alterations and gene expression patterns that provide important insights into the genetic features that distinguish the properties of tumor cells associated with similar histologies. We have established a large repository of human tumor cell lines (>1000) corresponding to a wide variety of tumor types, and we have developed a methodology for profiling the collection for sensitivity to putative anticancer compounds. The rationale for examining tumor cell lines on this relatively large scale reflects accumulating evidence indicating that there is substantial genetic heterogeneity among human tumor cells—even those derived from tumors of similar

Center for Molecular Therapeutics, Massachusetts General Hospital Cancer Center and Harvard Medical School, Charlestown, Massachusetts

Methods in Enzymology, Volume 438
ISSN 0076-6879, DOI: 10.1016/S0076-6879(07)38023-3

histologies. Thus, to develop an accurate picture of the molecular determinants of tumorigenesis and response to therapy, it is essential to study the nature of such heterogeneity in a relatively large sample set. Here, we describe the methodologies used to conduct such screens and we describe a "proof-of-concept" screen using the EGFR kinase inhibitor, erlotinib (Tarceva), with a panel of lung cancer lines to demonstrate a correlation between EGFR mutations and drug sensitivity.

1. INTRODUCTION

A new era of targeted molecular therapeutics has emerged, and consequently, the prospects for more effective and less toxic cancer treatments have improved tremendously. These new targeted drugs, many of which are selective kinase inhibitors, include the Food and Drug Administration–approved agents Herceptin, Erbitux, Tarceva, Iressa, Gleevec, Sutent, Sorafanib, and Avastin, and are capable of producing dramatic clinical responses in a subset of treated patients (Ross *et al.*, 2004). Importantly, paralleling the discovery of such drugs has been the realization that the molecular genetic features of an individual's tumor play a critical role in determining whether that patient will respond to a particular targeted drug. Human cancers vary enormously in their genetic constitution, and it is now becoming widely accepted that these genetic differences, even in tumors with the same basic histologic features, are the key determinants of response to any particular drug. For example, the HER2 receptor–directed therapeutic antibody Herceptin exhibits clinical efficacy specifically in a subset of breast cancers that exhibit amplification of the *HER2* gene (Slamon *et al.*, 2001). Similarly, the multi-kinase inhibitor, Gleevec, exhibits clinical activity in patients with gastrointestinal stromal tumors that harbor activating c-KIT receptor mutations (Slamon *et al.*, 2001). Finally, the selective EGFR kinase inhibitors, Iressa and Tarceva, are particularly effective in lung cancers that harbor mutationally activated alleles of the EGF receptor (Lynch *et al.*, 2004; Paez *et al.*, 2004; Pao *et al.*, 2004).

We have undertaken efforts to model sensitivity to such selective kinase inhibitors using a large panel of human cancer cell lines, with the goal of identifying genotypes that correlate well with drug sensitivity and can be used to guide clinical treatment. Our collection includes more than 1000 cancer cell lines derived from tumors of virtually every tissue type. We have optimized technologies to maintain and propagate this large collection of cell lines, as well as to screen the collection for sensitivity to drugs and other targeted inhibitory compounds. The cell lines are being tested using various assays of cell growth and survival following exposure to a variety of experimental compounds to determine whether a subset of the lines exhibit exceptional drug sensitivity. Sensitive lines are further analyzed to identify molecular features that are shared among the responsive cell lines, which may serve as "biomarkers" that can be used to predict treatment outcomes.

Here, we describe the methodologies required to conduct such high-throughput human tumor cell line screening, and we describe findings using a panel of 116 non–small-cell lung cancer (NSCLC) lines in a screen for sensitivity to the EGFR kinase inhibitor, erlotinib. Erlotinib sensitivity is seen clinically in ~10% of non–small-cell lung cancers (NSCLC), and is well correlated with the presence of somatic EGFR kinase domain mutations in tumors. Activating K-Ras mutations are found in ~35% of NSCLCs, and these patients typically do not respond to erlotinib (Pao *et al.*, 2005). EGFR and K-Ras mutations are nonoverlapping in NSCLC, and it has been suggested that this may reflect the fact that Ras is a downstream mediator of EGFR function, and therefore, such mutations would be redundant. Our cell line screening data demonstrate that the high-throughput cell line screening approach can be used to recapitulate the EGFR mutation–correlated sensitivity that has been observed clinically.

2. METHODS

2.1. Human cancer cell lines

Approximately 1100 human cancer cell lines were obtained from the American Type Culture Collection (ATCC, Rockville, MD), the Deutsche Sammlung von Mikroorganismen und Zellkulturen GmbH (DSMZ, Braunschweig, Germany), the Japanese Collection of Research Bioresources (JHSF, Osaka, Japan), or the European Collection of Cell Cultures (ECACC, Porton Down, UK) (Fig. 23.1). Cells were initially grown in RPMI medium supplemented with 5% FBS, and maintained at 37° in a humidified atmosphere at 5% CO_2. They were only switched to the richer DMEM/F12 growth medium if they did not grow optimally in this medium. Cell lines were routinely propagated in these two media in order to minimize the potential effect of varying the media on their sensitivity to therapeutic compounds in our assay, and to facilitate high-throughput screening. A considerable challenge when dealing with a cell line collection of this size is minimizing the potential risks of bacterial or fungal infections and, perhaps more significantly, eliminating the chances of cross-contamination of cell lines. Studies from the Sanger Institute (Cambridge, UK) have already highlighted the considerable problem of cell lines with common ancestry (http://www.sanger.ac.uk/genetics/CGP/Genotyping/synlinestable.shtml). Therefore, we have obtained whole genome SNP (single nucleotide polymorphism) arrays for each of the cell lines, which essentially provide a "fingerprint" that can be used to compare lines for potential cross-identity. In addition, to reduce the

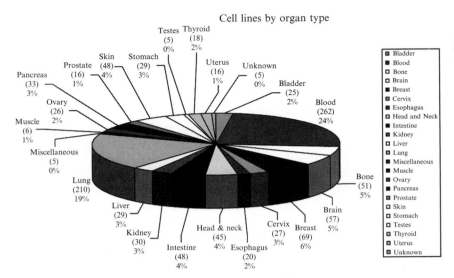

Figure 23.1 Pie chart representation of the 1100 human cancer cell lines used for high-throughput screening. They are divided according to the relevant tissue of origin and percentage of the total number of lines. The actual number of cell lines in each group is indicated in parentheses.

likelihood of cell line contamination, we elected to use Opticell® plates (Opticell, Westerville, OH) for the growth and maintenance of cell lines. In this contained system, the cells are grown between two gas–permeable membranes held in a rigid plastic frame, and accessed through two self-sealing rubber ports. The ports allow medium to be changed and cells to be trypsinized while eliminating any risks of cross–contamination. In addition, the cells can be stored at −140° in the Opticells when a DMSO-containing medium is used. Prior to screening a cell line, the Opticell is defrosted and the freezing medium replaced with fresh medium. When 60 to 70% confluent, the Opticell is trypsinized to a 10-cm plate where the cells are grown for an additional 2 to 3 days before plating on 96-well plates for screening. In our experience, many cell lines adhere and grow more rapidly on the plates compared to the Opticells.

2.2. Assay compounds

The selective EGFR kinase inhibitor, erlotinib, was kindly provided by Genentech, Inc. (San Francisco, CA). Additional compounds that we have screened in related studies were obtained through a variety of MTA agreements with pharmaceutical and biotechnology companies or are synthesized by collaborating chemists. The emphasis in our screen has been on molecules in preclinical development as potential anticancer drugs or molecules already being tested in clinical trials in humans in order for any findings to have the greatest potential clinical significance. All compounds are stored at a 10 mM concentration in DMSO at −80° unless specified otherwise by the source. Cells are treated in triplicate with three concentrations of each compound, to yield a 100-fold difference between the lowest and highest concentrations. The range of concentrations selected for each compound was based on *in vitro* data of the concentrations inhibiting relevant kinase activity and cell viability, as well as clinical data indicating peak and trough plasma concentrations in human subjects.

2.3. Cell viability assay

Cells were seeded in 96-well microplates (BD Biosciences, Franklin Lakes, NJ) at ~15% confluency in medium with 5% FBS. After overnight incubation, the cells were treated with three concentrations of each compound using the Sciclone ALH3000 multichannel liquid handling workstation (Caliper Lifesciences, Hopkinton, MA), and then returned to the incubator for assay at a 72-h time point. At this time point, the cells were processed using the Sciclone, which involved fixation in 4% formaldehyde in PBS (Boston Bioproducts, Worcester, MA) and staining in a 1:5000 solution of the fluorescent nucleic acid stain Syto60 (Molecular Probes, Carlsbad, CA). Although alternate "live cell" assays were quicker and more convenient, we also found that they were more likely to give misleading readings, depending on the particular pharmacologic inhibitor being tested. Quantitation of fluorescent signal intensity was carried out at excitation and emission wavelengths of 630 and 695 nM, respectively, using the SpectraMax M5 plate reader (Molecular Devices, Sunnyvale, CA). The mean of triplicate values for each drug concentration was compared to untreated wells, and a ratio corresponding to response was calculated (Fig. 23.2).

To validate this method of high-throughput tumor cell line screening as a strategy to identify genotype-correlated drug sensitivities, we screened 116 human NSCLC cell lines with the EGFR tyrosine kinase inhibitor erlotinib (Tarceva) (Fig. 23.3). The lines exhibited a wide range of sensitivity to erlotinib, with the vast majority of lines being refractory to treatment at clinically relevant drug concentration (2 micromolar). However, several lines exhibited extreme erlotinib sensitivity, with greater than 80% of cell growth being inhibited after a 72-hour exposure. Of the seven cell lines showing the most marked sensitivity to erlotinib, six were determined to harbor previously identified sensitizing kinase domain mutations upon sequencing EGFR exons 18 to 21. Interestingly, the only cell line in this group with wildtype EGFR exhibited high levels of Her2/Erbb2 gene amplification. Furthermore, as previously reported by Pao *et al.* (2005), mutations in EGFR and KRAS appear to be mutually exclusive.

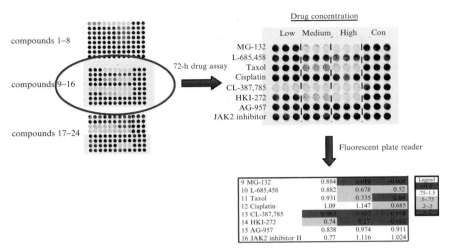

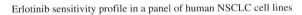

Figure 23.2 A schematic flow chart corresponding to the steps involved in assaying the effect of putative anticancer compounds on cell viability in the high-throughput screen. Following incubation with a fluorescent nuclear stain, the emission intensity of each well is read on a SectraMaxM5 plate reader, and the signal intensity of each well is calculated relative to the untreated cells. These data enable us to generate a profile of sensitivity across a panel of lines for each compound.

Erlotinib sensitivity profile in a panel of human NSCLC cell lines

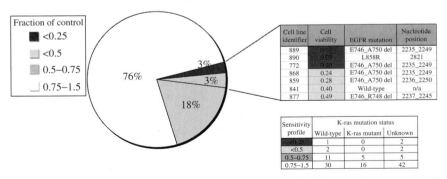

Figure 23.3 Pie chart representation of the sensitivity of 116 human lung cancer cell lines to treatment with 200 n*M* of the EGFR TKI erlotinib (Tarceva). The drug effect was calculated as the fraction of untreated cells at 72 h. Findings corresponding to the most sensitive cell lines are detailed. The bottom table depicts a sensitivity profile as correlated with K-Ras mutational status.

These methodologies were subsequently applied to a panel of 250 human cancer cell lines treated with a variety of kinase inhibitors thought to be important as either prosurvival or antiapoptotic factors in cancer cells (Fig. 23.4). This high-throughput screening platform is capable of identifying subsets of cell lines with marked sensitivities to specific agents; these lines will be further analyzed to identify those molecular features that may be "biomarkers" of response both *in vitro* and potentially *in vivo*.

2.4. EGFR sequencing

Genomic DNA was isolated from cell lines using the Gentra purification system according to the manufacturer's protocol. The EGFR kinase domain (exons 18 to 24) was amplified from genomic DNA by PCR with primers listed below. PCR products were purified using exonuclease I and shrimp alkaline phosphatase (United States Biochemical, Cleveland, OH) followed by bidirectional sequencing using BigDye v1.1 (Applied Biosystems, Foster City, CA) in combination with an ABI3100 sequencer (Applied Biosystems). Primers used for sequencing of exons 18 to 21 are listed in the following. For exons 22 to 24, sequencing primers were identical to PCR primers. Electropherograms were analyzed using Sequence Navigator software (Applied Biosystems). All mutations were confirmed by at least two independent PCR amplifications (Table 23.1).

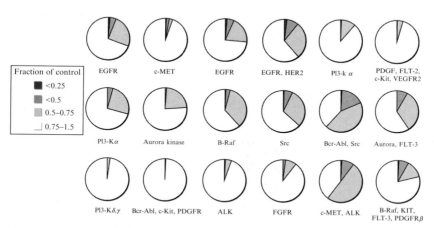

Figure 23.4 Pie chart representation of the sensitivity of 250 human cancer cell lines to a variety of kinase inhibitors following treatment for 72 h. The target kinase for various compounds that have been screened is indicated below the relevant pie chart, and the magnitude of effect is calculated as the fraction of viable cells relative to untreated controls. The concentration selected for each compound was based on *in vitro* data of the concentrations inhibiting relevant kinase activity and cell viability, as well as clinical data indicating peak and trough plasma concentrations in human subjects.

2.5. Western blot analysis of signaling pathways

Some of the cell lines exhibiting differential erlotinib sensitivity were subsequently examined for biochemical signaling responses to drug treatment. To harvest protein lysates, cells were washed with PBS, scraped in lysis buffer (150 mM NaCl, 1% NP-40, 50 mM Tris, 2 mM EDTA, 10% glycerol, 5 μg/ml each of aprotinin, leupeptine, and pepstatin, and 1 mM each of NaF, Na_3VO_4 and PMSF) and incubated on ice for 40 min. The lysates were centrifuged at 14,000 RPM for 20 min and the supernatant collected. Protein concentration was measured with the bicinchoninic acid protein assay (Pierce, Rockford, IL) and resolved by SDS–polyacrylamide gel. The gels were electroblotted onto PVDF membranes (Hybond-P, Amersham). Antibody detection was performed with a chemiluminescence

Table 23.1   PCR amplification primers

PCR Primers		
Designation	Primer Sequence (5'-3')	Orientation
18F	CAAGTGCCGTGTCCTGGCACCCAAGC	Sense
19F	GCAATATCAGCCTTAGGTGCGGCTC	Sense
20F	CCATGAGTACGTATTTTGAAACTC	Sense
21F	CTAACGTTCGCCAGCCATAAGTCC	Sense
22F	GAGCAGCCCTGAACTCCGTCAGACTG	Sense
23F	CAGGACTACAGAAATGTAGGTTTC	Sense
24F	GACTGGAAGTGTCGCATCACCAATG	Sense
18R	CCAAACACTCAGTGAAACAAAGAG	Antisense
19R	CATAGAAAGTGAACATTTAGGATGTG	Antisense
20R	CATATCCCCATGGCAAACTCTTGC	Antisense
21R	GCTGCGAGCTCACCCAGAATGTCTGG	Antisense
22R	CTCAGTACAATAGATAGACAGCAATG	Antisense
23R	GTGCCTGCCTTAAGTAATGTGATGAC	Antisense
24R	GGTTTAATAATGCGATCTGGGACAC	Antisense
Sequencing Primers		
18F.SEQ	GCACCCAAGCCCATGCCGTGGCTGC	Sense
19F.SEQ	CCTTAGGTGCGGCTCCACAGC	Sense
20F.SEQ	GAAACTCAAGATCGCATTCATGC	Sense
21F.SEQ	CGTGGAGAGGCTCAGAGCCTGGCATG	Sense
18R.SEQ	GAAACAAAGAGTAAAGTAGATGATGG	Antisense
19R.SEQ	CATTTAGGATGTGGAGATGAGC	Antisense
20R.SEQ	GCAAACTCTTGCTATCCCAGGAG	Antisense
21R.SEQ	CATCCTCCCCTGCATGTGTTAAAC	Antisense

Note: For exons 22,23,24 sequencing primers are identical to PCR primers.

kit (Supersignal, Pierce). Equal lane loading was assessed using a β-tubulin mouse monoclonal primary antibody (Sigma). The Akt, Erk1/2, phospho-Erk1/2(T202/Y204), and phospho-STAT3 (S727) antibodies were from Cell Signaling Technology (Beverly, MA). The phospho-EGFR (Y1068) antibody was from Abcam (Cambridge, MA). The phospho-Akt (S473) antibody was from BioSource International (Camarillo, CA), and the β-tubulin antibody was from Sigma (St. Louis, MO). All antibodies were used at a 1:1000 dilution, except for the β-tubulin antibody, which was used at 1:10,000 dilution.

The effect of erlotinib on the various prosurvival signaling pathways downstream of the receptor tyrosine kinase was determined in an EGFR mutant, KRAS wildtype cell line (PC9) and in an EGFR wildtype, KRAS mutant cell line (NCI-H1573). Sensitivity to erlotinib was associated with a dramatic abrogation of phosphorylation of EGFR and its signaling partners STAT3, Akt, and Erk1/2 (Fig. 23.5). In contrast, in the KRAS mutant cell line, there was no effect of erlotinib on any of these pathways.

2.6. Database management

Data analysis was executed using the FileMaker Pro software package (FileMaker Inc, Santa Clara, CA) and dedicated software programming. Each cell line was assigned a unique identifier number to facilitate recovery

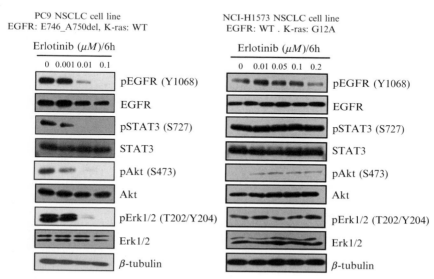

Figure 23.5 Effect of a range of erlotinib treatment at 6 h on downstream pro-survival signaling pathways in an EGFR mutant, K-Ras wildtype cell line (PC9) or an EGFR wildtype, K-Ras mutant (NCI-H1573) cell line.

from liquid nitrogen storage facilities, and also to track how often cell lines were defrosted/frozen. The cell line sensitivity data to the compounds tested are stored in a dedicated database that allows us to rapidly sort cell lines by magnitude of sensitivity to various compounds as well as by tissue type, growth conditions, origin, and so on.

3. CONCLUSIONS

High-throughput screening of a large number of tumor cell lines with a variety of putative anticancer compounds can be readily achieved using automated technology. Moreover, this strategy has the potential to identify small subsets of cancer cell lines showing marked sensitivity to clinically relevant kinase inhibitors. In our test case, using NSCLCs that are known to harbor EGFR- and Ras-activating mutations, we were able to identify drug-sensitive and -insensitive lines that correlate with genotypes that have been associated with clinical responses. Identification of compound-sensitive tumor lines, when using a compound whose target is poorly defined or is not yet validated, can potentially be the first step in determining the molecular characteristics that underlie sensitivity, and can provide essential preclinical data to guide clinical trials of genotype-directed therapy in patients with such novel therapeutics. Overall, these findings validate the strategy of screening large numbers of tumor cell lines for sensitivity to putative anticancer compounds, and suggest that human cancer cells can be informative models for therapeutic responses to signaling pathway inhibitors, and that large numbers may be required to fully capture the genetic heterogeneity in human cancers.

REFERENCES

Lynch, T. J., Bell, D. W., Sordella, R., Gurubhagavatula, S., Okimoto, R. A., Brannigan, B. W., Harris, P. L., Haserlat, S. M., Supko, J. G., Haluska, F. G., Louis, D. N., Christiani, D. C., *et al.* (2004). Activating mutations in the epidermal growth factor receptor underlying responsiveness of non–small-cell lung cancer to gefitinib. *N. Engl. J. Med.* **350,** 2129–2139.

Paez, J. G., Janne, P. A., Lee, J. C., Tracy, S., Greulich, H., Gabriel, S., Herman, P., Kaye, F. J., Lindeman, N., Boggon, T. J., Naoki, K., Sasaki, H., *et al.* (2004). EGFR mutations in lung cancer: correlation with clinical response to gefitinib therapy. *Science* **304,** 1497–1500.

Pao, W., Miller, V., Zakowski, M., Doherty, J., Politi, K., Sarkaria, I., Singh, B., Heelan, R., Rusch, V., Fulton, L., Mardis, E., Kupfer, D., et al. (2004). EGF receptor gene mutations are common in lung cancers from "never smokers" and are associated with sensitivity of tumors to gefitinib and erlotinib. Proc. Natl. Acad. Sci. U S A. 101, 13306–13311.

Pao, W., Wang, T. Y., Riely, G. J., Miller, V. A., Pan, Q., Ladanyi, M., Zakowski, M. F., Heelan, R. T., Kris, M. G., and Varmus, H. E. (2005). KRAS mutations and primary resistance of lung adenocarcinomas to gefitinib or erlotinib. PLoS Med. 2, e17.

Ross, J. S., Schenkein, D. P., Pietrusko, R., Rolfe, M., Linette, G. P., Stec, J., Stagliano, N. E., Ginsburg, G. S., Symmans, W. F., Pusztai, L., and Hortobagyi, G. N. (2004). Targeted therapies for cancer 2004. Am. J. Clin. Pathol. 122, 598–609.

Slamon, D. J., Leyland-Jones, B., Shak, S., Fuchs, H., Paton, V., Bajamonde, A., Fleming, T., Eiermann, W., Wolter, J., Pegram, M., Baselga, J., and Norton, L. (2001). Use of chemotherapy plus a monoclonal antibody against HER2 for metastatic breast cancer that overexpresses HER2. N. Engl. J. Med. 344, 783–792.

CHAPTER TWENTY-FOUR

In Vitro Signaling by MAPK and NFκB Pathways Inhibited by *Yersinia* YopJ

Sohini Mukherjee *and* Kim Orth

Contents

Abstract

A procedure for an *in vitro* signaling assay is described for the MAPK and NFκB pathways. The method uses a membrane-cleared lysate that contains all the soluble components required for activating these signaling cascades. The pathways can be activated by variety of molecules, including kinases, G-proteins, and E3 ligases. We demonstrate that YopJ inhibits downstream of all these activators. The *in vitro* signaling assay is ideal for initial biochemical studies on activators and inhibitors of the MAPK and NFκB pathways.

1. Introduction

Yersinia pestis is the infectious agent that caused the Black Death in the Middle Ages, and *Y. pseudotuberculosis* and *Y. enterocolitica* (two closely related food-borne pathogens) are causal agents of gastrointestinal disorders (Cornelis, 2000). *Yersinia* spp. harbor a 70-kb plasmid that encodes a type III secretion system (TTSS) that secretes Yops (*Yersinia* outer proteins, also referred to as virulence factors or effectors) (Cornelis, 2002). The type III secretion system is a transport vehicle used by the microbial pathogen to

University of Texas Southwestern Medical Center, Dallas, Texas

Methods in Enzymology, Volume 438
ISSN 0076-6879, DOI: 10.1016/S0076-6879(07)38024-5

deliver effectors from the bacterial cytosol to the target host cytosol (Viboud and Bliska, 2005). Upon translocation of the effectors into the target host cell, these molecules cripple the target cell by (1) blocking phagocytosis, (2) destroying the host defense system, and (3) inducing apoptosis (Orth, 2002). Each effector appears to mimic or capture the activity of a eukaryotic protein, which alters the signaling machinery in the target cell and gives an advantage to the pathogen during infection. Paradoxically, the effectors remain in a quiescent state inside the pathogen due to the lack of a substrate or an activator or possibly the presence of a chaperone (Feldman and Cornelis, 2003; Orth, 2002).

Early studies on one of the *Yersinia* effectors, YopJ, demonstrate that this 32-kDa effector inhibits cytokine production and induces apoptosis in a target cell during infection (reviewed by Orth, 2002). YopJ, with no obvious homology to any known eukaryotic protein domain, possesses the remarkable capacity to block multiple signaling pathways including the MAPK and the NFκB pathways. Interruption of these signaling pathways disrupts expression of TNFα and IL-8 by the infected target cell (Monack *et al.*, 1997; Palmer *et al.*, 1998; Ruckdeschel *et al.*, 1998).

Later studies revealed that YopJ interacted specifically with the family of mitogen activated protein kinase kinases (MKKs), and did not interact with downstream kinases (MAPKs: ERK, JNK and p38) or with upstream kinases (B-Raf) (Orth *et al.*, 1999). Although the MAPK and NFκB pathways are distinct, insights into how YopJ could inhibit both of these signaling systems are gained when the pathways are viewed in parallel as depicted in (Fig. 24.1) (Orth *et al.*, 1999). Herein, a representative MAPK pathway is shown that includes a cascade of three kinases (Raf, MKK, and

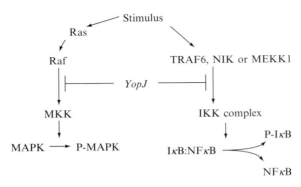

Figure 24.1 MAPK and NFκB signaling pathways. The MAPK pathway is activated *in vitro* by Ras or Raf and the phosphorylation of MAPKinase (P-MAPK) is observed to indicate activation of the pathway. The NFκB pathway is activated by TRAF6, NIK, or MEKK1, and the phosphorylation of IκB (P-IκB) is observed to indicate activation of the pathway.

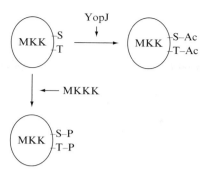

Figure 24.2 YopJ inhibits activation MAPK kinase by acetylation of serine and threonine residues, thereby preventing phosphorylation by upstream kinase (MKKK).

MAPK) resulting in the phosphorylation of the MAPK. The NFκB pathway is activated by a number of stimuli that converge on the activation of the IκB kinase (IKK) complex. When the IKK complex is activated, IκB from the NFκB:IκB complex is phosphorylated, which provides a signal for its own ubiquitination and subsequent degradation. Removal of IκB exposes the NLS on the NFκB transcription factor, resulting in translocation of NFκB to the nucleus where it activates transcription (see Fig. 24.1). Studies demonstrating that YopJ binds IKKβ, the MKK equivalent in the NFκB pathway, support the hypothesis that YopJ blocks the pathways at a similar point. In total, these observations identified a common critical point in the MAPK pathways and the NFκB pathway that is targeted for disruption by a single *Yersinia* virulence factor, YopJ (Orth *et al.*, 1999).

Further studies revealed that YopJ contains a catalytic triad with a cysteine, histidine, and aspartate, similar to those found in hydrolases. Mutation of the catalytic cysteine results in a mutant of YopJ (C172A) that is unable to inhibit the signaling pathways. It was discovered that YopJ uses its catalytic site to acetylate serine and threonine residues on the activation loop of MKKs, thereby directly competing with the post-translational modification phosphorylation (Mukherjee *et al.*, 2006) (Fig. 24.2). The finding of this newly discovered post-translational modification on how YopJ inhibited these signaling pathways was made, in part, by development of the *in vitro* signaling assay described in this article.

2. THE *IN VITRO* SIGNALING ASSAY

The cell-free signaling system uses concentrated, membrane-free, soluble lysate, an ATP regenerating system, and various partially purified or purified proteins to activate the various signaling pathways. The readout for this assay is the production of phosphorylated proteins: for the MAPK pathways the readout is the phosphorylated MAPKs, including ERK, p38, and JNK; for the NFκB pathway the readout is phosphorylated IκB. Each of these phosphorylated proteins can be detected by specific antibodies in immunoblot analysis.

3. MATERIALS AND METHODS

3.1. Materials

3.1.1. Plasmids and reagents

Expression constructs used in this study are pSFFV-YopJ-Flag and pSFFV-YopJ-Flag-C172A (Orth *et al.*, 1999), pGEX-TEV-YopJ (Mukherjee *et al.*, 2006), pRK7-Flag-NIK (kindly provided by Zhijian J. Chen), and pDCR-HA-RasV12 (kindly provided by Mike White).

Antibodies for phospho-ERK1/2, phospho-MKK, phospho-IκB, phospho-IKK, and phospho-MKK6 and MKK1/2, are purchased from Cell Signaling; for IKKβ and aldolase, Santa Cruz Biotechnology; for anti-HA antibody, Covance; and for anti-FLAG antibody, SIGMA.

Recombinant B-Raf is purchased from Upstate. EGF and Okadaic acid are purchased from Alexis; anti-Flag M2 resin from SIGMA; and protease inhibitor cocktail tablet from Roche.

3.1.2. Solutions

Phosphate buffered saline (PBS): 8 mM Na$_2$HPO$_4$, 1.1 mM NaH$_2$PO$_4$, 150 mM NaCl

HTX: 10 mM HEPES, pH 7.4, 0.5% Triton-X 100, 10 mM MgCl$_2$, 1 mM MnCl$_2$, 0.1 mM EGTA

HNT: 10 mM HEPES, pH7.4, 50 mM NaCl, and 1% TritonX-100

10$\times$ ATP Regenerating Solution (ARS): 10 mM ATP, 350 mM creatine phosphate, 20 mM Hepes, pH 7.2, 10 mM MgCl$_2$, 500μg/ml creatine kinase

5$\times$ SDS-PAGE sample buffer

3.1.3. Cell culture

HEK293 cells are used in all experiments. The cells are cultured in Dulbecco's modified Eagle's medium (Invitrogen) with 10% cosmic calf serum (Gemini) in 5% CO$_2$. Large-scale transfections are performed with cells cultured on 150-mm plates with 10 μg of plasmid DNA using calcium-phosphate-based transfection method (Russell, 2006). In brief, cells are transfected in the morning, media is changed 8 h after transfection, and then 24 h later, cells are harvested for NFκB experiments, or cells are starved with serum-free media overnight and then harvested for MAPK experiments.

3.2. Methods

3.2.1. Membrane-cleared lysates

The protocol for lysate preparation is a modified version of a protocol developed by Sturgill and colleagues (Dent *et al.*, 1993).

1. One 150-mm plate of HEK293 cells, transfected with pSFFV and pSFFV-YopJ-Flag or pSFFV-YopJ-Flag-C172A as described above, is placed on ice and gently washed with 2.5 ml of cold PBS + 1 mM EDTA. After removing the cold wash, cells are popped loose from the bottom of the plate by tapping sharply on the side of the dish. Scraping off cells from the plate is avoided during harvesting to avoid shearing of cells. Loose cells are harvested by gentle washing of the plate with 5 ml of cold PBS + 1mM EDTA, followed by centrifugation at 400×g for 5 min at 4°.

2. Each cell pellet is gently resuspended in 1 ml of cold PBS and transferred into a 1.5-ml eppendorf tube. It is important to keep the volume of cells in each tube below ~250 μl so that the cells are efficiently lysed after step 3.

3. Cells are recentrifuged at 400×g for 5 min, and after removal of the liquid above the cell pellet, a volume of HTX lysis buffer (10 mM HEPES, pH 7.4, 0.5% Triton-X 100, 10 mM MgCl$_2$, 1 mM MnCl$_2$, 0.1 mM EGTA, and protease inhibitor cocktail tablet [Roche]) equal to the volume of the cell pellet is added to the eppendorf tube. The cells are gently resuspended and then incubated on ice for 20 to 60 min. Close to the end of the incubation period the cells are dounced with an eppendorf pestle.

4. The cell lysate is subjected to differential centrifugation. At first it is centrifuged for 5 min at 800×g to separate the nuclear fraction. The cloudy liquid above the pellet is transferred to a prechilled eppendorf (if more than one 150-mm plate has been used to prepare a large lysate preparation, the lysates can be combined at this step into a single eppendorf). The lysate is next spun at 16,000×g for 10 min to remove various cellular organelles such as lysosomes and mitochondria. To clear the lysate of membranes, the supernatant is transferred to a prechilled 1.5-ml polyallomer microfuge tube (Beckman) for centrifugation at 100,000×g for 30 to 60 min in a table-top Beckman ultracentrifuge (TL-100).

5. In the final step, the supernatant is transferred to a prechilled 1.5-ml eppendorf tube. The protein concentration of the lysate is measured using Coomassie Plus Protein Assay reagent (Pierce). Protein concentration

should be ~10 mg/ml. For long-term storage, the lysate is snap-frozen on dry ice and transferred to a -80° freezer.

6. Stability of cell lysates can be sensitive to freeze/thaw. Background levels of phosphorylated MAP kinases may increase with samples that are freeze/thawed multiple times. Based on our experience with this assay, the NFκB signaling pathway is less sensitive to freeze/thaw storage.

3.2.2. Isolation of EGF- and RasV12-activated membranes

The preparation of stimulated membranes is essentially the same as the preparation of lysate with the following exceptions (Dent *et al.*, 1993):

1. For RasV12-activated membranes, one 150-mm plate of HEK293 cells is transfected with 10 μg of pDCR-HA-RasV12 plasmid DNA using calcium-phosphate-based transfection method (Russell, 2006) in the morning, media is changed 8 h after transfection, and then 24 h later cells are harvested after overnight starvation with serum-free media.

   For EGF-stimulated membranes, HEK293 cells are plated at 80% confluency and are starved with serum-free media overnight. Cells are then incubated with 50 ng/ml EGF (2 ml/150-mm plate) for 5 min at room temperature.

   For both RasV12- and EGF-treated cells, one 150-mm plate of transfected HEK293 cells is placed on ice and gently washed with 2.5 ml of cold PBS + 1 mM EDTA. After removing the cold wash, the cells are popped loose from the bottom of the plate by tapping sharply on the side of the dish. Scraping off cells from the plate is avoided during harvesting to avoid shearing of cells. Loose cells are harvested by gentle washing of the plate with 5 ml of cold PBS + 1mM EDTA, followed by centrifugation at 400×g for 5 min at 4°.

2. Follow steps 2 to 4 in section 3.2.1.

3. The supernatant after the 100,000×g spin is discarded and the membrane pellet is resuspended in 100 μl of HTX buffer, forming a crude suspension of membranes containing activated signaling components. Protein concentrations can be estimated using the Coomassie Plus protein assay reagent (Pierce).

3.2.3. Purification of recombinant TRAF6

pFast-Bac-T6RZC virus (kindly provided by Zhijian J. Chen and Jun-ichiro Inoue) was used to purify the recombinant TRAF6 protein from Sf9 cells using Ni^{2+} affinity purification (Qiagen). Infected cells were lysed using Emulsiflex C-5 cell homogenizer (Avastin) in lysis buffer containing 20 mM Tris (pH 7.5), 10 mM KCl, 1.5 mM MgCl$_2$, 10 mM NaF, 40 mM EGTA, 0.05% β—mercaptoethanol, and a protease inhibitor cocktail tablet (Roche). Cell lysate was allowed to bind to Ni-NTA resin for 1 h at 4°, and then eluted from beads using 250 mM imidazole-containing buffer.

3.2.4. Isolation of activated FLAG-NIK beads

1. One 100-mm plate of HEK293 cells, at 60 to 70% confluency, is transfected with 3.5 μg PRK7-FLAG-NIK using Fugene transfection reagent (Roche) and harvested 24 h post-transfection.
2. Cells are placed on ice and gently washed with 2 ml of cold PBS. After removing the cold wash, the cells are incubated with 1 ml of HNT lysis buffer (10 m*M* HEPES, pH 7.4, 50 m*M* NaCl, 1% TritonX-100, 1 m*M* DTT, 20 m*M* NaF, 20 m*M* β-glycerophosphate, 0.5 m*M* sodium vanadate, 0.5 m*M* EGTA and protease inhibitor cocktail tablet) for 10 min on ice.
3. Cells are scraped from the plate using a scraper, and the lysate is centrifuged at 16,000×*g* for 10 min at 4° to get rid of the cell debris.
4. The cleared lysate is subjected to immunoprecipitation using anti–FLAG M2 resin (30 μl beads for 1 ml lysate) for 1 h at 4°, following which the beads are washed three times with HNT buffer. The beads are next used in the assay described in the following section to activate the NFκB pathway.

3.2.5. *In Vitro* signaling assays

3.2.5.1. In Vitro *signaling assay for activation of ERK/MAPK signaling pathway*
The results of the following assay are shown in Fig. 24.3, where control lysates and YopJ lysates have been activated by the addition

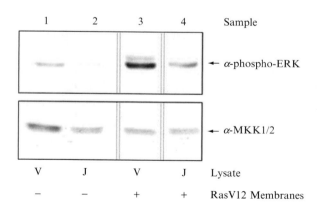

Figure 24.3 *In vitro* activation of the ERK/MAPK pathways by RasVl2–activated membranes. Lysates from cells transfected with a control plasmid (V) (lanes 1, 2) or YopJ expression plasmid (J) (lanes 3, 4) are incubated in the absence (lanes 1, 3) or presence (lanes 2, 4) of RasV12–activated membranes. Samples are assayed by immunoblot analysis with anti–phospho-ERK antibody (used as an indicator for activation of pathway) and anti-MKK1,2 antibody (used as a load control for lysates). (Note slightly more V lysate ias added to sample #1, resulting in a slight increase in background for the anti–phospho-ERK blot.)

of RasV12-activated membranes. The MAPK pathway in control lysates is activated as demonstrated by the appearance of phosphorylated ERK/MAP-Kinase (sample 3). By contrast, the MAPK pathway in the YopJ lysates is inhibited, as demonstrated by the lack of phosphorylated ERK/MAPKinase (sample 4).

Materials used in this assay include:

Membrane-cleared lysates for MAPK signaling from cells transfected with pSFFV (V) and pSFFV-YopJ-Flag (J). All lysates should be adjusted to 10 mg/ml protein as measured by the Coomassie Plus protein assay reagent (Pierce).
RasV12-activated membranes
$10\times$ ARS
HTX

Samples are prepared as described below in prechilled eppendorf tubes. Solutions are added in order as described in Table 24.1.

1. Reactions are initiated by addition of $2\,\mu l$ of RasV12 activated membranes to samples 3 and 4.
2. All samples are incubated at 37° for 10 min.
3. Reactions are terminated by the addition of 20 μl of $5\times$ SDS-PAGE sample buffer, followed by incubation at more than 90° for 5 min.
4. Samples are separated by SDS-PAGE and analyzed by immunoblot analysis with anti–phospho-ERK antibody followed by immunoblot analysis with anti–MKK antibody (for equal loading of lysate). The phosphorylated molecules should always be analyzed first as the post-translational modification appears labile after multiple immunoblot analyses.

Similar experiments can be designed with other activators, such as activated membranes from cells treated with EGF or recombinant B-Raf as demonstrated by Mukherjee et al. (2006). In this protocol, $2\,\mu l$ of the activator is added in step1, followed by steps 2 to 4 as described (see Fig. 24.3).

Table 24.1 *In vitro* MAPK signaling assay

Sample	1	2	3	4
V	16 μl	—	16 μl	—
J	—	16 μl	—	16 μl
$10\times$ARS	2 μl	2 μl	2 μl	2 μl
HTX	2 μl	2 μl	—	—

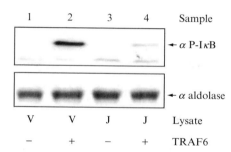

Figure 24.4 *In vitro* activation of the NFκB pathway by recombinant TRAF6. Lysates from cells transfected with a control plasmid (V) (lanes 1, 3) or YopJ expression plasmid (J) (lanes 2, 4) are incubated in the absence (lanes 1, 2) or presence (lanes 3, 4) of TRAF6. Samples are assayed by immunoblot analysis with anti-phospho-IκB antibody (used as an indicator for activation of pathway) and antialdolase antibody (used as a load control for lysates).

3.2.5.2. In Vitro *signaling assay for activation of NFκB signaling pathway by TRAF6*

The results of the following assay are shown in Fig. 24.4, where control lysates and YopJ lysates have been activated by the addition of TRAF6. The NFκB pathway in control lysates is activated as demonstrated by the appearance of phosphorylated IκB. By contrast, the NFκB pathway in the YopJ lysates is inhibited, as demonstrated by the lack of phosphorylated IκB.

Materials used in this assay include:

Membrane-cleared lysates for MAPK signaling from cells transfected with pSFFV (V) and pSFFV-YopJ-Flag (J). All lysates should be adjusted to ~10 mg/ml protein as measured by the Coomassie Plus protein assay reagent (Pierce).

Recombinant TRAF6 (TRAF6)
10× ARS
HTX

Samples are prepared as described below in prechilled eppendorf tubes. Solutions are added in order as described in Table 24.2.

1. Reactions are initiated by addition of 2 μl of TRAF6 (final 0.5 μM) to samples 2, 4, and 6.
2. All samples are incubated at 37° for 10 min.
3. Reactions are terminated by the addition of 20 μl of 5×SDS-PAGE sample buffer, followed by incubation at more than 90° for 5 min.
4. Samples are separated by SDS-PAGE and analyzed by immunoblot analysis with anti-phospho-IκB antibody followed by immunoblot analysis with antialdolase antibody or anti-IKKβ (for equal loading of lysate).

Table 24.2 *In vitro* NFκB signaling assay

Sample	1	2	3	4
V	16 μl	16 μl	—	—
J	—	—	16 μl	16 μl
10×ARS	2 μl	2 μl	2 μl	2 μl
HNT	2 μl	—	2 μl	—

The phosphorylated molecules should always be analyzed first as the post-translational modification appears labile after multiple immunoblot analyses.

Similar experiments can be designed by addition of activated FLAG-NIK beads, prepared as described above. In the protocol shown above, 2 μl of a 1:1 bead slurry are added in step 1 instead of TRAF6. Steps 2 to 4 are then performed as described. The addition of equal amounts of activated FLAG-NIK beads is confirmed by immunoblot analysis with anti-FLAG antibody.

Alternatively, another recombinant kinase, such as MEKK1, can be added to start the *in vitro* signaling reaction, with confirmation that equal amounts of the kinase are added by performing Western blot analysis (see Fig. 24.4).

4. Conclusion

We have described an *in vitro* signaling assay for MAPK and NFκB signaling pathways. Diverse types of activators, including kinases, activated G-proteins, and E3 ligases, are used to trigger the activation of signaling cascades within a concentrated lysate. The time course for the assay is relatively short (<10 min) and the readout for activation of a pathway is simply an immunoblot. Herein, we demonstrate that this assay can be used to study inhibitors, such as the *Yersinia* effector, YopJ. This *in vitro* signaling assay is ideal for initial studies on novel activators and inhibitors of the MAPK and NFκB signaling pathways. Immunoblot analysis of intermediate components of the signaling pathway can be analyzed using phospho-specific antibodies (such as anti-phospho-MKK or anti-phospho-IKK). The observations from the *in vitro* signaling system are consistent with previous genetic, microbial and cellular studies on the inhibitory activity of YopJ on the MAPK and NFκB signaling pathways. The *in vitro* signaling assay is a method for analyzing signaling pathways in a test tube and provides a means to manipulate the signaling system by the addition of extraneous proteins or chemicals.

ACKNOWLEDGMENTS

We thank the members of the Orth Lab for their generous support and helpful suggestions. This work is supported by grants from the National Institutes of Health (R01-AI056404 and R21-DK072134) and the Welch Research Foundation (I-1561). K. O. is a Burroughs Wellcome Fund investigator in the pathogenesis of infectious disease.

REFERENCES

Cornelis, G. R. (2000). Molecular and cell biology aspects of plague. *Proc. Natl. Acad. Sci. U. S. A.* **97,** 8778–8783.

Cornelis, G. R. (2002). The *Yersinia* Ysc-Yop "type III" weaponry. *Nat. Rev. Mol. Cell. Biol.* **3,** 742–752.

Dent, P., Wu, J., Romero, G., Vincent, L. A., Castle, D., and Sturgill, T. W. (1993). Activation of the mitogen-activated protein kinase pathway in Triton X-100 disrupted NIH-3T3 cells by p21 ras and *in vitro* by plasma membranes from NIH 3T3 cells. *Mol. Biol. Cell* **4,** 483–493.

Feldman, M. F., and Cornelis, G. R. (2003). The multitalented type III chaperones: All you can do with 15 kDa. *FEMS Microbiol. Lett.* **219,** 151–158.

Monack, D. M., Mecsas, J., Ghori, N., and Falkow, S. (1997). *Yersinia* signals macrophages to undergo apoptosis and YopJ is necessary for this cell death. *Proc. Natl. Acad. Sci. USA.* **94,** 10385–10390.

Mukherjee, S., Keitany, G., Li, Y., Wang, Y., Ball, H. L., Goldsmith, E. J., and Orth, K. (2006). *Yersinia* YopJ acetylates and inhibits kinase activation by blocking phosphorylation. *Science* **312,** 1211–1214.

Orth, K. (2002). Function of the *Yersinia* effector YopJ. *Curr. Opin. Microbiol.* **5,** 38–43.

Orth, K., Palmer, L. E., Bao, Z. Q., Stewart, S., Rudolph, A. E., Bliska, J. B., and Dixon, J. E. (1999). Inhibition of the mitogen-activated protein kinase kinase superfamily by a *Yersinia* effector. *Science* **285,** 1920–1923.

Palmer, L. E., Hobbie, S., Galan, J. E., and Bliska, J. B. (1998). YopJ of *Yersinia* pseudotuberculosis is required for the inhibition of macrophage TNF-alpha production and downregulation of the MAP kinases p38 and JNK. *Mol. Microbiol.* **27,** 953–965.

Ruckdeschel, K., Harb, S., Roggenkamp, A., Hornef, M., Zumbihl, R., Kohler, S., Heesemann, J., and Rouot, B. (1998). *Yersinia* enterocolitica impairs activation of transcription factor NF-kappaB: Involvement in the induction of programmed cell death and in the suppression of the macrophage tumor necrosis factor alpha production. *J. Exp. Med.* **187,** 1069–1079.

Russell, J. S.a.D. (2006). In "Calcium-phosphate mediated transfection of eukaryotic cells with plasmid DNAs." Cold Spring Harbor Laboratory Press, Cold Spring Harbor, NY.

Viboud, G. I., and Bliska, J. B. (2005). *Yersinia* outer proteins: Role in modulation of host cell signaling responses and pathogenesis. *Annu. Rev. Microbiol.* **59,** 69–89.

CHAPTER TWENTY-FIVE

Biological and Biochemical Characterization of Anthrax Lethal Factor, a Proteolytic Inhibitor of MEK Signaling Pathways

Jennifer L. Bromberg-White *and* Nicholas S. Duesbery

Contents

Abstract

The secretion of factors that block critical intracellular signaling pathways is a common strategy used by pathogenic bacteria for disabling host defenses and causing disease. Anthrax lethal toxin (LeTx) has been shown to cleave and inactivate mitogen-activated protein kinase (MAPK) kinases (MKKs or MEKs) and to inhibit MKK signaling. Cleavage of MKKs by LeTx prevents activation of their downstream substrates, the MAPKs. Because MAPK pathways regulate a variety of crucial cellular functions including proliferation, survival, differentiation, adhesion, and motility, LeTx has become a focus of study as an investigative tool as well as for the treatment and prevention of diseases due to malfunctions in MAPK signaling. This chapter describes methods for expressing and purifying the components of LeTx and focuses on techniques available for assessing its activity.

Laboratory of Cancer and Developmental Cell Biology, Van Andel Research Institute, Grand Rapids, Michigan

Methods in Enzymology, Volume 438
ISSN 0076-6879, DOI: 10.1016/S0076-6879(07)38025-7

1. INTRODUCTION

Macrophages play crucial roles in immune response, acting as first responders and as sentinels, sending out a general alarm that alerts host defenses to threats posed by invading organisms. Activation of mitogen-activated protein kinase (MAPK) signaling pathways is essential for macrophage function, not only for phagocytosis but also for the release of cytokines that potentiate immune response (Rosenberger and Finlay, 2003). Not surprisingly, pathogenic bacteria have evolved diverse and fascinating mechanisms to inhibit MAPK activation. *Yersinia pestis*, the causative agent of bubonic plague, uses a type III secretion system to introduce YopJ into cells. YopJ inactivates MAPK kinase (MKK) by O-acylation of phosphorylation sites in the activation segment (Mittal *et al.*, 2006). Species of *Shigella*, which cause dysentery, use a similar secretion system to introduce a factor called OspF into cells. OspF is a phosphothreonine lyase that modifies phosphorylation sites in the activation segment of MAPK, preventing its phosphorylation and activation (Li *et al.*, 2007). *Bacillus anthracis*, which causes anthrax, secretes lethal toxin, a binary toxin consisting of a binding moiety, protective antigen (PA), and a catalytic moiety, lethal factor (LF). LF is a Zn^{2+}-metalloprotease that cleaves the NH2 termini of mitogen-activated protein kinase kinases (MKKs) 1 to 4, 6, and 7, but not MKK 5 (Duesbery *et al.*, 1998; Pellizzari *et al.*, 1999; Vitale *et al.*, 1998 Vitale *et al.*, 2000). The NH2 termini of MKKs contain docking domains that are required for substrate association, so their loss after proteolysis by LF prevents MKK signaling (Bardwell *et al.*, 2004; Chopra *et al.*, 2003; Duesbery *et al.*, 1998).

Among these models, anthrax in particular has attracted the attention of the scientific and lay communities, in part because of its long history as a model of microbial disease. Studies of anthrax by notable scientists such as Robert Koch and Louis Pasteur led to the development of the germ theory of disease and laid the foundation for modern microbiology. More recently, its use as an agent of bioterrorism has heightened interest in developing therapies for this disease.

While vaccination is an effective prophylactic measure for individuals with a known risk of exposure to anthrax, it is not viewed as a practical measure for protecting large populations due to the limited supply of vaccine (Working Group on Civilian Biodefense *et al.*, 2002). Vaccination in conjunction with antibiotic administration is considered to be the most effective way of treating exposed individuals following a biological attack (Working Group on Civilian Biodefense *et al.*, 2002). However, such measures become less effective as the interval between initial exposure and the onset of treatment widens. For these reasons, therapeutic strategies that emphasize toxin inactivation are seen as worthy of pursuit.

Smith and Keppie (1954) first demonstrated conclusively that the effects of *B. anthracis* were mediated by an exotoxin when they showed that sterile-filtered serum from anthrax-infected guinea pigs could induce edema when injected subcutaneously or death when injected intravenously. Subsequently, it was shown that this toxin was composed of three proteins: PA, LF, and edema factor (EF) (reviewed by Singh *et al.*, 2005). PA is the central receptor-binding component that delivers the catalytic effector molecules, LF or EF, to the cytosol. Pairwise combination of PA with EF or LF produces two toxins with different pathologies. EF in combination with PA induces edema when injected subcutaneously, but is nontoxic upon intravenous injection into animals and hence is referred to as edema toxin, or EdTx. PA in combination with LF does not induce edema when injected subcutaneously, but is toxic upon intravenous injection into animals and hence is referred to as lethal toxin, or LeTx.

Lethal toxin is the dominant virulence factor produced by *B. anthracis*. This was elegantly demonstrated by Mock and colleagues (Cataldi *et al.*, 1990; Pezard *et al.*, 1991, 1993), who showed that strains of *B. anthracis* deficient in the production of LF or PA are not lethal in mice, whereas strains deficient in the production of EF are lethal. For these reasons, efforts to develop an antitoxin have largely focused on understanding and inhibiting LF activity, although it should be noted that edema toxin contributes significantly to the pathogenesis of anthrax (Firoved *et al.*, 2005, 2007).

In the following discussion, we will outline several techniques that can be used to assay LF activity. Fluorescent peptide-based assays for LF activity are commercially available (List Biological Labs, Campbell, CA). However, since sites outside the active site complex on both LF and MEK are required for efficient proteolysis of MEK (Chopra *et al.*, 2003; Liang *et al.*, 2004; Vitale *et al.*, 1998), we favor *in vitro* and cell-based assays that use full-length substrate and proceed under conditions that mimic cellular physiology.

2. Materials and Methods

2.1. Protein expression and purification

Although commercial sources of both PA and LF are available (e.g., List Biological Labs), their cost can make many experiments prohibitively expensive. In such cases, it makes sense to isolate and purify your own proteins. To make both PA and LF, we follow an optimized procedure developed by Park and Leppla (2000). Although fermentation may be carried out using Erlenmeyer flasks and a shaker incubator, we use a New Brunswick Bio-Flo 110 fermentor equipped with a 5.6-l (working volume) glass culture vessel (New Brunswick Scientific Co., Edison, NJ) that allows greater control of oxygenation, temperature, agitation, and pH. An attenuated strain of *B. anthracis* (BH445) expresses either PA or LF on a shuttle vector under the control of the PA promoter. Fermentation is initiated by the addition of a 50-ml overnight starter culture to 5 liters of modified FA medium (35 g/l tryptone, 5 g/l yeast extract, 3.2 g/l Na_2HPO_4, 1 g/l KH_2PO_4, 5.5 g/l NaCl, 0.04 g/l L-tryptophan, 0.04 g/l L-methionine, 5 mg/l thiamine-HCl, 25 mg/l uracil, pH 7.4), 0.1 ml antifoam, and 10 μg/ml kanamycin. Fermentation is allowed to proceed at 39° while sparging air at 4 l/min, and increasing agitation from 100 rpm to 400 rpm as the level of dissolved oxygen drops below 50% saturation. After 17 to 18 h of growth, the cells are removed by centrifugation ($3500 \times g$ for 30 min, 4°). The supernatant is sterile-filtered and concentrated by tangential flow filtration using a Millipore prep/scale-TFF cartridge with 1 ft^2 of 30-kDa MWCO polyethersulfone membrane (Millipore Corp., Billerica, MA), collecting the filtrate at approximately 50 ml/min under a 1-bar back pressure. Expressed protein is then purified by ammonium sulfate fractionation and fast-pressure liquid chromatography (FPLC) using phenyl sepharose and Q sepharose. The concentration of protein may be estimated using the bicinchoninic acid method (Smith *et al.*, 1985) and by densitometric analyses of coomassie blue-stained polyacrylamide gels.

Recombinant human MEK1 protein is expressed in *Spodoptera frugiperda* (Sf9) cells that have been infected with baculovirus containing human MEK1 ligated into the pVL1393 vector backbone (pKM636). The protein is isolated from supernatants of lysed cells and eluted over 10 column volumes in a linear gradient of 0 to 500 mM NaCl from a 20-ml Q-Sepharose column. The peak fractions containing MEK are pooled and loaded directly onto a 10-ml Ni-NTA column. After washing the column with 30 mM imidazole, MEK is eluted with 100 mM imidazole. At this point, the eluate is adjusted to 3 mM EDTA, 3 mM $MnCl_2$, and 2 mM DTT, and then 25 units of protein phosphatase 1 (New England Biolabs, Beverly, MA) are added to the reaction, and it is incubated at 30° for 4 h. Samples are then concentrated and applied to a 320-ml Sephacryl 200 column in 25 mM HEPES (pH 8.0), 100 mM NaCl, 2 mM DTT, and 10% glycerol.

His_6-tagged ERK2 protein is expressed in *Escherichia coli* and purified by FPLC using a HiTrap chelating column as described in detail in an earlier publication (Robbins *et al.*, 1993). Reagents for chromatography were obtained from GE Healthcare (Pittsburgh, PA). All other reagents were obtained from Sigma-Aldrich (St. Louis, MO) unless otherwise indicated.

2.2. Assays of LF activity

2.2.1. *In vitro* MKK cleavage assays

LF activity may be assayed *in vitro* using purified recombinant protein. Although this method clearly is nonphysiologic, it is invaluable for demonstrating that an inhibitor directly affects LF activity. In this assay,

LF activity is indirectly measured by reacting a constant concentration of MEK protein with varying amounts of LF, using MEK activity (i.e., ERK phosphorylation) as a readout (Fig. 25.1A). Briefly, 0.35 μg MEK1 is added to 3 μl of cleavage buffer (20 mM 3-(N-morpholino)propanesulfonic acid (pH 7.2), 25 mM β-glycerophosphate, 5 mM ethylene glycol-bis (2-aminoethylether)-N,N,N',N'-tetraacetic acid, 1 mM sodium orthovanadate, and 1 mM dithiothreitol) in the presence of LF (0.002 μg to 10 μg) and in a total volume of 10 μl. These cleavage reactions are incubated at 30° for 10 min. After cooling on ice for 2 min, 10 μl of kinase buffer (0.5 mM ATP [diluted 9:1 with [γ-^{32}P]ATP, Amersham Biosciences, Piscataway NJ,

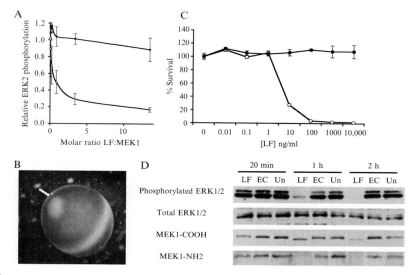

Figure 25.1 Measuring lethal factor (LF) activity. (A) *In vitro* MKK cleavage assay. Increasing concentrations of LF (empty circle) or inactive LF (E687C) (filled circle) are incubated with recombinant MEK1 *in vitro*. MEK1 activity after cleavage is measured using an *in vitro* kinase assay with ERK2 as a substrate. (B) Oocyte maturation assay. A mature oocyte of *Xenopus laevis* (1.2-mm diameter) is shown. The arrow indicates the white spot. (C) *In vitro* cell viability assay. Increasing concentrations of LF (empty circle) or inactive LF (E687C) (filled circle) are incubated with J774A.1 macrophages. Cell viability is measured by CellTiter 96® Aqueous Nonradioactive Cell Proliferation Assay (Promega Corp.) after 3 h of incubation. (D) NH$_2$-terminal MKK cleavage by immunoblotting. Lysates of V12 H-Ras-expressing NIH3T3 cells were treated with LeTx (LF) or PA plus inactive LF (E687C, EC) for 20 min, 1 h, or 2 h. Then they were separated by SDS polyacrylamide gel electrophoresis and immunoblotted with antibodies against phosphorylated ERK1/2, total ERK1/2, the COOH-terminus of MEK1, or the NH$_2$-terminus of MEK1. (Panel D, from Duesbery, Ahn, N. G., Copeland, T. D., Fukasawa, K., Gordon, V. M., Klimpel, K. R., Leppla, S. H., Oskarsson, M. K., Paull, K. D., Vande Woude, G. F. N. S., and Webb, C. P. (1998). Proteolytic inactivation of MAP-kinase-kinase by anthrax lethal factor. *Science* **280**, 734–737.)

10 mCi/ml, 3000 mCi/mmol], 75 mM MgCl$_2$, and 0.4 μg of ERK2) is added and samples are incubated for 10 min at 30°. After cooling on ice for 2 min, one volume of 2× SDS-buffer (200 mM Tris-HCl, 5% glycerol, 6% SDS, 10% β-mercaptoethanol, bromophenol blue to color) is added, and samples are incubated in a boiling water bath for 3 min. Proteins (10 μl) are then separated by SDS polyacrylamide electrophoresis on 10% gels. When electrophoresis is complete, the gels are fixed and the background radiation removed by repeated (6×) immersion in 40% methanol, 10% acetic acid, 1% glycerol for 15 min. Gels are then dried onto filter paper and ERK2 phosphorylation is quantified using a phosphorimager.

2.2.2. Oocyte maturation assays

The enzymatic activity of LF was first demonstrated in oocytes of the African clawed frog *Xenopus laevis* (Duesbery *et al.*, 1998). Immature oocytes are arrested at the first prophase of meiosis. In response to hormonal stimulation, the oocytes resume meiosis, complete the first meiotic cell cycle, and arrest again at metaphase II as a "mature" oocyte or egg. Progression through meiosis is easily measured by scoring for the presence of a "white spot" that is at the position of the meiotic spindle at the cortex of the animal pole, or by fixing and scoring for the breakdown of the oocyte nucleus (also called the germinal vesicle) (see Fig. 25.1B). Maturation requires polyadenylation and translation of c-mos mRNA. c-mos is a serine/threonine protein kinase that activates the MEK/MAPK signaling pathway. Since injection of LF inhibits MEK activity and thus prevents hormone-stimulated maturation, this system may be used to assay LF activity.

A detailed description of the procedure for isolating and preparing oocytes for injection has been recently published (Bodart and Duesbery, 2006). Briefly, ovarian lobes are isolated from healthy adult frogs. Individual oocytes are separated from these lobes by enzymatic digestion with collagenase or by manual defolliculation with fine forceps. Defolliculated oocytes are incubated in Masui's oocyte medium (MOM) (Zhang and Masui, 1992): 85 mM NaCl, 0.8 mM KCl, 0.5 mM CaCl$_2$, 0.6 mM MgSO$_4$, 5.4 mM Tris-base, and 4.0 mM sulfadiazine. *X. laevis* oocytes do not express a functional anthrax toxin receptor, so LF must be injected into the oocytes. We use a PMI-200 nitrogen pressure microinjector (Dagan Corp., Minneapolis, MN) coupled to glass needles that have been pulled to a fine point using a PC-10 needle puller (Narishige International, East Meadow, NY). Oocytes are incubated in MOM (20 to 30 per treatment), and are injected with between 20 and 40 nl of buffered saline solution (pH 7.2) containing 0.1 to 10 ng LF per oocyte. After 0.5 h of incubation, oocytes are treated with 10 μg/ml progesterone (from a 10 mg/ml stock in ethanol) to initiate maturation. White spot formation is then assessed at 0.5-h intervals for up to 12 h. In untreated and control injected oocytes, the first meiotic division (white spot formation) will typically occur 4 to 8 h after progesterone addition.

At the end of the experiment, oocytes should be fixed in one volume of 20% formalin (v/v), 60% ethanol, 10% acetic acid, and 10% glycerol. After about 12 h, oocytes will have hardened to a point where they may be dissected without losing their shape. Oocytes fixed under these conditions may be stored for weeks; preparations which have dried out may be rehydrated and examined months after fixation. A dissecting microscope should be used to examine each oocyte and record the presence or absence of a white spot. Each oocyte is then bisected along the animal–vegetal axis with a sharp scalpel and examined for the presence or absence of a germinal vesicle, which is easily distinguished from the yolky cytoplasm by its pale color and consistency.

The rate and extent of maturation varies among oocytes obtained from different females and under various environmental conditions. Moreover, it is not unusual to observe maturation in "unstimulated" oocytes, the most likely cause being contamination by forceps that have come in contact with progesterone-containing medium. These difficulties may be overcome by including appropriate positive (hormone-treated) and negative (untreated) controls in every experiment, and by expressing the timing of meiotic events in experimental treatments relative to that in oocytes treated with progesterone alone. For instance, timing is frequently expressed relative to when control oocytes complete 50% germinal vesicle breakdown ($GVBD_{50}$). If 50% of hormone-treated control oocytes complete GVBD by 4 h, and 50% of experimentally treated oocytes complete GVBD within 5 h, then the experimental group should be reported as having $GVBD_{50}$ equal to 1.25. A similar approach may be used to express the timing of white spot formation.

2.2.3. *In vitro* cell viability assays

LeTx is cytotoxic to some mouse macrophage-derived cell lines, so their survival may be used to measure LF activity (Friedlander, 1986) (see Fig. 25.1C). Other cell types (such as melanoma-derived cell lines) may be used instead, but their cytotoxicity may take longer to become manifest (Abi-Habib *et al.*, 2005). J774A.1 cells or RAW264.7 macrophage cells (American Type Culture Collection, Manassas, VA) are cultured in Dulbecco's Modified Eagle Medium supplemented with 10% fetal bovine serum and 1% penicillin/streptomycin. Cells are maintained at 37° in a humidified 5% CO_2 incubator. For *in vitro* cell viability assays, cells are seeded in 96-well dishes in 100 μl culture medium per well. When cells reach 80 to 90% confluence (30 to 40% for melanoma cells), the medium is replaced with fresh medium containing a range of LeTx concentrations (1 μg/ml PA plus 1 to 10,000 ng/ml LF in 10-fold increments). Cells are then incubated for 3 or 4 h (or up to 72 h for melanoma cells), at the end of which cell viability is assayed by using the CellTiter 96® Aqueous Non-Radioactive Cell Proliferation Assay (Promega Corp., Madison, WI) according to the manufacturer's instructions.

Results should be normalized to 100% survival with no LF present, and then plotted as a logarithmic function of the concentration of LF versus the proportion of viable cells. These values may then be used to determine the concentrations of LF required to cause 50% cell death (IC_{50}) or 50% maximal response (EC_{50}). Appropriate calculations may be performed using commercial software (e.g., Sigmaplot [Systat Software Inc., San Jose, CA] or GraphPad Prism [GraphPad Software, San Diego, CA]).

2.2.4. NH₂-terminal MKK cleavage by immunoblotting

Cleavage of the MKK NH_2-terminus may be directly assayed by measuring changes in molecular weight. However, the differences may be subtle and not readily determined except by making tagged constructs or by using sensitive assays such as mass spectrometry. Fortunately, there is an abundance of antibodies that have been generated against different regions of MKK. Thus, pairwise use of antibodies that recognize the cleaved NH_2-terminal fragment of MKK and those that recognize the remaining COOH-terminus may be applied to detecting cleaved versus intact protein in cell lines of interest (see Fig. 25.1D).

Typically, cells are cultured in 10-cm dishes containing 10 ml culture medium as described previously. When cells reach 60 to 70% confluence, the medium is replaced with fresh medium containing either PA alone (1 μg/ml), PA + 100 ng/ml inactive LF (E687C), or PA + LF (100 ng/ml). After 4 to 8 h in culture, the medium is removed and cells are washed once with PBS (pH 7.2). Cells are lysed in 0.5 ml RIPA buffer (50 mM Tris-HCl, 150 mM NaCl, 1 mM EDTA, 1 mM EGTA, 2 mM Na$_3$VO$_4$, 20 mM Na-pyrophosphate, 1% Triton X-100, 1% sodium deoxycholate, and 0.1% SDS, with Complete EDTA-free Protease Tablets [Roche Corp., Palo Alto, CA]), followed by sonication. The concentration of protein lysates is determined by the bicinchoninic acid method (Smith *et al.*, 1985). Protein samples are mixed in a 1:1 ratio with 2×SDS sample buffer, and proteins are separated (10 μg/lane) by denaturing SDS-PAGE in Tris–glycine–SDS running buffer (2.5 mM Tris-HCl, 19.2 mM glycine, 0.1% SDS, pH 8.3). Separated proteins are then transferred to PVDF membrane (Immobilon-P transfer membrane, Millipore) in Tris–glycine transfer buffer (12 mM Tris-HCl, 96 mM glycine, and 20% methanol, pH 8.3). Membranes are blocked using 10% nonfat milk in TBS-T (100 mM Tris-HCl, 150 mM NaCl, 0.05% Tween-20) for 30 min, and then immunostained with an antibody against the NH_2-terminus of MEK1 (07-641, 1:1000) (Upstate Biotechnology, Lake Placid, NY) overnight at 4°. Membranes are washed in TBS-T and then immunostained with a goat antirabbit IgG HRP-conjugated antibody (1:2000) (KPL Inc., Gaithersburg, MD) for 30 min.

Immunoblots are analyzed by chemiluminescence (LumiGLO Chemilumi-nescent Substrate; Cell Signaling Technology, Danvers, MA) followed by exposure to Kodak BioMax Light Film (Eastman Kodak Company, Roche-ster, NY). To verify equal loading and MEK1 cleavage, membranes are stripped by immersion in stripping buffer (62.5 mM Tris-HCl, 20 mM DTT, and 2% SDS, pH 6.7) for 30 min at 55° and reprobed with an antibody against the COOH-terminus of MEK1 (sc-219, 1:1000) (Santa Cruz Bio-technology, Santa Cruz, CA). Membranes can also be probed for α-tubulin (T9026, Sigma-Aldrich, St. Louis, MO) to verify equal loading.

3. CONCLUDING REMARKS

Although the mechanism by which anthrax kills its host is still unknown, the demonstration that LF proteolytically modifies and inactivates MKKs emphasizes the importance of MKK signaling in this disease. Assays for LF activity are invaluable for studies involving disease pathogenesis, as well as in the development of therapeutic strategies for this disease. We have described four common methods for assaying LF activity. Although the *in vitro* MKK cleavage assay is nonphysiologic and only indirectly measures LF activity, it can be extremely useful for therapeutic targeting of LF, as the direct effects of inhibitors on LF can be assayed in the absence of potentially confounding cellular effects. The oocyte maturation assay provides a conve-nient *in vivo* model system for LF activity, as the involvement of MAPK signaling in oocyte maturation has been well defined and there are clear endpoints. However, these endpoints (germinal vesicle breakdown and white spot formation) are only indirect measurements of LF activity, as opposed to direct measurement of the proteolytic activity of LF on MKKs. Furthermore, this assay requires a certain amount of expertise in oocyte manipulation and the developmental processes involved in oocyte matura-tion. Similarly, the *in vitro* cell viability assay does not measure LF activity directly, but assays for a consequence of this activity on cell viability. How-ever, this assay is quantitative, extremely sensitive, quick to perform, and relatively inexpensive. The NH$_2$-terminal MKK cleavage assay is a direct measurement of LF activity and employs widely used immunoblotting tech-niques. On the other hand, this assay may be influenced indirectly by other cellular activities and cannot detect subtle differences in LF activity.

ACKNOWLEDGMENTS

We wish to thank John J. Young and David E. Nadziejka for their helpful comments. This publication was made possible by a grant (CA109308-01) from the National Cancer

Institute. Its contents are solely the responsibility of the authors and do not necessarily represent the official views of the National Institutes of Health.

REFERENCES

Abi–Habib, R. J., Urieto, J. O., Liu, S., Leppla, S. H., Duesbery, N. S., and Frankel, A. E. (2005). BRAF status and mitogen–activated protein/extracellular signal–regulated kinase kinase 1/2 activity indicate sensitivity of melanoma cells to anthrax lethal toxin. *Mol. Cancer Ther.* **4,** 1303–1310.

Bardwell, A. J., Abdollahi, M., and Bardwell, L. (2004). Anthrax lethal factor–cleavage products of MAPK (mitogen–activated protein kinase) kinases exhibit reduced binding to their cognate MAPKs. *Biochem. J.* **378,** 569–577.

Bodart, J. F., and Duesbery, N. S. (2006). *Xenopus tropicalis* oocytes: More than just a beautiful genome. *Methods Mol. Biol.* **322,** 43–53.

Cataldi, A., Labruyere, E., and Mock, M. (1990). Construction and characterization of a protective antigen–deficient Bacillus anthracis strain. *Mol. Microbiol.* **4,** 1111–1117.

Chopra, A. P., Boone, S. A., Liang, X., and Duesbery, N. S. (2003). Anthrax lethal factor proteolysis and inactivation of MAPK kinase. *J. Biol. Chem.* **278,** 9402–9406.

Duesbery, N. S., and Vande Woude, G. F. (1999). Anthrax toxins. *Cell. Mol. Life Sci.* **55,** 1599–1609.

Firoved, A. M., Miller, G. F., Moayeri, M., Kakkar, R., Shen, Y., Wiggins, J. F., McNally, E. M., Tang, W.–J., and Leppla, S. H. (2005). *Bacillus anthracis* edema toxin causes extensive tissue lesions and rapid lethality in mice. *Am. J. Pathol.* **167,** 1309–1320.

Firoved, A. M., Moayeri, M., Wiggins, J. F., Shen, Y., Tang, W. J., and Leppla, S. H. (2007). Anthrax edema toxin sensitizes DBA/2J mice to lethal toxin. *Infect. Immun.* **75,** 2120–2125.

Friedlander, A. M. (1986). Macrophages are sensitive to anthrax lethal toxin through an acid–dependent process. *J. Biol. Chem.* **261,** 7123–7126.

Li, H., Xu, H., Zhou, Y., Zhang, J., Long, C., Li, S., Chen, S., Zhou, J.–M., and Shao, F. (2007). The phosphothreonine lyase activity of a bacterial type III effector family. *Science* **315,** 1000–1003.

Liang, X., Young, J. J., Boone, S. A., Waugh, D. S., and Duesbery, N. S. (2004). Involvement of domain II in toxicity of anthrax lethal factor. *J. Biol. Chem.* **279,** 52473–52478.

Mittal, R., Peak–Chew, S. Y., and McMahon, H. T. (2006). Acetylation of MEK2 and I kappa B kinase (IKK) activation loop residues by YopJ inhibits signaling. *Proc. Natl. Acad. Sci. USA* **103,** 18574–18479.

Park, S., and Leppla, S. H. (2000). Optimized production and purification of *Bacillus anthracis* lethal factor. *Protein Expr. Purif.* **18,** 293–302.

Pellizzari, R., Guidi–Rontani, C., Vitale, G., Mock, M., and Montecucco, C. (1999). Anthrax lethal factor cleaves MKK3 in macrophages and inhibits the LPS/IFNgamma–induced release of NO and TNFalpha. *FEBS Lett.* **462,** 199–204.

Pezard, C., Berche, P., and Mock, M. (1991). Contribution of individual toxin components to virulence of *Bacillus anthracis*. *Infect. Immun.* **59,** 3472–3477.

Pezard, C., Duflot, E., and Mock, M. (1993). Construction of *Bacillus anthracis* mutant strains producing a single toxin component. *J. Gen. Microbiol.* **139,** 2459–2463.

Robbins, D. J., Zhen, E., Owaki, H., Vanderbilt, C. A., Ebert, D., Geppert, T. D., and Cobb, M. H. (1993). Regulation and properties of extracellular signal–regulated protein kinases 1 and 2 *in vitro*. *J. Biol. Chem.* **268,** 5097–5106.

Rosenberger, C. M., and Finlay, B. B. (2003). Phagocyte sabotage: Disruption of macrophage signalling by bacterial pathogens. *Nat. Rev. Mol. Cell. Biol.* **4,** 385–396.

Singh, Y., Liang, X., and Duesbery, N. S. (2005). Microbial toxins. *In* "Molecular and Cellular Biology." (T. Proft, ed.), pp. 285–312. Horizon Bioscience, Norfolk, England.

Smith, H., and Keppie, J. (1954). Observations on experimental anthrax: Demonstration of a specific lethal factor produced *in vivo* by *Bacillus anthracis*. *Nature* **173**, 869–870.

Smith, P. K., Krohn, R. I., Hermanson, G. T., Mallia, A. K., Gartner, F. H., Provenzano, M. D., Fujimoto, E. K., Goeke, N. M., Olson, B. J., and Klenk, D. C. (1985). Measurement of protein using bicinchoninic acid. *Anal. Biochem.* **150**, 76–85.

Vitale, G., Bernardi, L., Napolitani, G., Mock, M., and Montecucco, C. (2000). Susceptibility of mitogen–activated protein kinase kinase family members to proteolysis by anthrax lethal factor. *Biochem. J.* **352**(Pt 3), 739–745.

Vitale, G., Pellizzari, R., Recchi, C., Napolitani, G., Mock, M., and Montecucco, C. (1998). Anthrax lethal factor cleaves the N–terminus of MAPKKs and induces tyrosine/threonine phosphorylation of MAPKs in cultured macrophages. *Biochem. Biophys. Res. Commun.* **248**, 706–711.

Working Group on Civilian Biodefense (2002). Anthrax as a biological weapon, 2002: Updated recommendations for management. *JAMA* **287**, 2236–2252.

Zhang, S. C., and Masui, Y. (1992). Activation of *Xenopus laevis* eggs in the absence of intracellular Ca activity by the protein phosphorylation inhibitor, 6–dimethylaminopurine (6–DMAP). *J. Exp. Zool.* **262**, 317–329.

CHAPTER TWENTY-SIX

Genetic Analyses of the Role of RCE1 in RAS Membrane Association and Transformation

Martin O. Bergo,[‡] Annika M. Wahlstrom,[‡] Loren G. Fong,[†] and Stephen G. Young[*,†]

Contents

* Corresponding author
† Division of Cardiology, Department of Internal Medicine, University of California, Los Angeles, California
‡ Wallenberg Laboratory, Department of Medicine, Sahlgrenska University Hospital, Gothenburg, Sweden

Methods in Enzymology, Volume 438

ISSN 0076-6879, DOI: 10.1016/S0076-6879(07)38026-9

Abstract

Proteins terminating with a *CAAX* motif, such as the nuclear lamins and the RAS family of proteins, undergo post-translational modification of a carboxyl-terminal cysteine with an isoprenyl lipid—a process called protein prenylation. After prenylation, the last three residues of *CAAX* proteins are clipped off by an endoprotease of the endoplasmic reticulum. RCE1 is responsible for the endoproteolytic processing of the RAS proteins and is likely responsible for endoproteolytic processing of the vast majority of *CAAX* proteins. Prenylation has been shown to be essential for the proper intracellular targeting and function of several *CAAX* proteins, but the physiologic importance of the endoprotease step has remained less certain. Here, we will review methods that have been used to define the physiologic importance of the endoproteolytic processing step of *CAAX* protein processing.

1. INTRODUCTION

Eukaryotic proteins that terminate with a $CAAX$ sequence typically undergo a series of post-translational processing reactions, beginning with protein prenylation, which is the attachment of a farnesyl or geranylgeranyl lipid to the thiol group of the cysteine residue (the "C" of the $CAAX$ motif) (Zhang and Casey, 1996). Second, the last three amino acids of the protein (i.e., the $-AAX$) are removed by a prenylprotein-specific endoprotease of the endoplasmic reticulum (ER) (Ashby, 1998; Boyartchuk *et al.*, 1997; Young *et al.*, 2000). Third, the carboxyl group of the newly exposed isoprenylcysteine is methylated by isoprenylcysteine carboxyl methyltransferase (ICMT) (Dai *et al.*, 1998; Hrycyna *et al.*, 1991; Young *et al.*, 2000). These three processing steps have been studied intensively with yeast **a**–factor, a farnesylated yeast mating pheromone, and with the yeast and mammalian RAS proteins, small GTP-binding proteins involved in signal transduction. However, large numbers of cellular proteins with diverse biological functions contain a $CAAX$ motif and undergo these modifications.

The enzymes responsible for the prenylation of the $CAAX$ proteins, protein farnesyltransferase and protein geranylgeranyltransferase type I, have been characterized extensively in both yeast and higher organisms (Zhang and Casey, 1996). Progress in understanding the importance of the endoprotease step has lagged behind, but during the past few years some progress has been made in understanding the $CAAX$ endoproteases and their functional importance.

Two genes from *Saccharomyces cerevisiae*, $RCE1$ and $STE24$ ($AFC1$), were identified as playing a role in the endoproteolytic removal of the "$-AAX$"

from Ras2p and the precursor to the yeast mating pheromone **a**-factor (Boyartchuk et al., 1997). Rce1p (for Ras and a-factor converting enzyme) is essential for cleaving the -*AAX* from Ras2p. In addition, Rce1p is quite capable of cleaving the -*AAX* from **a**-factor. Ste24p is a zinc metalloproteinase that plays dual roles in the endoproteolytic processing of the precursor to **a**-factor. Along with Rce1p, Ste24p is capable of carrying out the carboxyl-terminal cleavage reaction (i.e., the release of the -*AAX* from **a**-factor) (Boyartchuk et al., 1997; Tam et al., 1998). In addition, however, it cleaves seven amino acids from the amino terminus of the protein (Tam et al., 1998). Ste24p plays no role in the endoproteolytic processing of the yeast Ras proteins (Boyartchuk et al., 1997). In fact, **a**-factor is the only well-established substrate for Ste24p in yeast (Boyartchuk et al., 1997).

Both yeast Rce1p and Ste24p are ER proteins with multiple transmembrane domains (Young et al., 2000). Rce1p is 315 amino acids in length and is predicted to contain multiple transmembrane helices (Boyartchuk et al., 1997). The Rce1p amino acid sequence does not contain sequence motifs characteristic of the well-defined classes of proteases, but there are remote similarities with the type IIb signal peptidase, which cleaves signal sequences from proteins containing nearby lipid modifications (Boyartchuk et al., 1997). Ste24p contains 453 amino acids, as well as multiple transmembrane domains (Boyartchuk et al., 1997). Ste24p contains a HEXXH (H, His; E, Glu) motif between amino acids 297 and 301, a sequence shared by many zinc-dependent metalloproteases. Mutating either of the conserved histidines in the HEXXH motif within Ste24p blocks **a**-factor processing (Boyartchuk et al., 1997; Fujimura-Kamada et al., 1997).

In yeast, Rce1p is essential for Ras processing and function (Boyartchuk et al., 1997). Heat shock sensitivity elicited by a mutationally activated Ras2p is suppressed by *RCE1* deficiency. However, *RCE1* deficiency in yeast almost certainly does not completely block all Ras functions, given that rce1Δ yeast are viable (Boyartchuk et al., 1997) and yeast lacking the Ras proteins are not (Kataoka et al., 1984). *RCE1* deficiency in yeast mislocalizes Ras2p away from the plasma membrane to the cytosol (Boyartchuk et al., 1997). In wildtype yeast, a green fluorescent protein (GFP)-*RAS2* fusion is targeted to the periphery of the cells (i.e., the plasma membrane). In contrast, the GFP-*RAS2* fusion in rce1Δ yeast is located inside the cell (i.e., either in the cytosol or associated with internal membrane compartments) (Boyartchuk et al., 1997).

The identification of *RCE1* and *STE24* in yeast made it possible to identify the human and mouse orthologs, *RCE1* and *ZMPSTE24* (Boyartchuk et al., 1997; Fujimura-Kamada et al., 1997; Kumagai et al., 1999; Leung et al., 2001; Young et al., 2000). Human RCE1 is required for normal processing of the RAS proteins; the *RCE1* gene is expressed ubiquitously, with the highest levels of expression in placenta, heart, and skeletal muscle (Kim et al., 1999; Otto et al., 1999). Human RCE1 is 329 amino acids in length, 27.6% identical to Rce1p in *S. cerevisiae*, and contains multiple transmembrane helices (Young *et al.*, 2000). The human ortholog for yeast *STE24*, *ZMPSTE24*, is also expressed ubiquitously (Bergo *et al.*, 2002b; Leung *et al.*, 2001). It is 475 amino acids in length, with 36% identity to the yeast protein (Young *et al.*, 2000). Like yeast Ste24p, human ZMPSTE24 contains the zinc metalloprotease motif HEXXH and multiple transmembrane helices (Young *et al.*, 2000). Human and mouse ZMPSTE24 faithfully carries out the processing of yeast **a**–factor (Leung *et al.*, 2001); however, no **a**-factor orthologs have been identified in mammals. During the past few years, several laboratories have shown that ZMPSTE24 is critical for the endoproteolytic processing of prelamin A, a farnesylated *CAAX* protein that is the precursor to lamin A, a structural protein of the nuclear envelope (Bergo *et al.*, 2002b; Pendás *et al.*, 2002). Akin to Ste24p-processing of **a**-factor, ZMPSTE24 is almost certainly capable of cleaving the -*AAX* from prelamin A (Bergo *et al.*, 2002b; Corrigan *et al.*, 2005). In addition, it cleaves prelamin A a second time, 15 residues upstream from the farnesylcysteine, releasing mature lamin A (Corrigan *et al.*, 2005; Dalton and Sinensky, 1995; Kilic *et al.*, 1997). Remarkably, prelamin A and yeast **a**-factor are apparently cleaved twice by ZMPSTE24 in a highly specific fashion, despite very little amino acid similarity surrounding the four different sessile bonds.

2. ESTABLISHING THAT HUMAN *RCE1* IS A *CAAX* ENDOPROTEASE

The laboratory of Patrick Casey (Duke University) expressed human RCE1 protein *Sf9* insect cells and performed biochemical assays to show that it was a *CAAX* endoprotease (Kim *et al.*, 1999; Otto *et al.*, 1999). Extracts from the *Sf9* cells were mixed with recombinant isoprenylated *CAAX* proteins. In this assay, recombinant RCE1 cleaves the -*AAX* from the isoprenylated proteins, rendering the proteins susceptible to methylation with recombinant Ste14p (the yeast isoprenylcysteine carboxyl methyltrans-ferase) and [*methyl*-^{3}H]-*S*-adenosylmethionine. With this "coupled endo-proteolysis/methylation assay," Casey's laboratory demonstrated that human RCE1 is capable of processing farnesylated K-RAS, farnesylated H-RAS, farnesylated N-RAS, the farnesylated heterotrimeric G-protein $G_{\gamma 1}$ subunit, geranylgeranylated K-RAS, and geranylgeranyl-RAP1B. Both farnesylated and geranylgeranylated K-RAS exhibited a K_m value of approximately 0.5 μM and similar K_{cat} values. Prenylated *CAAX* peptides, but not nonisoprenylated peptides, were able to compete for the processing of isoprenylated K-RAS (Otto *et al.*, 1999). Also, a previously identified inhibitor of endoproteolytic processing, RPI, was shown to be an effective inhibitor of RCE1, with an IC_{50} of ~5 n*M* (Otto *et al.*, 1999).

3. *RCE1* KNOCKOUT MICE

The laboratory of Stephen Young (University of California, Los Angeles) created *Rce1*-deficient mice with standard gene-targeting techniques (Kim *et al.*, 1999). Heterozygous knockout mice ($Rce1^{+/-}$) were normal. However, almost all of the homozygous embryos ($Rce1^{-/-}$) died, beginning at about embryonic day 15 (E15). $Rce1^{-/-}$ mice were born alive very rarely, and those mice were invariably quite small and died within a few weeks. The precise reason for the death of $Rce1^{-/-}$ mice is not known; the viable $Rce1^{-/-}$ embryos at E15.5 revealed no consistent differences in morphology, organogenesis, stage of development, color, or size.

4. ANALYZING RAS FUNCTION IN *RCE1*$^{-/-}$ MEFs

Primary mouse embryonic fibroblasts (MEFs) could be cultured from the $Rce1^{-/-}$ embryos (Kim *et al.*, 1999). To generate fibroblasts, $Rce1^{+/-}$ mice (which can be obtained from the authors of this chapter), are intercrossed, and pregnant females are sacrificed at E13.5. Genotyping of embryos is performed by Southern blot analysis of DNA prepared from fetal membranes (Kim *et al.*, 1999). The embryos are incubated in 5.0 ml of 0.25% trypsin-EDTA (Life Technologies, Gaithersburg, MD) for 8 h at 4°, followed by a 20-min incubation at 37°. The embryos are mechanically disrupted by repeated pipetting in 5 ml of DME-FBS medium (Dulbecco's modified Eagle's medium supplemented with 10% v/v fetal bovine serum [FBS], L-glutamine, nonessential amino acids, penicillin-streptomycin [all from Life Technologies], and 2-mercaptoethanol [Sigma, 100 μM final concentration]). The debris is allowed to settle, and the cell suspension, diluted to 25 ml, is plated in T150 tissue culture flasks. The cells are grown in an incubator at 37° with 7% CO_2 ($\sim 1 \times 10^7$ cells are obtained from each embryo). To immortalize the MEFS, cells are passaged according to a 3T3 protocol (Todaro, 1969), in which cells are harvested and counted every 3 days and replated at a density of 2×10^6 cells per 100-mm dish.

5. ABSENCE OF *CAAX* ENDOPROTEASE ACTIVITY IN *RCE1*-DEFICIENT CELLS

CAAX endoprotease activity in $Rce1^{-/-}$ and $Rce1^{+/+}$ MEFs can be measured with a coupled endoprotease/methylation assay (Kim *et al.*, 1999).

Substrate proteins (e.g., K–RAS, N–RAS, and RAP1B) containing a *CAAX* sequence are expressed in *Escherichia coli* and then prenylated *in vitro* with either recombinant protein farnesyltransferase or protein geranylgeranyltransferase (Otto *et al.*, 1999). To prepare membranes from *Rce1*^{+/+} and *Rce1*^{−/−} embryonic fibroblasts, cells were lysed by sonication. After centrifugation at $500 \times g$ for 10 min (to remove debris and cell nuclei), the membrane fraction is prepared by spinning at $200,000 \times g$ for 1.5 h. The membranes are then resuspended in 50 mM Tris–HCl at protein concentrations of 10 to 25 mg/ml. To measure the prenylprotein-specific *CAAX* endoprotease activity, the prenylated protein substrates and membrane fractions are mixed in 50 μl of a 100-mM Hepes buffer (pH 7.5) containing 5 mM MgCl$_2$. After a 30-min incubation at 37°, the proteolysis reactions are stopped, and the methyltransferase reaction is initiated by adding a cocktail (20 μl) containing 5 mM NaHPO$_4$, 87.5 mM EDTA, several protease inhibitors (300 μM N-tosyl-L-phenylalanine chloromethylketone, 1 mM phenylmethylsulfonyl fluoride [PMSF], 10 mM 1,10-phenanthroline), 20 μg of *Sf9* cell membranes expressing yeast Ste14p, and 17.5 μM [^{3}H]AdoMet (1.5 Ci/mmol). After a 20-min incubation at 37°, the methylation reaction is terminated by adding 0.5 ml of 4% sodium dodecyl sulfate (SDS). Bovine brain cytosol (50 μg protein) is added as a carrier, and the mixture is incubated for 20 min at room temperature, after which the proteins were precipitated with 0.5 ml of 30% trichloroacetic acid. Precipitated proteins are collected on glass fiber or nitrocellulose filters, and the extent of proteolysis is determined by quantifying ^{3}H incorporation into the protein by scintillation counting. With this assay, *Rce1*^{−/−} membranes were shown to lack *CAAX* processing activity for farnesylated heterotrimeric G$_{\gamma1}$ subunit, farnesylated H–RAS, farnesylated N–RAS, geranylgeranylated K–RAS, and geranylgeranyl-RAP1B (Kim *et al.*, 1999; Otto *et al.*, 1999).

6. Documenting an Accumulation of Uncleaved *RCE1* Substrates in *RCE1*-Deficient Cells

Rce1^{−/−} MEFs accumulate uncleaved RCE1 substrates, and this accumulation of RCE1 substrates can be detected with a coupled endoproteolysis/methylation assay (Leung *et al.*, 2001). Whole-cell extracts (150 μg) from *Rce1*^{+/+} and *Rce1*^{−/−} MEFs are incubated with 10 μM S-adenosyl-L-[*methyl*-¹⁴C]methionine (55 Ci/mol, Amersham Pharmacia), *Sf9* cells expressing high levels of RCE1 (100 μg), and *Sf9* cells expressing high levels of yeast Ste14p (100 μg). The reaction is incubated for 2 h at 37°. The reaction is stopped by adding 50 μl of 1.0 M NaOH containing 0.1%

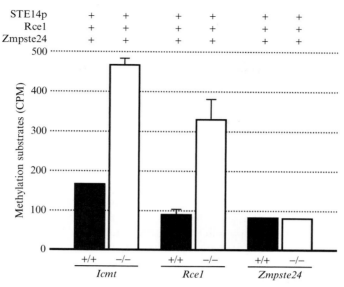

Figure 26.1 An accumulation of "methylatable" substrates in $Rce1^{-/-}$ cells. Methylation of whole-cell extracts from primary fibroblasts derived from $Zmpste24^{+/+}$, $Zmpste24^{-/-}$, $Icmt^{+/+}$, $Icmt^{-/-}$, $Rce1^{+/+}$, and $Rce1^{-/-}$ embryos. The cell extracts were incubated with S-adenosyl-L-[$methyl$-^{14}C]methionine, membranes containing high levels of Ste14p, yeast membranes containing high levels of mouse RCE1, and yeast membranes containing high levels of mouse ZMPSTE24. The relative level of methylatable substrates in wild-type and knockout cells was assessed with a base hydrolysis/methanol diffusion assay. For the $Zmpste24^{+/+}$ and $Zmpste24^{-/-}$ cells, we observed identical results when membranes expressing RCE1 were left out of the reaction mixture. (From Leung, G. K., Schmidt, W. K., Bergo, M. O., Gavino, B., Wong, D. H., Tam, A., Ashby, M. N., Michaelis, S., and Young, S. G. (2001). Biochemical studies of $Zmpste24$-deficient mice. *J. Biol. Chem.* **276**, 29051–29058, with permission.)

SDS. The reaction mixture (90 μl) can then be spotted onto a pleated 2 × 8-cm filter paper wedged in the neck of a 20-ml scintillation vial containing 5 ml of scintillation fluid (ScintiSafe Econo 1, Fisher). The vials are capped and incubated at room temperature for 5 h to allow the [^{14}C] methanol (formed by base hydrolysis of methyl esters) to diffuse into the scintillation fluid. The amount of base-releasable methanol is far higher in the $Rce1^{-/-}$ fibroblasts than in the $Rce1^{+/+}$, reflecting the accumulation of uncleaved $RCE1$ substrates (substrates that can be readily cleaved and methylated in the setting of high levels of $RCE1$ and Ste14p) (Fig. 26.1). Importantly, there was no accumulation of "methylatable" protein substrates in fibroblasts lacking $Zmpste24$ (Leung *et al.*, 2001) (see Fig. 26.1).

7. ALTERED ELECTROPHORETIC MIGRATION OF RAS PROTEINS IN *RCE1*⁻/⁻ FIBROBLASTS

The electrophoretic migration of the RAS proteins in $Rce1^{-/-}$ cells is abnormally slow, reflecting an absence of endoproteolytic processing (Kim *et al.*, 1999). To study the electrophoretic migration of RAS proteins, it is first necessary to prepare lysates of MEFs (Kim *et al.*, 1999). Confluent cultures of MEFs are harvested from a 100-mm plate in 0.5 ml of RIPA lysis buffer (50 mM Tris-HCl, pH 8.0, 150 mM NaCl, 5 mM MgCl$_2$, 1% Triton X-100, 0.5% sodium deoxycholate, 0.1% SDS, 0.5 mM PMSF, 10 μg/ml leupeptin, and 10 μg/ml of aprotinin), and disrupted by sonication for 10 s in a Branson model 400 sonifier (duty cycle constant, output control 0.3) (Danbury, CT). The samples are then heated to 95° for 5 min and loaded onto an SDS/10 to 20% gradient polyacrylamide gel. After electrophoretic transfer of the proteins to a nitrocellulose membrane, Western blotting is performed with the "pan-RAS" monoclonal antibody Ab-4 (i.e., binds to N-, K-, and H-RAS) (Oncogene Science, Uniondale, NY) or the K-RAS-specific monoclonal antibody Ab-1 (Oncogene Science), followed by an incubation with horseradish peroxidase-conjugated sheep antimouse IgG (Amersham). The binding of the secondary antibodies is detected by enhanced chemiluminescence (Amersham). The electrophoretic mobility of the RAS proteins from $Rce1^{-/-}$ MEFs (or $Rce1^{-/-}$ embryos) is distinctly abnormal, even on a minigel. In fact, RAS proteins with a normal electrophoretic mobility are undetectable in $Rce1^{-/-}$ cells (Fig. 26.2). When equal amounts of $Rce1^{-/-}$ and $Rce1^{+/+}$ MEF lysates are mixed, one observes a doublet RAS band, consisting of the processed RAS proteins in $Rce1^{+/+}$ MEFs and the unprocessed RAS proteins in $Rce1^{-/-}$ MEFs (Kim *et al.*, 1999).

8. METHYLATION OF RAS PROTEINS IN *RCE1*⁻/⁻ FIBROBLASTS

The absence of RAS endoproteolysis means that the RAS proteins cannot be methylated. The absence of a methyl ester in the RAS proteins from $Rce1^{-/-}$ MEFs can be documented by a metabolic labeling assay (Clarke *et al.*, 1988; Kim *et al.*, 1999). In this assay, 100-mm dishes containing approximately 2.5 × 10⁶ $Rce1^{-/-}$ or $Rce1^{+/+}$ MEFs are incubated in the presence of 1 mCi of L-[*methyl*-³H]methionine (80 Ci/mmol) (Amersham-Pharmacia, Chicago, IL) and [³⁵S]cysteine (>1000 Ci/mmol) (Amersham-Pharmacia) in methionine- and cysteine-free DMEM/10% FBS in a 7% CO$_2$ incubator at 37° for 24 h. The medium is removed, and the cells are washed with 5 ml of ice-cold PBS, scraped

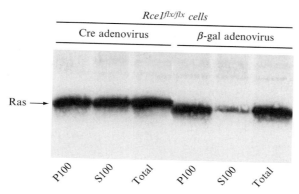

Figure 26.2 Intracellular localization of RAS proteins in immortalized fibroblasts. $Rce1^{flx/flx}$ cells express normal levels of $Rce1$, whereas $Rce1^{\Delta/\Delta}$ fibroblasts are completely deficient in $Rce1$ expression. Cre adenovirus fully converted the $Rce1^{flx/flx}$ fibroblasts into $Rce1^{\Delta/\Delta}$ fibroblasts. Fibroblasts were fractionated into cytosolic (S100) and membrane (P100) fractions; RAS proteins were immunoprecipitated and analyzed on Western blots of sodium dodecyl sulfate-polyacrylamide gels with antibody Ab-4. Note higher molecular weight of the RAS proteins in $Rce1^{\Delta/\Delta}$ fibroblasts. (From Bergo, M. O., Ambroziak, P., Gregory, C., George, A., Otto, J. C., Kim, E., Nagase, H., Casey, P. J., Balmain, A., and Young, S. G. (2002a). Absence of the $CAAX$ endoprotease $Rce1$: Effects on cell growth and transformation. $Mol. Cell. Biol.$ **22**, 171–181, with permission.)

from the dish into 1 ml of ice-cold PBS, and centrifuged at $500 \times g$. Cell pellets are resuspended in 1.0 ml of RIPA lysis buffer and disrupted by sonication, as described above. RAS proteins are then immunoprecipitated with antibody Y13–259, and the immunoprecipitate is size-fractionated on an SDS/10 to 20% polyacrylamide gel. The gel is fixed in isopropanol, water, and acetic acid (25:65:10, v/v), soaked for 30 min in Amplify (Amersham–Pharmacia), and dried. The gel is imaged with a phosphorimager, and the region of the gel corresponding to the RAS band is excised. The amount of [³H]methanol released from the gel fragment by base hydrolysis of the farnesylcysteine methyl esters is quantified by scintillation counting. The gel slice is placed in a 1.5-ml capless microcentrifuge tube and mixed with 200 μl of 2 M NaOH; the tube is lowered into 20-ml scintillation vial containing 6 ml of Safety-Solve II counting fluor (RPI) fluid. The scintillation vial is then capped and incubated at 55°. The [³H]methanol released from the farnesylcysteine methyl esters diffuses into the fluor within the capped vial, while the ³⁵S-labeled protein remains in the gel slice. After 24 h, the microcentrifuge tubes are removed, and the radioactivity in the fluor is counted. As judged by the [³H]methanol release, the RAS proteins from $Rce1^{+/+}$ cells contain a ³H-labeled farnesylcysteine methyl ester, whereas RAS proteins from the $Rce1^{-/-}$ cells do not (Kim *et al.*, 1999).

9. SUBCELLULAR FRACTIONATION AND LOCALIZATION OF RAS PROTEINS IN *RCE1⁻/⁻* MEFs

To determine the degree of membrane association of RAS proteins in *Rce1⁻/⁻* MEFs, subcellular fractionation studies can be performed (Kim *et al.*, 1999). In this study, 100-mm dishes containing confluent MEFs are washed with ice-cold PBS, and cells are collected in 1.0 ml of PBS and centrifuged at 500×*g* for 10 min. Cell pellets are incubated with 1225 μl of hypotonic buffer (10 m*M* Tris-HCl, pH 7.5, 1.0 m*M* MgCl$_2$, 0.5 m*M* PMSF, 10 μg/ml leupeptin, 10 μg/ml of aprotinin, 1 μ*M* DTT) and placed on ice for 10 min. Cells are disrupted with an ice-cold dounce tissue homogenizer, after which 225 μl of 1 *M* NaCl is added. A total of 450 μl of this solution ("total lysate") is transferred to a microfuge tube and set aside. The remaining 1000 μl is transferred to a polycarbonate ultracentrifuge tube and spun at 100,000×*g* for 30 min at 4°. The supernatant fluid (S100, representing the cytosolic fraction) is transferred to a new microcentrifuge tube; the pellet (P100, representing the membrane fraction) is resuspended in 850 μl of hypotonic buffer and 150 μl of 1 *M* NaCl. Next, 50 μl of 10× RIPA buffer is added to the "total lysate" sample, and 110 μl is added to the S100 and P100 fractions. After incubation on ice for 10 min, the lysates are clarified by centrifugation at 25,000×*g* for 30 min at 4°. Supernatants are transferred to new microfuge tubes, and immunoprecipitation, gel fractionation, and Western blot detection of the RAS proteins are performed as described earlier. A substantial difference in the intracellular distribution of the RAS proteins is apparent by immunoblot analysis of P100 and S100 fractions of *Rce1⁻/⁻* and *Rce1⁺/⁺* fibroblasts (Kim *et al.*, 1999). Virtually all of the RAS proteins in *Rce1⁺/⁺* cells are located in the P100 fraction, whereas about one-half of the RAS proteins in *Rce1⁻/⁻* cells are located in the S100 fraction (see Fig. 26.2).

To further assess the association of the RAS proteins with membranes of *Rce1⁻/⁻* and *Rce1⁺/⁺* embryonic fibroblasts, enhanced GFP-RAS fusion protein constructs can be expressed in MEFs and then visualized by fluorescence microscopy (Bergo *et al.*, 2002a; Kim *et al.*, 1999). *Rce1⁻/⁻* and *Rce1⁺/⁺* MEFs are transfected with 2.0 μg of GFP-RAS fusion plasmids with SuperFect Reagent (Qiagen, Valencia, CA). Cells are fixed in 4% formalin 24 to 28 h after transfection, mounted, and viewed by confocal microscopy. In *Rce1⁺/⁺* cells, the fluorescence is localized almost entirely to the plasma membrane. In contrast, the vast majority of the fluorescence in the *Rce1⁻/⁻* cells is cytosolic or associated with internal membranes. However, very small amounts of fluorescence can be observed along the plasma membrane in the *Rce1⁻/⁻* cells.

The importance of RCE1 for the localization of GFP-RAS fusion proteins is dependent on the fact that the RAS proteins are farnesylated.

When mutations are introduced into the *CAAX* motif so as to create GFP–RAS fusions that are geranylgeranylated, the impact of *Rce1* deficiency on intracellular localization is no longer detectable (Michaelson *et al.*, 2005). Thus, geranylgeranylated GFP-N-RAS and geranylgeranylated GFP-K-RAS are largely localized to the plasma membrane in *Rce1*⁻/⁻ MEFs and not mislocalized to the cytosol (i.e., a pattern indistinguishable from that in *Rce1*⁺/⁺ cells). Changing the isoprenyl chain length of RAC1 and RHOB (proteins that are normally geranylgeranylated) had the opposite effect. The intracellular localization of geranylgeranylated GFP-RAC1 and geranylgeranylated GFP-RHOB is unaffected by *Rce1* deficiency. However, the intracellular localization of farnesylated GFP-RAC1 and farnesylated GFP-RHOB is sensitive to *Rce1* deficiency (Michaelson *et al.*, 2005).

10. DEFINING THE IMPACT OF *RCE1* DEFICIENCY ON MAP KINASE ACTIVATION IN *RCE1*⁻/⁻ FIBROBLASTS

We suspected that absent endoproteolysis of the RAS proteins might result in impaired growth factor-mediated activation of the RAS effector Erk1/2. Serum-stimulated activation of Erk1/2 was assessed by seeding 1×10^5 *Rce1*⁺/⁺ and *Rce1*⁻/⁻ cells on 60-mm dishes followed by overnight serum starvation (Bergo *et al.*, 2004). The next morning, medium containing 10% serum was added to the cells. Cells were harvested at various time points after serum stimulation and total cell lysates were analyzed by immunoblotting with an antibody recognizing phosphorylated ERK1/2 (phospho-p44/42 MAP kinase E10 monoclonal, Cell Signaling Technology, Beverly, MA), and total ERK1/2 (p44/42 MAP kinase polyclonal, Cell Signaling Technology). Contrary to our expectation, we observed no effect of *Rce1* deficiency on the activation of ERK1/2 (i.e., activation was no different in *Rce1*⁻/⁻ and *Rce1*⁺/⁺ cells). We were concerned that intrinsic genetic differences between independent lines of immortalized *Rce1*⁻/⁻ and *Rce1*⁺/⁺ fibroblasts might conceivably have prevented us from observing subtle differences in ERK1/2 activation. Accordingly, we repeated the ERK1/2 activation experiments in *Rce1*ᶠˡˣ/ᶠˡˣ cells (which have normal levels of *Rce1* expression) and the *Rce1*-deficient *Rce1*Δ/Δ cells that were derived from them by *Cre* expression (i.e., two fibroblast cell lines that were identical except for *Rce1* expression). Once again, however, we observed no differences in *Rce1*-expressing and *Rce1*-deficient cells in ERK1/2 activation, either in response to serum or epidermal growth factor (Bergo *et al.*, 2004).

11. IMPACT OF ENDOPROTEOLYSIS ON THE BINDING OF K-RAS TO MICROTUBULES

The group at the laboratory of Pat Casey identified a specific and prenylation-dependent interaction between tubulin/microtubules and K-RAS, but not H-RAS or several other small GTPases (Thissen *et al.*, 1997). In a follow-up study, they found that the interaction between K-RAS and microtubules is highly dependent both on the polylysine domain with K-RAS, as well as endoproteolysis and methylation at the carboxyl terminus (Chen *et al.*, 2000). Partially processed K-RAS that was farnesylated (but retained the -*AAX* residues) bound microtubules. However, surprisingly, endoproteolytic removal of the -*AAX* from K-RAS abolished all binding. Even more surprising, the binding of K-RAS to microtubules was restored by methylation of the C-terminal prenylcysteine. Consistent with these results, localization of the green fluorescent protein–K-RAS fusion was paclitaxel-sensitive in cells lacking *Rce1*, while no paclitaxel effect was observed in cells lacking the methyltransferase ICMT. These studies show that the polylysine domain is critical for the interaction between K-RAS and tubulin/microtubules and provide the first evidence for a functional consequence of *CAAX*-processing steps (Chen *et al.*, 2000).

12. DEFINING THE IMPACT OF *RCE1* DEFICIENCY ON CELL GROWTH AND ON TRANSFORMATION BY ACTIVATED FORMS OF RAS

To study the role of *Rce1* on cell growth and RAS-transformation, we examined *Rce1*$^{+/+}$ and *Rce1*$^{-/-}$ MEFs. We also constructed a conditional knockout allele in which the entire *Rce1* coding sequence was "floxed" (flanked by loxP sites). Mice homozygous for the conditional allele (*Rce1*$^{flx/flx}$) were viable, healthy, and fertile (Bergo *et al.*, 2002a). *Rce1*$^{flx/flx}$ MEFs were produced, and it was possible to completely delete the *Rce1* gene from these cells by treating the cells with *Cre* adenovirus (thereby creating *Rce1*$^{\Delta/\Delta}$ MEFs). Treating *Rce1*$^{flx/flx}$ MEFs with very small amounts of *Cre* adenovirus resulted in mixed cultures of *Rce1*$^{flx/flx}$ and *Rce1*$^{\Delta/\Delta}$ MEFs (Bergo *et al.*, 2002a).

To define the effect of the *Rce1* excision on cell growth, we examined mixed cultures of *Rce1*$^{flx/flx}$ and *Rce1*$^{\Delta/\Delta}$ cells by transfecting them with large amounts of a *lacZ* adenovirus and small amounts of *Cre* adenovirus (so as to delete *Rce1* from only ~50% of the cells in the culture). The cells were then allowed to grow for multiple passages, and the *Rce1* genotype was checked by Southern blotting at frequent intervals (Bergo *et al.*, 2002a).

Experiment 1

Experiment 3

Experiment 2

Experiment 4

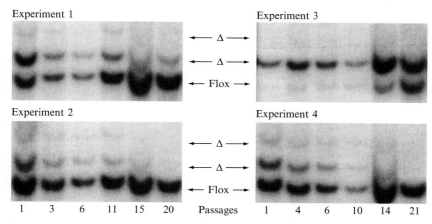

| 1 | 3 | 6 | 11 | 15 | 20 | Passages | 1 | 4 | 6 | 10 | 14 | 21 |

Figure 26.3   Southern blot assessment of the ratio of $Rce1^{flx}$ and $Rce1^{\Delta}$ alleles during the growth of mixed cultures of $Rce1^{flx/flx}$ and $Rce1^{\Delta/\Delta}$ fibroblasts. Mixed cultures of $Rce1^{flx/flx}$ and $Rce1^{\Delta/\Delta}$ cells were obtained by infecting immortalized $Rce1^{flx/flx}$ fibroblasts with *lacZ* adenovirus and *Cre* adenovirus. Southern blots showing the ratio of $Rce1^{\Delta}$ and $Rce1^{flx}$ alleles at different passages in four independent experiments. These experiments show that the $Rce1^{flx/flx}$ cells, which express $Rce1$, grow more rapidly than the $Rce1^{\Delta/\Delta}$ fibroblasts, which lack $Rce1$ expression. (From Bergo, M. O., Ambroziak, P., Gregory, C., George, A., Otto, J. C., Kim, E., Nagase, H., Casey, P. J., Balmain, A., and Young, S. G. (2002a). Absence of the *CAAX* endoprotease Rce1: Effects on cell growth and transformation. *Mol. Cell. Biol.* **22,** 171–181, with permission.)

If the $Rce1$ excision had no effect on cell growth (i.e., if there were no differences in the growth of $Rce1^{flx/flx}$ and $Rce1^{\Delta/\Delta}$ cells), one would expect that the Southern blots would reveal, over multiple passages, stability in the relative intensities of the 5.0-kb $Rce1^{flx}$ band and the 6.5-kb $Rce1^{\Delta}$ band. This was not the case. In each of multiple experiments, the ratio of the 5.0-kb $Rce1^{flx}$ to the 6.5-kb $Rce1^{\Delta}$ band increased steadily, indicating a competitive growth advantage of the $Rce1^{flx/flx}$ cells over $Rce1^{\Delta/\Delta}$ cells (Fig. 26.3). Control experiments established that the competitive growth advantage of $Rce1^{flx/flx}$ cells over $Rce1^{\Delta/\Delta}$ cells was due to the excision of the $Rce1$ gene rather than to the loss of the neo (a drug selection marker in the $Rce1^{flx}$ allele) (Bergo *et al.*, 2002a).

13. EFFECTS OF AN *RCE1* EXCISION ON THE GROWTH OF RAS-TRANSFECTED CELLS IN SOFT AGAR AND IN NUDE MICE

In mammalian cells, the expression of activated RAS proteins allow fibroblasts the ability to grow in soft agar (anchorage-independent growth).

To test whether deletion of *Rce1* would influence this phenotype, *Rce1*$^{flx/flx}$ fibroblasts were first infected with a retrovirus coding for a mutationally activated K-RAS (Bergo *et al.*, 2002b). After selection for retroviral infection with puromycin (4 μg/ml), the overexpression of K-RAS was documented by Western blotting. The RAS-transfected *Rce1*$^{flx/flx}$ fibroblasts were incubated with a *lacZ* adenovirus or a *Cre* adenovirus (to produce RAS-transfected *Rce1*$^{\Delta/\Delta}$ cells) (Bergo *et al.*, 2002a). Equal numbers (3000 cells/well) of RAS-transfected *Rce1*$^{flx/flx}$ or *Rce1*$^{\Delta/\Delta}$ cells were mixed with medium containing 0.35% agarose and poured onto wells of 12-well plates that contained a 0.7% agarose base. The plates were incubated in a humidified culture incubator at 37°, 7% CO_2, for 14 to 21 days. Colonies were stained with MTT (3-[4,5-dimethylthiazol-2-yl]2,5-diphenyltetrazolium bromide; thiazolyl blue, Sigma, 1 mg/ml in phosphate-buffered saline). Each well was photographed with a digital camera. The images were imported into Adobe PhotoShop, and colony numbers were determined with an image-processing tool kit. The deletion of *Rce1* resulted in a $\sim$30% reduction in colonies in each of four independent experiments ($p = 0.029$; $p = 0.09$; $p = 0.005$; $p = 0.008$) (Bergo *et al.*, 2002b) (Fig. 26.4). A $\sim$75% reduction in transformed colonies ($p < 0.001$) was observed when using a second pair of *Rce1*$^{flx/flx}$ and *Rce1*$^{\Delta/\Delta}$ cells (not shown). In parallel, we generated H-RAS–transformed *Rce1*$^{flx/flx}$ and *Rce1*$^{\Delta/\Delta}$ cells, and found that deletion of *Rce1* resulted in a 65% reduction in colonies ($p = 0.002$). We also obtained K-RAS-transformed *Rce1*$^{-/-}$ and *Rce1*$^{+/+}$ cells by transfecting primary embryonic fibroblasts with E1A and K-RAS. We observed a $\sim$50% reduction in numbers of colonies from *Rce1*$^{-/-}$ cells, compared with *Rce1*$^{+/+}$ cells in three independent experiments ($p = 0.003$, $p < 0.001$, $p < 0.001$) (Bergo *et al.*, 2002a).

The reduced ability of K-RAS-transfected *Rce1*$^{\Delta/\Delta}$ cells to grow in soft agar was associated with a reduced capacity to contribute to the growth of tumors in nude mice (Bergo *et al.*, 2002a). Nude mice were injected with a mixed culture of K-RAS-transfected *Rce1*$^{flx/flx}$ and *Rce1*$^{\Delta/\Delta}$ cells. The *Rce1*$^{flx/flx}$ cells manifested an increased capacity to contribute to the formation of tumors relative to the *Rce1*$^{\Delta/\Delta}$ cells. As judged by Southern blots, the ratio of *Rce1*$^{\Delta}$ to *Rce1*^{flx} alleles in the injected cells was 1.99. The ratio increased to 11.8 ± 2.9 (mean $\pm$ standard deviation) in the DNA prepared from established tumors, reflecting more rapid growth of *Rce1*$^{flx/flx}$ cells relative to *Rce1*$^{\Delta/\Delta}$ cells.

A *Rce1*$^{flx/flx}$ skin carcinoma cell line harboring a mutationally activated H-RAS was produced by treating the skin of *Rce1*$^{flx/flx}$ mice with mutagens (Bergo *et al.*, 2002a). We treated the *Rce1*$^{flx/flx}$ cells with 10^8 pfu of *Cre* adenovirus, successfully deleting *Rce1* from 90% of the cells. However, we found that the small fraction of *Rce1*$^{flx/flx}$ cells overgrew the culture within a few days, strongly suggesting that *Rce1*$^{flx/flx}$ skin carcinoma cells grow more rapidly than the *Rce1*$^{\Delta/\Delta}$ cells.

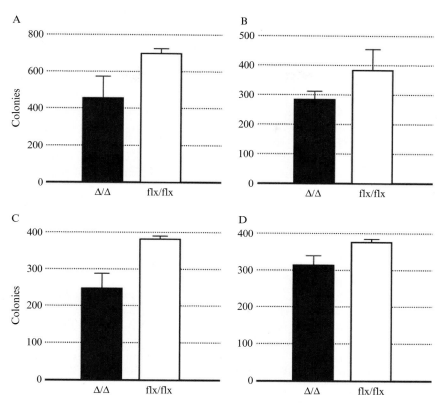

Figure 26.4 Comparison of the ability of K-RAS–transformed immortalized $Rce1^{\Delta/\Delta}$ and $Rce1^{flx/flx}$ fibroblasts to form colonies in soft agar. $Rce1^{flx/flx}$ fibroblasts were infected with a K-RAS retrovirus, and then treated with either *lacZ* adenovirus or *Cre* adenovirus. In $Rce1^{flx/flx}$ cells treated with *Cre* adenovirus, the *Rce1* was fully deleted, generating $Rce1^{\Delta/\Delta}$ fibroblasts. In each experiment, the excision of *Rce1* resulted in fewer colonies in soft agar ($p = 0.029$, $p = 0.005$, $p = 0.09$, $p = 0.008$). (From Bergo, M. O., Ambroziak, P., Gregory, C., George, A., Otto, J. C., Kim, E., Nagase, H., Casey, P. J., Balmain, A., and Young, S. G. (2002a). Absence of the *CAAX* endoprotease *Rce1*: Effects on cell growth and transformation. *Mol. Cell. Biol.* **22**, 171–181, with permission.)

14. ANALYZING THE FUNCTIONAL IMPORTANCE OF *RCE1* IN THE LIVER AND HEART

To inactivate *Rce1* in the liver, $Rce1^{flx/flx}$ mice were given an intravenous injection of a *Cre* adenovirus (2×10^{11} plaque-forming units). After 3 days, livers were removed, and the levels of *CAAX* endoprotease activity in the tissues were determined, and the electrophoretic mobilities of the RAS proteins were analyzed (Bergo *et al.*, 2004). Western blot analysis of the RAS proteins in the liver revealed that approximately one-half of the RAS proteins had retarded electrophoretic mobility, indicating a loss of endoproteolytic processing in a large fraction of the liver cells. Despite the loss of *Rce1* in the liver, the mice remained quite healthy. Mice lacking *Rce1* in the liver were also generated by breeding *Rce1*$^{flx/flx}$ mice carrying the interferon-inducible Mx1-*Cre* transgene, and then inducing *Cre* expression with intraperitoneal injections of polyinosinic-polycytidylic ribonucleic acid (pI-pC, Sigma, St. Louis, MO). As judged by Southern blots, the excision of *Rce1* in the livers of the pI-pC–treated mice was complete, and RCE1 activity levels were dramatically reduced. Again, despite the absence of *Rce1* in the liver, the pI-pC–treated *Rce1*$^{flx/flx}$Mx1-*Cre*$^{+/+}$ mice gained weight and exhibited normal vitality over 2 to 3 months of observation. During this time, transaminase levels remained normal, and the histological appearance of the liver on hematoxylin and eosin-stained sections was indistinguishable from that of wildtype mice (Bergo *et al.*, 2004).

Mice lacking *Rce1* in the heart were created by breeding *Rce1*$^{flx/\Delta}$ mice carrying a *Cre* transgene driven by the α-myosin heavy-chain (αMyhc) promoter (Bergo *et al.*, 2004). The cardiac myocytes in these mice would be expected to lack *Rce1* expression in cardiac myocytes. Indeed, quantitative PCR studies revealed lower *Rce1* mRNA levels in the hearts of *Rce1*$^{flx/\Delta}\alpha$Myhc-*Cre*$^{+/o}$ mice than in the hearts of *Rce1*$^{flx/\Delta}$ controls ($p < 0.01$) (Bergo *et al.*, 2004). Similarly, using the coupled endoprotease/methylation assay, we observed lower *CAAX* endoprotease activity levels in the hearts of *Rce1*$^{flx/\Delta}\alpha$Myhc-*Cre*$^{+/o}$ mice than in the hearts of *Rce1*$^{flx/\Delta}$ control mice ($p < 0.01$). As expected, we observed retarded electrophoretic mobility of the RAS proteins in hearts of *Rce1*$^{flx/\Delta}\alpha$Myhc-*Cre*$^{+/o}$ mice, and we were able to document, with the coupled endoprotease/methylation assay, a significant accumulation of RCE1 protein substrates in the hearts of *Rce1*$^{flx/\Delta}\alpha$Myhc-*Cre*$^{+/o}$ mice.

The *Rce1*$^{flx/\Delta}\alpha$Myhc-*Cre*$^{+/o}$ mice and control *Rce1*$^{flx/flx}$ mice were bred and observed for 10 months. The *Rce1*$^{flx/\Delta}\alpha$Myhc-*Cre*$^{+/o}$ mice appeared healthy at 1 month of age, but started dying by 3 to 5 months of age (Bergo *et al.*, 2004). By 7 months of age, 50% of the *Rce1*$^{flx/\Delta}\alpha$Myhc-*Cre*$^{+/o}$ mice had died; by 10 months, 70% had died. In contrast, none of the *Rce1*$^{flx/flx}$ mice or *Rce1*$^{+/+}\alpha$Myhc-*Cre*$^{+/o}$ mice died during the 10-month follow-up period. Similarly, no premature deaths were observed in *Rce1*$^{flx/\Delta}$ mice. Several weeks before their deaths, most of the heart-specific *Rce1* knockout mice (*Rce1*$^{fl/\Delta}\alpha$Myhc-*Cre*$^{+/o}$) appeared listless and had ruffled fur. The hearts of *Rce1*$^{fl/\Delta}\alpha$Myhc-*Cre*$^{+/o}$ mice were invariably enlarged (Fig. 26.5), and echocardiography showed dilated left ventricles (Bergo *et al.*, 2004). Histological sections revealed dilatation of all four chambers of the heart, and organized thrombi were occasionally noted within the left atrium. The left ventricular musculature was thin, and there were increased amounts

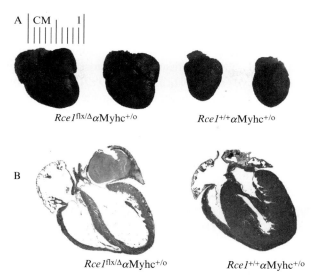

Figure 26.5 Dilated cardiomyopathy in $Rce1^{flx/\Delta}\alpha$Myhc-$Cre^{+/o}$ mice. (A) Increased size of hearts from $Rce1^{flx/\Delta}\alpha$Myhc-$Cre^{+/o}$ mice, compared with $Rce1^{+/+}\alpha$Myhc-$Cre^{+/o}$ controls. (B) H&E-stained sections of an $Rce1^{flx/\Delta}\alpha$Myhc-$Cre^{+/o}$ heart and an $Rce1^{+/+}\alpha$ Myhc-$Cre^{+/o}$ heart, both from 7-month-old mice. The hearts from $Rce1^{flx/\Delta}\alpha$Myhc-$Cre^{+/o}$ mice were invariably dilated. Note the organized left atrial thrombus in the heart from the $Rce1^{flx/\Delta}\alpha$Myhc-$Cre^{+/o}$ mouse. (From Bergo, M. O., Lieu, H. D., Gavino, B. J., Ambroziak, P., Otto, J. C., Casey, P. J., Walker, Q. M., and Young, S. G. (2004). On the physiological importance of endoproteolysis of *CAAX* proteins: heart-specific *RCE1* knockout mice develop a lethal cardiomyopathy. *J. Biol. Chem.* **279**, 4729–4736, with permission.)

of collagen between myocytes. Heart tissue from $Rce1^{flx/flx}$, $Rce1^{+/+}$ αMyhc-$Cre^{+/o}$, and $Rce1^{flx/\Delta}$ control mice was normal. Thus, in contrast to the situation with the liver and hematopoietic cells, $Rce1$ deficiency in the heart had significant adverse consequences.

It is tempting to speculate that the effects of $Rce1$ deficiency in the heart are somehow related to the absence of RAS endoproteolytic processing, particularly given that an activated H-RAS transgene in mice causes myocardial hypertrophy (Gottshall *et al.*, 1997) and the expression of a dominant-negative H-RAS causes a dilated cardiomyopathy (Seigo Izumo, personal communication; see also http://cardiogenomics.med. harvard.edu/groups/proj1/pages/ras_home.html). However, despite the attractiveness of the "RAS hypothesis," there are clearly reasons for caution. First, we did not observe any effect of $Rce1$ deficiency in fibroblasts on the activation of RAS effectors by growth factors (Bergo *et al.*, 2004). Second, RCE1 has dozens of protein substrates, and it is conceivable that the elimination of endoproteolysis of some of the non-RAS substrates underlies the pathology in the heart-specific *Rce1* knockout mice. For example, the endoproteolytic processing of lamin B1 (a key structural component of the nuclear lamina) does not occur in the absence of *Rce1* (Maske *et al.*, 2003). Missense mutations in lamin A/C clearly cause dilated cardiomyopathy (Fatkin *et al.*, 1999), so one could imagine that absent endoproteolytic processing of lamin B1 might affect heart function. One could make similar arguments about the potential involvement of many other *CAAX* proteins in mediating the cardiomyopathy of *Rce1* deficiency.

15. MINIMAL EFFECTS OF *RCE1* DEFICIENCY ON THE HEMATOPOIETIC SYSTEM IN MICE

Initially, we considered the possibility that $Rce1^{-/-}$ embryos might die from defective hematopoiesis. However, this did not appear to be the case. Lethally irradiated mice were successfully rescued with hematopoietic stem cells from the livers of $Rce1^{-/-}$ embryos (Kim *et al.*, 1999). Both the red and white cell counts remained stable for more than 6 months of follow-up, suggesting that there was no significant impact of *Rce1* deficiency on hematopoiesis. ERK activities were measured in bone marrow collected 3 to 6 months after adoptive transfer. In multiple experiments, $Rce1^{-/-}$ and $Rce1^{+/+}$ bone marrow demonstrated equivalent basal and GM-CSF-stimulated ERK activities.

We considered the possibility that the potential of $Rce1^{-/-}$ stem cells might have a relatively reduced capacity to repopulate the bone marrow when compared directly with stem cells from $Rce1^{+/+}$ bone marrow. However, this was not the case (Aiyagari *et al.*, 2003). Wildtype or $Rce1^{-/-}$ fetal liver cells were injected into irradiated hosts with the same reference population of BoyJ competitor cells to directly compare their potential to repopulate the bone marrow. The $Rce1^{-/-}$ and $Rce1^{+/+}$ hematopoietic cells exhibited equivalent bone-marrow repopulating potential over a dose range that produced 10 to 70% donor cell chimerism.

To further analyze the consequences of *Rce1* deficiency on the growth of hematopoietic cells, we harvested bone marrow cells from $Rce1^{flx/flx}$Mx1-$Cre^{+/o}$ and control $Rce1^{flx/+}$Mx1-$Cre^{+/o}$ mice, 5 weeks after injection of pI-pC and compared their ability to form colonies in methylcellulose. As expected, bone marrow cells from both groups of mice formed similar numbers and types of colonies (Wahlstrom *et al.*, unpublished observations). These data strongly suggest that inactivation of *Rce1* does not impair the proliferative capacity of hematopoietic cells.

16. *RCE1* DEFICIENCY ACCELERATES THE DEVELOPMENT OF A K-RAS–INDUCED MYELOPROLIFERATIVE DISEASE

As described earlier, the inactivation of *Rce1* resulted in mislocalization of the RAS proteins within cells and reduced RAS-induced transformation of fibroblasts. Therefore, we hypothesized that inactivating *Rce1* would inhibit the development of RAS-induced malignancies *in vivo*. To approach this issue, we bred *Rce1*$^{flx/flx}$Mx1-*Cre*$^{+/o}$ mice with mice harboring a latent oncogenic K-RAS allele (*Kras2*^{LSL}). Injection of pI–pC into *Rce1*$^{flx/flx}$Mx1-*Cre*$^{+/o}$*Kras2*$^{LSL/+}$ mice simultaneously inactivated the expression of *Rce1* and activated the expression of oncogenic K-RAS^{G12D} in hematopoietic cells. Normally, activation of the oncogenic K-RAS allele in hematopoietic cells leads to a rapidly progressing and lethal myeloproliferative disease with leukocytosis, splenomegaly, tissue infiltration, and an ability of hematopoietic cells to form colonies in methylcellulose in the

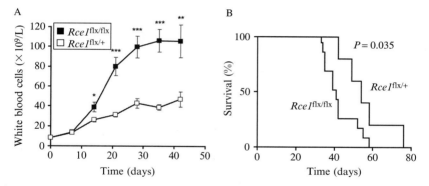

Figure 26.6 Inactivation of *Rce1* accelerates the development of a K-RAS–induced myeloproliferative disease. Groups of *Rce1*$^{flx/flx}$Mx1-*Cre*$^{+/o}$*Kras2*$^{LSL/+}$ mice and control *Rce1*$^{flx/+}$Mx1-*Cre*$^{+/o}$*Kras2*$^{LSL/+}$ mice were injected at weaning with pI–pC to induce a myeloproliferative disease. (A) The number of white blood cells was elevated in *Rce1*$^{flx/flx}$ *Kras2*$^{LSL/+}$Mx1-*Cre*$^{+/o}$ compared with *Rce1*$^{flx/+}$*Kras2*$^{LSL/+}$Mx1-*Cre*$^{+/o}$ mice with myeloproliferative disease. Statistically significant changes at each time point are indicated: $\star p <$ 0.05; $\star\star p < 0.01$; and $\star\star\star p < 0.001$; $n = 11$ to 27 per time point (except for the values at 42 days where $n = 3$ [*Rce1*$^{flx/flx}$*Kras2*$^{LSL/+}$Mx1-*Cre*$^{+/o}$] and $n = 5$ (*Rce1*$^{flx/+}$*Kras2*$^{LSL/+}$Mx1-*Cre*$^{+/o}$]). (B) Kaplan-Meier curve demonstrating reduced survival of pI–pC–treated *Rce1*$^{flx/}$ ^{flx}*Kras2*$^{LSL/+}$Mx1-*Cre*$^{+/o}$ mice (median survival 40 days, $n = 12$) compared with *Rce1*$^{flx/+}$ *Kras2*$^{LSL/+}$Mx1-*Cre*$^{+/o}$ mice (median survival 54 days, $n = 5$). (From Wahlstrom, A. M., Cutts, B. A., Karlsson, C., Andersson, K. M., Liu, M., Sjogren, A. K., Swolin, B., Young, S. G., and Bergo, M. O. (2007). *Rce1* deficiency accelerates the development of K-RAS-induced myeloproliferative disease. *Blood* **109**, 763–768, with permission.)

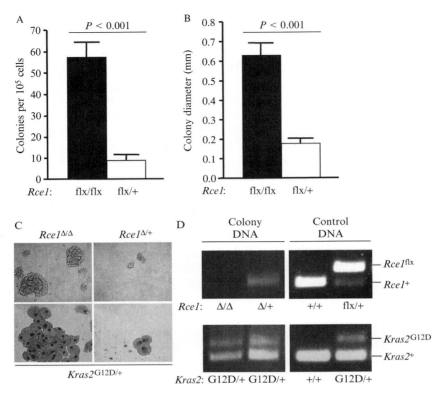

Figure 26.7 Inactivation of *Rce1* increases colony formation of K-RAS^{G12D}–expressing splenocytes. (A and B) Growth factor–independent colony growth of splenocytes from *Rce1*$^{flx/flx}$Mx1-*Cre*$^{+/o}$*Kras2*$^{LSL/+}$ ($n = 3$) and *Rce1*$^{flx/+}$Mx1-*Cre*$^{+/o}$*Kras2*$^{LSL/+}$ ($n = 3$) mice. A, colony number; B, colony size. Splenocytes from control mice were incapable of colony growth in this assay (not shown). (C) Upper panel: Photographs showing *Rce1*$^{\Delta/\Delta}$Mx1-*Cre*$^{+/o}$*Kras2*^{G12D} and *Rce1*$^{\Delta/+}$Mx1-*Cre*$^{+/o}$*Kras2*^{G12D} splenocyte colonies from a typical experiment in A and B. Lower panel: May-Grünwald-Giemsa–stained cytospins of individual colonies. (D) PCR amplification of genomic DNA from individual colonies to detect the *Rce1*^{flx} and *Rce1*$^+$ alleles (upper panel) and the *Kras2*$^+$ and *Kras2*^{G12D} alleles (lower panel). Lane 1, *Rce1*-deficient *Kras2*^{G12D} colony; lane 2, heterozygous *Rce1*-deficient *Kras2*^{G12D} colony; lanes 3 and 4, control DNA from mouse tails. (From Wahlstrom, A. M., Cutts, B. A., Karlsson, C., Andersson, K. M., Liu, M., Sjogren, A. K., Swolin, B., Young, S. G., and Bergo, M. O. (2007). *Rce1* deficiency accelerates the development of K-RAS-induced myeloproliferative disease. *Blood* **109**, 763–768, with permission.) (See color insert.)

absence of exogenous growth factors (Braun *et al.*, 2004; Chan *et al.*, 2004; Wahlstrom *et al.*, 2007). Contrary to our hypothesis, the inactivation of *Rce1* resulted in an up to 2.7-fold increase in white blood cell counts ($p < 0.0001$) (see Fig. 26.6A), a 5.3-fold increase in the percentage of immature cells in the circulation ($p < 0.0001$), increased infiltration of cells into liver and spleen, and reduced survival ($p < 0.035$) (see Fig. 26.6B) (Wahlstrom et al., 2007). Moreover, inactivation of Rce1 in hematopoietic cells expressing oncogenic K-RAS resulted in a 6.6-fold increase in colony number and a 3.4-fold increase in colony size compared to control when grown in methylcellulose in the absence of exogenous growth factors ($p < 0.001$) (Fig. 26.7A to C). Genotyping of genomic DNA isolated from individual colonies confirmed the absence of Rce1 in K-RAS^{G12D}-expressing cells (see Fig. 26.7D). Thus, we were forced to conclude that inactivation of Rce1 dramatically worsens essentially all phenotypes associated with the K-RAS-induced myeloproliferative disease.

This study does not support the concept that RCE1 inhibitors would be useful for the treatment of RAS-induced hematological malignancies. Indeed, such a strategy might even be dangerous. At this point, the mechanism for the worsened myeloproliferative disease and the increased proliferation of Rce1-deficient hematopoietic cells is unknown. The simplest potential explanation would be that RCE1 normally processes an isoprenylated CAAX protein that suppresses cell proliferation and that this protein is dysfunctional in the absence of RCE1-mediated endoproteolytic processing. If this explanation were true, than Rce1 deficiency should accelerate the development of myeloid leukemia caused by other genetic interventions (e.g., Nf1 deficiency [Le et al., 2004]).

REFERENCES

Aiyagari, A. L., Taylor, B. R., Aurora, V., Young, S. G., and Shannon, K. M. (2003). Hematologic effects of inactivating the Ras processing enzyme Rce1. Blood 101, 2250–2252.

Ashby, M. N. (1998). CaaX converting enzymes. Curr. Opin. Lipidol. 9, 99–102.

Bergo, M. O., Ambroziak, P., Gregory, C., George, A., Otto, J. C., Kim, E., Nagase, H., Casey, P. J., Balmain, A., and Young, S. G. (2002a). Absence of the CAAX endoprotease Rce1: Effects on cell growth and transformation. Mol. Cell. Biol. 22, 171–181.

Bergo, M. O., Gavino, B., Ross, J., Schmidt, W. K., Hong, C., Kendall, L. V., Mohr, A., Meta, M., Genant, H., Jiang, Y., Wisner, E. R., Van Bruggen, N., et al. (2002b). Zmpste24 deficiency in mice causes spontaneous bone fractures, muscle weakness, and a prelamin A processing defect. Proc. Natl. Acad. Sci. USA 99, 13049–13054.

Bergo, M. O., Lieu, H. D., Gavino, B. J., Ambroziak, P., Otto, J. C., Casey, P. J., Walker, Q. M., and Young, S. G. (2004). On the physiological importance of endoproteolysis of CAAX proteins: heart-specific RCE1 knockout mice develop a lethal cardiomyopathy. J. Biol. Chem. 279, 4729–4736.

Boyartchuk, V. L., Ashby, M. N., and Rine, J. (1997). Modulation of Ras and a-factor function by carboxyl-terminal proteolysis. Science 275, 1796–1800.

Braun, B. S., Tuveson, D. A., Kong, N., Le, D. T., Kogan, S. C., Rozmus, J., Le Beau, M. M., Jacks, T. E., and Shannon, K. M. (2004). Somatic activation of oncogenic Kras in hematopoietic cells initiates a rapidly fatal myeloproliferative disorder. Proc. Natl. Acad. Sci. USA 101, 597–602.

Chan, I. T., Kutok, J. L., Williams, I. R., Cohen, S., Kelly, L., Shigematsu, H., Johnson, L., Akashi, K., Tuveson, D. A., Jacks, T., and Gilliland, D. G. (2004). Conditional expression of oncogenic K-ras from its endogenous promoter induces a myeloproliferative disease. *J. Clin. Invest.* **113,** 528–538.

Chen, Z., Otto, J. C., Bergo, M. O., Young, S. G., and Casey, P. J. (2000). The C-terminal polylysine domain and methylation of K-Ras are critical for the interaction between K-Ras and microtubules. *J. Biol. Chem.* **275,** 41251–41257.

Clarke, S., Vogel, J. P., Deschenes, R. J., and Stock, J. (1988). Posttranslational modification of the Ha-ras oncogene protein: Evidence for a third class of protein carboxyl methyltransferases. *Proc. Natl. Acad. Sci. USA* **85,** 4643–4647.

Corrigan, D. P., Kuszczak, D., Rusinol, A. E., Thewke, D. P., Hrycyna, C. A., Michaelis, S., and Sinensky, M. S. (2005). Prelamin A endoproteolytic processing *in vitro* by recombinant *Zmpste*24. *Biochem. J.* **387,** 129–138.

Dai, Q., Choy, E., Chiu, V., Romano, J., Slivka, S. R., Steitz, S. A., Michaelis, S., and Philips, M. R. (1998). Mammalian prenylcysteine carboxyl methyltransferase is in the endoplasmic reticulum. *J. Biol. Chem.* **273,** 15030–15034.

Dalton, M., and Sinensky, M. (1995). Expression systems for nuclear lamin proteins: Farnesylation in assembly of nuclear lamina. *Methods Enzymol.* **250,** 134–148.

Fatkin, D., MacRae, C., Sasaki, T., Wolff, M. R., Porcu, M., Frenneaux, M., Atherton, J., Vidaillet, H. J., Jr., Spudich, S., de Girolami, U., Seidman, J. G., Seidman, C., *et al.* (1999). Missense mutations in the rod domain of the lamin A/C gene as causes of dilated cardiomyopathy and conduction-system disease. *N. Engl. J. Med.* **341,** 1715–1724.

Fujimura-Kamada, K., Nouvet, F. J., and Michaelis, S. (1997). A novel membrane-associated metalloprotease, Ste24p, is required for the first step of NH2-terminal processing of the yeast a-factor precursor. *J. Cell Biol.* **136,** 271–285.

Gottshall, K. R., Hunter, J. J., Tanaka, N., Dalton, N., Becker, K. D., Ross, J., Jr., and Chien, K. R. (1997). Ras-dependent pathways induce obstructive hypertrophy in echo-selected transgenic mice. *Proc. Natl. Acad. Sci. USA* **94,** 4710–4715.

Hrycyna, C. A., Sapperstein, S. K., Clarke, S., and Michaelis, S. (1991). The *Saccharomyces cerevisiae* STE14 gene encodes a methyltransferase that mediates C-terminal methylation of a-factor and Ras proteins. *EMBO J.* **10,** 1699–1709.

Kataoka, T., Powers, S., McGill, C., Fasano, O., Strathern, J., Broach, J., and Wigler, M. (1984). Genetic analysis of yeast *RAS1* and *RAS2* genes. *Cell* **37,** 437–445.

Kilic, F., Dalton, M. B., Burrell, S. K., Mayer, J. P., Patterson, S. D., and Sinensky, M. (1997). *In vitro* assay and characterization of the farnesylation-dependent prelamin A endoprotease. *J. Biol. Chem.* **272,** 5298–5304.

Kim, E., Ambroziak, P., Otto, J. C., Taylor, B., Ashby, M., Shannon, K., Casey, P. J., and Young, S. G. (1999). Disruption of the mouse *Rce1* gene results in defective Ras processing and mislocalization of Ras within cells. *J. Biol. Chem.* **274,** 8383–8390.

Kumagai, H., Kawamura, Y., Yanagisawa, K., and Komano, H. (1999). Identification of a human cDNA encoding a novel protein structurally related to the yeast membrane-associated metalloprotease, Ste24p. *Biochim. Biophys. Acta* **1426,** 468–474.

Le, D. T., Kong, N., Zhu, Y., Lauchle, J. O., Aiyigari, A., Braun, B. S., Wang, E., Kogan, S. C., Le Beau, M. M., Parada, L., and Shannon, K. M. (2004). Somatic inactivation of *Nf1* in hematopoietic cells results in a progressive myeloproliferative disorder. *Blood* **103,** 4243–4250.

Leung, G. K., Schmidt, W. K., Bergo, M. O., Gavino, B., Wong, D. H., Tam, A., Ashby, M. N., Michaelis, S., and Young, S. G. (2001). Biochemical studies of *Zmpste*24-deficient mice. *J. Biol. Chem.* **276,** 29051–29058.

Maske, C. P., Hollinshead, M. S., Higbee, N. C., Bergo, M. O., Young, S. G., and Vaux, D. J. (2003). A carboxyl-terminal interaction of lamin B1 is dependent on the *CAAX* endoprotease Rce1 and carboxymethylation. *J. Cell Biol.* **162,** 1223–1232.

Michaelson, D., Ali, W., Chiu, V. K., Bergo, M., Silletti, J., Wright, L., Young, S. G., and Philips, M. (2005). Postprenylation CAAX processing is required for proper localization of Ras but not Rho GTPases. Mol. Biol. Cell **16,** 1606–1616.

Otto, J. C., Kim, E., Young, S. G., and Casey, P. J. (1999). Cloning and characterization of a mammalian prenyl protein-specific protease. J. Biol. Chem. **274,** 8379–8382.

Pendás, A. M., Zhou, Z., Cadiñanos, J., Freije, J. M. P., Wang, J., Hultenby, K., Astudillo, A., Wernerson, A., Rodríguez, F., Tryggvason, K., and Lopéz-Otín, C. (2002). Defective prelamin A processing and muscular and adipocyte alterations in Zmpste24 metalloproteinase-deficient mice. Nat. Genet. **31,** 94–99.

Tam, A., Nouvet, F. J., Fujimura-Kamada, K., Slunt, H., Sisodia, S. S., and Michaelis, S. (1998). Dual roles for Ste24p in yeast a-factor maturation: NH2-terminal proteolysis and COOH-terminal CAAX processing. J. Cell Biol. **142,** 635–649.

Thissen, J. A., Gross, J. M., Subramanian, K., Meyer, T., and Casey, P. J. (1997). Prenylation-dependent association of Ki-Ras with microtubules. Evidence for a role in subcellular trafficking. J. Biol. Chem. **272,** 30362–30370.

Todaro, G. J. (1969). Transformation assay using cell line 3T3. In "Fundamental Techniques in Virology." (K. Habel and N. P. Salzman, eds.), pp. 220–228. Academic Press, New York.

Wahlstrom, A. M., Cutts, B. A., Karlsson, C., Andersson, K. M., Liu, M., Sjogren, A. K., Swolin, B., Young, S. G., and Bergo, M. O. (2007). Rce1 deficiency accelerates the development of K-RAS-induced myeloproliferative disease. Blood **109,** 763–768.

Young, S. G., Ambroziak, P., Kim, E., and Clarke, S. (2000). Postisoprenylation protein processing: CXXX (CaaX) endoproteases and isoprenylcysteine carboxyl methyltransferase. In "The Enzymes" (F. Tamanoi and D. S. Sigman, eds.), pp. 155–213. Academic Press, San Diego, CA.

Zhang, F. L., and Casey, P. J. (1996). Protein prenylation: Molecular mechanisms and functional consequences. Annu. Rev. Biochem. **65,** 241–269.

CHAPTER TWENTY-SEVEN

CHARACTERIZATION OF KRAS-MEDIATED PANCREATIC TUMORIGENESIS IN ZEBRAFISH

Jon M. Davison,* Seung Woo Park,[†] Jerry M. Rhee,[‡] and Steven D. Leach[‡]

Contents

* Department of Pathology, Johns Hopkins University School of Medicine, Baltimore, Maryland
† Department of Internal Medicine, Yonsei University College of Medicine, Seoul, South Korea
‡ Department of Surgery, Oncology and Cell Biology, Johns Hopkins University School of Medicine, Baltimore, Maryland

Methods in Enzymology, Volume 438
ISSN 0076-6879, DOI: 10.1016/S0076-6879(07)38027-0

Abstract

Activating Kras mutations are a pervasive and characteristic feature of human pancreatic cancer. In order to examine the earliest *in vivo* effects of oncogenic Kras expression in the exocrine pancreas, we generated two lines of zebrafish expressing eGFP alone or eGFP fused to human Kras with an activating mutation in codon 12 (Kras^{G12V}) driven by *ptf1a* regulatory elements using a BAC recombineering strategy (Park *et al.*, 2008). In this review, we describe the techniques that we used to observe the effects of eGFP-Kras^{G12V} expression in pancreatic progenitor cells of the zebrafish embryo, as well as techniques used to characterize malignant pancreatic tumors in the adult zebrafish. This zebrafish model of pancreatic neoplasia provides a unique view of the effects of oncogenic Kras in the embryonic pancreas and suggests that the zebrafish will be a useful model organism in which to study the biology of Kras-initiated pancreatic neoplasia.

1. INTRODUCTION

Almost 20 years have passed since the seminal observation that the vast majority (85 to 90%) of pancreatic adenocarcinomas harbor activating point mutations in the Kras proto-oncogene (Almoguera *et al.*, 1988). Sadly, the prognosis for a patient with this diagnosis has remained entirely unchanged during this span of time (Jemal *et al.*, 2006). This is true in spite of remarkable insights into the molecular genetics underlying pancreatic carcinogenesis and the mechanisms of ras signal transduction (Hezel *et al.*, 2006). The dismal prognosis for pancreatic cancer is due, at least in part, to the advanced stage of disease at which most tumors are initially detected. This fact has given impetus to the study of early pancreatic lesions that might still be curable.

There is a large body of evidence that implicates lesions known as "pancreatic intraepithelial neoplasia" or PanINs as the *de facto* precursors of invasive ductal adenocarcinoma (Hruban *et al.*, 2000). At the genetic level, there are similarities between invasive pancreatic adenocarcinomas and PanIN lesions, including a high frequency of oncogenic mutations in Kras (Caldas *et al.*, 1994; Yanagisawa *et al.*, 1993). These observations suggest that oncogenic Kras mutations represent an early event—possibly an initiating event—in a model of progressive accumulation of genetic mutations during pancreatic carcinogenesis. Furthermore, murine models of pancreatic adenocarcinoma in which oncogenic Kras expression is induced from its endogenous locus, either in pancreatic progenitor cells or in adult acinar/centroacinar cells strongly buttress the concept that oncogenic Kras mutations play an initiating role in pancreatic neoplasia (Leach, 2004; Murtaugh and Leach, 2007).

Animal models of pancreatic adenocarcinoma involving the initiation of oncogenic ras expression during embryonic life display a spectrum of embryonic and neonatal phenotypes, which in some instances, morphologically resemble those appearing in adult transgenic animals (Grippo et al., 2003; Quaife et al., 1987; Tuveson et al., 2004, 2006). These findings suggest that some aspects of Kras-initiated neoplasia may be effectively observed and studied in the embryo. A relevant embryonic model of early pancreatic neoplasia would be advantageous due to the reduced latency to expression of the phenotype. In the mouse, this advantage may not be apparent, but in a vertebrate organism such as the zebrafish (Danio rerio), which is amenable to forward genetic and chemical modifier screens, an embryonic model of early pancreatic neoplasia would provide a powerful living system to identify compounds which modulate oncogenic Kras activity as well as novel genetic modifiers of the early neoplastic phenotype.

In recent years, the zebrafish has emerged as a useful model system for the study of cancer biology due to the existence of many tractable techniques for analysis of gene function as well as amenability of the species to forward genetic and chemical modifier screens (Stern et al., 2005; Stern and Zon, 2003). In addition, the optical clarity of early zebrafish embryos which develop externally provides an opportunity for real-time, in vivo examination of cell behavior. This capability can be further enhanced by cell type–specific expression of fluorescent transgenes. Zebrafish tumor models have been generated by isolation of zebrafish with inactivating mutations in the TP53 and APC tumor suppressor genes (Berghmans et al., 2005; Haramis et al., 2006), by expression of human oncogenes such as MYCN, BRAF or the TEL-AML1 fusion protein utilizing tissue-specific promoter elements (Langenau et al., 2003; Patton et al., 2005; Sabaawy et al., 2006) and through the use of cre-lox technology for conditional expression of murine c-Myc (Langenau et al., 2005). Moreover, forward genetic screens have identified a number of mutations associated with increased susceptibility to cancer in the fish (Amsterdam et al., 2004; Moore et al., 2006; Shepard et al., 2005, 2007).

Previous studies have demonstrated significant anatomic and histologic similarity between the mammalian and zebrafish pancreas (Chen et al., 2007; Wallace and Pack, 2003). In addition, zebrafish have been shown to develop pancreatic cancer after exposure to chemical mutagens (Moore et al., 2006). Among several transcription factors which have conserved function between zebrafish and mammals, the basic helix-loop-helix transcription factor Ptf1a is known to play a critical role in the specification, proliferation and differentiation of pancreatic progenitor cells (Kawaguchi *et al.*, 2002; Lin *et al.*, 2004; Zecchin *et al.*, 2004). Ptf1a–expressing progenitor cells may also play an important role in cancer pathogenesis, as activation of oncogenic Kras in Ptf1a-expressing cells induces the formation of mouse PanIN lesions and invasive pancreatic carcinoma (Hingorani *et al.*, 2003).

In order to create a zebrafish system for the study of exocrine pancreatic cancer, we generated stable transgenic lines expressing either eGFP alone or eGFP fused to oncogenic human Kras^{G12V} (Niv *et al.*, 1999) in exocrine pancreatic progenitor cells under *ptf1a* regulatory elements (Park *et al.*, 2008). Exocrine pancreas progenitor cells expressing eGFP-Kras^{G12V} failed to undergo primary exocrine differentiation, leading to an accumulation of undifferentiated progenitor cells expressing oncogenic Kras. This embryonic phenotype was associated with the subsequent appearance of malignant pancreatic carcinomas. These results provide a unique view of the early effects of oncogenic Kras in a living vertebrate organism, and suggest that zebrafish models of pancreatic cancer may prove useful in advancing our understanding of the human disease.

The techniques described in this review were central to the initial characterization of the phenotype of transgenic *ptf1a*:eGFP-Kras^{G12V} fish. For other useful protocols for working with zebrafish which are beyond the scope of this paper, we refer the reader to several standard texts (Detrich III *et al.*, 2004a, 2004b; Nusslein-Volhard and Dahm, 2002; Westerfield, 2000).

2. Targeted Expression of eGFP and eGFP-Kras^{G12V} in the Developing and Adult Zebrafish Exocrine Pancreas

In order to stably express the eGFP and eGFP-Kras^{G12V} proteins in exocrine pancreas progenitor cells in the zebrafish, we utilized a BAC recombineering strategy (Datsenko and Wanner, 2000) to replace the coding sequences of the zebrafish *ptf1a* gene encoded in a genomic BAC clone (CH211–142H2, Genbank accession number AL845362) which spans the zebrafish *ptf1a* locus (Fig. 27.1). The completely sequenced BAC clone is 160 Kbp in size and spans the *ptf1a* coding sequence, encompassing approximately 110 Kbp of 5′ noncoding sequence and 50 Kbp of 3′ noncoding sequence. The genomic regulatory sequences of the *ptf1a* gene have not been fully characterized and large genomic clones such as BACs are likely to encompass both proximal and distal regulatory elements as well as sequences which may insulate against epigenetic silencing of transgene expression, potentially yielding transgene expression patterns which more faithfully recapitulate expression of the gene of interest (Jessen *et al.*, 1998).

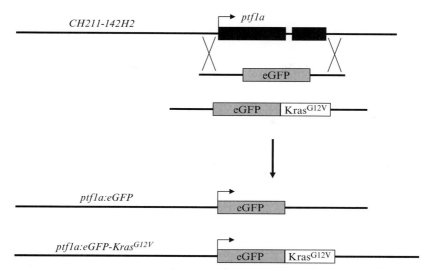

Figure 27.1 Schematic diagram of the BAC recombineering strategy utilized by Park *et al.* (2008). PCR products encoding the *eGFP* and *eGFP-Kras*^{G12V} transgenes are flanked by sequences homologous to the CH211–142H2 BAC that spans the *ptf1a* locus. Homologous recombination results in precise replacement of the *ptf1a* coding sequences (two exons, black rectangles) with the *eGFP* and *eGFP-Kras*^{G12V} transgenes.

Generation of BAC transgenes is unnecessary, however, when discrete promoter elements can be identified for effective targeting of transgene expression to cell types of interest. Although, as mentioned above, there are theoretical advantages associated with the use of large DNA constructs for the expression of transgenes, a major disadvantage is the low germline transgenesis efficiency associated with large DNA constructs (see following discussion). If it is feasible to utilize smaller plasmid constructs that may be synthesized with traditional cloning methods, germline transgenesis efficiency can be improved several-fold, particularly if coupled with the use of transposable elements (Urasaki *et al.*, 2006).

The λ-Red recombinase system that we used to generate the recombinant BAC transgenes for this study (Datsenko and Wanner, 2000) has been superseded by other protocols and reagents. Protocols that we have subsequently used in our lab can be found at http://recombineering.ncifcrf.gov/. BAC recombineering methods as applied to zebrafish transgenesis have also been recently reviewed in detail elsewhere (Yang *et al.*, 2006).

Recombineering of the *ptf1a* BAC resulted in precise replacement of the *ptf1a* coding sequences with eGFP or eGFP-Kras^{G12V} coding sequences (see Fig. 27.1). This was confirmed by direct sequencing of the recombineered BACs.

3. GENERATION OF TRANSGENIC ZEBRAFISH WITH RECOMBINANT BAC DNA CONSTRUCTS

Exogenous DNA constructs may be directly injected into single-cell–stage embryos and incorporated into zebrafish chromosomal DNA at low frequency. BAC DNA is prepared for injection with standard plasmid maxiprep kits (e.g., Qiagen) and then diluted to a concentration of 100 to 150 ng/μl in 0.1 M potassium chloride and 0.03% phenol red for injection (Yang *et al.*, 2006). Protocols for mating adult fish, and for injecting embryos are beyond the scope of this review and are described in detail in standard technical manuals (Nusslein-Volhard and Dahm, 2002).

Injected embryos can be raised and visually screened for transient expression of a fluorescent marker up to 5 days postfertilization. This is an important functional confirmation of the BAC construct that has been previously validated by sequencing or restriction analysis. In order to increase the likelihood of recovering F0 adults that have incorporated the transgene into the germline, we selectively raise F0 embryos which express BAC transgenes in somatic tissues of the pancreas, hindbrain, and retina in a mosaic fashion.

Injected F0 embryos that incorporate BAC DNA into the germline invariably show evidence of germline mosaicism. This is reflected in lower than Mendelian ratios of transgenic offspring produced when F0 adults are mated to wildtype fish. In our experience with BAC transgenesis, only 1 to 3% of injected F0 fish produce transgenic offspring, reflecting relatively inefficient incorporation of BAC DNA into the zebrafish germline. When a germline transgenic F0 fish is identified, 1 to 3% of embryos from a given outcross to wildtype fish express the transgene. This necessitates screening of 100 to 150 embryos from each F0 adult (typically a single cross). The process of screening for transgenic offspring is greatly facilitated by the incorporation of fluorescent proteins into the transgene. The F1 embryos which express the transgene can be identified by visual inspection under the fluorescent dissecting microscope and raised to adulthood and subsequently mated to wildtype fish to recover the line in the F2 generation.

The relative inefficiency of this process when compared to other forms of zebrafish transgenesis requires a large number of injected F0 embryos be raised to adulthood in order to guarantee the recovery of multiple lines of transgenic fish. We recommend injecting enough embryos to recover at least 200 adult fish. The number of injections will depend on the baseline survival rates of zebrafish in the nursery as well as any inherent toxicity associated with expression of the transgene or the injected DNA preparation itself. We raised approximately 300 injected F0s to adulthood in order to recover the *ptf1a*:eGFP and the *ptf1a*:eGFP-Kras^{G12V} lines (Park *et al.*, 2008).

4. ANALYSIS OF THE EFFECTS OF EGFP-KRAS^G12V ON CELLULAR DIFFERENTIATION IN THE ZEBRAFISH EMBRYONIC PANCREAS

The *ptf1a*:eGFP-Kras^{G12V} and *ptf1a*:eGFP transgenes appropriately target protein expression to the pancreas, hindbrain, and retina, faithfully recapitulating expression of endogenous *ptf1a* (Lin *et al.*, 2004; Zecchin *et al.*, 2004). However, pancreatic expression of the *ptf1a*:eGFP-Kras^{G12V} transgene is mosaic in the line that we recovered (Fig. 27.2B). In a subset of transgenic *ptf1a*:eGFP-Kras^{G12V} embryos, we initially observed that transgene expression was appropriately initiated in the left lateral endoderm, and that cells expressing the eGFP-Kras^{G12V} migrated across the midline as

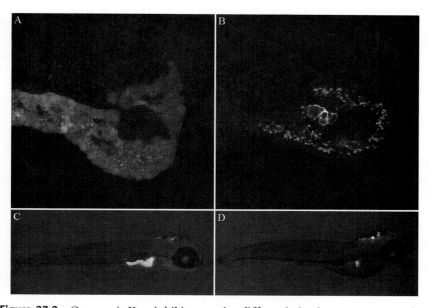

Figure 27.2 Oncogenic Kras inhibits exocrine differentiation in pancreas progenitor cells. (A) Carboxypeptidase A (a marker of exocrine differentiation) is uniformly immunolabeled (red, Cy3) in apical secretory granules of *eGFP*-positive pancreas progenitor cells at 76 h postfertilization (hfp) in *ptf1a:eGFP* embryos. (B) In contrast, *ptf1a: eGFP-Kras*^{G12V} embryos exhibit mosaic expression of the eGFP fusion protein. In cells that express the eGFP-Kras^{G12V} fusion protein, CPA expression is not detected, consistent with a block in exocrine differentiation. (C) *In situ* hybridization detects *eGFP* RNA expression in the pancreas and hindbrain, in a pattern that is identical to protein expression. (D) The pattern of *eGFP-Kras*^{G12V} RNA expression (detected with an antisense probe to *eGFP*) in *ptf1a:eGFP-Kras*^{G12V} embryos is similar to the focal, mosaic pattern of protein expression seen in most embryos (as depicted in B). (See color insert.)

previously described for *ptf1a*-positive pancreas progenitor cells (Lin *et al.*, 2004). In a majority of embryos, however, expression of the eGFP-Kras^{G12V} protein was subsequently downregulated, and by 72 h postfertilization (hpf) was only detected by confocal microscopy in manually dissected viscera. Confocal analysis revealed expression of eGFP-Kras^{G12V} in isolated cells and groups of cells within the pancreatic primordium (see Fig. 27.2B).

We evaluated the effects of eGFP-Kras^{G12V} expression on exocrine differentiation in the zebrafish pancreatic anlage, as prior reports have suggested a block in differentiation occurring in the mouse embryonic pancreas when activated ras (Hras^{G12V}) was expressed under the transcriptional control of the rat elastase promoter (Quaife *et al.*, 1987). Carboxypeptidase A (CPA) is a digestive enzyme that is a marker of exocrine differentiation. CPA protein expression is first detected in the pancreas between 56 and 60 hpf by immunolabeling. We evaluated CPA expression in the *ptf1a*:eGFP and the *ptf1a*:eGFP-Kras^{G12V} lines by immunofluorescent labeling between 76 and 96 hpf when robust expression is expected in the vast majority of *ptf1a*-positive cells (see Fig. 27.2A). We observed that expression of eGFP-Kras^{G12V} was associated with a lack of CPA expression, consistent with a block in differentiation in the subset of cells expressing oncogenic Kras (see Fig. 27.2A and B).

Immunofluorescent staining is a widely used protocol and several variables can affect the quality of staining, including fixation, age of the embryo, and the nature of the antigen and tissues that are to be labeled. The following is a standard protocol used in our lab that works with many of the antibodies we routinely use. The reader is referred to other resources for additional protocols (Nusslein-Volhard and Dahm, 2002; Westerfield, 2000).

4.1. Protocol: Whole-mount immunofluorescent labeling of zebrafish embryos

All steps can be performed in standard 1.5-ml microcentrifuge tubes at room temperature with constant gentle agitation (e.g., using a rotating nutator or rocker platform) unless otherwise indicated. Another option when working with multiple groups of embryos is to perform incubations and washes in 24-well tissue culture plates.

1. Raise embryos in embryo media containing 0.003% phenylthiourea (PTU) starting at 24 h postfertilization to inhibit pigmentation.
2. Once embryos are the appropriate age, dechorionate embryos with forceps as required and rinse in fresh phosphate buffered saline (PBS), pH 7.4.

3. Fix embryos by incubating in 4% paraformaldehyde overnight at 4°. Paraformaldehyde is made by dissolving the compound in PBS by heating to 50 to 60° and adding 1N sodium hydroxide to obtain pH 7.0. Store at −20° until used.

4. Wash embryos with PBST (PBS, 0.1% Tween-20) three times, 5 min each. Other protocols advise removing the yolk for staining of gastro-intestinal tract structures. We have not found this necessary for staining the pancreas using this protocol.

5. Replace one-half the volume of PBST with ice-cold methanol. Incubate on rotator 2 min at room temperature. Repeat this procedure. Last, replace as much of the fluid as possible with 100% methanol. Place the embryos at −20° for 30 min. Embryos may be stored at −20° for several months before proceeding.

6. Remove one-half volume of methanol and replace with equal amount of PBST. Incubate at room temperature for 5 min.

7. Rinse two more times with PBST, 5 min each.

8. Treat embryos with collagenase (Sigma) diluted in PBST from 10% glycerol stock solution stored at −20°. The age of the embryo determines the duration of treatment and the concentration of collagenase. The following are starting points which will vary depending on the activity of each batch of collagenase: 96 h postfertilization (hpf) embryos are treated with 0.2% collagenase for 45 min; 72 to 80 hpf embryos are treated for 45 min in 0.1% collagenase; 48 hpf embryos are treated for 30 min in 0.1% collagenase; and 24 hpf embryos are treated for 10 to 15 min in 0.1% collagenase.

9. Remove collagenase by washing with PBST, three times, 5 min each.

10. Remove as much PBST as possible from embryos and lid of the tube, being careful not to expose embryos to air.

11. Prepare appropriate dilution of primary antibody in PBST supplemented with 10% normal serum. Add 200 μl of the diluted primary antibody to each microcentrifuge tube or enough to completely cover all embryos in a well of a 24-well tissue culture plate (approximately 250 μl). Incubate at 4° overnight.

12. Wash at least three times in PBST for 15 min each.

13. Incubate the embryos with fluorescent secondary antibody raised against the species and antibody isotype of the primary antibody diluted in PBST supplemented with 10% normal serum overnight at 4°.

14. Wash off the secondary antibody in PBST at least 3 times for 15 min each. Longer and more numerous washes will improve the signal to background ratio.

15. For fluorescent microscopy, embryos can be incubated overnight at room temperature in Hoechst 33342 dye (0.1 μg/ml) diluted in PBS for a nuclear counterstain.

4.2. Dissection of endodermal organs for optimal confocal imaging

There are several ways to visualize the results of whole-mount immunofluorescent staining in zebrafish embryos. In order to obtain optimal confocal images of the pancreas, it should be mounted in contact with the glass coverslip. To accomplish this, we have found it helpful to manually dissect the endodermal organs away from the ventral yolk sac and dorsal anatomic structures (e.g., dorsal aorta, pronephric ducts, myotomes, etc.). Alternatives include mounting the whole embryo, cutting cryosections, or cutting vibratome sections. Each approach has its specific advantages and disadvantages. We will discuss a manual dissection technique that we have found particularly useful. We refer the reader to other resources for helpful embryo sectioning techniques: Nusslein-Volhard and Dahm (2002) and Westerfield (2000).

4.2.1. Protocol: Manual dissection of the embryonic gastrointestinal organs

1. Manual dissection is facilitated by transferring the embryo into PBS and then into 4% paraformaldehyde. Fixation after antibody staining will facilitate the dissection as it makes the tissues more brittle and more easily teased apart.
2. To dissect the endodermal organs *en bloc*, transfer an embryo onto a 1.5% agarose–coated Petri dish that is filled with enough PBS to cover the embryo. The agarose base prevents excessive needle breakage and damage to forceps during dissections.
3. Prior to dissection, prepare a 22 × 22–mm glass coverslip by washing the surface with distilled water and placing four small beads of high-vacuum grease in each corner to serve as spacers. Also have ready a small Petri dish with PBST. A dissecting microscope is used to visualize the embryo during dissection.
4. To dissect out the gastrointestinal organs, remove the anterior portion of the embryo including the pectoral fins and head. This can be accomplished by grasping the embryo anterior to the yolk and immediately posterior to the pectoral fins with forceps. Use a second pair of forceps to scrape off the head immediately anterior to where the embryo is grasped. This method preserves the pancreas from being crushed (Fig. 27.3B).
5. Use forceps to remove the tail distal to the main mass of yolk in order to preserve the pancreas.
6. Next remove the yolk. We prefer to use microinjection needles for this part of the dissection. We start with 4- or 6-inch, 1.5-mm diameter glass capillary tubes (World Precision Instruments) and pull them into fine-tip microinjection needles. For manual dissection, 3- to 3.5-inch needles are

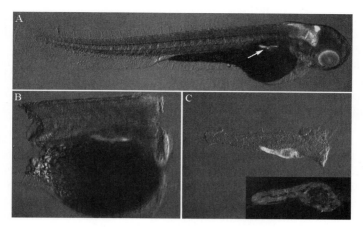

Figure 27.3   Stages of manual dissection of a single *ptf1a:eGFP* embryo. Anterior is to the right in all images. (A) An 80-h postfertilization embryo prior to dissection. eGFP protein is expressed in the retina, hindbrain, and pancreas (arrow). (B) Dissection begins with amputation of the head and tail followed by (C) removal of the yolk and separation of the remaining endodermal organs from ventral anatomic structures. The pancreas may be isolated by careful dissection from the gut (C inset). (See color insert.)

    ideal for manual dissection. Prepare a needle for dissection by breaking off the tip with fine-tipped forceps so that the needle is suitably rigid, but fine enough to accomplish delicate dissection tasks. Some experimentation with different tips will be necessary to make ideal instruments for manual dissection. If the tissues have been adequately fixed prior to dissection, the yolk often dissociates with relative ease using the tips of the needles to gently pry the yolk away from the rest of the embryo.

7. The agarose may be used to hold the embryo in place during dissection. It is sometimes helpful to cut a small hole in the agarose using forceps or a fairly blunt needle and position a portion of the embryo in this hole to limit embryo movement during the dissection.

8. After the yolk is removed, next begin at the caudal end to separate the posterior gut from the dorsal musculature. Working anteriorly, the entire gut with attached pancreas, swim bladder, liver remnants, and biliary structures may be separated from the dorsal embryonic structures (see Fig. 27.3C).

9. Using fine-tipped needles, the pancreas can be dissected away from the other gastrointestinal structures, if needed, for optimal imaging (see Fig. 27.3C, inset).

10. Carefully transfer the dissected tissues using a glass pipette onto the center of the glass coverslip while visualizing the process under the dissecting microscope. Dissected tissues will stick to plastic pipettes, thus they should be avoided.

11. If the tissues adhere to the glass pipette, transfer the tissue into the PBST buffer while visualizing the transfer with the dissecting microscope, and then back into PBS and attempt to transfer to the coverslip once again. PBST may prevent the tissue from adhering to the glass coverslip, so care should be taken to avoid placing tissues on the coverslip in PBST. If this cannot be avoided, transfer the tissues onto the coverslip in PBST and carefully remove the excess using the capillary action of a blunt microinjection needle rather than a plastic pipette tip. Use of a blunt microinjection needle will prevent overly vigorous aspiration of fluid and inadvertent aspiration of the specimen associated with the use of plastic micropipette tips. Blow on the end of the needle to expel excess PBST from the tip, and reuse the needle until all excess PBST has been removed. Replace with a drop of PBS.

12. When the tissue is on the coverslip in a drop of PBS, draw away all excess PBS using the capillary action of a blunt-ended microinjection needle. Then use a fine-tipped microinjection needle to position the tissue such that the pancreas (or other tissue of interest) is in direct contact with the coverslip.

13. Allow the specimen to dry very briefly (less than 30 s) while placing a glass slide on top of the prearranged vacuum grease spacers. Orient the coverslip such that the long axis of the specimen is parallel to the long axis of the glass slide (this will orient and position the specimen well for subsequent imaging). Carefully depress the glass slide onto the coverslip without crushing the specimen, leaving approximately 0.5 to 1.0 mm of space between the glass slide and coverslip.

14. Invert the glass slide so that the coverslip is now on top. Slowly add approximately 200 to 400 μl of PBS under the coverslip to hydrate the specimen, filling the entire space under the coverslip. If the PBS is added too quickly, the specimen may be washed off of the coverslip or moved. This will necessitate repositioning or dissection of another embryo. If there is too much space between the coverslip and the glass slide, the PBS will leak out and this may disrupt the orientation of the specimen.

15. The specimen is now ready for imaging. Bear in mind that the PBS will evaporate unless a seal has been created. The specimen embedded in this fashion must, therefore, be imaged without significant delay. There is no need to create a seal if the specimen can be imaged without significant delay.

16. Techniques for using a conventional fluorescence microscope or a confocal microscope are beyond the scope of this review. In general, though, it is advisable to acquire a low-magnification image of the entire specimen and then capture additional images using higher magnification objectives and zoom factors. It is best to capture a set of images with a standard set of objectives and zoom factors so that images from separate specimens can be compared.

4.3. Whole-Mount Fluorescent *in situ* hybridization with antisense riboprobes

In situ hybridization with labeled antisense RNA riboprobes can be used to examine RNA expression patterns in whole embryos. Protocols for hybridization and detection using chromogenic reagents are standard techniques (Thisse and Thisse, 1998). We have used fluorescent detection of antisense riboprobes as an alternative to chromogenic detection in order to permit visualization with fluorescent microscopy (see Fig. 2C and D).

Antisense riboprobes can be generated from linearized plasmid DNA or directly from PCR products if the target sequence is amplified with a primer encoding an SP6 or T7 RNA polymerase promoter.

4.3.1. Protocol: Antisense riboprobe preparation

1. PCR products or DNA cloned into plasmids may be used for *in vitro* RNA synthesis to generate antisense riboprobes. To generate a template for the *in vitro* RNA synthesis reaction, plasmid DNA should be linearized with appropriate restriction enzymes. The linearized DNA spanning the RNA polymerase promoter and template must be purified from small amounts of contaminating circularized plasmid by gel purification in order to prevent synthesis of large, run-on RNA transcripts from circular templates.

2. To generate a PCR product for *in vitro* RNA synthesis, set up a standard PCR reaction in which one of the primers encodes the T7 or SP6 RNA polymerase promoter sequence (i.e., the primer annealing to the 3-prime end of the sequence of interest). For instance, we generated a probe to detect eGFP expression using the following primers: forward 5′CAAGGGCGAGGAGCTGTT3′ and reverse 5′TGT<u>AATACGA</u><u>CTCACTATAGG</u>GCTCGTCCATGCCGAGAGT3′ (T7 RNA polymerase promoter is underlined).

3. Gel purify the appropriate-sized PCR product. For a probe with an unknown pattern of expression, it will be necessary to verify the sequence of the PCR product.

4. Use the purified PCR product or linearized plasmid DNA as a template for an *in vitro* RNA synthesis reaction according to the manufacturer's protocol (Boehringer DIG RNA Labeling Kit SP6/T7). It is generally safe to assume that 5 to 10 μg of RNA will be synthesized from 1 μg of linear template DNA using this kit.

5. Purify the synthesized RNA using the Qiagen RNeasy Micro kit (RNA cleanup protocol), incorporating the DNAse treatment.

6. After a final elution in 100 μl of RNAse free water, it is worthwhile to verify the synthesis of RNA by running 2 to 5 μl of purified RNA on a gel. The RNA should run as a single robust band on a denaturing gel.

A high-molecular-weight band may be due to contaminating circular plasmid DNA. A low-molecular-weight smear indicates an inefficient reaction or RNA degradation. Neither of these products should be used in subsequent steps.

7. Dilute the RNA to 1 ml in prehybridization solution (50% formamide, 5× SSC, 0.1% Tween-20, 5 mg/ml torula RNA, 50 μg/ml heparin; stored at −20° after preparation) and store at −20° as a stock solution. The working probe concentration is determined empirically, but generally ranges from 0.1 to 1.0 ng/μl. (A useful starting point is to dilute the stock solution 20-fold to obtain the working riboprobe solution).

4.3.2. Protocol: Whole-mount fluorescence *in situ* hybridization with antisense riboprobe

In general, it is optimal to perform the in situ hybridization in 1.5-ml microcentrifuge tubes with locks to prevent the solutions from leaking out of the tubes. All washes and incubations can be performed at room temperature unless otherwise specified.

1. Fix embryos as per above immunofluorescent staining protocol (steps 1 to 5).
2. Rehydrate fixed embryos which have been stored in methanol by replacing 50% of the volume of methanol with PBST and incubate for 5 min, followed by two washes in PBST for 5 min each.
3. Permeabilize embryos with proteinase K digestion for 25 min at room temperature using an empirically determined concentration of proteinase K (will vary with each stock and the age of the embryo). The following will be useful starting points: no treatment for embryos younger than 34 hpf; 0.25 μg/ml for a 34 hpf; 0.5 μg/ml for 36 hpf; 1.0 μg/ml for 40 hpf; 2.5 μg/ml for 44 hpf; 5 μg/ml for a 48 hpf; 10 μg/ml for 56 hpf; 20 μg/ml for 72 hpf; 25 μg/ml for 80 hpf; and 50 μg/ml for 9 days postfertilization embryos. Dilute a 5-mg/ml stock of proteinase K solution in PBST to the appropriate concentration.
4. Stop the digestion by rinsing briefly with PBST, and then fix the embryos in 4% paraformaldehyde for 20 min.
5. Rinse three times in PBST, 5 min each. Remove PBST and replace with pre-hybridization solution (see step 7 of riboprobe preparation). Prehybridize at 68° for at least 4 h.
6. When ready to add the riboprobe, dilute the riboprobe stock to the working concentration in pre-hybridization solution and denature it at 70° for 10 min. Remove the pre-hybridization solution from the embryos and add enough of the denatured riboprobe solution to completely cover the embryos (approximately 200 μl). Incubate at 68° overnight. Embryos should be handled with extreme care as they are quite fragile when heated to 68° in solutions containing formamide.

7. Prepare a series of five low-stringency wash solutions starting with 50% formamide, 5× SSC and finishing with 2× SSC. The intermediate washes can be made by combining these two solutions in 3:1, 1:1, and 1:3 ratios. After hybridizing with the probe overnight, remove the riboprobe solution and perform the series of low stringency washes at 68° for 10 min each. All wash buffers (low and high stringency, see below) can be prepared the night before and prewarmed in the hybridization oven for use the next day. *Note*: After hybridization, the riboprobe solution can be saved and stored at −20° for reuse at least twice. It has been observed that some probes are associated with less background signal on reuse.

8. Perform two high-stringency washes in 0.2× SSC, 0.01% Tween-20 for 30 min each at 68°.

9. Equilibrate the embryos back into room temperature PBST by removing half the volume of the high-stringency wash and replacing with PBST for 10 min. Next, wash in PBST for 10 min.

10. Quench endogenous peroxidase activity by incubating embryos in 0.3% hydrogen peroxide made in PBS for 1 h. Altering the concentration of hydrogen peroxide at this point may alter the amount of background peroxidase activity in the tissue and this step may need optimization (try a range of 0.1 to 1%).

11. Wash three times in PBST at room temperature.

12. Block for 1 h at room temperature with PBST, 10% goat serum, or other suitable blocking reagent.

13. Replace blocking solution with 250 μl antibody solution containing antidigoxigenin antibody conjugated to peroxidase (Roche) diluted 1:1000 in blocking buffer. Incubate overnight at 4°.

14. Wash in PBST a minimum of five times for 15 min each to minimize background.

15. Carefully transfer embryos to a 24-well tissue culture dish. Use the TSA-Plus Cyanine 3 System (Perkin Elmer) or the TSA-Plus Fluorescein System (Perkin Elmer) to detect peroxidase activity. Dilute the fluorophore tyramide stock solution which has been prepared according to manufacturer's instructions 1:50 in the 1× Plus Amplification Diluent and apply to the embryos. A minimum of 250 μl is needed to completely cover the embryos.

16. Develop in the dark for 1 to 10 min. Fluorescence can be monitored using a fluorescent microscope. Specific signal is usually apparent before 4 h. High background can be reduced substantially by extensive washing in PBST (we have washed up to 2 days) after signal development, by reducing probe concentration, lowering the antibody concentration, using a different blocking reagent or by trying different conditions for quenching endogenous peroxidase activity. The protocol

recommendations in the Perkin Elmer manual include many useful suggestions for troubleshooting and for incorporation of appropriate controls.

17. Fluorescent signals can be visualized in whole embryos or sectioned embryos. Counterstain with Hoechst dye ($0.1\ \mu g/\mu l$) overnight at room temperature prior to confocal imaging.

This protocol can be combined with the antibody immunofluorescent staining protocol by following the immunofluorescent labeling protocol above, leaving out the collagenase treatment and proceeding with incubation in the primary antibody and all subsequent steps. It is likely that some epitopes will not survive after *in situ* hybridization.

5. IDENTIFICATION OF TUMORS EXPRESSING EGFP-KRAS^{G12V}

One of the advantages of working with zebrafish embryos is their translucency, which greatly facilitates the visualization of fluorescent transgenes. Although adult wildtype zebrafish are extensively pigmented, visceral expression of fluorescent transgenes can be detected transcutaneously. Because oncogenic Kras is fused to eGFP in the *ptf1a*:eGFP-Kras^{G12V} fish, and because pancreatic eGFP expression is at extremely low levels in surviving fry, we are able to detect the development of early tumors by screening for the re-emergence eGFP expression in adult fish (Park *et al.*, 2008). Even very small eGFP-positive tumors (approximately 1 mm in size) can be visualized with a fluorescent microscope (Fig. 27.4A).

5.1. Protocol: Identifying fluorescent tumors in adult fish

1. Transfer a fish into a 250 ml beaker containing 95.8 ml of fish water and 4.2 ml of 0.4% tricaine stock solution (0.4% tricaine, 10 mM Tris, pH 7.0).
2. Anesthetize the fish until it begins to lose its balance. Prolonged exposure to tricaine can kill the fish, so it is important to monitor the fish carefully.
3. Briefly remove the fish from the anesthetic and place it on a Petri dish lid. Quickly scan the fish for expression of the transgene using a dissecting microscope equipped with a fluorescent light source and appropriate filters. It is important to evaluate both sides of the fish to ensure that small lateralized tumors are not missed.
4. Return the fish to a tank of fresh system water for recovery. It should be borne in mind that in a population of fish some are more sensitive to tricaine than others; it is therefore advisable to monitor all fish carefully during this process and avoid the temptation to anesthetize large numbers of fish simultaneously.

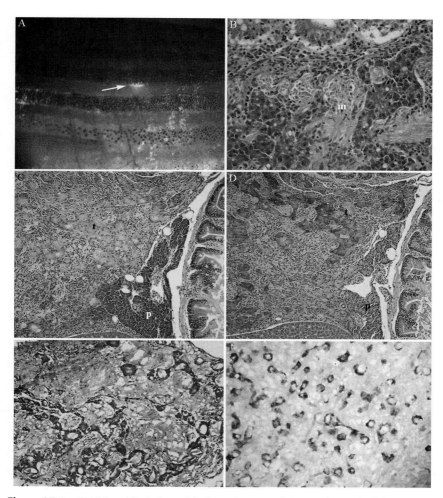

Figure 27.4 (A) Magnified view of the lateral aspect of an anesthetized adult zebrafish with a focal lesion (arrow) expressing $eGFP\text{-}Kras^{G12V}$ within the abdominal viscera (measuring approximately 1 mm) detected by fluorescent transillumination. (B) Histologic section of a malignant pancreatic carcinoma exhibiting predominantly acinar differentiation invading through the muscularis propria (m) of the gut (400×). (C) Histologic section of a tumor (t) arising in the pancreas (p) of a $ptf1a\text{:}eGFP\text{-}Kras^{G12V}$ fish. The carcinoma shows mixed acinar and ductal morphology (200×). (D) Immunohistochemical detection of the $eGFP\text{-}Kras^{G12V}$ fusion protein with an anti-GFP antibody (mouse, Chemicon, 1:300) in a section adjacent to the one depicted in C. Note that expression of the fusion protein is detected in the tumor (t), but not in the morphologically normal pancreas (p) (200×). (E) Frozen section of a pancreatic adenocarcinoma arising in a $ptf1a\text{:}eGFP\text{-}Kras^{G12V}$ fish with predominantly glandular differentiation and abundant extracellular mucin (200×). (F) *In situ* hybridization with an anti-$eGFP$ riboprobe detects expression of $eGFP\text{-}Kras^{G12V}$ in a pancreatic adenocarcinoma histologically similar to that shown in E (200×). (See color insert.)

6. Histologic Evaluation of Solid Tumors

In order to establish a correlation between eGFP expression and the presence of a neoplasm, fish are examined histologically. Tissues can be embedded in paraffin or frozen and processed for cryosections. Paraffin embedding gives superior histologic detail, and immunolabeling and special stains can be performed. Moreover, the entire fish can be embedded in paraffin and sectioned in order to evaluate the extent of tumor invasion. The disadvantage is that embedding tissue and sectioning paraffin blocks is more time consuming and does require some specialized expertise, although many universities have centralized histology core labs to assist with processing of tissues. For paraffin embedding, fish can be fixed for 24 h in 10% neutral buffered formalin at 4°. A 10-fold volume excess of formalin is recommended. To prevent autolysis of internal organs, fish should be kept on ice and the formalin should be prechilled. To ensure adequate fixation of internal organs, a small amount of fixative can be injected into the abdominal cavity with a 1-ml syringe fitted with a 26-gauge needle, or the coelomic cavity can be opened by making an incision along the ventral surface of the adult fish (without damaging internal organs). Fish can be stored in 70% ethanol after fixation or one can proceed directly with decalcification and embedding. Decalcify adult fish by incubation in 0.5 M EDTA, pH 7.8, for 7 days at room temperature. This is recommended to obtain optimal paraffin sections, although young fish aged less than 1 month do not require this (Moore *et al.*, 2004). After decalcification, cut the fish into 3- to 5-mm thick transverse sections with a fresh razor blade. If there is a mass, the first section should be through the center of the mass. Transverse sections, in our experience, are best to identify foci where the tumor is invading adjacent structures (an important finding in establishing that the tumor is malignant). The alternative is to bisect the fish longitudinally along the midline and process both halves for histologic evaluation. Protocols for paraffin embedding and sectioning will not be discussed.

Although some histologic detail is lost with frozen sections, tissue embedding is very simple and the quality of histology is generally adequate. It may also facilitate immunolabeling of tissues due to better preservation of epitopes. Moreover, the native fluorescence of proteins expressed from transgenes is usually preserved in frozen tissues, which are fixed in paraformaldehyde.

6.1. Protocol: Preparing frozen sections of dissected visceral organs

1. Sacrifice the adult fish by inducing hypothermia by placing the fish in an ice bath.
2. Assign each fish a unique designation. A database which tracks the genotype, parental founders, age, gross and histologic findings, results

of any other ancillary assays, and so on should be established to organize the specimens. The unique identifier will also help organize images and slides that are saved.

3. In our experience, it is difficult to cut frozen sections of complete cross-sections of the zebrafish due to the fact that the skin of the fish does not adhere well to the embedding matrix. We have had greater success cutting sections of the dissected viscera.

4. Place the fish on a moistened paper towel.

5. Open the coelomic cavity by grasping the skin with forceps and make an incision along the ventral surface of the fish from the cloacal opening to the gills.

6. Reflect back the skin of the fish and grasp the posterior end of the viscera with forceps. Gently dissect the viscera away from the skin and along the spinal column with microdissection scissors , starting posteriorly and working anteriorly. Make a final cut anterior to the liver at the level of the gills to remove the organs en bloc.

7. eGFP-positive tumor tissue can be harvested at this point under the fluorescent dissecting microscope. The remaining tissue can be fixed for histologic evaluation.

8. Fix the tissue in 4% paraformaldehyde for 24 h at 4°. A 10-fold volume excess is recommended for fixation. Some antigens may not tolerate paraformaldehyde fixation; in these instances, follow the antibody manufacturer's recommendations for fixation. One option is to embed the tissue in OCT compound (TissueTek) without prior fixation (see steps 10 and 11 of this protocol), and then cut frozen sections and fix the tissues in the appropriate fixative (e.g., acetone, methanol, etc.) after sections are cut.

9. Cryoprotect the fixed tissue by immersing it in a solution of 30% sucrose (made in PBS) to limit the formation of tissue-distorting ice crystals during the freezing process. Incubate the tissue in 30% sucrose at 4° overnight.

10. To embed the tissue for cryosectioning, cover the bottom of an appropriate size plastic cryomold (TissueTek) with a layer of OCT Compound (TissueTek). Remove the tissue from the sucrose solution and rinse in PBS. Blot the tissue dry with a Kimwipe® and place the tissue on top of the layer of OCT and orient the tissue for sectioning. Tissues should be oriented in order to obtain complete cross-sections. Overlay the tissue with OCT.

11. Rapidly freeze the tissues in OCT by partial submersion in liquid nitrogen. Another option is to place the cryomold at −80°, although this results in slower rate of freezing. More rapid freezing serves to prevent the formation of large ice crystals that artifactually distort the tissue. If using liquid nitrogen, care should be taken not to overcool

the block× which will cause the tissue and OCT block to crack. Frozen tissue blocks can be stored at −80° until ready for use.

12. Equilibrate the frozen OCT block to −20° in the cryostat prior to sectioning. Remove the OCT block with the embedded tissues from the plastic mold and affix it to a clean, prechilled cryostat chuck by placing a drop of OCT onto the chuck and lightly depressing the frozen OCT block onto the chuck. Allow this to cool to −20° before attempting to cut sections. The chuck should be affixed to the microtome and oriented so that complete, even cross-sections are obtained. Many helpful suggestions for cutting frozen sections can be found at http://www.ihcworld.com/_protocols/histology/frozen_section_technique_1.htm.

13. Cut 5- to 10-μm thick sections onto plus–charged glass slides if subsequent immunolabeling experiments will be performed. Slides should be air dried briefly and then stored at −80° until needed.

14. It is helpful to cut a series of adjacent sections and number the slides sequentially. Stain one or two of the sections with hematoxalin and eosin in order to evaluate tumor histology and identify surrounding structures. Sections that are to be stained with hematoxalin and eosin should be immediately placed into 95% ethanol in a coplin jar upon sectioning for optimal staining.

6.2. Protocol: Rapid evaluation of eGFP-Kras^{G12V} expression in tumors

It may be helpful to obtain a rapid confirmation of the presence of tumor expressing eGFP-Kras^{G12V} in frozen sections to ensure that sections are being taken at the appropriate level within the block.

1. Wash a frozen section three times in PBS.
2. Blot off excess PBS and mount a glass coverslip using an fluorescent mounting medium (Dako Fluorescent Mounting Medium) supplemented with 4',6-diamidino-2-phenylindole (DAPI) nuclear counterstain (1 μg/ml concentration). Alternatively, the coverslip may be mounted after a 1-min treatment with DAPI diluted in PBS (1 μg/ml) and two brief washes in PBS. The slide is immediately ready to review with a fluorescent microscope.

7. BRIEF REMARKS ON THE DIAGNOSIS OF MALIGNANT PANCREATIC TUMORS IN ZEBRAFISH

When evaluating zebrafish pancreatic neoplasms, we believe that documenting invasion of surrounding organs (e.g., invasion of the liver, lateral wall, gut, gonad, etc.) provides unequivocal evidence of malignancy.

In the absence of detailed knowledge of the natural history of small localized tumors, a conservative approach to characterizing tumors that may appear to invade the pancreatic parenchyma is warranted. Histologic proof of hematogenous metastasis has been difficult to obtain in the zebrafish because tumors are theoretically capable of "seeding" other coelomic organs or spreading contiguously in an insidious fashion. Such insidious patterns of invasion could give a false impression of a separate mass that could be mistakenly interpreted as a metastasis.

8. CHARACTERIZATION OF PROTEIN EXPRESSION AND RNA EXPRESSION IN ZEBRAFISH TUMORS BY IMMUNOLABELING AND *IN SITU* HYBRIDIZATION

Immunolabeling and *in situ* hybridization are standard techniques that can be applied to frozen sections or paraffin sections to further characterize the phenotype of solid tumors. The methods below are basic protocols that we have used. Optimal results are achieved by attention to fixation protocols, cryosectioning technique, and titration of antibody dilutions. For the protocols below, all steps should be performed at room temperature unless otherwise indicated in a humidified chamber.

8.1. Protocol: Immunofluorescent labeling of frozen sections

1. Frozen section slides that have been stored at −80° should be warmed to room temperature prior to use.
2. Permeabilize in PBS, 0.2% Triton X-100, 15 min.
3. Wash three times in PBS, 2 min each.
4. Block by incubating slide with blocking buffer (5% normal serum from same species as secondary antibody in PBS, 0.1% Triton X-100) for 60 min at room temperature.
5. Replace blocking buffer with primary antibody diluted in blocking buffer. Incubate overnight at 4°.
6. Wash in PBS three times for 10 min each.
7. Incubate with secondary antibody-fluorophore conjugate for 2 h in blocking buffer. The secondary antibody is raised against the species and antibody isotype of the primary antibody.
8. After washing in PBS three times, 10 min each, nuclei are counterstained with DAPI (1 μg/ml diluted in PBS) for 1 min.
9. Wash in PBS 3 times, 5 min each and mount a glass coverslip with fluorescence mounting medium (Dako Fluorescence Mounting Medium or Vectashield Mounting Medium, Vector Laboratories).

8.2. Protocol: *In situ* hybridization to detect RNA expression in frozen tissue sections

1. Prior to *in situ* hybridization, sections are cut onto plus–charged glass slides and dehydrated through a series of ethanol washes (50 to 100%), dried and stored at $-80°$ until use.

2. Prepare antisense riboprobe as described in the antisense riboprobe preparation protocol (above). The final working probe concentration is determined empirically, but will generally range from 0.1 to 1.0 ng/μl. It will be useful, therefore, to save a stock dissolved in the prehybridization buffer (described below) and stored at $-20°$ until used.

3. Defrost sections for 30 min at room temperature.

4. Incubate sections in a solution of proteinase K (10 μg/ml in 50 mM Tris-HCl, pH 7.5, 5 mM EDTA) for 2 min at room temperature.

5. Stop proteinase K digestion by washing in PBS containing 0.2% glycine.

6. Rinse slides twice in fresh PBS for 30 s each.

7. Incubate slides for 15 min in 4% paraformaldehyde.

8. Rinse slides twice in PBS, 30 s each.

9. Prehybridize with hybridization solution for 1 h in a hybridization chamber at 50°. Hybridization solution consists of 4× SSC, 55% deionized formamide, 10% dextran sulfate, 1 mg/ml yeast tRNA, 1× Denhardt's solution (0.02% Ficoll 400, 0.02% polyvinyl pyrolidone, 0.02% bovine serum albumin) and 2 mM EDTA. A hybridization chamber can be constructed from any plastic container. Put absorbent paper on the bottom of the chamber and soak the paper with humidifying solution (50% formamide, 4× SSC) to prevent evaporation of the hybridization solution from the surface of the slide.

10. It is convenient to place plastic pipettes cut to the appropriate length on top of the paper towels to prevent slides from sticking to the paper.

11. During prehybridization, prepare probe hybridization solution. Approximately 100 to 300 μl of probe solution at approximately 0.1 to 1.0 ng/μl will be needed per slide. The optimal probe concentration should be determined empirically. Denature the working probe hybridization solution by heating for 10 min at 70°.

12. Remove prehybridization solution and add enough hybridization solution to sufficiently cover the section. Cover with a glass coverslip, remove air bubbles, and seal the coverslip using rubber cement. Incubate in the hybridization chamber overnight at 50°. Make sure to add humidifying solution to the absorbent paper to minimize evaporation of the probe solution.

13. Prepare wash solutions (see steps 15 and 16) and prewarm them in the hybridization oven overnight.

14. Remove slide from hybridization chamber and peel off rubber cement and gently remove the coverslip.

15. Perform a series of five low-stringency washes for 10 min each at 50° starting with 50% formamide/5× SSC and ending with 2× SSC. Intermediate washes can be performed by combining the 50% formamide/5× SSC solution with the 2× SSC solutions in 3:1, 1:1, and 1:3 ratios.

16. Wash with 0.2× SSC twice for 30 min each at 50°.

17. Remove as much of the 0.2× SSC wash solution as possible, and incubate with 50% 0.2× SSC, 50% PBST at room temperature for 10 min.

18. Rinse with PBST twice, 5 min each.

19. Drain off PBST as much as possible and add blocking solution (PBST, 5% goat serum), covering the whole section. Incubate the slide in moist chamber for 1 h.

20. Drain off as much blocking solution as possible without allowing slide to dry. Add 200 μl alkaline phosphatase-conjugated anti–digoxigenin Fab antibody (Roche; diluted 1:2000 in blocking solution) and incubate the slide in moist chamber for 2 h.

21. Remove antibody solution and wash with PBST four times for 15 min each.

22. Rinse twice with alkaline phosphatase buffer (100 mM sodium chloride, 50 mM magnesium chloride, 100 mM Tris hydrochloride, 0.1% Tween-20, pH 9.5) for 5 min each.

23. Replace buffer with freshly prepared BCIP (5-Bromo-4-Chloro-3'-Indolyphosphate p-Toluidine Salt, Roche) and NBT (Nitro-Blue Tetrazolium Chloride, Roche) solution (prepared by combining 4.5 μl NBT, 3.5 μl BCIP in 1 ml of alkaline phosphatase buffer) and incubate in dark at room temperature until color development is suitable.

24. Stop the reaction by brief rinsing with PBST.

25. Fix by replacing PBST with 4% paraformaldehyde. Incubate 15 min at room temperature.

26. Rinse in PBST twice, 5 min each.

27. Mount with an aqueous mounting medium.

8.3. Protocol: Immunohistochemical labeling of paraffin sections

Immunohistochemical labeling is performed on 5 μm of formalin-fixed paraffin sections that have been heated at 60° for 30 min to ensure complete dehydration. All steps may be performed at room temperature in a humidified chamber unless otherwise specified. A standard protocol used in our lab follows:

1. Deparaffinize slides in two xylene washes, 5 min each.

2. Incubate in 100% ethanol twice, 10 min each, and then in 95% ethanol twice, 10 min each. Then wash twice in distilled water, 5 min each.

3. Antigen retrieval may be required for a given antigen in formalin-fixed tissues. Consult the antibody manufacturer's recommendations for the specific antigen. A standard protocol used in our lab follows. First boil distilled water in a rice cooker. Place the slides in a slide rack and immerse them in a plastic staining dish (TissueTek) with 250 ml of 10 mM citrate buffer, pH 6.0. Place the staining dish containing the citrate buffer and slides in the boiling water and cover for 10 min. Remove the plastic staining dish with citrate buffer and slides from the boiling water and place them on ice, allowing the citrate buffer and slides to cool to room temperature.

4. Rinse slides in distilled water three times for 5 min each.

5. Quench endogenous peroxidase activity by treating slides with 3% hydrogen peroxide diluted in distilled water for 10 min.

6. Incubate slides with blocking buffer (5% normal serum diluted in PBS, 0.1% Tween-20) for 60 min. The choice of serum should correspond to the species in which the secondary antibody-horse radish peroxidase (HRP) conjugate is raised.

7. Remove the blocking buffer and replace it with the primary antibody diluted in blocking buffer and incubate overnight at 4°. Consult the antibody manufacturer for recommended dilution. When first working with an antibody, it may be necessary to try multiple dilutions to establish the optimal concentration of antibody.

8. Wash with PBST three times for 5 min each.

9. Incubate with secondary antibody-HRP conjugate diluted in blocking buffer at the manufacturer's recommended dilution for 1 h. The secondary antibody is raised against the species and antibody isotype of the primary antibody.

10. Peroxidase activity is detected using 3, 3'-diaminobenzidine (DAB) available in kit form (Dako Corporation) according to the manufacturer's protocol. Note that DAB is a suspected carcinogen and appropriate care should be exercised with its use.

11. Lightly counterstain slides with hematoxylin by rinsing in distilled water, then hematoxylin for 15 s, rinse in distilled water, then dehydrate in 95% ethanol, 100% ethanol, and two washes in xylene, each for 1 min.

Mount and coverslip with nonaqueous mounting medium (e.g., Histomount, Zymed; Permount, Fisher Scientific).

REFERENCES

Almoguera, C., Shibata, D., Forrester, K., Martin, J., Arnheim, N., and Perucho, M. (1988). Most human carcinomas of the exocrine pancreas contain mutant c-K-ras genes. *Cell* **53,** 549–554.

Amsterdam, A., Sadler, K. C., Lai, K., Farrington, S., Bronson, R. T., Lees, J. A., and Hopkins, N. (2004). Many ribosomal protein genes are cancer genes in zebrafish. *PLoS Biol.* **2**, E139.

Berghmans, S., Murphey, R. D., Wienholds, E., Neuberg, D., Kutok, J. L., Fletcher, C. D., Morris, J. P., Liu, T. X., Schulte-Merker, S., Kanki, J. P., Plasterk, R., Zon, L. I., *et al.* (2005). tp53 mutant zebrafish develop malignant peripheral nerve sheath tumors. *Proc. Natl. Acad. Sci. USA* **102**, 407–412.

Caldas, C., Hahn, S. A., Hruban, R. H., Redston, M. S., Yeo, C. J., and Kern, S. E. (1994). Detection of K-ras mutations in the stool of patients with pancreatic adenocarcinoma and pancreatic ductal hyperplasia. *Cancer Res.* **54**, 3568–3573.

Chen, S., Li, C., Yuan, G., and Xie, F. (2007). Anatomical and histological observation on the pancreas in adult zebrafish. *Pancreas* **34**, 120–125.

Datsenko, K. A., and Wanner, B. L. (2000). One-step inactivation of chromosomal genes in *Escherichia coli* K-12 using PCR products. *Proc. Natl. Acad. Sci. USA* **97**, 6640–6645.

Detrich, III, H. W., Westerfield, M., and Zon, L. I. (2004a). "The zebrafish: Cellular and developmental biology." Boston: Elsevier Academic Press, Boston.

Detrich, III, H. W., Westerfield,M., and Zon, L. I. (2004b). The zebrafish: Genetics, genomics and informatics. *Methods Cell Biol.* **77**.

Grippo, P. J., Nowlin, P. S., Demeure, M. J., Longnecker, D. S., and Sandgren, E. P. (2003). Preinvasive pancreatic neoplasia of ductal phenotype induced by acinar cell targeting of mutant Kras in transgenic mice. *Cancer Res.* **63**, 2016–2019.

Haramis, A. P., Hurlstone, A., van der Velden, Y., Begthel, H., van den Born, M., Offerhaus, G. J., and Clevers, H. C. (2006). Adenomatous polyposis coli-deficient zebrafish are susceptible to digestive tract neoplasia. *EMBO Rep.* **7**, 444–449.

Hezel, A. F., Kimmelman, A. C., Stanger, B. Z., Bardeesy, N., and Depinho, R. A. (2006). Genetics and biology of pancreatic ductal adenocarcinoma. *Genes Dev.* **20**, 1218–1249.

Hingorani, S. R., Petricoin, E. F., Maitra, A., Rajapakse, V., King, C., Jacobetz, M. A., Ross, S., Conrads, T. P., Veenstra, T. D., Hitt, B. A., Kawaguchi, Y., Johann, D., *et al.* (2003). Preinvasive and invasive ductal pancreatic cancer and its early detection in the mouse. *Cancer Cell* **4**, 437–450.

Hruban, R. H., Wilentz, R. E., and Kern, S. E. (2000). Genetic progression in the pancreatic ducts. *Am. J. Pathol.* **156**, 1821–1825.

Jemal, A., Siegel, R., Ward, E., Murray, T., Xu, J., Smigal, C., and Thun, M. J. (2006). Cancer statistics, 2006. *CA Cancer J. Clin.* **56**, 106–130.

Jessen, J. R., Meng, A., McFarlane, R. J., Paw, B. H., Zon, L. I., Smith, G. R., and Lin, S. (1998). Modification of bacterial artificial chromosomes through chi-stimulated homologous recombination and its application in zebrafish transgenesis. *Proc. Natl. Acad. Sci. USA* **95**, 5121–5126.

Kawaguchi, Y., Cooper, B., Gannon, M., Ray, M., MacDonald, R. J., and Wright, C. V. (2002). The role of the transcriptional regulator Ptf1a in converting intestinal to pancreatic progenitors. *Nat. Genet.* **32**, 128–134.

Langenau, D. M., Traver, D., Ferrando, A. A., Kutok, J. L., Aster, J. C., Kanki, J. P., Lin, S., Prochownik, E., Trede, N., Zon, L. I., and Look, A. T. (2003). Myc-induced T cell leukemia in transgenic zebrafish. *Science* **299**, 887–890.

Langenau, D. M., Feng, H., Berghmans, S., Kanki, J. P., Kutok, J. L., and Look, A. T. (2005). Cre/lox-regulated transgenic zebrafish model with conditional myc-induced T cell acute lymphoblastic leukemia. *Proc. Natl. Acad. Sci. USA* **102**, 6068–6073.

Leach, S. D. (2004). Mouse models of pancreatic cancer: The fur is finally flying! *Cancer Cell* **5**, 7–11.

Lin, J. W., Biankin, A. V., Horb, M. E., Ghosh, B., Prasad, N. B., Yee, N. S., Pack, M. A., and Leach, S. D. (2004). Differential requirement for ptf1a in endocrine and exocrine lineages of developing zebrafish pancreas. *Dev. Biol.* **274**, 491–503.

Moore, J. L., Gestl, E. E., and Cheng, K. C. (2004). Mosaic eyes, genomic instability mutants, and cancer susceptibility. *Methods Cell Biol.* **76,** 555–568.

Moore, J. L., Rush, L. M., Breneman, C., Mohideen, M. A., and Cheng, K. C. (2006). Zebrafish genomic instability mutants and cancer susceptibility. *Genetics* **174,** 585–600.

Murtaugh, L. C., and Leach, S. D. (2007). A case of mistaken identity? Nonductal origins of pancreatic "ductal" cancers. *Cancer Cell* **11,** 211–213.

Niv, H., Gutman, O., Henis, Y. I., and Kloog, Y. (1999). Membrane interactions of a constitutively active GFP-Ki-Ras 4B and their role in signaling. Evidence from lateral mobility studies. *J. Biol. Chem.* **274,** 1606–1613.

Nusslein-Volhard, C., and Dahm, R. (2002). "The zebrafish: A practical approach." New York: Oxford University Press, New York.

Park, S. W., Davison, J., Rhee, J., Pisharath, H., Hruban, R. H., Maitra, A., and Leach, S. D. (2008). Oncogenic KRAS induces progenitor cell expansion and malignant transformation in zebrafish exocrine pancreas. *Gastroenterology.* in press.

Patton, E. E., Widlund, H. R., Kutok, J. L., Kopani, K. R., Amatruda, J. F., Murphey, R. D., Berghmans, S., Mayhall, E. A., Traver, D., Fletcher, C. D., Aster, J. C., Granter, S. R., *et al.* (2005). BRAF mutations are sufficient to promote nevi formation and cooperate with p53 in the genesis of melanoma. *Curr. Biol.* **15,** 249–254.

Quaife, C. J., Pinkert, C. A., Ornitz, D. M., Palmiter, R. D., and Brinster, R. L. (1987). Pancreatic neoplasia induced by ras expression in acinar cells of transgenic mice. *Cell* **48,** 1023–1034.

Sabaawy, H. E., Azuma, M., Embree, L. J., Tsai, H. J., Starost, M. F., and Hickstein, D. D. (2006). TEL-AML1 transgenic zebrafish model of precursor B cell acute lymphoblastic leukemia. *Proc. Natl. Acad. Sci. USA* **103,** 15166–15171.

Shepard, J. L., Amatruda, J. F., Stern, H. M., Subramanian, A., Finkelstein, D., Ziai, J., Finley, K. R., Pfaff, K. L., Hersey, C., Zhou, Y., Barut, B., Freedman, M., *et al.* (2005). A zebrafish bmyb mutation causes genome instability and increased cancer susceptibility. *Proc. Natl. Acad. Sci. USA* **102,** 13194–13199.

Shepard, J. L., Amatruda, J. F., Finkelstein, D., Ziai, J., Finley, K. R., Stern, H. M., Chiang, K., Hersey, C., Barut, B., Freeman, J. L., Lee, C., Glickman, J. N., *et al.* (2007). A mutation in separase causes genome instability and increased susceptibility to epithelial cancer. *Genes Dev.* **21,** 55–59.

Stern, H. M., Murphey, R. D., Shepard, J. L., Amatruda, J. F., Straub, C. T., Pfaff, K. L., Weber, G., Tallarico, J. A., King, R. W., and Zon, L. I. (2005). Small molecules that delay S phase suppress a zebrafish bmyb mutant. *Nat. Chem. Biol.* **1,** 366–370.

Stern, H. M., and Zon, L. I. (2003). Cancer genetics and drug discovery in the zebrafish. *Nat. Rev. Cancer* **3,** 533–539.

Thisse, B., and Thisse, C. (1998). High resolution whole-mount *in situ* hybridization. *Zebrafish Sci. Monitor* **5** (online).

Tuveson, D. A., Shaw, A. T., Willis, N. A., Silver, D. P., Jackson, E. L., Chang, S., Mercer, K. L., Grochow, R., Hock, H., Crowley, D., Hingorani, S. R., Zaks, T., *et al.* (2004). Endogenous oncogenic K-ras(G12D) stimulates proliferation and wide-spread neoplastic and developmental defects. *Cancer Cell* **5,** 375–387.

Tuveson, D. A., Zhu, L., Gopinathan, A., Willis, N. A., Kachatrian, L., Grochow, R., Pin, C. L., Mitin, N. Y., Taparowsky, E. J., Gimotty, P. A., Hruban, R. H., Jacks, T., *et al.* (2006). Mist1-KrasG12D knock-in mice develop mixed differentiation metastatic exocrine pancreatic carcinoma and hepatocellular carcinoma. *Cancer Res.* **66,** 242–247.

Urasaki, A., Morvan, G., and Kawakami, K. (2006). Functional dissection of the Tol2 transposable element identified the minimal cis-sequence and a highly repetitive sequence in the subterminal region essential for transposition. *Genetics* **174,** 639–649.

Wallace, K. N., and Pack, M. (2003). Unique and conserved aspects of gut development in zebrafish. *Dev. Biol.* **255,** 12–29.

Westerfield, M. (2000). "The zebrafish book. A guide for the laboratory use of zebrafish (*Danio rerio*)." Eugene: University of Oregon Press, Eugene.

Yanagisawa, A., Ohtake, K., Ohashi, K., Hori, M., Kitagawa, T., Sugano, H., and Kato, Y. (1993). Frequent c-Ki-ras oncogene activation in mucous cell hyperplasias of pancreas suffering from chronic inflammation. *Cancer Res.* **53,** 953–956.

Yang, Z., Jiang, H., Chachainasakul, T., Gong, S., Yang, X. W., Heintz, N., and Lin, S. (2006). Modified bacterial artificial chromosomes for zebrafish transgenesis. *Methods* **39,** 183–188.

Zecchin, E., Mavropoulos, A., Devos, N., Filippi, A., Tiso, N., Meyer, D., Peers, B., Bortolussi, M., and Argenton, F. (2004). Evolutionary conserved role of ptf1a in the specification of exocrine pancreatic fates. *Dev. Biol.* **268,** 174–184.

CHAPTER TWENTY-EIGHT

Genetic Modeling of Ras-Induced Human Rhabdomyosarcoma

Corinne M. Linardic*,† *and* Christopher M. Counter†,‡

Contents

Abstract

Rhabdomyosarcoma is the most common soft tissue sarcoma of childhood and adolescence. Historically, rhabdomyosarcoma has been studied by the manipulation of human cell lines derived from primary rhabdomyosarcoma tumor tissue adapted to grow in culture. Recently, mouse models have been added to the arsenal of tools to study this disease *in vivo*. However, given the emerging understanding of the genetic variability and mutability of human tumor–derived cell lines, and the existing differences between human and murine tumorigenesis, we sought to uniformly dissect the genetic events required to generate rhabdomyosarcoma from primary human skeletal muscle precursors. To this end, primary human skeletal muscle cells were transformed with defined genetic elements to corrupt the p53, Rb, Myc, telomerase, and Ras pathways, resulting in cells that, when assayed as subcutaneous xenografts in immunocompromised mice, formed tumors indistinguishable at the immunohistochemical level from the embryonal histologic variant of rhabdomyosarcoma. This chapter will discuss the techniques used to transform primary human skeletal muscle cells, the assays used to verify expression of the ectopically expressed genetic elements, and the methods used to evaluate the tumorigenic capacity of the resulting cell lines.

* Department of Pediatrics, Duke University, Durham, North Carolina
† Department of Pharmacology and Cancer Biology, Durham, North Carolina
‡ Department of Radiation Oncology, Durham, North Carolina

Methods in Enzymology, Volume 438
ISSN 0076-6879, DOI: 10.1016/S0076-6879(07)38028-2

1. INTRODUCTION

Historically, the study of rhabdomyosarcoma has relied on cultured cell lines created from patient-derived tumor samples that have been adapted to grow *in vitro*. Beginning in the 1990s, several transgenic mouse mutants intended to study the role of specific genes were serendipitously found to develop tumors resembling embryonal rhabdomyosarcoma (Merlino and Helman, 1999; Nanni *et al.*, 2003; Sharp *et al.*, 2002; Tsumura *et al.*, 2006). In 2004, a mouse model for alveolar rhabdomyosarcoma was described (Keller *et al.*, 2004). While both cell line and mouse genetic modeling approaches have been invaluable in dissecting rhabdomyosarcoma, both also have limitations. Cell lines derive from genetically diverse individuals, may accumulate additional mutations with prolonged culturing, and represent the culmination of genetic mis-events that eventually result in cancer, thus limiting the ability to define early and intermediate tumorigenic events. Mouse models accurately reflect murine rhabdomyosarcoma formation, but given the recent observed differences between human and murine tumorigenesis (Hamad *et al.*, 2002; Rangarajan *et al.*, 2004), cannot precisely model the human disease. Efforts were therefore directed toward developing *in vitro* models in human cells that could dissect the discrete genetic events required to stepwise convert primary human skeletal muscle cells into their transformed counterparts.

The first genetically defined model of human cancer showed that in primary human cells, it was necessary and sufficient to inactivate the p53 and Rb pathways, upregulate Myc pathways, reactivate telomerase in order to stabilize telomeres, and activate oncogenic Ras in order to provide self-sufficiency in growth signaling and angiogenesis (Hahn *et al.*, 1999; Kendall *et al.*, 2005; Yeh *et al.*, 2004). This was accomplished by the stable serial introduction into primary human cells of four transgenes encoding SV40 large-T and small-t oncoproteins, the hTERT catalytic subunit of telomerase, and an oncogenic version of the G-protein H-RasG12V. In this chapter, we will describe the adaptation of this approach toward modeling human embryonal rhabdomyosarcoma (Linardic *et al.*, 2005). Importantly, we have chosen to use primary human skeletal muscle cells as the starting cell type for this approach, as rhabdomyosarcoma is defined by its skeletal muscle features. We have also validated the choice of the four described transgenes by assuring that rhabdomyosarcoma indeed demonstrates lesions in the p53, Rb, Myc, telomerase, and Ras pathways (Linardic *et al.*, 2005; Xia *et al.*, 2002). Regarding oncogenic H-Ras, although this mutation is not common in human embryonal rhabdomyosarcoma, it adequately represents the inappropriate upregulation of the Ras pathway found in rhabdomyosarcoma, manifest as gain-of-function mutations in proteins that drive Ras signaling, such as receptor tyrosine kinases (Merlino and Helman, 1999), or loss-of-function mutations that normally suppress Ras signaling, such as neurofibromin (Reed and Gutmann, 2001). The recent identification of germline mutations of H-RAS in Costello syndrome, a genetic disorder that predisposes affected individuals to embryonic tumors including embryonal rhabdomyosarcoma, further validates the role of H-Ras as an important genetic mutation in the development of rhabdomyosarcoma (Aoki *et al.*, 2005). Following a detailed explanation of the techniques used to transform skeletal muscle cells, there will be a description of the assays used to verify expression of the transgenes introduced into the cell lines, and the methods used to evaluate their tumorigenic capacity.

 ## 2. MATERIALS AND METHODS

2.1. Overview

This genetically defined model of rhabdomyosarcoma is based on the serial stable introduction of the transgenes encoding SV40 DNA tumor virus early region, hTERT, and H-RasG12V into primary human skeletal muscle myoblasts (HSMM cells) obtained from Lonza (formerly Cambrex Corp.), using amphotrophic retrovirus. The resulting transformed cells are abbreviated as "HSMM T-H-R" cells. It is important to note that although another type of primary human skeletal muscle cell (SkMC cells, Lonza catalog no. CC-2561) is available from this supplier, when transformed by the T-H-R approach these cells yield tumors of varying phenotype, suggesting that the stock culture of primary SkMC cells is a heterogeneous population that cannot reproducibly generate tumors resembling rhabdomyosarcoma. HSMM cells are therefore recommended. Because the plasmids and methodology used to generate infectious amphotrophic retrovirus are as published previously (O'Hayer and Counter, 2005), when the necessary reagents or methods are divergent from those published, the precise alternative approach is detailed.

2.2. Creating HSMM T-H-R cells

2.2.1. Materials

- Low-passage HSMM cells confirmed by the supplier to be of skeletal muscle origin, as assessed by desmin staining 90% or greater, and capable of forming skeletal muscle myotubes when cultured in differentiation media (Lonza catalog no. CC-2580)
- SKBM-2 media bullet kit (Lonza catalog no. CC-3245)

- Subculturing reagent pack including HEPES-buffered saline solution, trypsin/EDTA, and trypsin-neutralizing solution (Lonza catalog no. CC-5034)
- Monoclonal antibodies to desmin (DakoCytomation catalog no. M0760) and skeletal muscle–specific actin (DakoCytomation catalog no. M0635, clone HHF35)
- Amphotrophic retrovirus encoding neomycin-SV40 early region (T/t-Ag), hygromycin-hTERT, and puromycin-FLAG epitope-tagged-H-RasG12V (O'Hayer and Counter, 2005)
- Neomycin, diluted to stock 50 mg/ml (Gibco catalog no. 10131–035)
- Hygromycin B, diluted to stock 50 mg/ml (Invitrogen catalog no. 10687–010)
- Puromycin, reconstituted in 1× PBS to stock 1 mg/ml (Sigma catalog no. P8833)
- Tissue culture plastic ware

2.2.2. Methods

1. Thaw, count, and seed HSMM cells into 6-cm tissue culture dishes per supplier instructions; allow cells to attach overnight. Subculture as suggested by the supplier.
2. Prior to beginning infection with amphotrophic retrovirus, verify expression of desmin and skeletal muscle-specific actin in the HSMM starting cell population by subjecting 100-μg HSMM whole-cell lysates to standard separation on SDS-PAGE and immunoblotting with anti–desmin monoclonal antibody at a dilution of 1:200, and anti-skeletal muscle–specific actin antibody at a dilution of 1:200, followed by a secondary HRP-conjugated antimouse antibody and chemiluminescence exposure. On standard SDS-PAGE, desmin and skeletal muscle–specific actin migrate at approximately 53 and 38 kD, respectively. In combination with a positive assay for myotube formation, expression of these protein markers verifies the skeletal muscle origin of the HSMM cells.
3. To begin the process of retroviral infection, when HSMM cells are 40% confluent, aspirate SKBM-2 media and replace with media containing amphotrophic retrovirus encoding SV40 T/t antigen. Amphotrophic virus will have been made in DMEM with 10% FBS; HSMM cells will tolerate a 24-h exposure to this media. See O'Hayer and Counter (2005) on the preparation of amphotrophic retrovirus and related safety measures when working with this method of transgene delivery.
4. After 24 h, remove viral media and replace with SKBM-2. Washing is not necessary.
5. After 24 h, initiate antibiotic selection for neomycin-expressing HSMM cells, indicating that they have incorporated the retrovirus, by adding

neomycin to working concentration of 250 μg/ml. Selection should take 10 days or less. If there is excessive cell death, replace with fresh neomycin selection media so that dead or dying cells do not interfere with the growth of the remaining adherent cells. Once the HSMM cells are stably expressing the SV40 early region, they will be capable of bypassing tissue culture–induced senescence, and will be amenable to further transduction with hTERT and H-RasG12V. If the HSMM cells become more than 70% confluent during selection process, subculture using the reagent pack as per supplier instructions so that cells maintain log-phase growth. There will be a very low background rate of spontaneous differentiation to myotubes if this practice is followed.

6. In a similar manner to the SV40 early-region transgene, transduce cells with amphotrophic retrovirus encoding hTERT, using hygromycin B at a working concentration of 50 μg/m for 10 days, and then amphotrophic retrovirus encoding FLAG-H-RasG12V, using puromycin at a working concentration of 0.25 μg/ml for 7 days. Because HSMM cells will not tolerate culturing in more than one selection antibiotic at a time, perform each retroviral infection and its associated selection separately. See O'Hayer and Counter (2005) on the preparation of this amphotrophic retrovirus.

7. Following the introduction of each transgene, cells surviving selection should be frozen per supplier instructions and stored in liquid nitrogen so that future genetic manipulations can be performed as needed. Minimal cell passages should occur between transgene introductions.

8. Following final infection and selection, confirm that the HSMM T-H-R cells do not shed amphotrophic retrovirus, using the horizontal spread assay (Bielicki et al., 1996). Once confirmed, the cells may be manipulated outside of a BSL-2 facility.

2.2.3. Validating transgene expression

Prior to using the HSMM T-H-R cells for experimental inquiry, it is critical to examine the expression level of each of the introduced transgenes (Linardic et al., 2005). To prove stable expression of the SV40 early region, separate 100 μg of an HSMM T-H-R whole-cell lysate by standard SDS-PAGE and immunoblot using a monoclonal antibody against large-T antigen (Santa Cruz catalog no. SC-147) at a dilution of 1:1000, followed by a secondary HRP–conjugated antimouse antibody and chemiluminescence exposure. To prove expression of FLAG-epitope–tagged H-RasG12V, similarly separate and immunoblot an HSMM T-H-R whole-cell lysate using anti-FLAG-M2 monoclonal antibody (Sigma catalog no. F3165) at a dilution of 1:4000). SV40 large-T antigen and H-Ras migrate at 94 and 21 kD, respectively.

Since hTERT will not be detectable at the protein level, use RT-PCR to confirm expression of the hTERT cDNA (Kendall et al., 2005). In brief, following the isolation of total RNA from HSMM T-H-R cells using the RNAzol B (TEL-TEST catalog no. CS-104) or similar reagent, reverse transcribe 2 μg of total RNA using the Omniscript RT (Qiagen catalog no. 205111) or similar kit with OligodT (Life Technologies Invitrogen catalog no. 18418–012) primer. Input 4 μl of each reaction into PCR amplification reactions for hTERT cloned into a pBABE plasmid (5'-GAGGTG CAGAGCGACTAC and 5'-GCTGTTCACCTGCAAATCCA), or GAP DH (5'-GAGAGACCCTCACTGCTG and 5'-GATGGTACATGACAA GGTGC). Separate each PCR reaction on a standard 2% agarose gel.

Retroviral transduction may yield variable levels of stable transgene expression in target cells, depending upon the efficiency of retroviral production and infection. In this regard, a low level of FLAG-H-RasG12V protein expression within the HSMM T-H-R cells will result in decreased tumorigenic potential (see next section for a discussion of the assays used to assess tumorigenic potential). That is, soft agar colony count will be low and tumorigenesis *in vivo* will be slow. Therefore if needed, the HSMM T-H-R cells can be reinfected with and selected for retroviral-transduced FLAG-H-RasG12V expression. Once the level of transgene expression has been examined and is satisfactory, further culturing of the HSMM T-H-R cell lines in selection antibiotics to promote transgene retention is not necessary, since the combination of introduced transgenes is not only tolerated in HSMM cells, but confers a growth advantage.

2.2.4. Soft agar and tumorigenic growth assays

The soft agar assay is an *in vitro* method to measure cellular transformation, as assessed by the ability of cells to grow under anchorage-independent conditions. HSMM T-H-R cells are tested for anchorage-independent growth as previously described (O'Hayer and Counter, 2005), with the following modifications:

1. HSMM T-H-R cells can be assayed in 2x SKBM-2, or to reduce cost, convert them to RPMI-1640 (or DMEM) with 10% FBS, as most human rhabdomyosarcoma cell lines are grown in these common media types.
2. Count colonies after 4, rather than 3, weeks.

Although the soft agar assay measures anchorage-independent growth, a more rigorous assessment of tumorigenesis is to assay whether cells can form tumors *in vivo*. To determine whether the transformed HSMM T-H-R cells possess this property, these cells are injected as subcutaneous xenografts in immunocompromised SCID-beige mice *in vivo* as previously described (O'Hayer and Counter, 2005), with the following modifications:

1. For subcutaneous injection, 7 million, rather than 10 million cells, are adequate.

2. When preparing cells for injection, HSMM T-H-R cells require additional rinsing with PBS in order to effectively trypsinize them off of the tissue culture plate. Therefore, rinse each 15-cm dish of confluent HSMM T-H-R cells twice with 10 ml of 1× PBS, and then trypsinize with 4 ml of trypsin-EDTA and neutralize with 4 ml trypsin–neutralizing solution.

3. Prior to injection into SCID-beige mice, cell lines should be tested to prove that they are bacteria, fungus, and mycoplasma-free. Bacterial and fungal contamination can be detected by visual inspection, but mycoplasma infection may be occult and therefore requires specific testing, such as with the Gen-Probe nucleic acid hybridization mycoplasma detection kit (Fisher Scientific catalog no. GP-1591).

2.2.5. Verification of tumor xenograft pathology characteristics

To verify that the tumors generated from subcutaneous injection of HSMM cells transformed with T-H-R are appropriately phenocopying embryonal rhabdomyosarcoma, it is imperative to examine these tumors for morphology by standard H&E methods, and immunohistochemical markers of skeletal muscle. Harvested tumors are fixed in formalin, and then placed in 75% ethanol pending embedding in paraffin. The first tier of markers traditionally used to immunohistochemically identify rhabdomyosarcoma includes desmin, myoglobin, and skeletal muscle–specific actin. Recently, a second tier of markers, the muscle-specific transcription factors myogenin and MyoD1, was incorporated into the diagnostic algorithm, although the number of institutions that have developed these assays into high-throughput mode is low. A description of the immunohistochemical techniques used to perform these assays is beyond the scope of this chapter, but we refer the reader to Linardic et al. (2005) and Morotti et al. (2006). For further proof of skeletal muscle origin, transmission electron microsocopy can be performed on the tumor xenografts to verify the presence of skeletal muscle–specific myofilaments (Linardic et al., 2005). A pathologist with experience in the analysis of soft tissue sarcomas should be involved in the interpretation of all of these analyses.

3. CONCLUDING REMARKS

To complement traditional cancer models employing tumor-derived cell lines and genetically manipulated mice, this method now provides a third approach in which to dissect the genetic events necessary to transform primary human skeletal muscle cells into cells that form tumors mimicking embryonal rhabdomyosarcoma. The benefits of this method include its ability to flexibly evaluate the role of starting cell type in rhabdomyosarcoma histogenesis, and the role of specific genes in the discrete aspects of the pathology of rhabdomyosarcoma. Indeed, our exploration of two different starting primary human skeletal muscle cell types, SkMC and HSMM cells, illustrates the importance of choosing a cell type that is as homogeneous as possible, and once transformed accurately represents the tumor type being studied. In addition to its ability to accommodate a variety of cell types and genes, because this model is based on the stepwise accumulation of genes that initiate, as well propagate, the tumorigenic phenotype, it may be used to investigate both early and late events in human rhabdomyosarcoma formation.

REFERENCES

Aoki, Y., Niihori, T., Kawame, H., Kurosawa, K., Ohashi, H., Tanaka, Y., Filocamo, M., Kato, K., Suzuki, Y., Kure, S., and Matsubara, Y. (2005). Germline mutations in HRAS proto-oncogene cause Costello syndrome. *Nat. Genet.* **37,** 1038–1040.

Bielicki, J., Hopwood, J. J., and Anson, D. S. (1996). Correction of Sanfilippo A skin fibroblasts by retroviral vector-mediated gene transfer. *Hum. Gene Ther.* **7,** 1965–1970.

Hahn, W. C., Counter, C. M., Lundberg, A. S., Beijersbergen, R. L., Brooks, M. W., and Weinberg, R. A. (1999). Creation of human tumour cells with defined genetic elements. *Nature* **400,** 464–468.

Hamad, N. M., Elconin, J. H., Karnoub, A. E., Bai, W., Rich, J. N., Abraham, R. T., Der, C. J., and Counter, C. M. (2002). Distinct requirements for Ras oncogenesis in human versus mouse cells. *Genes Dev.* **16,** 2045–2057.

Keller, C., Arenkiel, B. R., Coffin, C. M., El Bardeesy, N., DePinho, R. A., and Capecchi, M. R. (2004). Alveolar rhabdomyosarcomas in conditional Pax3:Fkhr mice: Cooperativity of Ink4a/ARF and Trp53 loss of function. *Genes Dev.* **18,** 2614–2626.

Kendall, S. D., Linardic, C. M., Adam, S. J., and Counter, C. M. (2005). A network of genetic events sufficient to convert normal human cells to a tumorigenic state. *Cancer Res.* **65,** 9824–9828.

Linardic, C. M., Downie, D. L., Qualman, S. J., Bentley, R. C., and Counter, C. M. (2005). Genetic modeling of human rhabdomyosarcoma. *Cancer Res.* **65,** 4490–4495.

Merlino, G., and Helman, L. J. (1999). Rhabdomyosarcoma—working out the pathways. *Oncogene* **18,** 5340–5348.

Morotti, R. A., Nicol, K. K., Parham, D. M., Teot, L. A., Moore, J., Hayes, J., Meyer, W., and Qualman, S. J. (2006). An immunohistochemical algorithm to facilitate diagnosis and subtyping of rhabdomyosarcoma: the Children's Oncology Group experience. *Am. J. Surg. Pathol.* **30,** 962–968.

Nanni, P., Nicoletti, G., De Giovanni, C., Croci, S., Astolfi, A., Landuzzi, L., Di Carlo, E., Iezzi, M., Musiani, P., and Lollini, P. L. (2003). Development of rhabdomyosarcoma in HER–2/neu transgenic p53 mutant mice. *Cancer Res.* **63,** 2728–2732.

O'Hayer, K. M., and Counter, C. M. (2005). A genetically defined normal somatic human cell system to study ras oncogenesis *in vitro* and *in vivo*. *Methods Enzymol.* **407,** 637–647.

Rangarajan, A., Hong, S. J., Gifford, A., and Weinberg, R. A. (2004). Species- and cell type-specific requirements for cellular transformation. *Cancer Cell* **6,** 171–183.

Reed, N., and Gutmann, D. H. (2001). Tumorigenesis in neurofibromatosis: new insights and potential therapies. *Trends Mol. Med.* **7,** 157–162.

Sharp, R., Recio, J. A., Jhappan, C., Otsuka, T., Liu, S., Yu, Y., Liu, W., Anver, M., Navid, F., Helman, L. J., DePinho, R. A., and Merlino, G. (2002). Synergism between

INK4a/ARF inactivation and aberrant HGF/SF signaling in rhabdomyosarcomagenesis. *Nat. Med.* **8,** 1276–1280.

Tsumura, H., Yoshida, T., Saito, H., Imanaka-Yoshida, K., and Suzuki, N. (2006). Cooperation of oncogenic K-ras and p53 deficiency in pleomorphic rhabdomyosarcoma development in adult mice. *Oncogene* **25,** 7673–7679.

Xia, S. J., Pressey, J. G., and Barr, F. G. (2002). Molecular pathogenesis of rhabdomyosarcoma. *Cancer Biol. Ther.* **1,** 97–104.

Yeh, E., Cunningham, M., Arnold, H., Chasse, D., Monteith, T., Ivaldi, G., Hahn, W. C., Stukenberg, P. T., Shenolikar, S., Uchida, T., Counter, C. M., Nevins, J. R., *et al.* (2004). A signalling pathway controlling c-Myc degradation that impacts oncogenic transformation of human cells. *Nat. Cell Biol.* **6,** 308–318.

Author Index

Subject Index

A

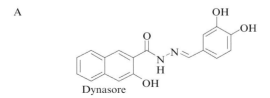

Dynasore

B

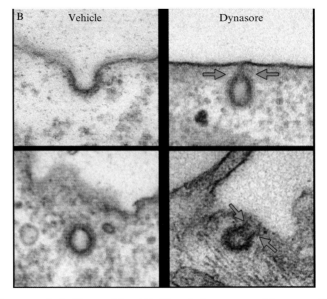

Tom Kirchhausen *et al.*, Figure 6.1 (A) Chemical structure of dynasore. (B) Effect of dynasore on clathrin-coated structures. The figure shows representative images of clathrin-coated structures of cells treated with DMSO (vehicle) or with 80 μM dynasore. The upper and lower left panels illustrate the appearance of endocytic coated pits and coated vesicles; the upper and lower right panels show the appearance of "U" and "O" shape–coated pits associated with the plasma membrane in cells treated with dynasore. The gray arrows highlight the extent of the constriction states observed upon treatment with dynasore. (From Macia, E., Ehrlich, M., Massol, R., Boucrot, E., Brunner, C., and Kirchhausen, T. (2006). Dynasore, a cell-permeable inhibitor of dynamin. *Dev. Cell* **10**, 839–850.)

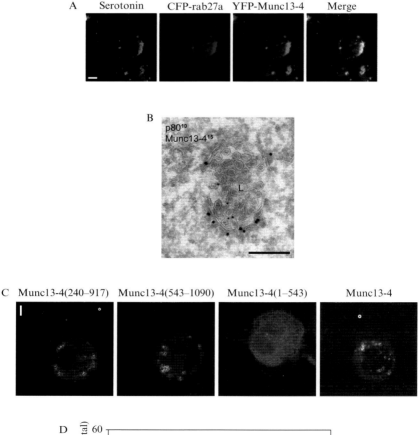

A   Serotonin   CFP-rab27a   YFP-Munc13-4   Merge

B

p80[10]
Munc13-4[15]

L

C Munc13-4(240–917)   Munc13-4(543–1090)   Munc13-4(1–543)   Munc13-4

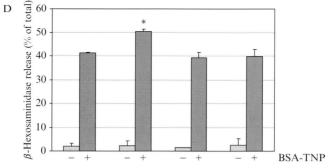

D

β-Hexosaminidase release (% of total)

BSA-TNP

Peter van der Sluijs *et al.*, Figure 13.2   Localization of munc13-4/rab27a complex on secretory lysosomes. (A) Munc13-4/rab27a colocalize on granular structures in the cytoplasm. RBL-2H3 cells were transfected with pEYFP-munc13-4 (green) and pECFP-rab27a (blue), and labeled with an antibody against serotonin (red), a content marker of secretory lysosomes. The merged image of this triple-label experiment shows extensive colocalization of munc13-4/rab27a with the marker on discrete cytoplasmic granules. Bar denotes 5 μm. (B) Munc13-4 localizes to the limiting membrane of multi-vesicular granules. Double labeling of YFP-munc13-4 (15 nm gold) and the lysosomal membrane glycoprotein p80 (10 nm gold) on cryosections prepared from RBL-2H3 transfectants. Bar denotes 200 nm. (C) Membrane localization of munc13-4 requires the MHDs. RBL-2H3 cells transfected with various GFP-tagged munc13-4 truncations were labeled with an antibody against serotonin. Note that only the truncations containing the region encompassing the MHDs localize to secretory lysosomes. Bar denotes 5 μm. (D) Munc13-4 is a positive regulator of secretory lysosome degranulation. RBL-2H3–expressing GFP-tagged munc13-4 constructs were assayed for their ability to release β-hexosaminidase in resting state and after stimulation. Full-length, munc13-4 enhanced antibody induced β-hexosaminidase secretion, while truncation mutants were without effect.

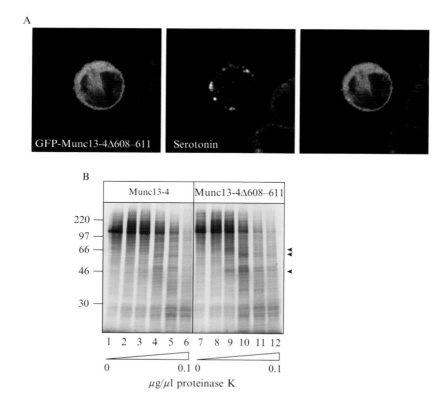

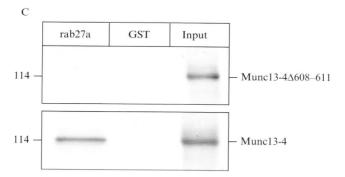

Peter van der Sluijs *et al.*, Figure 13.3 Analysis of FHL3 munc13-4Δ608–611 mutant. (A) Munc13-4Δ608–611 does not localize to secretory lysosomes. The FHL3 mutant (and control wildtype munc13-4) were transfected as GFP-tagged constructs in RBL-2H3 cells. Deletion of four amino acids in the first MHD creates a mutant that localizes to the cytoplasm, and is not recruited to serotonin containing secretory lysosomes. (B) Deletion of aa 608–611 causes misfolding of munc13-4. Conformational alterations of munc13-4Δ608–611 were investigated in a limited proteolysis assay. ^{35}S-labeled His$_6$-munc13-4 were produced in a coupled transcription–translation reaction and submitted

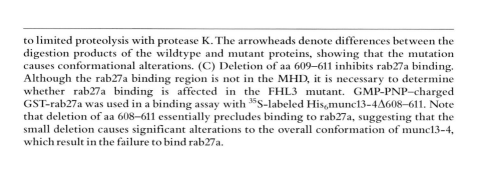

to limited proteolysis with protease K. The arrowheads denote differences between the digestion products of the wildtype and mutant proteins, showing that the mutation causes conformational alterations. (C) Deletion of aa 609–611 inhibits rab27a binding. Although the rab27a binding region is not in the MHD, it is necessary to determine whether rab27a binding is affected in the FHL3 mutant. GMP-PNP–charged GST-rab27a was used in a binding assay with ^{35}S-labeled His$_6$munc13-4Δ608–611. Note that deletion of aa 608–611 essentially precludes binding to rab27a, suggesting that the small deletion causes significant alterations to the overall conformation of munc13-4, which result in the failure to bind rab27a.

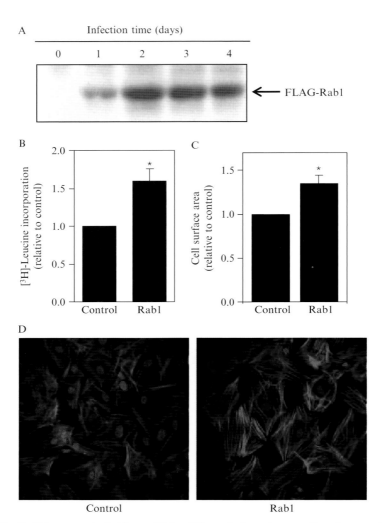

Catalin M. Filipeanu *et al.*, **Figure 15.1**   Effect of adenovirus-mediated expression of Rab1a on hypertrophic growth in neonatal cardiomyocytes. (A) Adenovirus-mediated expression of FLAG-Rab1a in neonatal rat ventricular myocytes. Neonatal myocytes were infected with empty adenoviral vector (control) or recombinant FLAG-Rab1a adenovirus for 1, 2, 3, and 4 days at 20 MOI. FLAG-Rab1a expression was detected by immunoblotting with anti-FLAG antibody M2 following SDS-PAGE. (B) The effect of adenovirus-driven FLAG-Rab1 expression on total protein synthesis. Cardiomyocytes were infected with control or Rab1 adenoviruses (20 MOI), and incubated with 1 μCi [3H]-leucine for 12 h. The data are shown as the x-fold increase over the control and represent the means (plus or minus standard error) of five separate experiments, each performed in duplicate. (C) The effect of Rab1a on the cell surface area. Cardiomyocytes were stained with phalloidin for F-actin after a 2-day infection. The cell surface area was measured by using the NIH Image program. At least 100 myocytes from 20 randomly selected fields in three separate experiments were measured. $\star p < 0.05$ versus cardiomyocytes infected with control adenovirus. (D) The effect of Rab1 on sarcomeric organization. The myocytes were stained with phalloidin and fluorescence visualized by a confocal microscope. (Figures adapted from Filipeanu, C. M., Zhou, F., Claycomb, W. C., and Wu, G. (2004). Regulation of the cell surface expression and function of angiotensin II type 1 receptor by Rab1-mediated endoplasmic reticulum-to-Golgi transport in cardiac myocytes. *J. Biol. Chem.* **279**, 41077–41084.)

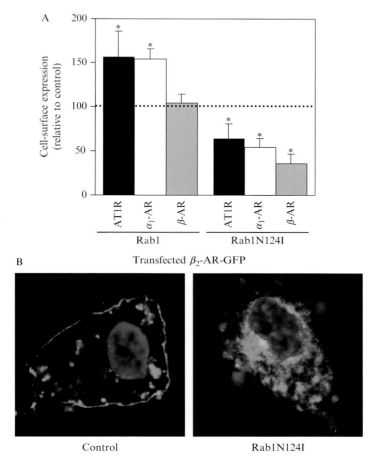

Guangyu Wu Figure 16.1 Effect of adenovirus-mediated expression of Rab1 and its dominant negative mutant Rab1N124I on the cell surface expression (A) and subcellular localization (B) of GPCRs in neonatal cardiomyocytes. (A) Cardiomyocytes were cultured and infected with control, Rab1, or Rab1N124I adenovirus for 2 days at a multiplicity of infection of 20. The cell-surface expression of AT1R, α_1-AR, and β-AR was determined by the radioligand [^{3}H]-Ang II, [^{3}H]-prazosin, and [^{3}H]-CGP12177, respectively. The data shown are the percentage of the mean value obtained from the cardiomyocytes infected with control adenovirus (the dotted line). $\star p < 0.05$ versus control. (B) Cardiomyocytes were grown on coverslips, infected with control or Rab1N124I adenoviruses, and transiently transfected with GFP-tagged β_2-AR. Two days after infection, the myocytes were fixed and stained with 4,6–diamidino-2-phenylindole (nuclear). Blue, nuclear stained by 4,6–diamidino-2-phenylindole; green, GFP-β_2-AR. (From Filipeanu, C. M., Zhou, F., Claycomb, W. C., and Wu, G. (2004). Regulation of the cell surface expression and function of angiotensin II type 1 receptor by Rab1–mediated endoplasmic reticulum–to–Golgi transport in cardiac myocytes. *J. Biol. Chem.* **279**, 41077–41084; and Filipeanu, C. M., Zhou, F., Fugetta, E. K., and Wu, G. (2006a). Differential regulation of the cell-surface targeting and function of β- and α1-adrenergic receptors by Rab1 GTPase in cardiac myocytes. *Mol. Pharmacol.* **69**, 1571–1578.)

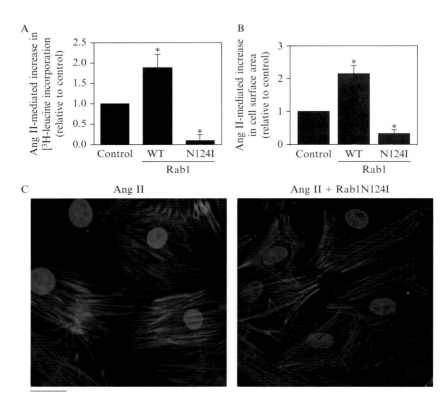

Guangyu Wu Figure 16.2 Effect of adenovirus-mediated expression of Rab1 and its dominant negative mutant Rab1N124I on Ang II–stimulated hypertrophic response in neonatal cardiomyocyte. To reflect the effect of Rab1 on Ang II–mediated stimulation, the contribution of Rab1 itself to protein synthesis and cell surface area was subtracted. Cardiomyocytes were cultured and infected with control, Rab1, and Rab1N124I adenoviruses for 2 days, and then stimulated with Ang II (100 nM) for 24 h. (A) Effect of Rab1 on Ang II–stimulated total protein synthesis measured by [^{3}H]-leucine incorporation. (B) Effect of Rab1 on Ang II–mediated increase in cell surface area. (C) Effect of Rab1N124I on Ang II–stimulated sarcomeric organization revealed by staining with phalloidin for F-actin. $\star p < 0.05$ versus control. (Adapted from From Filipeanu, C. M., Zhou, F., Claycomb, W. C., and Wu, G. (2004). Regulation of the cell surface expression and function of angiotensin II type 1 receptor by Rab1-mediated endoplasmic reticulum-to-Golgi transport in cardiac myocytes. *J. Biol. Chem.* **279,** 41077–41084.)

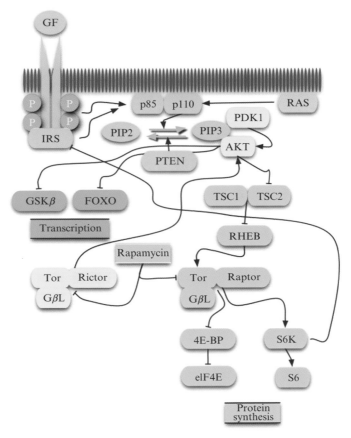

Adam Denley *et al.*, Figure 20.1 Example of canonical PI3K signaling pathway. The pathway is divided into four functionally distinct, color-coded modules. The basic enzymatic activity of PI3K is shown in green. Binding of an extracellular growth factor to a receptor tyrosine kinase leads to recruitment and activation of PI3K. Activated PI3K phosphorylates PIP$_2$ to PIP$_3$; the phosphatase PTEN catalyzes the reverse reaction. PIP$_3$ acts as a second messenger and initiates the singal by recruiting the serine threonine kinase Akt and its activating kinase, PDK1, shown in yellow. Akt is additionally activated by the TOR/rictor complex. Rapamycin is a specific inhibitor of TOR with a stronger effect on the TOR/raptor complex than on the TOR/rictor complex. Activated Akt controls gene transcription and translation, shown in magenta and orange, respectively. Akt phosphorylation negatively regulates the growth–attenuating targets FOXO and GSKβ. Activated Akt affects protein synthesis through TOR singlaing. Akt disrupts the TSC1/TSC2 complex, resulting in activation of Rheb and of the TOR/raptor complex. This complex controls the initiation of protein synthesis by activating S6 kinase and inhibiting 4E-BP. Several of the PI3K signaling components have oncogenic potential: the p85 and p110 subunits of PI3K, Akt, Rheb, and elF4E. The interactions of Ras and PI3K are described in the text.

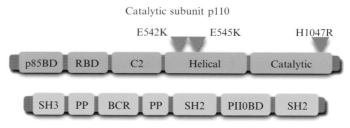

Adam Denley *et al.*, Figure 20.2 Domain structure of catalytic subunit p110 and regulatory subunit p85 of PI3K. The catalytic subunit contains the p85-binding domain (p85BD), Ras-binding domain (RBD), C2 domain (C2), helical domain, and catalytic domain. The three hot-spot mutations identified in human cancer are marked. The regulatory subunit contains three src homology domains (two SH2 and one SH3), polyproline regions (PP), BCR homology domain (BCR), and p110-binding domain (p110BD). Dimensions of the domains are not to scale.

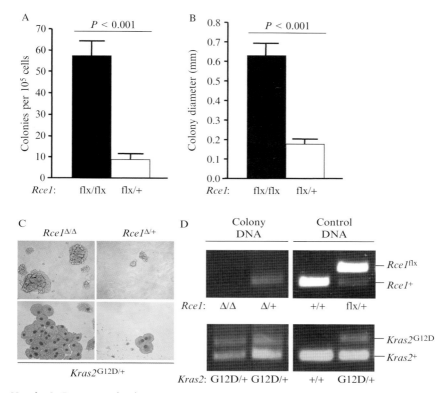

Martin O. Bergo *et al.*, Figure 26.7 Inactivation of *Rce1* increases colony formation of K-RAS^G12D–expressing splenocytes. (A and B) Growth factor–independent colony growth of splenocytes from *Rce1*^flx/flx^Mx1-*Cre*^+/o^*Kras2*^LSL/+^ ($n = 3$) and *Rce1*^flx/+^Mx1-*Cre*^+/o^*Kras2*^LSL/+^ ($n = 3$) mice. A, colony number; B, colony size. Splenocytes from control mice were incapable of colony growth in this assay (not shown). (C) Upper panel: Photographs showing *Rce1*^Δ/Δ^Mx1-*Cre*^+/o^*Kras2*^G12D^ and *Rce1*^Δ/+^Mx1-*Cre*^+/o^*Kras2*^G12D^ splenocyte colonies from a typical experiment in A and B. Lower panel: May-Grü nwald-Giemsa–stained cytospins of individual colonies. (D) PCR amplification of genomic DNA from individual colonies to detect the *Rce1*^flx^ and *Rce1*^+^alleles (upper panel) and the *Kras2*^+^ and *Kras2*^G12D^ alleles (lower panel). Lane 1, *Rce1*-deficient *Kras2*^G12D^ colony; lane 2, heterozygous *Rce1*-deficient *Kras2*^G12D^ colony; lanes 3 and 4, control DNA from mouse tails. (From Wahlstrom, A. M., Cutts, B. A., Karlsson, C., Andersson, K. M., Liu, M., Sjogren, A. K., Swolin, B., Young, S. G., and Bergo, M. O. (2007). *Rce1* deficiency accelerates the development of K-RAS-induced myeloproliferative disease. *Blood* **109**, 763–768, with permission.)

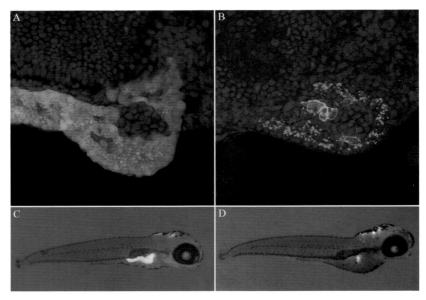

Jon M. Davison *et al.*, Figure 27.2   Oncogenic Kras inhibits exocrine differentiation
in pancreas progenitor cells. (A) Carboxypeptidase A (a marker of exocrine differentiation) is uniformly immunolabeled (red, Cy3) in apical secretory granules of *eGFP*-posi-
tive pancreas progenitor cells at 76 h postfertilization (hfp) in *ptf1a:eGFP* embryos.
(B) In contrast, *ptf1a:eGFP-Kras*[G12V] embryos exhibit mosaic expression of the eGFP
fusion protein. In cells that express the eGFP-Kras[G12V] fusion protein, CPA expression is
not detected, consistent with a block in exocrine differentiation. (C) *In situ* hybridization detects *eGFP* RNA expression in the pancreas and hindbrain, in a pattern that is
identical to protein expression. (D) The pattern of *eGFP-Kras*[G12V] RNA expression
(detected with an antisense probe to *eGFP*) in *ptf1a:eGFP-Kras*[G12V] embryos is similar to
the focal, mosaic pattern of protein expression seen in most embryos (as depicted in B).

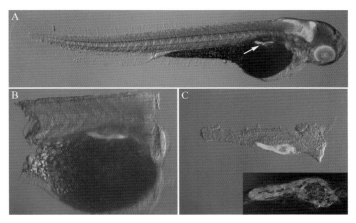

Jon M. Davison *et al.*, **Figure 27.3**   Stages of manual dissection of a single *ptf1a:eGFP* embryo. Anterior is to the right in all images. (A) An 80-h postfertilization embryo prior to dissection. eGFP protein is expressed in the retina, hindbrain, and pancreas (arrow). (B) Dissection begins with amputation of the head and tail followed by (C) removal of the yolk and separation of the remaining endodermal organs from ventral anatomic structures. The pancreas may be isolated by careful dissection from the gut (C inset).

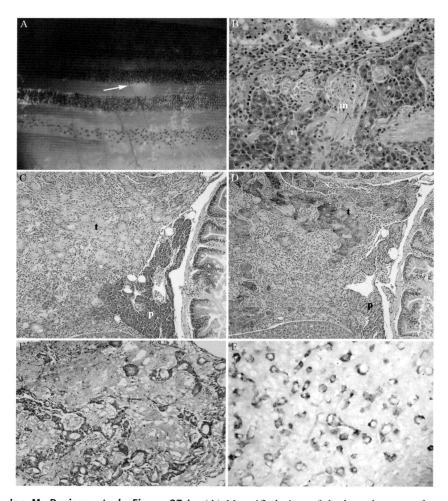

Jon M. Davison *et al.*, Figure 27.4   (A) Magnified view of the lateral aspect of an anesthetized adult zebrafish with a focal lesion (arrow) expressing *eGFP-Kras*^{G12V} within the abdominal viscera (measuring approximately 1 mm) detected by fluorescent transillumination. (B) Histologic section of a malignant pancreatic carcinoma exhibiting predominantly acinar differentiation invading through the muscularis propria (m) of the gut (400×). (C) Histologic section of a tumor (t) arising in the pancreas (p) of a *ptf1a:eGFP-Kras*^{G12V} fish. The carcinoma shows mixed acinar and ductal morphology (200×). (D) Immunohistochemical detection of the *eGFP-Kras*^{G12V} fusion protein with an anti–GFP antibody (mouse, Chemicon, 1:300) in a section adjacent to the one depicted in C. Note that expression of the fusion protein is detected in the tumor (t), but not in the morphologically normal pancreas (p) (200×). (E) Frozen section of a pancreatic adenocarcinoma arising in a *ptf1a:eGFP-Kras*^{G12V} fish with predominantly glandular differentiation and abundant extracellular mucin (200×). (F) *In situ* hybridization with an anti–*eGFP* riboprobe detects expression of *eGFP-* Kras^{G12V} in a pancreatic adenocarcinoma histologically similar to that shown in E (200×).